Carbon Nanotubes: Reinforced Metal Matrix Composites

Nanomaterials and Their Applications
Series Editor: M. Meyyappan

Inorganic Nanowires
Applications, Properties, and Characterization
M. Meyyappan, Mahendra K. Sunkara

Inorganic Nanoparticles
Synthesis, Applications, and Perspectives
Claudia Altavilla, Enrico Ciliberto

Carbon Nanotubes
Reinforced Metal Matrix Composites
Arvind Agarwal, Debrupa Lahiri, Srinivasa Rao Bakshi

Graphene
Synthesis and Applications
Wonbong Choi, Jo-won Lee

Upconverting Nanomaterials
Perspectives, Synthesis, and Applications
Claudia Altavilla

Plasma Processing of Nanomaterials
R. Mohan Sankaran

Biosensors Based on Nanomaterials and Nanodevices
Jun Li, Nianqiang Wu

Carbon Nanotubes
Reinforced Metal Matrix Composites, Second Edition
Andy Nieto, Arvind Agarwal, Debrupa Lahiri, Ankita Bisht and Srinivasa Rao Bakshi

For more information about this series, please visit:
https://www.routledge.com/Nanomaterials-and-their-Applications/book-series/CRCNANTHEAPP

Carbon Nanotubes
Reinforced Metal Matrix Composites

Andy Nieto
Arvind Agarwal
Debrupa Lahiri
Ankita Bisht
Srinivasa Rao Bakshi

CRC Press
Taylor & Francis Group
Boca Raton London New York

CRC Press is an imprint of the
Taylor & Francis Group, an **informa** business

Second edition published 2021
by CRC Press
6000 Broken Sound Parkway NW, Suite 300, Boca Raton, FL 33487-2742

and by CRC Press
2 Park Square, Milton Park, Abingdon, Oxon, OX14 4RN

First edition published by CRC Press 2010

CRC Press is an imprint of Taylor & Francis Group, LLC

Library of Congress Cataloging-in-Publication Data
Names: Agarwal, Arvind, author. | Nieto, Andy, author. | Lahiri, Debrupa, author. | Bisht, Ankita, author. | Bakshi, Srinivasa Rao, author.
Title: Carbon nanotubes : reinforced metal matrix composites / Andy Nieto, Arvind Agarwal, Debrupa Lahiri, Ankita Bisht, Srinivasa Rao Bakshi.
Description: Second edition. | Boca Raton : CRC Press, 2021. | Series: Nanomaterials and their applications | Includes bibliographical references and index.
Identifiers: LCCN 2021001760 (print) | LCCN 2021001761 (ebook) | ISBN 9780367274924 (hbk) | ISBN 9780429299582 (ebk)
Subjects: LCSH: Nanotubes. | Electrochemical sensors–materials.
Classification: LCC TA481 .A37 2021 (print) | LCC TA481 (ebook) | DDC 620.1/6–dc23
LC record available at https://lccn.loc.gov/2021001760
LC ebook record available at https://lccn.loc.gov/2021001761

ISBN: 978-0-367-27492-4 (hbk)
ISBN: 978-0-367-28087-1 (pbk)
ISBN: 978-0-429-29958-2 (ebk)

Typeset in Times
by SPi Global, India

Dedication

This book is dedicated to the authors' families, mentors, teachers, and fellow researchers

Contents

Foreword to the 1st Edition xiii
Foreword to the 2nd Edition xv
Preface to the 1st Edition xvii
Preface to the 2nd Edition xix
Acknowledgments xxi
About the Authors xxiii

Chapter 1 Introduction 1

1.1 Composite Materials 1
1.2 Development of Carbon Fibers 3
1.3 Carbon Nanotubes: Synthesis and Properties 3
1.4 Carbon Nanotube – Metal Matrix Composites (CNT-MMCs) 7
1.5 Chapter Highlights 10
References 11

Chapter 2 Dispersion of Carbon Nanotubes in the Metal Matrix 15

2.1 Significance of Carbon Nanotube Dispersion 15
2.2 Methods for Carbon Nanotube Dispersion 16
2.2.1 Blending 17
2.2.2 Ball Milling 17
2.2.2.1 High Energy Ball Milling (HEBM) 17
2.2.2.2 Cryogenic Mechanical Milling 19
2.2.2.3 Solution Ball Milling (SBM) 20
2.2.2.4 Rotational Milling Modeling 21
2.2.2.5 Shift-Speed Ball Milling (SBBM) 23
2.2.3 Flake Powder Metallurgy (FPM) 24
2.2.4 Nano-scale Dispersion (NSD) 27
2.2.5 Molecular Level Mixing 28
2.2.6 In Situ Growth on Metal Particles 29
2.2.7 Physical Vapor Deposition 31
2.2.8 Spray Drying 31
2.3 Quantification of Carbon Nanotube Dispersion 33
2.4 Chapter Highlights 39
References 39

Chapter 3 Composite Processing Techniques 43

3.1 Powder Metallurgy Routes 43
3.1.1 Conventional Sintering 45
3.1.2 Hot Pressing 47

3.1.3 Spark Plasma Sintering ... 49
3.1.4 Deformation Processing ... 51
3.2 Melt Processing ... 54
3.2.1 Casting ... 55
3.2.2 Melt Infiltration ... 57
3.2.3 Induction Melting ... 59
3.3 Thermal Spraying ... 60
3.3.1 Plasma Spraying ... 60
3.3.2 High Velocity Oxy-Fuel Spraying ... 66
3.3.3 Cold Spraying ... 67
3.4 Electrochemical Routes ... 71
3.5 Novel Techniques ... 75
3.5.1 Sputtering ... 75
3.5.2 Sandwich Processing ... 76
3.5.3 High Pressure Torsion (HPT) ... 77
3.5.4 Friction Stir Processing ... 77
3.5.5 Laser Deposition ... 81
3.6 Conclusion ... 81
3.7 Chapter Highlights ... 81
References ... 82

Chapter 4 Characterization Techniques for CNT-MMC Materials ... 91

4.1 X-Ray Diffraction ... 91
4.2 Raman Spectroscopy ... 92
4.3 Scanning Electron Microscopy with Energy Dispersive Spectroscopy ... 94
4.4 High Resolution Transmission Electron Microscopy ... 96
4.5 Electron Energy Loss Spectroscopy ... 97
4.6 X-Ray Photoelectron Spectroscopy ... 98
4.7 Mechanical Properties Evaluation ... 99
4.7.1 Nanoscale Mechanical Testing ... 99
4.7.1.1 Nano-Indentation ... 99
4.7.1.2 Nano Dynamic Modulus Analysis ... 102
4.7.1.3 Modulus Mapping ... 103
4.7.1.4 Nanoscratch ... 104
4.7.2 Macroscale/Bulk Mechanical Testing ... 105
4.7.2.1 Tensile/Compression Test ... 106
4.7.2.2 Tribological Property Evaluation ... 107
4.7.3 In Situ Mechanical Testing ... 108
4.8 Thermal Properties ... 109
4.9 Electrical Properties ... 110
4.10 Electrochemical Properties ... 111
4.11 Chapter Highlights ... 113
References ... 114

Chapter 5 Carbon Nanotube Reinforced Metal Matrix Composite (CNT-MMC) Systems 121

5.1 Aluminum-Carbon Nanotube System 122
5.2 Copper-Carbon Nanotube System 133
5.3 Nickel-Carbon Nanotube System 143
5.4 Magnesium-Carbon Nanotube System 151
5.5 CNT-MMC Foams 155
5.6 CNT-Metal Nanolaminates 161
5.7 Other Metal-Carbon Nanotube Systems 164
5.8 Chapter Highlights 173
References 174

Chapter 6 Mechanics of CNT-MMCs 187

6.1 Elastic Modulus of Metal Matrix-Carbon Nanotube Composites 188
6.1.1 Modified Rule of Mixtures 189
6.1.2 Cox Model 191
6.1.3 Halpin-Tsai Model 192
6.1.4 Hashin-Shtrikman Model 192
6.1.5 Modified Eshelby Model 193
6.1.6 Dispersion-based Model 194
6.2 Strengthening Mechanisms in Metal Matrix-Carbon Nanotube Composites 199
6.2.1 Shear Lag Models 204
6.2.2 Strengthening by Interphase 206
6.2.3 Strengthening by Carbon Nanotube Clusters 208
6.2.4 Halpin-Tsai Equations 209
6.2.5 Strengthening by Dislocations 210
6.2.6 Strengthening by Grain Refinement 212
6.2.7 In Situ Observation of CNT Strengthening Mechanisms 214
6.3 Chapter Highlights 221
References 222

Chapter 7 Interfacial Phenomena in CNT-MMCs 227

7.1 Significance of Interfacial Phenomena 227
7.2 Energetics of Carbon Nanotube-Metal Interaction 230
7.3 Carbon Nanotube-Metal Interaction in Various Systems 233
7.3.1 Formation and Effects of Carbides 233
7.3.2 Effect of CNT Defects on Al_4C_3 Nucleation 241
7.3.3 Effect of Temperature on Al_4C_3 Formation 243
7.3.4 Degree of Interfacial Reaction at CNT/Metal Interface 246

7.4 Incorporation of Interphases 248
7.5 Chapter Highlights 249
References 250

Chapter 8 Tribological, Transport, and Functional Properties & Behavior 253

8.1 Tribological Properties 253
8.2 Thermal Properties 258
8.3 Elevated Temperature Mechanical Behavior 262
8.4 Electrical Properties 270
8.5 Corrosion Properties 273
8.6 Hydrogen Storage 275
8.7 Sensors and Catalytic Properties 276
8.8 Chapter Highlights 278
References 279

Chapter 9 Synergistic Interactions with Secondary Reinforcements 285

9.1 Ceramic Particle-CNT Dual Reinforcement 285
9.2 Nanocarbon-CNT Dual Reinforcement 294
9.3 Chapter Highlights 300
References 300

Chapter 10 Computational Studies in CNT-MMCs 303

10.1 Overview of Early Computational Studies on CNT-MMCs 303
10.2 Thermodynamic Prediction of Carbon Nanotube-Metal Interface 306
10.3 Microstructural Simulation 307
10.4 Mechanical and Thermal Property Prediction by the Object-Oriented Finite Element Method 310
10.5 Chapter Highlights 312
References 312

Chapter 11 Patents and Commercialization 315

11.1 Patents 315
11.1.1 US Patents 315
11.1.2 International Patents 319
11.2 Prospects & Challenges for Commercialization 321
11.3 Chapter Highlights 326
References 326

Chapter 12 Summary and Future Directions 329

12.1 Summary of CNT-MMC Research 329
12.2 Future Directions 334
12.2.1 Purity & Quality of CNTs 334
12.2.2 Challenges Related to Processing 335
12.2.3 Optimal CNT Dimensions & Aligned CNT-MMCs 335
12.2.4 In Situ Characterization and Modeling 336
12.2.5 Dynamic Mechanical Behavior 336
12.2.6 Interphase Development 337
12.2.7 Extreme Environment Stability 337
12.2.8 Environmental and Toxicity Aspects of CNT-MMCs 338
12.2.9 CNTs as Key Technology Enablers 339
References 340

Index 345

Foreword to the 1st Edition

Carbon nanotubes (CNTs) have created an enormous amount of excitement in research laboratories across the world due to their extraordinary mechanical properties and unique electronic properties. This is evident from the exponential growth over the last decade in journal publications and conference presentations on all subjects related to CNTs. Even the price has begun to inch downward for single-walled CNTs. All of these developments point toward a bright road ahead for CNTs in various applications. One promising area is preparation of composites where CNTs can be loaded into a host matrix of a polymer, metal, or a ceramic. Of these, the subject of CNT-polymer matrix composites is the most widely investigated, primarily due to the relative ease of preparation techniques compared to the other two. The results appear to be promising and some small-scale commercial products have started to become available, for example, thin films for shielding electromagnetic interference. CNT-metal matrix composites have much promise in aerospace, automotive, and many other industries, and in thin film form, they are considered for electrical and electronic applications, corrosion resistance, wear resistance, and other protective coatings, to name a few.

In order to realize the full promise of CNT-metal matrix composites, certainly much more work is needed in all aspects: various composite preparation techniques, characterization, evaluation of properties, large-scale processing, and application development. Interest in this field has been steadily increasing, drawing more researchers into the arena. At this point, there is a critical need for a book to summarize the status of the field but more importantly to lay out the principles behind the technology. This is what Professor Arvind Agarwal and his co-workers from Florida International University have done here. The book discusses various processing techniques, characterization methods, models for metal matrix composite systems, properties of CNT composites, and their applications. The book is not only a useful reference for researchers in this area but can also serve as a textbook for a one-semester course. I am sure that the academic and industrial communities will find this book a valuable resource.

M. Meyyappan
NASA Ames, Moffett Field, CA

Foreword to the 2nd Edition

- The field of nanotechnology has begun to mature into technological applications over the last decade or so, and carbon nanotube reinforced composites are no exception. Myriad patents for potential products have been filed and many of the challenges outlined in the 1st edition are being addressed and in some cases new research directions have sprung forward as a result.
- A decade after helping pioneer the field of carbon nanotube reinforced metal matrix composites and the 1st edition, Prof. Agarwal and his mentees and collaborators have provided a much needed update on the state-of-the-art of CNT-MMC technology, at an opportune time when commercialization is on the horizon.
- The expertise of the authors spans across the vast field of CNT-MMCs, from basic powder metallurgy and composite science, to novel processing techniques, additive processing, in situ characterization, multi-scale mechanical properties, and material responses in high temperature and corrosive environments, to name a few.
- This book is a critical reference for new scientists and engineers embarking on the journey to study and develop CNT-MMCs. The book is laid out logically to guide a practitioner from synthesis and processing of CNT-MMCs, to materials characterization, to advanced techniques such as in situ mechanics and performance-driven evaluation such as wear and elevated temperature testing, corrosion, transport phenomena, and functional properties.
- This book promises to be a valuable resource to faculty and students covering nanotechnology, carbon nanotubes, and nanocomposites. The book is suitable for a one-quarter or one-semester book for an advanced graduate course on such topics, or as a complimentary textbook in composite and nanotechnology courses.

Prof. Sudipta Seal
University of Central Florida
Orlando, FL

Preface to the 1st Edition

The last two decades have seen exponential growth in the field of nanoscience and nanotechnology. This has been possible due to major developments in the processing equipment, characterization, and imaging equipment with improved resolution and, of course, the availability of research funding through federal, state, and private agencies. Consequently, nanoscience has made tremendous progress. However, its benefits have yet to reach the masses at a large scale. This has also caused some concern, if nanotechnology is the hype. The major hurdle in advancing the benefits of nanoscience is the lack of suitable "nano-manufacturing" techniques. Engineers, scientists, and researchers are working to develop techniques where devices and structures with nano-scale features could be manufactured at a large-scale and affordable cost.

Nanocomposite is one such application where nano-scale features, nano-particles, or nanofibers are introduced into the structure to obtain excellent mechanical, thermal, electrical, and functional properties. The carbon nanotube (CNT) is one of the nanomaterials that has been actively researched due to its excellent mechanical, electrical, chemical, and thermal properties for a variety of applications. It has also been used as a filler or reinforcement for nanocomposites. However, the majority of CNT-reinforced composite research has been focused on the polymer matrix composites. Primarily, this can be attributed to the relative ease of polymer processing, which often does not require high temperature for the consolidation, as needed for metals and ceramic matrices. Studies on CNT-reinforced ceramic matrices are few as compared to those on polymer matrices, whereas those on CNT-reinforced metal matrix components are even fewer. This is quite surprising considering the fact that most of the structural materials used in today's world are metals. Metal matrix composites with CNT reinforcement can be designed to possess qualities such as lightweight, high strength, low coefficient of thermal expansion, and high thermal conductivity for use in automobile, aerospace, and electronic packaging applications. The need for stronger and lightweight materials for energy efficiency is never ending and is recognized more in today's scenario where the world is seeing global warming and climatic changes. The Japanese Automobile Manufacturing Association (JAMA) has recognized that using aluminum-CNT and magnesium-CNT composites in automobiles can reduce by 50% CO_2 emission per year caused by vehicles. Furthermore, the cost of multiwall carbon nanotubes has come down significantly in the last few years, with large-scale production facilities in Japan (Mitsui) and Germany (Bayer) that can produce several metric tons of nanotubes. The desire to bring the benefits of nanotechnology to the masses by developing CNT-MMCs for large-scale applications, in synergy with reducing the cost of multiwall carbon nanotubes, motivated us to write this book. The need was further strengthened by the fact that no book or resource on such topic was available in 2008, when we conceptualized the idea. This book provides a comprehensive knowledge about the research performed on different CNT-MMCs. Our approach has been to highlight the critical issues in developing CNT-MMCs, instead of merely summarizing the current research. Accordingly, we have divided the book into 10 chapters. Some of the salient features of these chapters are as follows:

- Comprehensive analysis on advantages, limitations, and evolution of the processing techniques used for CNT-MMCs with a critical discussion on the scope of further development is included. Processing techniques have been classified for easy understanding. The merits as well as limitations of each process with respect to CNT-MMCs are discussed.
- The characterization techniques that are unique to study CNT-MMCs are discussed with their limitations.
- This book provides a comprehensive summary of research work on different CNT-MMCs in tabular form that includes composition, processing method, quality of CNT dispersion, and properties. Such tables will be of great importance to any researcher or student of CNT-MMCs. These tables would serve as a single source of compiled information on a majority of the research on CNT-MMCs.
- The strengthening due to CNT addition has been studied in detail. CNTs being a few microns in length have a load-sharing ability over small-length scales. The translation of the micro-mechanical strengthening effect to the macro level is discussed with insight into the applicability of micro-mechanics models.
- The significance of chemical stability of carbon nanotubes in the metal matrix as a function of processing and its impact on CNT/metal interface and mechanical properties is critically addressed.
- The most critical issue of CNT *dispersion* in the metal matrix is addressed. A unique way to quantify CNT distribution has been discussed. Quantification of CNT dispersion would enable researchers to improve the control of the processing parameters for obtaining improved properties.
- The computational studies on metal-matrix composites are almost non-existent in the literature. The computational studies that are of greater significance to the development of CNT-MMCs are described, with an emphasis on proposing some of the computational methods for future research.
- The potential applications of CNT-MMCs are discussed and the need for finding newer applications is emphasized. The future research direction and a possible roadmap to develop CNT-MMCs are elucidated.

This book would serve as the most comprehensive source for an expert as well as a beginner in the area of CNT-reinforced metal matrix composites and nanotechnology. This book will also be beneficial for graduate students and researchers in nanotechnology, materials engineers, mechanical engineers, and chemical engineers.

Preface to the 2nd Edition

The field of carbon nanotube reinforced metal matrix composites (CNT-MMCs) has continued to grow over the last decade, 11 years from the publication of the first edition. It has been a pleasure seeing many of the suggested research paths laid out in the first edition followed and investigated in the last decade. These investigations and new unforeseen directions of research and tools for studying these novel materials have warranted a second edition. The primary goals of this second edition are to update the state-of-the-art and provide researchers new goal posts to aim for as the journey of discovery continues and begins to overlap with a trek toward technological maturation and commercialization. Two new chapters, along with several other notable changes and sections, were added in this second edition:

- Processing of CNT-MMCs has been re-organized into Chapter 2 *Dispersion of Carbon Nanotubes in the Metal Matrix*, and Chapter 3 *Composite Processing Techniques*. Chapter 2 focuses on dispersion techniques including the advent of flake powder metallurgy and improved analytical models for understanding ball milling. Previous discussion on quantification of dispersion (previously in Chapter 7 of 1st edition) have all been incorporated in Chapter 2. Chapter 3 solely focuses on consolidation techniques for CNT-MMCs.
- A section on in situ *Mechanical Testing* was added to Chapter 4 to discuss techniques for real-time observation of deformation mechanisms in CNT MMCs and the new discoveries made as a result.
- New types of CNT-MMC systems, to include *foams* and *nanolaminates*, are discussed in new sections in Chapter 5.
- Chapter 9 *Synergistic Interactions with Secondary Reinforcements* is a new chapter devoted to dual reinforcements in CNT-MMCs. These new systems yield the prospect of harnessing properties from multiple nanomaterials and enabling materials by design thru tailoring of properties.
- Chapter 11 *Patents and Commercialization* presents a survey of the patent literature in order to take a glimpse into what the first big wave of CNT-MMC commercialization might look like. In addition, issues surrounding patent laws and invention rights as related to nanotechnology and CNT-MMCs are discussed.

It is the authors' sincere hope that readers will find this updated edition useful and insightful, and that it will serve as a key reference for practitioners in academia, government, and industry in further developing and harnessing the full potential of these still exciting materials – *Carbon Nanotubes: Reinforced Metal Matrix Composites*.

Acknowledgments

The authors are very thankful to all the students and researchers who have contributed to the Plasma Forming Laboratory at Florida International University at different stages over the past 18 years. They are Prof. Tapas Laha, Prof. Kantesh Balani, Prof. Yao Chen, Prof. Anup Kumar Keshri, Dr. Ruben Galiano Batista, Dr. Sybille Facca, Dr. Cheng Zhang, Dr. Pranjal Nautiyal, Dr. Archana Loganathan, Dr. Adeyinka Idowu, Dr. Chris Rudolf, Dr. Tony Thomas, Gabriela Gonzalez, Dayan Paez, Sunil Musali, Jorge Tercero, Riken R. Patel, Di Wang, Sara Rengifo, Ana Exime, Leslie Embrey, Jenniffer Bustillos, Tyler Dolmetsch, Melanie Andara, Tanisha Richard, David Axel Virzi, Suvrat Bhargava, Jonathan Solomon, Akanksha Bhargava, and Juan Puerta.

Dr. Lawrence Kabacoff (Office of Naval Research) and Dr. Kevin Llyons (NIST) are greatly acknowledged for their initial support on CNT-reinforced composite research. Arvind Agarwal is also thankful to the National Science Foundation CAREER Award, which enabled research on CNT-MMCs at Florida International University. The support of our research collaborators, namely Prof. Sudipta Seal (University of Central Florida), Prof. Graham McCartney (University of Nottingham, UK), and Prof. Wenzhi Li (Florida International University) is greatly appreciated.

Andy Nieto greatly acknowledges his closest family for their endearing support and love, from his loving wife Richa Agrawal, to his proud parents Hernando and Janeth who were the pillars of his upbringing and ultimately education, and his always-supportive brother Hernando Jr. Andy also acknowledges the many mentors he's had along his academic journey, all of whose insights and styles on learning, research, and mentoring have contributed to his own views and practices. The most influential are his postdoc mentor Anindya Ghoshal, doctoral advisor Julie Schoenung, and lastly but certainly not least, fellow coauthor, friend, and mentor Arvind Agarwal.

Arvind Agarwal thanks Florida International University, which provided enthusiastic support in this endeavor. Arvind Agarwal is very thankful to his wife Anuradha Godavarty and daughter Diya Agarwal for their continuous and endless support in this effort.

Debrupa Lahiri acknowledges valuable discussions with her husband Indranil Lahiri and continuous support and encouragement from her family.

Ankita Bisht acknowledges the unparalleled love and support of her husband Sanjay Singh Samant, her parents Prakash Chandra Bisht and Vimla Bisht. She is also very thankful to her sisters Ms. Monika Bisht and Ms. Niharika Bisht and her niece Khanak for pouring in the positive energy and enthusiasm. Ankita Bisht extends her sincere thanks to her supervisor Debrupa Lahiri, for valuable guidance, constant encouragement, support and providing multiple opportunities that shaped her research skills.

Srinivasa Rao Bakshi would also like to thank his family and wife for their unconditional support and understanding during the writing of this manuscript.

About the Authors

Dr. Andy Nieto is an assistant professor in the Department of Mechanical and Aerospace Engineering at the Naval Postgraduate School (NPS) in Monterey, California, USA. Current research interests focus on the development of advanced ceramic and metallic matrix composites and coatings for use in extreme environments. Additive processing of these systems thru cold spray and selective laser melting are a key focus. Prior to joining NPS, Dr. Nieto was an Oak Ridge Associated Universities postdoctoral fellow at the U.S. Army Research Laboratory, where he worked on the development of high temperature ceramics and coatings, and the understanding of environmental particulate attack mechanisms. Other past and present research interests include nanomaterials (synthesis, mechanical/functional behavior), novel biomaterials (structural, tissue engineering), tribology (low wear/friction materials), nanomechanics, UHTCs and MAX phases, and nanoparticle reinforced composites (polymer/metal/ceramic matrix). Dr. Nieto has served as a reviewer for over 35 international journals. He has published over 40 technical articles and his research has received over 1300 citations. Dr. Nieto was a co-author on the journal publication awarded the 2017 Nanomaterials & Energy Prize. Dr. Nieto earned his Ph.D. in Materials Science & Engineering (MSE) from the University of California Davis in 2016, and his M.S. in MSE from Florida International University in 2013. Dr. Nieto received an undergraduate degree in Aerospace engineering from the University of Miami.

Dr. Arvind Agarwal is a Distinguished University professor and chair of the Department of Mechanical and Materials Engineering at Florida International University (FIU), Miami, FL, USA. Prof. Agarwal obtained his B.S from the Indian Institute of Technology (IIT) Kanpur in Materials and Metallurgical Engineering and Ph.D. in Materials Science and Engineering from the University of Tennessee at Knoxville. After his Ph.D., Prof. Agarwal worked in the industry as a materials scientist at Plasma Processes Inc., Huntsville, AL, USA, for three (3) years, before embarking on an academic career at FIU in 2002.

Prof. Agarwal's current research interests include advanced materials processing, thermal and cold spray, surface engineering, spark plasma sintering, ultrahigh temperature ceramics, carbon nanotube (CNT), boron nitride nanotube (BNNT), and graphene reinforced composites and coatings, nanoindentation, and mechanical properties of low dimensional and biological materials. Prof. Agarwal has published 300 technical articles including 3 co-authored books, 7 edited books, and 8 patents.

Prof. Agarwal has received several honors and awards including National Science Foundation's (NSF) prestigious CAREER award, Defense Tech Connect Award for Top 15% Technologies, and Nanomaterials and Energy Prize (Best Journal Paper). Prof. Agarwal was honored as an elected Fellow of ASM International in 2012 and senior member of the National Academy of Inventors in 2020.

Dr. Debrupa Lahiri is an associate professor in the Department of Metallurgical and Materials Engineering, IIT Roorkee, India. She has received her Ph.D. in Materials Science and Engineering from Florida International University, USA, M. Tech from IIT Kanpur, India and B.E. from Bengal Engineering College, India (IIEST). In addition to her eight years' experience in academia, she has eight years of experience in industry and research environment as well, which includes serving as Scientist in Nuclear Fuel Complex (NFC, DAE) India. She authored more than 100 publications in peer reviewed journals, 5 book chapters, one book and four patents (applied to Indian Patent office). She is also the recipient of prestigious Zwick Science Award 2013, which is given for innovations in materials' testing. Currently, her group 'Biomaterials and Multiscale Mechanics Group' performs research in the areas of nano-phase reinforced composites with different functionalities. The aims of her different research activities are developing novel material systems for different applications, including structural, strategic and scaffolds and implants for hard and soft tissue engineering. She is also the co-founder of 'Heal-agnostic Innovations Pvt. Ltd.' (http://faculty.iitr.ac.in/~dlahifmt), a start-up working in the healthcare domain.

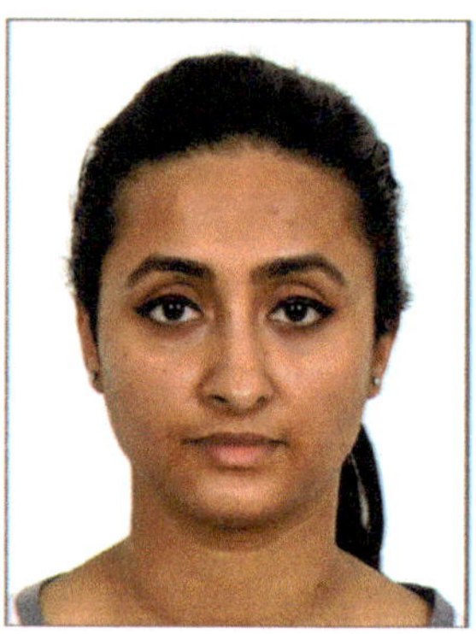

Dr. Ankita Bisht is an assistant professor in the Department of Mechanical Engineering, Women Institute of Technology (WIT), Dehradun, India. She completed her Ph.D. in Materials Science and Engineering from IIT Roorkee, India in February 2020. She did her M.Tech in Design and Production Engineering from GBPUAT, India and B.Tech in Production Engineering from GBPIET, India. Her areas of research actively involve fabrication and characterization of nano-phase and multiscale reinforcement composite for different functionalities. Her research includes multiscale-mechanics and nano-mechanics of composite materials with main focus on effective dispersion of nano phases along with its quantification, and synergistic effects of nano-fillers on mechanical and thermo-mechanical properties of composites. She has published 12 papers in peer reviewed journals.

Dr. Srinivasa Rao Bakshi is associate professor in the Department of Metallurgical and Materials Engineering at India Institute of Technology Madras, India. He has a Ph.D. Materials Science and Engineering from Florida International University, USA (2009). He completed his B.E. and M.E. in Metallurgical Engineering from NIT Rourkela (2001) and II.Sc. Bangalore (2003) respectively. His research interests include surface engineering, in situ composites, reactive sintering and ultra-high temperature ceramic composites. His research involves powder metallurgy (mechanical alloying and spark plasma sintering) and coating techniques (thermal spraying and weld cladding), wear and nanomechanical testing. Dr. Bakshi has published 84 *journal papers* (having more than 3500 *citations*) with an *h-index* of 30 and co-authored a book *Carbon Nanotube Reinforced Metal Matrix Composites* (CRC press) and a book chapter on *Effect of Carbon Addition on the Microstructure and Properties of CoCrFeMnNi High Entropy Alloys* (CRC Press). He has also applied for 3 Indian Patents. He has guided 6 Ph.D. and 1 M.S. scholar and 3 post-doctoral fellows. He has been active member in several professional societies including the Indian Institute of Metals Chennai Chapter, ASM International Chennai Chapter, and The Minerals Metals and Materials Society (TMS), USA. He has organized several workshops at IIT Madras and International symposiums. He was awarded a Certificate of Merit for International Activities by Division of Materials and Processing of Japan Society of Mechanical Engineers in 2012, Young Professional Award by ASM International Chennai Chapter in 2013, and Young Leader Professional Development Award by TMS, USA in 2014. He was awarded Institute Research and Development Award (Junior Level) by IIT Madras for 2016. He has served as head of Nanomaterials Laboratory in IIT Madras. He is currently head of the Surface Modifications and Additive Repair Technologies (SMART) Lab at IIT Madras.

1 Introduction

1.1 COMPOSITE MATERIALS

Composite materials consist of a base or matrix material and one or more physically distinct, distributed phases, known as reinforcements or fillers. The reinforcement/filler is added to the matrix in order to obtain desired properties like strength, stiffness, toughness, thermal conductivity, electrical conductivity, coefficient of thermal expansion, electromagnetic shielding, damping, and wear resistance. Composite materials can be seen everywhere, from airplanes to cars and sports equipment. They have become an essential part of our day-to-day life. In fact, the basic principles of composite materials were applied quite early in building mud houses, where the clay was reinforced with grass straws, and boat making in which wooden planks were held together with iron plates. The use of reinforced concrete in the construction and infrastructure industry is another example of a composite material. Nature is a great manufacturer and source of composite materials. Natural materials like wood and bone are composite materials with multi-scale microstructure and are quintessential examples of the synergistic principles behind the improvement of the properties. Nowadays, an entirely new field of study, namely biomimetics, is dedicated to the understanding and reproduction of the structure of the natural materials like nacre to enhance properties or to attain similar functionality.

The dimensional stability of the structure and the amount of material required to build it are determined by the mechanical properties of the material used, namely the strength and the elastic modulus. The stronger the material, the lesser the amount required and lighter is the structure. In some applications like aircrafts and automobiles, materials with low density and high strength are highly desirable for making them fuel-efficient. It is difficult to obtain a single homogeneous material having all the desirable properties. Although metals and alloys have very high strength and are tough, they have limited elastic modulus. Ceramics, on the other hand, have excellent elastic modulus but have low toughness and ductility. It is well known that metallurgical heat treatments can increase the strength of a material to appreciable extent, but cannot increase the elastic modulus significantly. One of the strategies to increase the strength of a material has been to decrease the grain size. This has led to the development of nanocrystalline materials [1]. However, the fabrication of bulk structural components with nanostructured grains is still a very big challenge due to severe grain growth, although several novel manufacturing methods have been developed [2]. The need for higher fuel efficiency and higher speed, demands lowering of the overall weight of an automobile. In applications such as space shuttles, space telescopes, and orbiter, employing lightweight and high-strength materials translates to lower cost of transportation as well as increased lifetime. Some applications like heat sinks in electronic circuits require increased strength and thermal conductivity while having a lower coefficient of thermal expansion (CTE). Fillers are added in order to achieve electronic conduction. Hence, the need for materials with tailored properties led to development of composite materials.

A lot of research has been carried out on particulate and fiber reinforced composites, which can be ascribed partly to the development of ceramic fibers and whiskers of high strength and stiffness. Due to the relatively lower amount of structural defects like dislocations and internal cracks in whiskers, strengths close to the theoretical cohesive strength can be achieved in this form. Fiberglass was invented in 1938 by Games Slayter of the Owens-Corning company [3,4] and was originally used for insulation purposes. In 1959, Claude P. Talley demonstrated the first boron fibers having stiffness of approximately 440 GPa and strength of approximately 2.4 GPa [5]. Another landmark was achieved in 1964, when Stephanie Kwolek discovered Kevlar fiber, which had up to 8 times the specific strength of aluminum alloy while having density less than 60% that of glass fiber [6]. A significant amount of research has been carried out during the last 40 years in fabrication and understanding of composite materials. Figure 1.1 shows the yearly cumulative number of research publications on various aspects of composites for different fiber reinforcements irrespective of the type of matrix. It is observed that fiberglass and boron fibers were very popular reinforcements in the composite industry in the late 1960s. Glass fiber reinforced plastics (GFRP) were used for structural applications like boats, storage tanks, houses, and even airplane interiors. The development and availability of high-quality and high-strength carbon fibers in the late 1970s fueled the research in carbon fiber reinforced composites as is seen by the rapid increase in the number of publications during the 1980s. Metal matrix composites having particulate as well as fibrous reinforcements have been developed that possess high-temperature capability, high

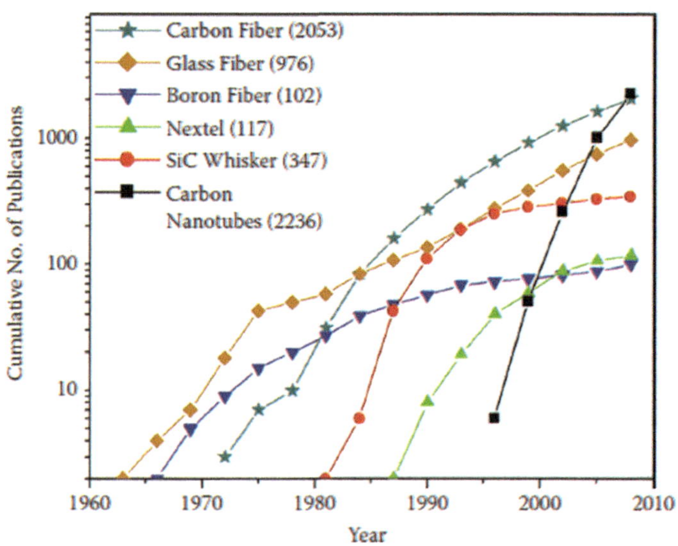

FIGURE 1.1 Year-wise cumulative number of publications on composites containing different kinds of fibrous reinforcements. (Data compiled using Scopus).

thermal conductivity, low CTE, and high specific stiffness and strength. They find applications in advanced automobiles, space antennas, aircraft brakes, sporting goods like tennis rackets and baseball bats, and heat dissipation and management in integrated circuits.

1.2 DEVELOPMENT OF CARBON FIBERS

Roger Bacon in 1958, working at the Union Carbide Corporation and studying the triple point of graphite, observed the formation of stalagmite-like structures caused by evaporation and condensation of the graphite from the anode during an arc discharge process under a high-pressure inert gas (approximately 92 atm, which is a little lower than the triple point pressure of graphite) [7]. The deposit contained whiskers of graphite from a fraction of a micron to a few microns in diameter and up to 3 cm in length. This was the first instance of synthesis of flexible fibers with strength up to 20 GPa and elastic modulus of up to 700 GPa, which was higher than any other fiber known during that time. Bacon [7] also proposed a scroll-like structure for the carbon whiskers. Subsequently, carbon fibers and woven mats were available, which were produced from the carbonization of rayon and polyacrylonitrile (PAN) fibers. Leonard Singer, also working at the Union Carbide Corporation, developed highly oriented graphitic fibers by carbonization of pitch during the 1970s [8]. These pitch-based fibers had a very high elastic modulus up to 1000 GPa and high thermal conductivity but had lower strength than PAN-based fibers. Vapor grown carbon fibers (VGCF) were produced by a catalytic chemical vapor deposition process (CCVD) in which a hydrocarbon/hydrogen mixture undergoes dissociation at high temperatures in the presence of catalyst particles with the result of formation of carbon fibers on the catalyst particles. Depending on the growth conditions, the fibers could be between 0.1 and 1.5 μm in diameter and up to 1 mm in length. For a comprehensive study on the fabrication and properties of carbon fibers and their composites, the readers are referred to a text by Peter Morgan [9]. Manufacture of carbon fibers of high strength in the 1960s and 1970s made them the first choice for the manufacture of advanced composites for use in rocket nozzle exit cones, missile nose tips, re-entry heat shields, packaging, and thermal management. Extensive research has been carried out in the area of carbon fiber reinforced metal matrix composites. Since 1970, carbon fiber reinforced composites have been extensively used in wide array of applications like aircraft brakes, space structures, military and commercial planes, lithium batteries, sporting goods, and structural reinforcement in construction.

1.3 CARBON NANOTUBES: SYNTHESIS AND PROPERTIES

The discovery of carbon nanotubes has been widely attributed to Iijima in 1991 [10]. However, this has been debated as several other researchers had synthesized and reported carbon structures similar to those reported by Iijima in 1991. Monthioux and Kuznetsov have compiled some of the earlier reports in a guest editorial of the journal *Carbon* [11]. Most notable are the filamentous tubes synthesized by Radushkevich et al. [12] in 1952, Bacon in 1960 [7], and Oberlin et al. [13] in 1976. Oberlin et al. had produced hollow tubes of carbon ranging between 2 and 50 nm in diameter by decomposition of a mixture of benzene and hydrogen and had described the structure as "turbostratic stacks of carbon layers, parallel to the fiber axis and arranged in concentric sheets like the annular rings of a tree." Although carbon nanotubes might have been synthesized earlier, it took the genius of Iijima to realize that these were made up of multiple seamless tubes arranged in a concentric manner as opposed to the scroll-like structure of filaments proposed by Bacon. Subsequent to discovery of multi-walled carbon nanotubes (referred to as CNT throughout this book), single-walled carbon nanotubes (hereafter referred to as SWNT throughout this book) were discovered independently by Iijima and Ichihashi [14] and Bethune et al. [15] and reported in the same issue of Nature in 1993.

A SWNT can be obtained by rolling a sheet of graphene (a single layer of sp^2 bonded carbon) to form a seamless tube. There could be many ways for doing this. As shown in Figure 1.2, when the graphene sheet is rolled along the chiral axis C_h, that is, by joining both ends of C_h, a nanotube would result with a circumference equal to the length of C_h. The chiral axis can be represented by the integers (n, m) where $C_h = n_{\mathbf{a1}} + m_{\mathbf{a2}}$; $\mathbf{a}_1$ and $\mathbf{a}_2$ being the lattice translation vectors as shown in Figure 1.2. The diameter of the nanotube would depend on the (n, m) and is given by

$$a\left(\sqrt{n^2 + m^2 + nm}\right) / \pi$$

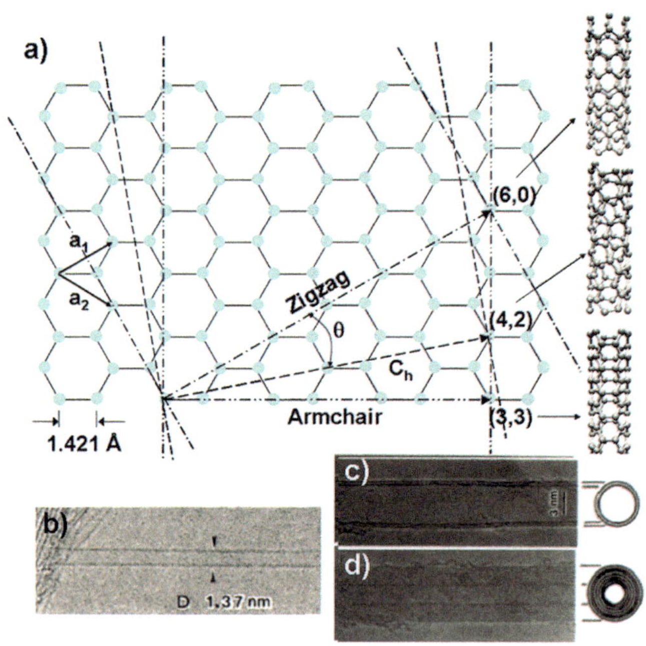

FIGURE 1.2 (a) Schematic showing the formation of an SWNT by rolling along different chiral vectors $\boldsymbol{C}_h$ and the resulting SWNTs, and (b)–(d) high-resolution TEM images showing a single, double, and seven-walled nanotube, respectively [10]. (From Nature Publishing Group. With permission).

where a is the lattice vector = 2.46 Å. "Armchair" nanotubes are formed when $n = m$ and a "zigzag" nanotube is formed when either n or $m = 0$. All armchair nanotubes and nanotubes with $n–m = 3k$ are metallic, whereas others are semiconducting. The physical properties of carbon nanotubes and related materials are tabulated in Table 1.1. SWNTs have excellent electrical and thermal conductivities owing to the ballistic nature of conduction of electrons and phonons that allows them to carry large current densities without significant heating. It is observed from Table 1.1 that although thermal conductivity of individual nanotubes is quite higher than metals (Cu = 400 W m^{-1} K^{-1}), aggregates have been shown to have lesser conductivity values. An excellent account of transport properties in CNTs is provided by Saito, Dresselhaus, and Dresselhaus [36].

The carbon nanotube synthesis set-up used by Iijima was an arc discharge apparatus similar to those used for carbon filament synthesis by Bacon, but operating at a

TABLE 1.1
Physical Properties of Carbon Materials

Property	Graphite	Diamond	Carbon Fiber	SWNT	CNT
Specific heat capacity (at 300 K), J kg^{-1} K^{-1}	710 [16]	486 [16]	–	~650 [17]	~480 [18]
Thermal conductivity at RT, W m^{-1} K^{-1}	165 [19]	3320 [20]	1900 for VGCF [21]	6600 for single SWNT [20], 35 W m^{-1} K^{-1} for disordered mat [22], 200 for aligned mats [23]	3000 for single CNT [24], 2.5 for bulk CNT sample [25]
Electrical conductivity	900–1700 S cm^{-1} [19]	Insulator	24 S cm^{-1} [26]	Resistivity of single rope $<10^{-4}$ohms cm^{-1} [27]	1850 S cm^{-1} with current density of 10^7 A cm^{-2} [28]
Magnetic susceptibility, emu g^{-1}	-30×10^{-6} when magnetic field is parallel to c-axis [29]	-4.9×10^{-7} [29]	–	Current densities up to 4×10^9 A cm^{-2} [30] -10.65×10^{-6} for bundles containing nanoparticles and magnetic field parallel to bundle axis [31]	Saturation magnetization of as grown Fe containing CNTs = 17.7 and pure CNT = 1.1 [32]
Thermoelectric power at 300 K, μV K^{-1}	−3.5 [33]	3500 for semiconducting diamond [34]	–	~50 [35]	~22 [33]

lower pressure of argon (100 torr). Multi-walled CNTs having 2–50 walls (or concentric tubes) were deposited by evaporation of carbon from the anode and condensation on the cathode. Ebbesen and Ajayan studied the arc discharge method further and found that the optimal pressure for CNT synthesis was 500 torr, which resulted in a ~75% conversion [37] thereby producing them in large quantities. SWNTs were formed when a small amount of iron was placed on a dimple in the cathode and a mixture of methane and argon atmosphere was used during arc discharge. Bethune et al. at IBM discovered the formation of SWNT on the cathode when a 2 at.% Co containing anode was used in the arc discharge apparatus under helium atmosphere. Guo et al. of Richard Smalley's group at Rice University were the first to synthesize SWNT by evaporation of a hot (1200 °C) transition metal containing carbon target by laser ablation method followed by the condensation on a cold finger [38]. Chemical vapor deposition has also been used to produce CNT and to some extent SWNT [39]. Wang et al. developed a large-scale fluidized bed CVD process for synthesis of CNT of up to 80% purity at the rate of 50 kg day^{-1} [40]. The temperature, gas compositions, and catalysts used are important parameters that determine quality of CNTs produced. CVD-grown CNTs are generally impure as compared to arc discharge CNTs due to the presence of nanometer-size catalyst particles unless purified. Presence of the catalyst sometimes impairs the formation of walls and leads to poor graphitization.

The worldwide interest in carbon nanotubes is evident from the fact that in a span of just 15 years, the number of publications on carbon nanotubes composites has exceeded that of the carbon fiber composites over the last 40 years (Figure 1.1). This is due to the near perfect structure of CNTs, which results in excellent properties [41]. The mechanical properties of SWNTs and CNTs have been measured using direct and indirect methods and have been tabulated in Table 1.2. Based on these results, it can be said that CNTs have an elastic modulus greater than carbon fibers

TABLE 1.2

Summary of Experimental Measurements of Mechanical Properties of CNTs

S. No.	Method	Remarks	Refs.
1	Amplitude of thermal vibrations of CNTs at different temperatures in a TEM	E = 0.4–4.15 TPa Avg. = 1.8 TPa	[42]
2	Same as 1 for SWNTs	E = 1.3–0.4/+0.6 TPa	[43]
3	Force-displacement curve of pinned CNT using AFM	E = 1.28 ± 0.59 TPa	[44]
5	Shifts in D* peaks of the Raman spectra of CNT in epoxy composites	E = 2.8–3.6 TPa for SWNT and 1.7–2.4 TPa for CNT	[45]
6	Frequency of electromechanical resonances	E = 0.1–1 TPa for CNT	[46]
7	Bend test of simply supported CNT	E = 870 GPa for arc CNT and 27 GPa for CVD CNT	[47]
8	Same as 7 for SWNT ropes	E = 1 TPa	[48]
9	Tensile test of CNT in SEM	E = 270–950 GPa Strength = 11–63 GPa	[49]
10	Same as 9 for SWNT ropes	E = 320–1470 GPa Strength = 13–52 GPa	[50]

and strength up to 5 times that of carbon fibers. Therefore, they are the strongest materials known to mankind. SWNTs have been found to have better physical and mechanical properties compared to CNTs due to the presence of defects in CNTs. Because of these reasons, as well as their superior thermal and electrical property, a lot of attention has been devoted to using carbon nanotubes as reinforcements for composite materials – the focus of this book.

1.4 CARBON NANOTUBE – METAL MATRIX COMPOSITES (CNT-MMCs)

Due to their extraordinary properties, be it experimentally measured or theoretically computed, CNTs caught the attention of researchers and work on development of CNT composites started at a tremendous pace as shown in Figure 1.1. Figure 1.3 shows the yearly number of publications on CNT reinforced metal, ceramic, and polymer composites for the first three decades after the discovery of CNTs. It is observed that most of the research is carried out on the development of CNT reinforced polymer matrix composites (PMCs). The idea was to replace graphite fiber with CNTs because the amount of CNTs required would be lower for achieving the same levels of strengthening. In fact, one of the early applications has been replacement conductive automotive fuel transmission lines, for which originally carbon black was employed. The main reason for a majority of the research focus on PMC can be attributed to the ease of polymer processing, which can be carried out at small stresses and low temperatures as compared to metal and ceramic matrices. Metal and ceramic processing requires high temperatures and/or high pressures, which requires more expansive and generally larger equipment. This makes startup costs and

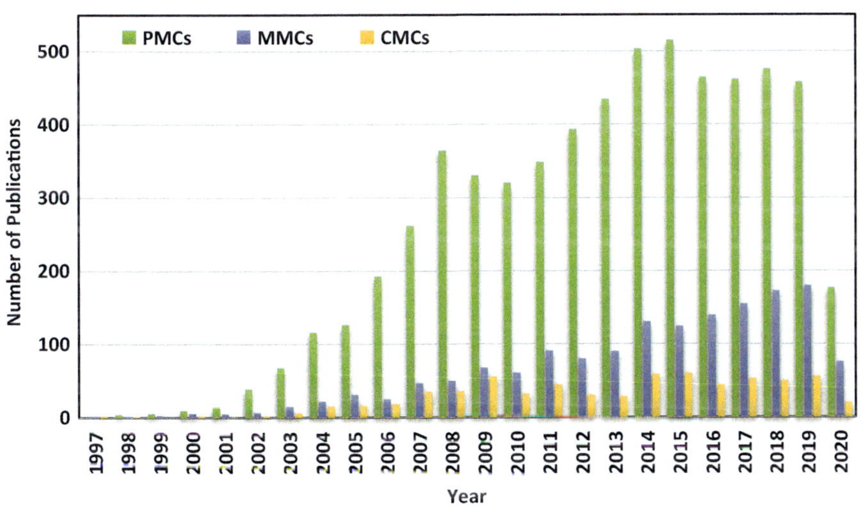

FIGURE 1.3 Published journal articles per year for PMCs, MMCs, and CMCs from 1997 till present. 2020 data is thru June. (Data compiled from Scopus using search criteria "carbon nanotube 'X' matrix composite, where X = polymer, metal, or ceramic).

accessibility higher, that is, the potential return on investment (especially for laboratories and industry) must be more evident. From the material standpoint itself, the high temperature and pressures make damage to the CNTs more likely, which in turn affects their efficacy as the reinforcement phase.

Metal matrix composite processing requires high temperatures and pressures, that can yield adverse effects on nanostructured materials. In addition, there are stringent requirements for metal's isolation from the atmosphere to avoid oxidation. Hence, this may require specially designed equipment, such as expensive vacuum systems. Carbon nanotubes might react with metals to form carbides and hence be destroyed. Some of these aspects initially restricted the interest in CNT reinforced metal matrix composites (MMCs). From Figure 1.3, it is seen that the interest in CNT reinforced MMCs has accelerated significantly over the last decade. With the demonstration of extraordinary increase in the strength and the elastic modulus [51], several groups have investigated a wide variety of metallic matrices. Figure 1.4 shows the plot of year-wise number of publications for major metal matrices that have been reinforced with CNTs. It is observed that, in general, interest in all matrices has continued increasing thru a second decade of CNT-MMC research. Figure 1.5 shows that a lot of research has been done in developing structural metals from light weight metals (e.g., Al, Mg, Ti), to heavier metals such as steel and tungsten. Many of these composites have envisioned dual increases in mechanical as well as functional properties. Cu has received attention for development of high thermal conductivity. Use in Co and W has pushed the application limits to higher temperatures, where CNTs have demonstrated surprising robustness.

There are several challenges in the fabrication of MMCs with CNT reinforcement. By far the most important challenge has been to obtain a uniform distribution of CNTs in the matrix. CNTs have large specific surface area up to 200 $m^2\ g^{-1}$ and hence they tend to agglomerate and form clusters due to van der Waals forces. In addition, the non-wetting nature of CNTs to most molten metals results in their clustering.

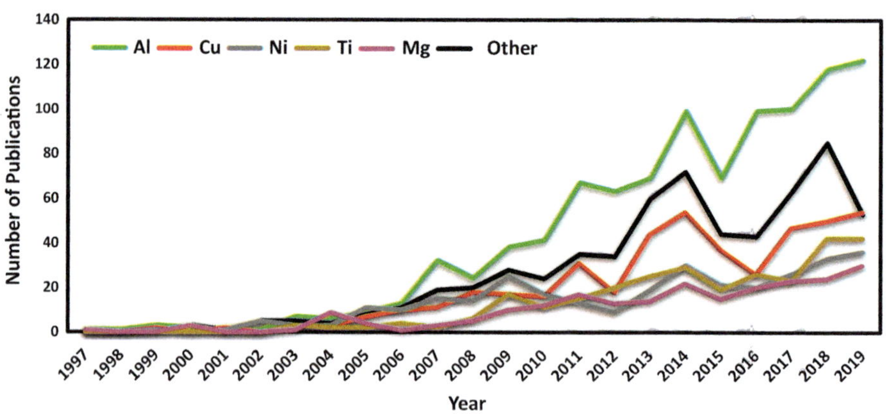

FIGURE 1.4 Journal articles published on different CNT-MMCs from 1997 thru 2019. (Data compiled from Scopus using search criteria "carbon nanotube 'Y' matrix composite, were Y is the specific metal).

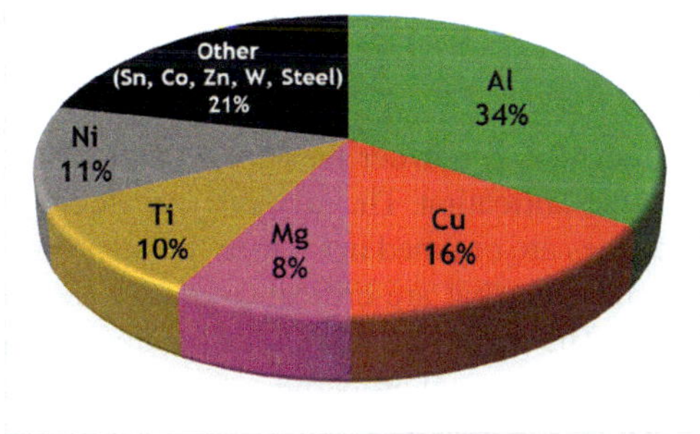

FIGURE 1.5 Total publications on different types of CNT-MMCs from 1997 to 2020. (Same Scopus search criteria used as in Figure 1.4).

Good dispersion of the reinforcement is a necessity for the efficient use of the properties as well as for obtaining homogeneous properties. CNT clusters have lower strength and higher porosity and serve as discontinuities and stress concentrators. The second important challenge is to ensure the structural and chemical stability of the CNTs in the metal matrix. Owing to the high temperatures and stresses involved in MMC processing, CNTs may be damaged or lost due to reaction. These aspects need special attention, which is not the case with PMCs. Carbon nanotubes surely have the potential to produce composites with extraordinary properties and combinations of properties not present in any materials known to man. Many applications have been projected for CNT metal matrix composites based on the mechanical and functional properties of CNTs. Much research is still underway for overcoming the challenges and understanding the behavior of these composites. Earlier research on carbon nanotube reinforced metal matrix composites (referred to as CNT-MMC throughout the book) was limited to miniature samples in the laboratory due to the high cost of carbon nanotubes. In the early 1990s, the cost of SWNT was almost \$1000 g^{-1}. Many new companies have started synthesizing carbon nanotubes, which has resulted in significant reduction in the cost of CNTs. The price of nanotubes depends on the level of purity desired and the specifications as well as on the quantity ordered. SWNTs are expensive because they are difficult to fabricate and purify. Nowadays one can obtain SWNTs for \$25,000 to over \$100,000 per pound and multi-walled CNTs for \$100–\$3000 per pound. These prices are still high when compared with carbon fibers, which depending on their form (free fiber or woven fabric) could be approximately \$10–\$100 per pound. Given the lower amount of CNTs required and the decreasing prices, CNTs might replace carbon fibers and carbon black in certain applications in the future. This book summarizes all the efforts on CNT reinforced metal matrix composites to date in this area. The novel processing methods developed and the idea behind them have been explained. Novel and emerging applications of CNT MMCs will be presented. This book is intended to address the challenges in CNT-MMC processing, the advantages and limitation of various existing processing techniques, and the design philosophy for novel methods of processing. This work will benefit new and existing researchers in this area by providing them all the information for getting started as well as pioneering in this field.

In the chapters that follow, multi-walled carbon nanotubes have been referred to as CNTs and single walled carbon nanotubes as SWNTs. Chapter 2 deals with the approaches and techniques for dispersing CNTs into metallic powders in order to synthesize CNT-MMC composite powder feedstocks, the critical first step to effective CNT-MMC composite fabrication. The methods and challenges in fabrication of bulk CNT-MMC composites are outlined in Chapter 3. Chapter 4 deals with the various characterization techniques that are critical to evaluating and investigating the behavior of CNT-MMCs. The techniques available for microstructural analysis and evaluation of mechanical and physical properties of CNT-MMCs are discussed, with emphasis on the advantages and disadvantages of the various techniques. Chapter 5 provides a comprehensive report of the research work on all metal matrix-CNT composites studied to date. This includes Al-CNT, Cu-CNT, Ni-CNT, Mg-CNT, Si-CNT, and other CNT-MMCs. The tables presented in Chapter 5 provide comprehensive information on the effect of processing techniques and CNT addition on the mechanical properties of the composite. Chapter 6 deals with understanding the strengthening mechanisms in CNT-MMCs, including the latest in situ testing results that confirm and reveal mechanisms unique to CNTs. The micromechanical models available from the fiber composites are outlined and their applicability in predicting properties of MM-CNT composites has been discussed with an example of the experimental data on Al-CNT composites. Chapter 7 deals with interfacial characteristics and phenomena of CNT-MMCs. The factors that influence interfacial reaction, product formation, and its consequence on the microstructure and properties of CNT-MMCs are presented. Chapter 8 summarizes the thermal, electrical, tribological, and corrosion properties of CNT-MMCs. The functional applications of CNT-MMCs include hydrogen storage, sensors, catalysts, and batteries. Chapter 9 touches on an exciting new field of dual nanoparticle reinforced composites. The synergistic use of secondary nanoparticles alongside CNTs is highlighted, and details on promising new composites such as CNT-Diamond- and CNT-Graphene-MMCs are presented. Chapter 10 summarizes the few studies on computational approach utilized in the design of CNT-MMCs and the microstructure and property evolution. Chapter 11 put the excitement surrounding CNT-MMCs into perspective, by highlighting the many patents on CNT-MMC technology filed, and increasing prospects for commercialization of CNT-MMC products. The conclusions from the research carried out on CNT-MMCs since 1997 have been outlined in Chapter 12. The scope and direction for the future work with a roadmap to develop CNT-MMCs is also discussed in Chapter 12.

1.5 CHAPTER HIGHLIGHTS

The idea of composite materials has emerged from the requirement of light-weight materials with improved mechanical and physical properties like strength, toughness, thermal and electrical conductivity, and lower CTE for targeted applications. Fiber reinforced composites are very suitable for structural applications for their high strength and stiffness. Carbon nanotubes are strong contenders in this category, due to their superior elastic modulus, tensile strength, and thermal and electrical conductivity rather than conventional carbon fibers. CNTs could be the nanomaterial of choice for designing and tailoring metallic composites with an array of multiple property combinations not previously thought possible. Being vigorously researched for now for nearly three decades, production cost of multiwall CNTs has come down considerably, where 1 kg of quality CNTs can be attained for less than $1000 USD. The main problem associated with the fabrication of composite structure is the agglomeration of CNTs due to their high surface tension, resulting in poor properties (strength, electrical and thermal conductivity, etc.) than expected. The last decade of CNT-MMC research has seen exciting advances in the in situ observation of CNT strengthening mechanisms, new possible applications of CNTs in high-temperature applications, and better and more scalable fabrication routes that yield uniform dispersion of CNTs.

REFERENCES

1. Mayers, M. A., Mishra, A., and D. J. Benson. 2006. Mechanical properties of nanocrystalline materials. *Prog. Mater. Sci.* 51: 427–556.
2. Viswanathan, V., Laha, T., Balani, K., Agarwal, A., and S. Seal. 2006. Challenges and advances in nanocomposite processing techniques. *Mater. Sci. Eng. R.* 54: 121–285.
3. Slayter, G. 1938. Method and apparatus for making glass wool. U.S.Patent 2133235.
4. Slayter, G. 1941. Method of producing glass fibers. U.S. Patent 2230272.
5. Talley, C. P. 1959. Mechanical properties of glassy boron. *J. Appl. Phys.* 30: 1114–1115.
6. Kwolek, S. 1968. South African Patent Application 6813051.
7. Bacon, R. 1960. Growth, structure, and properties of graphite whiskers. *J. Appl. Phys.* 31: 283–290.
8. Singer, L. S. 1978. The mesophase and high modulus carbon fibers from pitch. *Carbon* 16: 409–415.
9. Morgan, P. 2005. *Carbon Fibers and Their Composites*. Boca Raton, FL: Taylor and Francis, Inc.
10. IIjima, S. 1991. Helical microtubules of graphitic carbon. *Nature* 354: 56–58.
11. Monthioux, M., and V. L. Kuznetsov. 2006. Who should be given the credit for the discovery of carbon nanotubes? *Carbon* 44: 1621–1623.
12. Radushkevich, L. V., and V. M. Lukyanovich. 1952. Zurn. *Fisic. Chem.* 26: 88–95.
13. Oberlin, A., Endo, M., and T. Koyama. 1976. Filamentous growth of carbon through benzene decomposition. *J. Crys. Growth* 32: 335–349.
14. Iijima, S., and T. Ichihashi. 1993. Single-shell carbon nanotubes of 1-nm diameter. *Nature* 363: 603–605.
15. Bethune, D. S., Kiang, C. H., de Vries, M. S., Gorman, G., Savoy, R., Vasquez, J., and R. Beyers. 1993. Cobalt-catalyzed growth of carbon nanotubes with single atomic-layer walls. *Nature* 363: 605–607.
16. *FactSage 5.2*. GTyT Technologies, Kaiserstr. 100, 52134 Herzogenrath, Germany, 2003.
17. Hone, J., Batlogg, B., Benes, J., Johnson, A. T., and J. E. Fisher. 2000. Quantized phonon spectrum of single-wall carbon nanotubes. *Science* 289: 1730–1733.
18. Yi, W., Lu, L., Zhang, D. L., Pan, Z. W., and S. S. Xie. 1999. Linear specific heat of carbon nanotubes. *Phys. Rev. B.* 59. 9015–9018.
19. Buerschaper, R. A. 1944. Thermal and electrical conductivity of graphite and carbon at low temperatures. *J. Appl. Phys.* 15: 452–454.
20. Berber, S. Kwon, Y. K., and D. Tomanek. 2000. Unusually high thermal conductivity of carbon nanotubes. *Phys. Rev. Letts.* 84: 4613–4616.
21. Biercuk, M. J., Llaguno, M. C., Radosavljevic, M., Hyun, J. K., Johnson, A. T., and J. E. Fisher. 2002. Carbon nanotube composites for thermal management. *Appl. Phys. Letts.* 80: 2767–2769.

22. Hone, J., Whitney, M., Piskoti, C., and A. Zettl. 1999. Thermal conductivity of single-walled carbon nanotubes. *Phys. Rev. B.* 59: 2514–2516.
23. Hone, J., Llaguno, M. C., Nemes, N. M., Johnson, A. T., Fischer, J. E., Walters, D. A., Casavant, M. J., Schmidt, J., and R. E. Smalley. 2000. Electrical and thermal transport properties of magnetically aligned single wall carbon nanotubes films. *Appl. Phys. Letts.* 77: 666–668.
24. Kim, P., Shi, L., Majumdar, A., and P. L. McEuen. 2001. Thermal transport measurements of individual multiwalled nanotubes. *Phys. Rev. Letts.* 87: 215502 (1–4).
25. Qin, C., Shi, X., Bai, S. Q., Chen, L. D., and L. J. Wang. 2006. High temperature thermal and electrical properties of bulk carbon nanotube prepared by SPS. *Mater. Sci. Eng. A.* 420: 208–211.
26. Imai, J., and K. Kaneko. 1992. Electrical conductivity of a single micrographitic carbon fiber with a high surface area under various atmospheres. *Langmuir* 8: 1695–1697.
27. Thess, A., Lee, R., Nikolaev, P., Dai, H., Petit, P., Robert, J., Xu, C., Lee, Y. H., Kim, S. G., Rinzler, A. G., Colbert, D. T., Scuseria, G. E., Tomanek, D., Fischer, J. E., and R. E. Smalley. 1996. Crystalline ropes of metallic carbon nanotubes. *Science* 273: 483–487.
28. Ando, Y., Zhao, X., Shimoyama, H., Sakai, G., and K. Kaneto. 1999. Physical properties of multiwalled carbon nanotubes. *Int. J. Inorg. Mater.* 1: 77–82.
29. Issi, J. P., Langer, L., Heremans, J., and C. H. Olk. 1995. Electronic properties of carbon nanotubes: experimental results. *Carbon* 33: 941–948.
30. Dai, H., Javey, A., Pop, E., Mann, D., Kim, W., and Y. Lu. 2006. Electrical trans-port properties and field effect transistors of carbon nanotubes. *NANO: Brief Reports and Reviews* 1: 1–4.
31. Wang, X. K., Lin, X. W., Song, S. N., Dravid, V. P., Ketterson, J. P., and R. P. H. Chang. 1995. Properties of buckytubes and derivatives. *Carbon* 33: 949–958.
32. Gui, X. C., Wang, K. L., Wei, J. Q., Lu, R. T., Shu, Q. K., Jia, Y., Wang, C., Zhu, H. W., and D. H. Wu. 2009. Microwave absorbing properties and magnetic properties of different carbon nanotubes. *Sci China E: Technol. Sci.* 52: 227–231.
33. Tian, M., Li, F., Chen, L., and Z. Mao. 1998. Thermoelectric power behavior in carbon nanotubule bundles from 4.2 to 300 K. *Phys. Rev. B.* 58: 1166–1168.
34. Goldsmid, H. J., Jenns, C. C., and D. A. Wright. 1959. The thermoelectric power of semiconducting diamond. *Proc. Phys. Soc.* 73: 393–398.
35. Hone, J., Ellwood, I., Muno, M., Mizal, A., Cohen, M. L., Zettl, A., Rinzler, A. G., and R. E. Smalley. 1998. Thermoelectric power of single-walled carbon nanotubes. *Phys. Rev. Lett.* 80: 1042–1045.
36. Saito, R., Dresselhaus, M. S., and S. Dresselhaus. 1998. *Physical Properties of Carbon Nanotubes*. London: Imperial College Press.
37. Ebbesen, T. W., and P. M. Ajayan. 1992. Large-scale synthesis of carbon nanotubes. *Nature* 358: 220–222.
38. Guo, T., Nikolaev, P., Thess, A., Colbert, D. T., and R. E. Smalley. 1995. Catalytic growth of single-walled nanotubes by laser vaporization. *Chem. Phys. Lett.* 243: 49–54.
39. Yacaman, M. J., Yoshida, M. M., Rendon, L., and J. G. Santiesteban. 1993. Catalytic growth of carbon microtubules with fullerene structure. *Appl. Phys. Lett.* 62: 202–204.
40. Wang, Y., Wei, F., Luo, G., Yu, H., and G. Gu. 2002. The large-scale production of carbon nanotubes in a nano-agglomerate fluidized-bed reactor. *Chem. Phys. Lett.* 364: 568–572.
41. Baugman, R. H., Zakhidov, A. A., and W. A. de Heer. 2002. Carbon nanotubes: the route towards applications. *Science* 297: 787–792.
42. Treacy, M. M. J., Ebbesen, T. W., and J. M. Gibson. 1996. Exceptionally high Young's modulus observed for individual carbon nanotubes. *Nature* 381: 678–680.
43. Krishnan, A., Dujardin, E., Ebbesen, T. W., Yianilos, P. N., and M. M. J. Treacy. 1998. Young's modulus of single-walled nanotubes. *Phys. Rev. B.* 58: 14013–14019.

44. Wong, E. W., Sheehan, P. E., and C. M. Lieber. 1997. Nanobeam mechanics: elasticity, strength, and toughness of nanorods and nanotubes. *Science* 277: 1971–1974.
45. Lourie, O., and H. D. Wagner. 1998. Evaluation of Young's modulus of carbon nanotubes by micro-Raman spectroscopy. *J. Mater. Res.* 13: 2418–2422.
46. Poncharal, P., Wang, Z. L., Ugarte, D., and W. A. de Heer. 1999. Electrostatic deflections and electromechanical resonances of carbon nanotubes. *Science* 283: 1513–1516.
47. Salvetat, J. P., Kulik, A. J., Bonard, J-M., Briggs, G. A. D., Stockli, T., Metenier, K., Bonnamy, S., Beguin, F., Burnham, N. A., L. Forro. 1999. Elastic modulus of ordered and disordered multiwalled carbon nanotubes. *Adv. Mater.* 11: 161–165.
48. Salvetat, J. P., Briggs, G. A. D., Bonard, J. M., Bacsa, R. R., and A. J. Kulik. 1999. Elastic and shear moduli of single-walled carbon nanotube ropes. *Phys. Rev. Lett.* 82: 944–947.
49. Yu, M. F., Laurie, O., Dyer, M. J., Moloni, K., Kelly, T. F., and R. S. Ruoff. 2000. Strength and breaking mechanism of multiwalled carbon nanotubes. *Science* 287: 637–640.
50. Yu, M. F., Files, B. S., Arepalli, S., and R. S. Ruoff. 2000. Tensile loading of ropes of single wall carbon nanotubes and their mechanical properties. *Phys. Rev. Lett.* 84: 5552–5555.
51. Cha, S. I., Kim, K. T., Arshad, S. N., Mo, C. B., and S. H. Hong. 2005. Extraordinary strengthening effect of carbon nanotubes in metal-matrix nanocomposites processed by molecular-level mixing. *Adv. Mater.* 17: 1377–1381.

2 Dispersion of Carbon Nanotubes in the Metal Matrix

This chapter is intended to emphasize the significance of carbon nanotube dispersion and its quantification methods. Normally, carbon nanotubes are obtained in the form of a black powder similar to metallic powders. The powders are made up of lumps, which are made up of entangled CNTs. When looked inside a scanning electron microscope (SEM), one can see the CNTs twisted and entangled with each other. In case of chemical vapor deposition (CVD) grown CNTs, the catalyst particles can be seen in the transmission electron microscope (TEM) images. The uniform dispersion of CNTs in the metal matrix is a major challenge. The reasons for difficulty in dispersing CNTs in the metal matrix are discussed. A discussion of the various methods utilized for improving CNT dispersion is presented with their applicability, limitations, and drawbacks. Achieving uniform carbon nanotube dispersion in the metal matrix is the main criteria for successful processing. As such, novel methods are presented for the quantification of CNT dispersion in the composite. The necessity and the advantage of CNT distribution quantification are also discussed.

2.1 SIGNIFICANCE OF CARBON NANOTUBE DISPERSION

Uniform dispersion of CNTs has been by far the most significant challenge in the field of CNT-reinforced composites. This observation is applicable in all kinds of CNT composites – polymer, ceramic, or metal matrix. The tremendous surface area of CNTs of up to 200 $m^2\ g^{-1}$ leads to formation of clusters due to Van der Waals forces. In the subsequent chapters, it will be a recurring theme that adequate dispersion of CNTs is one of the most significant effects on the mechanical and functional properties of CNT reinforced metal matrix composites (CNT-MMCs). Different degrees of dispersion also cause scatter in the mechanical and functional properties for the same CNT content. Clustering and non-uniform dispersion of CNTs will lead to inhomogeneous property distribution in the structural component because the properties like thermal expansion coefficient, thermal conductivity, elastic modulus, and strength depend on the volume fraction of the CNTs. In addition, it will result in a less efficient utilization of the mechanical properties of the individual CNTs [1–3].

The primary challenges involved in adequate dispersion of CNTs within the metal matrix reaction are related to interfacial and often intrinsic incompatibilities between CNTs and metals. The high specific surface area of CNTs makes them intrinsically difficult to disperse as strong van der Waals attractive forces promote the formation of CNT bundles and clusters. The flexible tubular nature of many CNTs can exacerbate this issue by enabling string-like entanglement. CNTs often exhibit hydrophobic properties, while commonly used metals such as Al are hydrophilic due to the natural nanometric oxide film. This contributes to weak chemical attraction between CNT and metals such as Al. Lastly, many metal feed stocks have specific surface area that is orders of magnitude smaller than CNTs, and hence there can be insufficient space or area on the surface of metal particulates for CNTs to become attached onto.

Most of the early research on fabrication of CNT-MMC composites used blending techniques to add CNTs to metals [4–6]. Blending by mixing is not very effective in dispersing the CNTs. Consequently, the properties have been found to be poor in clustered systems. A decrease in the hardness has been observed in the case of silver matrix composites for CNT volume fractions higher than 10% [6]. The wear rate of Cu matrix composites is found to decrease with the addition of CNTs up to 12 vol.%, but the wear rate increases for 16 vol.% CNT composite [4]. In another study on Cu matrix composites, the hardness was found to increase and the wear rate was found to decrease with an increase in the CNT content up to 10 vol.% [7]. Further increase in the CNT content led to the deterioration of wear properties suggesting a critical CNT content, where clustering is avoided. A decrease in the hardness and strength of lead-free composite solders has been observed with larger CNT contents [8,9]. These reports indicate the necessity for having a uniform CNT distribution and the deterioration of properties due to clustering at higher CNT content.

On the other hand, improved dispersion has resulted in better properties. A sevenfold increase in the compressive yield strength was reported for Al-1.6 wt.% CNT composites prepared by powders on which CNTs were dispersed by nanoscale dispersion (NSD) [10]. A sample prepared by hot extrusion of SPS compacts of these powders showed an increase in the tensile yield strength by 128% for Al-5 vol.% CNT composite [11]. Good dispersion obtained by mechanical alloying has been shown to improve the compressive yield strength of Al-4 vol.% CNT composites by 46% [12]. Similar processing leads to an increase in the tensile yield strength and elastic modulus by 133% and 57% in Al-4.5 vol.% CNT composite [13]. Use of the CVD technique to grow CNTs in situ on Al powders has been found to be successful for obtaining powders with good dispersion [14]. Samples prepared by pressing-sintering of these powders were shown to have an increase in the tensile strength and elastic modulus by 130% and 30% for Al-8 vol.% CNT composite [15]. Improved dispersion attained by the molecular level mixing method has been found to lead to an increase in the compressive yield strength by two times in a Cu-10 vol.% CNT composite [16]. These results indicate the improvement of the properties due to CNT dispersion.

2.2 METHODS FOR CARBON NANOTUBE DISPERSION

Several methods have been suggested for the improvement of dispersion of CNTs in the starting powder feedstock and composite. These methods are summarized briefly in the following sections.

2.2.1 Blending

Most of the research efforts on the synthesis of CNT-MMCs have been by the powder metallurgical route. Thus, the dispersion of the CNTs in the precursor powders has received much attention. Mere blending by using a turbula shaker or by roller milling for small lengths of time was inefficient and the poor dispersion in the powders was carried over to the composite [5,17]. Figure 2.1 shows poorly dispersed Al-CNT powders and the fracture surface of the composite prepared by hot rolling of the powders. It can be seen that the CNT clustering in the powders (Figure 2.1a) is carried over to the final product (Figure 2.1b). The composite showed a decrease in the tensile strength by 20% due to the clustering phenomena [17].

2.2.2 Ball Milling

2.2.2.1 High Energy Ball Milling (HEBM)

The first and most extensively used method to disperse CNTs in the metal powder is ball milling or mechanical milling. The process is also referred to as mechanical alloying, however, in the composites field mixing rather than alloying is the objective. Mechanical milling can be carried out using various types of milling machines, the most common being a high energy ball mill (HEBM), of which there are several configurations. In a planetary ball mill configuration, several jars can be utilized at once, with the jars being arranged on a rotating sun wheel as shown in the schematic in Figure 2.2. During mechanical milling, the metal powder-CNT mixture is fed into a jar. Hardened steel balls or ceramic balls are added along with the powder mixture. Sometimes a mixture of balls of different sizes is utilized to ensure that the milling action continues in the voids between the larger balls. The rotation of the sun wheel causes the jars to rotate and finally the energy is transferred to the balls and powders in the jar. The ball-to-powder ratio (BPR) is an important parameter that affects the amount of energy transferred to the metal-CNT mixture. The CNT-metal mixture is

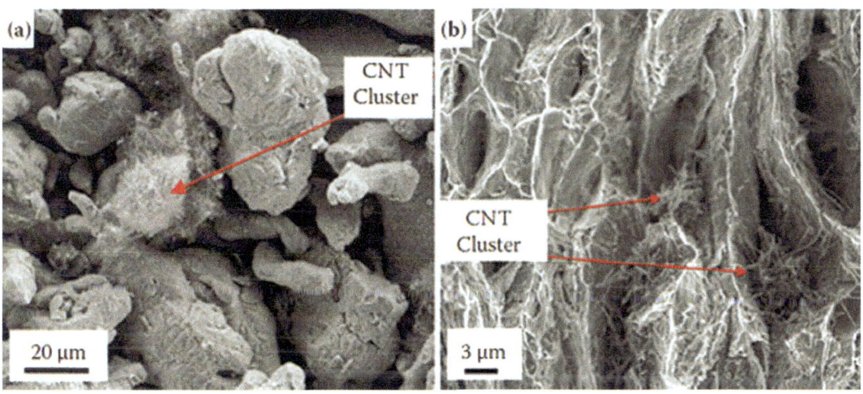

FIGURE 2.1 Micrographs showing (a) poor dispersion and clustering in Al-CNT powder prepared by improper milling, and (b) clusters observed in the fracture surface of the composite prepared by hot rolling of the powders [17]. (Reproduced with permission from Elsevier).

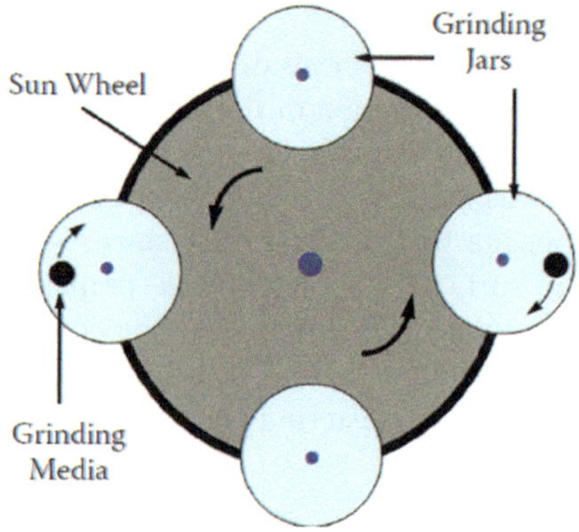

FIGURE 2.2 Schematic of the arrangement of the milling jars on the sun wheel

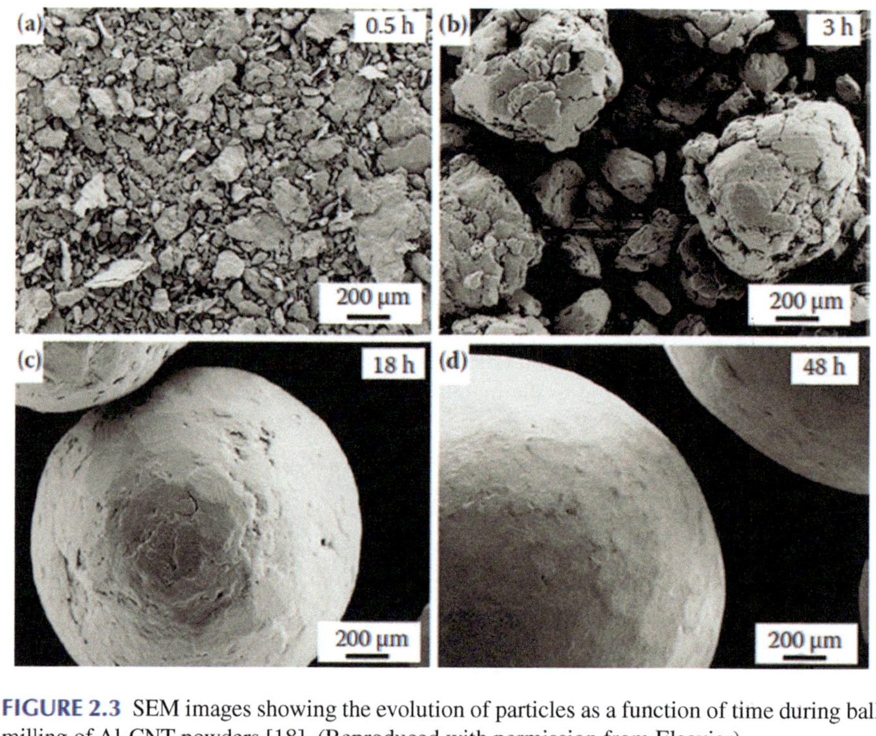

FIGURE 2.3 SEM images showing the evolution of particles as a function of time during ball milling of Al-CNT powders [18]. (Reproduced with permission from Elsevier).

subject to the repeated impact between the balls and the container walls, which can lead to the breakdown of CNT clusters and the localized welding of the CNTs to the metal. Addition of a small amount of stearic acid can prevent cold welding of the particles and inhibit coarsening. Figure 2.3 shows the SEM image of an Al-2 wt.% CNT mixture that has undergone mechanical milling for different amounts of time [18]. It is observed from Figure 2.3b that after 3 h of milling, the particle size starts

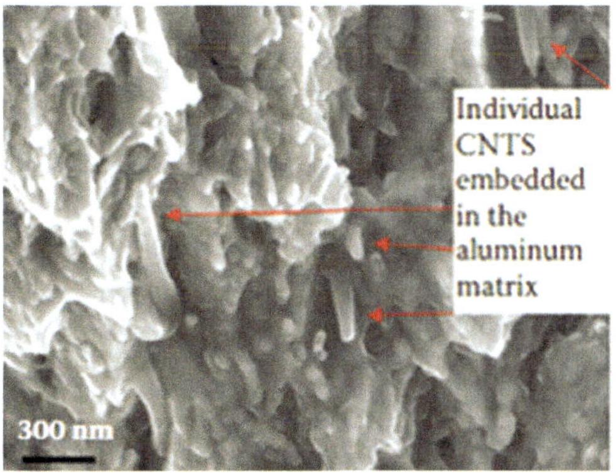

FIGURE 2.4 SEM of powder dispersion attained by ball milling [18]. (Reproduced with permission from Elsevier).

to increase due to cold welding. The particles start to become large and rounder due to the rolling action as shown in Figure 2.3c and 2.3d. The particle size after 48 h of rolling is more than 1 mm, which might not be suitable for use in several processing techniques. Figure 2.4 shows the SEM of the fracture surface of the particle after 48 h of milling with moderately good dispersion of the CNTs in the aluminum matrix. The effects and effectiveness of ball milling have also been found to be dependent on the length and morphology of CNTs used. Esawi et al. [19] investigated the interactions of small (30–50 nm diameter, 10–20 μm length) and bent CNTs, in comparison with larger (140 nm diameter, 7 μm length) straight CNTs in an aluminum matrix. Both Al/CNT mixtures were milled under argon atmosphere with methanol as a processing control agent using stainless steel balls, at rotational speeds of 400 rpm, and a BPR of 10:1. After consolidation, the composite with the long and straight CNTs showed no signs of agglomeration, in contrast to the composite using the small bent CNTs where CNT clusters were found. The smaller CNTs were also more prone to forming carbide secondary phases. Mechanical blending/milling has been a moderately effective technique to disperse CNTs in the metal powder but the parameters like BPR, rotation speed, milling duration, and starting powders size must be optimized to get better dispersion and bonding between CNTs and the metal powder [20].

2.2.2.2 Cryogenic Mechanical Milling

Cryogenic mechanical milling, also known as "cryomilling," is a modified ball milling technique that utilizes cryogenic media, typically liquid nitrogen, to provide cooling to the ball milling process. The liquid nitrogen can be either internal or external to the milling jars containing the powders to be mixed. Cryomilling is most widely used to develop nanostructured alloys as it provides several advantages over other ball milling techniques, especially HEBM. The low temperatures induced by cryomilling make it difficult for particles to agglomerate, sinter, or melt together, and make reactions slower or less likely. Oxidation is especially reduced during cryomilling. The low temperatures minimize cold welding or grain growth, thereby enabling the attainment of fine grain structures. The severe mechanical impacts along with the reduction of dislocation annihilation and recrystallization yield metallic powders with very high dislocation densities and hence very high strength.

Maite et al. [21] utilize cryomilling to synthesize Al-CNT composites using gas atomized spherical Al5083 powder and MWCNTs of 40–47 nm diameter and length of 0.2–0.5 μm. One set of Al/CNT mixtures were cryomilled together for 8 h, with a BPR of 32:1, and a speed of 180 rpm using 10 mm diameter stainless steel balls. Liquid nitrogen was maintained at a constant level within the milling attritor. A second mixture of Al/CNT consisted of Al powder cryomilled using the above parameters, followed by a second cryomilling run where MWCNTs were introduced. The second lower energy run was conducted using 10 mm stainless balls for 1 h, at a speed of 180 rpm, and a BPR of 10:1. Powders were consolidated using spark plasma sintering.

Agglomerates persisted using the two-step approach, resulting in poor densification, mechanical properties, and wear resistance. The combined mixing yielded improved CNT dispersion, better densification, and mechanical properties over the two-step mixing approach. However, CNTs exhibited more damage in the combined mixing approach and mechanical properties only matched that of cryomilled Al. Wear resistance did improve in the Al-MWCNT composite using the combined mixing, compared to the control cryomilled Al material [21]. Omidi et al. [22] used similar cryomilling parameters to synthesize Al6061-1 wt.% CNT composites. Omidi et al. [22] found that cryomilled Al-CNT composites exhibited greater wear resistance over ball milled composites.

2.2.2.3 Solution Ball Milling (SBM)

Solution ball milling (SBM) is an approach that combines wet chemistry with ball milling [23]. Figure 2.5 shows the process schematically. Chen et al. [23] first use an isopropanol solution laden with surfactants (1 wt.% zwitterionic surfactants

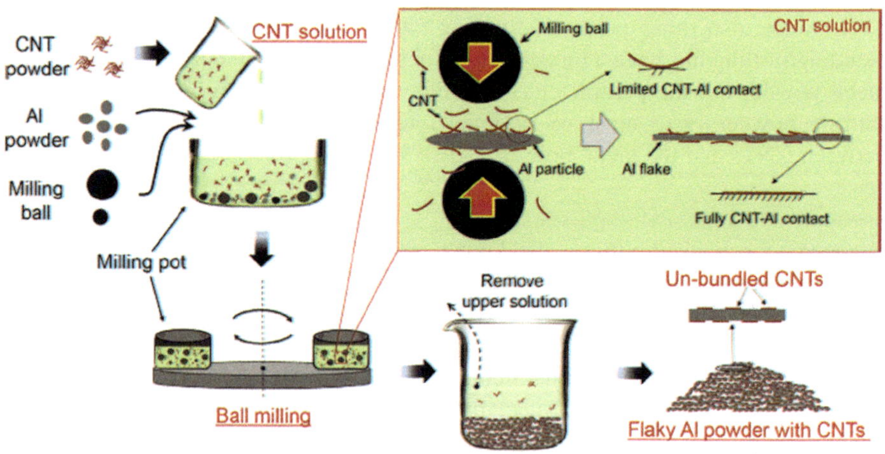

FIGURE 2.5 Schematic of solution ball milling (SBM) process for CNT dispersion [23]. (Reproduced with permission from Elsevier).

with both hydrophilic and hydrophobic groups) to create a CNT solution with 1 wt.% CNT. This solution is then introduced into the milling jars, along with the aluminum matrix powder and the milling media. The second step is to mill the combined powder and CNT solution; Chen et al. [23] utilize a planetary ball mill but other ball milling configurations could be used. 160 g of the 1 wt.% CNT solution was added to 160 g of Al powder, yielding Al-0.5 wt.% CNT composites. The milling media and jars were made of zirconia. Chen et al. [23] use milling parameters of 200 rpm for 60 min total milling time, with an intermittent 10 min "off" time used for every 10 min of milling in order to prevent over-heating. The milling process results in a flaky Al powder with CNTs distributed throughout the flaky powder particulate surfaces. The milled slurry is then transferred to a beaker where it is let to stand for ~15 min in order to allow the composite Al-CNT powders to settle. The upper solution, which likely contains some remaining suspended CNT particles, is then poured out. The settled slurry mixture is dried in an oven at 80 °C for 30 min.

Raman spectroscopy revealed that the CNTs in the SBM processed powder incurred less damage than those processed by high energy ball milling. SBM (717 nm CNTs after process) was able to prevent CNT rupturing and preserve the starting length of CNTs (1135 nm) much better than HEBM (503 nm after 2 h milling) [24]. The SBM processed Al-CNT composites exhibited greater elongation during tensile tests than Al-CNT composites processed by HEBM for 4–48 h of milling [24].

2.2.2.4 Rotational Milling Modeling

A quick peruse of the ball milling literature reveals that myriad milling parameters have been used to synthesize CNT-MMCs, as well as other nanocarbon-based metal matrix composites. Liu et al. [25] present a modeling effort to determine the best milling parameters to attain optimal CNT dispersion with minimal damage to the CNTs themselves. Experimental studies on variable speed milling utilizing stainless steel milling media with a 10:1 BPR, Al-5 Mg alloy as the matrix, and entangled CNTs with a diameter of 10 nm and a length of ~5 μm, were used to compliment and support the modeling assumptions. Milling parameters ranged from a mild condition of 300 rpm for 2 h, to 450 rpm for 8 h. The physical mechanisms observed in the experiments and underpinning the foundations of the model are as follows: (i) deformation stage where metal particulates becomes flattened, (ii) CNT cluster break-up and dispersion onto metal particle surface, (iii) cold welding state where flattened metal flakes weld together to form laminate structures, and (iv) CNT incorporation into the bulk of metal flakes welded together. CNT cluster break-up must sufficiently occur in the deformation stage in order to have a uniform distribution of individual CNTs on the metal flakes. Once cold welding occurs, further break up of CNT clusters would be minimal, and any remaining CNT clusters would be incorporating into the new coarsening or welded metallic particulates. CNT clusters would adversely affect performance by serving as stress concentrators and inhibiting adequate percolation to promote enhanced transport properties.

Aiken et al. [26] established a model for the fraction of powders ($f_{Al,0}$) in their initial undeformed state and the fraction of powders (f_{Al}) that have been deformed and cold welded, based on the probability of cold welding (a), and the time interval between two different ball collisions (τ), given by Eq. 2.1.

$$\frac{f_{Al}}{1-f_{Al}} = \frac{f_{Al,0}}{1-f_{(Al,0)}} \exp\left(-\frac{at}{\tau}\right) \tag{2.1}$$

As milling time increases from time "0" to time "t" the fraction of deformed and cold welded powder increases and at a certain fraction, it can be presumed that the cold welding mechanism is dominant. The time interval between different ball collisions is a function of the mean free path of balls during milling (λ), and the velocity of balls at the instant before collision (v), given by Eq. 2.2.

$$\tau = \frac{\lambda}{v} \tag{2.2}$$

The time interval for two different ball collisions directly correlates with the time for the metallic powder particulates to flatten, denoted as t_{flatten}. As previously stated, CNT clusters will break up simultaneously during the metallic powder deformation stage. As such, the break-up of CNT clusters can be quantified by the evolving size of CNTs clusters as milling time increases, as given by Eq. 2.3.

$$D^{-3} - D_0^{-3} = \xi \varepsilon t \tag{2.3}$$

Where D is the CNT cluster size at time t, D_0 is the initial CNT cluster size, ξ is a coefficient, and ε is the strain on the milled powders. As the cluster size decreases, at a sufficiently small size the cluster can be considered to "disappear," that is, no longer considered a cluster and instead when this happens it can be presumed that adequate CNT dispersion is achieved. As such, the time necessary for adequate CNT dispersion can be formulated as per Eq. 2.4.

$$t_{CNT-dispersion} \propto \varepsilon^{-1} \tag{2.4}$$

Furthermore, intuitively it is understood that high impact velocities would correlate to large strains, and this is quantified by Maurice et al. [27,28] as shown in Eq. 2.5.

$$\varepsilon \propto v^{0.4} \tag{2.5}$$

The time for CNT dispersion is hence proportional to impact velocity. A ratio between the time necessary for dispersing CNTs to the time necessary to flatten metallic powder particles can be thereby be established, as provided by Eq. 2.6.

$$\frac{t_{CNT-dispersion}}{t_{flatten}} \propto v^{0.6} \tag{2.6}$$

Equation 2.6 indicates that at very high impact velocities, which are driven by higher milling rotational speeds, the time for CNT dispersion will be greater than that needed to flatten metal powder particles. Once powders are flattened, cold welding begins, and hence CNT clusters that have not been broken down will become

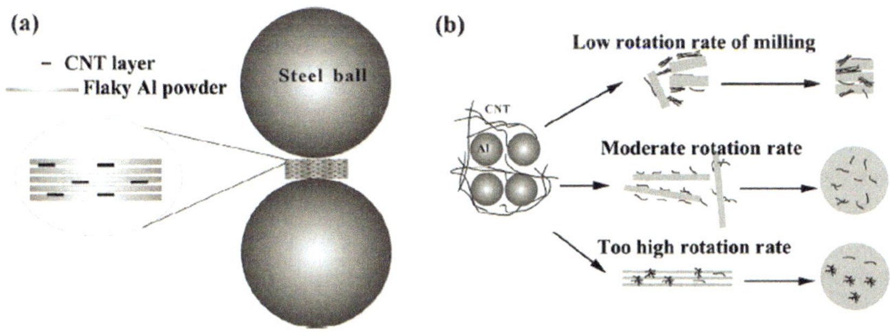

FIGURE 2.6 Schematic of (a) Al-CNT powders during ball impact during milling, and (b) CNT distributions during milling at different rotating speeds [25]. (Reproduced with permission from Elsevier).

embedded within the cold welded metal particles. In contrast, at low velocities or milling speeds, CNT clusters may be broken down before metallic particles are sufficiently flattened in order to provide the high particulate surface area needed to disperse individual CNTs on the surface. These conceptual extremes are shown schematically in Figure 2.6. At intermediate speeds, the onset of cold welding is sufficiently delayed to allow for metal powders to flatten and provide sufficient surface area for the proper dispersal of the broken up CNT clusters. Lie et al. [25] indicated that milling paramcters of 400 rpm for 8 h appeared to attain this intermediate condition where adequate dispersion of CNTs was achieved.

2.2.2.5 Shift-Speed Ball Milling (SSBM)

The shift-speed ball milling (SSBM) is a ball milling approach that integrates the implications of previously discussed rolling milling modeling, namely that optimal milling conditions for CNT-MMC mixing consist of both a high- and low-intensity milling step. SSBM utilizes both low-speed ball milling (LSBM) with a low rotational speed, and high-speed ball milling (HSBM) with a higher rotational speed. Xu et al. [29] present the first iteration of SSBM for synthesizing Al-CNT composites. A SSBM approach consisting of 8 h of LSBM and 1 h or HSBM was compared with LSBM-only and HSBM-only runs at 9 h each. For all runs, a planetary ball mill with stainless steel milling media was used. The jars were filled with Ar gas, the ball-to-powder ratio was kept at 20:1, and the instrument had a 300 mm revolution radius and a 90 mm rotational radius. The LSBM runs or steps utilized a rotational speed of 135 rpm, while the HSBM runs and steps had a rotational speed of 270 rpm.

During LSBM, the deformation mechanism of the powders transitions from flattening to fracturing after 6 h. Powder milled using only LSBM (for 9 h) produced flakes with 200 nm thickness and 40 μm diameter (starting powder was spherical with 30 μm diameter). During HSBM, the first 2 h consisting of powder flattening, after which cold welding occurs. Powder produced using only HSBM produced particles tens of microns thick and hundreds of microns thick. SSBM yielded intermediate results, with flakes typically <20 μm thick and <150 μm wide. CNTs were observed to be well embedded in the SSBM case. Raman spectroscopy was used to quantify damage on the starting CNTs (1.11) and milled CNTs using the ID/IG ratio.

It was revealed that damage incurred onto CNTs during SSBM (1.21) was relatively low and closer to that occurring during LSBM (1.13), than that caused by HSBM (1.69). Overall, the LSBM step during SSBM provides sufficient time for CNTs to be dispersed onto the gradually flattening Al powder particles. A slow flattening rate delays the onset of cold welding, where CNT dispersion and cluster break-up are no longer possible. Once proper dispersion occurs during LSBM, the short (1 h) HSBM step promotes between bonding between Al flakes and CNTs. Consolidated compacts from the SSBM powder exhibited similar strength to those made by only LSBM or only HSBM, but ductility that was 2–3 times higher [29].

2.2.3 Flake Powder Metallurgy (FPM)

The flake powder metallurgy (FPM) method developed by Jiang et al. [30,31] for CNT-MMC powder synthesis and consolidation consists of a hybrid slurry and ball milling process to form metallic flakes with dispersed CNTs, followed by consolidation using conventional powder metallurgy techniques. FPM has been used to develop Al-CNT composites with up to 20 vol.% of uniformly dispersed CNTs, a significant milestones in CNT dispersion in CNT-MMCs. The effectiveness of FPM has led to adoption of this technique by several research groups.

The first use of the FPM method (shown schematically in Figure 2.7) began with spherical Al powder that was ball milled under Ar atmosphere using stainless steel milling media, with a rotational speed of 423 rpm and a ball to powder ratio of 20:1. Various milling times were used to produce Al flakes with thicknesses varying from 0.2 to 10 μm. The dispersion technique was demonstrated using both the starting spherical Al powder, as well as the Al flakes (the flake case is discussed here). The Al flakes were then introduced into a 3 wt.% polyvinyl alcohol (PVA) aqueous solution in order to coat the Al particles with functional groups that will promote

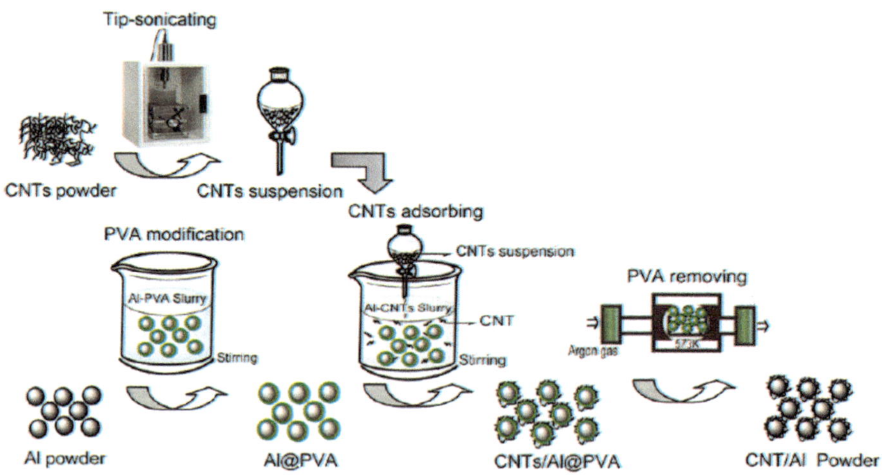

FIGURE 2.7 Schematic of powder processing steps during flake powder metallurgy [30]. (Reproduced with permission from Elsevier).

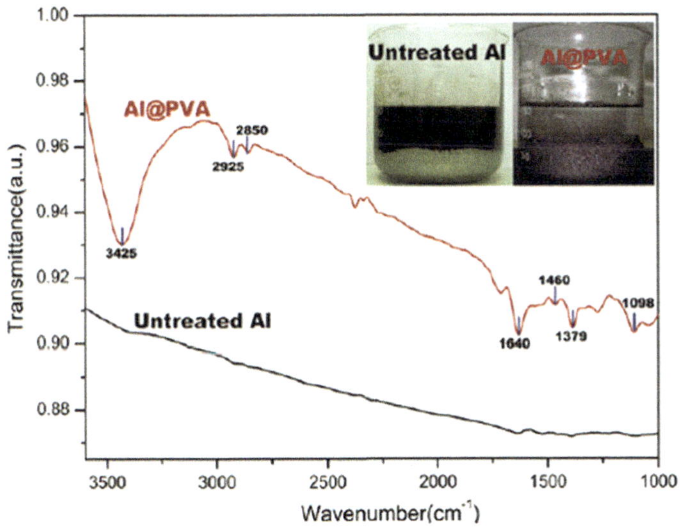

FIGURE 2.8 IR patterns of treated and untreated Al powders, insets are images of the suspensions after CNT adsorption onto Al matrix powder [30]. (Reproduced with permission from Elsevier).

adsorption of CNTs onto the surface of Al particles. This first solution is then filtered and rinsed with deionized (DI) water. Separately, tangled CNTs were introduced into an aqueous solution that was then sonicated for 2 h. Then, 100 g of the PVD functionalized Al powder was added to 1000 mL of DI water. The CNT suspension is then added dropwise to this suspension. The mixed Al/CNT suspension is stirred until the color visibly changes from black to transparent. The inset of Figure 2.8 shows the Al/CNT suspensions using Al flakes and PVA modified Al flakes. While the mixture using the PVA modified Al was transparent, that using untreated Al maintained the black color. The black color indicates the CNTs remain in suspension after the Al flakes sediment, indicating CNTs did not adhere to the Al flakes, in contrast to the case with PVA modified Al, where the CNTs coat the Al flakes and remain attach to settled Al flakes.

In addition, the IR transmission patterns in Figure 2.8 show the presence of various bonds owing to the functionalization occurring on the surfaces of Al and CNT. CNTs are functionalized with –COOH groups that are hydrophilic. In contrast, untreated Al powder maintains a surface oxide (Al_2O_3) that is hydrophobic. The incompatibility between hydrophilic and hydrophobic groups contributes to the poor adsorption of CNTs onto the untreated Al powder. Instead CNTs in the untreated Al solution will agglomerate back into clusters over time. However, the –OH groups on the Al_2O_3 film are useful for promoting adsorption of the PVA molecules via hydrogen bonding, and enabling cross-linking of PVA as shown schematically in Figure 2.9. This results in a stable and well-adhered PVA film on Al powder particulates. The PVA membrane acts two-fold to enhance CNT absorption onto Al powder: (1) the hydrophilic groups on PVA enhanced contact of the Al powder with the aqueous

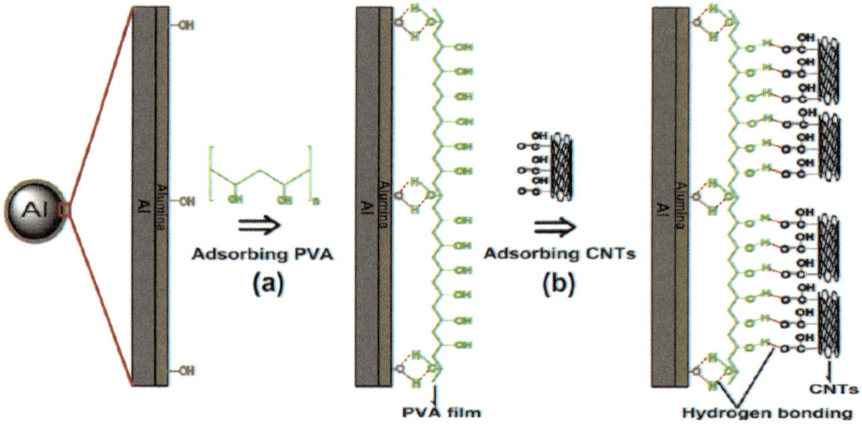

FIGURE 2.9 Schematics of CNT adsorption mechanism on the PVD-modified Al powder surface, (a) PVA membrane formation, and (b) CNT adsorption [30]. (Reproduced with permission from Elsevier).

solution, providing more opportunities for CNTs to interact with Al powders, and (2) PVA groups promote CNT adhesion via hydrogen bonding between –COOH groups on CNTs and –OH groups on PVA. The degree of CNT dispersion is further enhanced by increasing the area available on the powder particles. Flakes will have greater surface area than spherical powder, and thinner flakes will have greater surface area than thicker flakes. Figure 2.10 demonstrates the improved distribution of CNTs on thinner flakes [30].

After mixing functionalized Al spherical powders or flakes with CNTs, the mixture is then rinsed with DI water and vacuum dried. PVA is removed prior to consolidation in a flowing Ar atmosphere at 500 °C for 2 h. Powders are consolidated by uniaxial compaction at 500 MPa and hot extrusion at 440 °C with a 20:1 extrusion ratio. The resulting Al-2 vol.% CNT composites had a tensile strength of 435 MPa and failure strain of 6%.

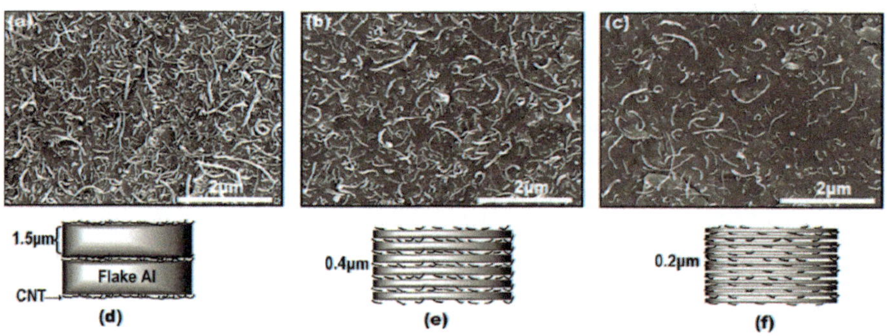

FIGURE 2.10 SEM of Al flakes with 2.5 vol.% CNT, (a) 1.5 μm thick flakes, (b) 0.4 μm thick flakes, (c) 0.2 μm flakes, and (d)–(f) corresponding schematic illustrating the distribution of CNTs between each type of Al flake [30]. (Reproduced with permission from Elsevier).

2.2.4 Nano-scale Dispersion (NSD)

The NSD route of CNT-MMC composite powder synthesis was developed by Noguchi and co-workers [10] for improvement of CNT dispersion. It is described in the schematic shown in Figure 2.11a. This method utilizes natural rubber (NR) to maintain dispersion of CNTs in metallic powder. A preform of CNT in NR and CNTs and metal mixture in NR is stacked alternatively. The stacks are then compression molded into slabs at 80 °C and heated in N_2 atmosphere at 800 °C. This treatment burns off the rubber and melts the Al, incorporating the CNTs into the composite in a dispersed manner. A sevenfold increase in compressive yield strength was reported by the addition of 1.6 vol.% CNTs [10]. The NSD process has been used to produce precursor Al powder on which CNTs were distributed uniformly as shown in Figure 2.11b. The quality of CNT dispersion yielded by NSD method is dependent on the powder size used. Since the CNTs are dispersed uniformly on the powder surface, the dispersion is better if smaller metallic powders are used. Use of large powders will result in the presence of CNTs at the grain boundaries only [11]. The powder can then be consolidated using conventional powder metallurgy techniques such as hot extrusion [32] and spark plasma sintering (SPS) [11]. Kwon et al. [11] utilized SPS followed by hot extrusion for synthesizing 5 vol.% CNT reinforced Al composite.

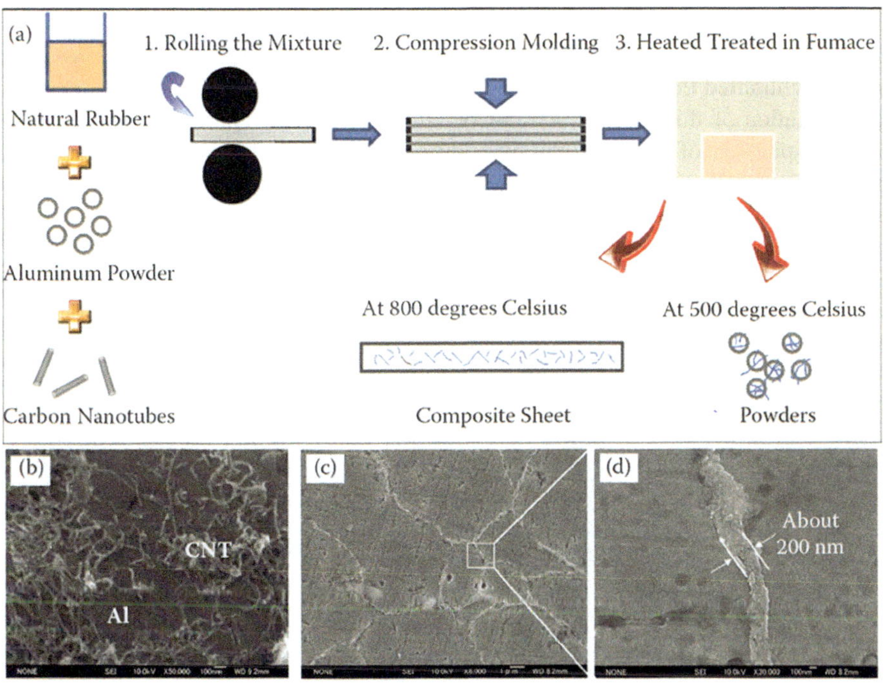

FIGURE 2.11 Schematic showing the (a) steps in the NSD method, (b) SEM image of Al powder with dispersed CNTs on the surface prepared by NSD, and (c) and (d) SEM images of the cross-section of Al-5 vol.% CNT composite prepared by SPS of the powders showing CNTs segregated to boundaries [11]. (Reproduced with permission from Elsevier).

As seen in Figure 2.11c and 2.11d, due to the nature of CNT dispersion, the CNTs were segregated at the grain boundaries after SPS consolidation. The composite prepared by hot extrusion of SPS compacts showed a tensile strength (194 MPa) twice that of pure aluminum [11].

2.2.5 Molecular Level Mixing

Molecular level mixing is a composite powder preparation technique that was developed by Cha and co-workers at Korea Advanced Institute of Science and Technology (KAIST) in Daejeon, Korea [33]. As the name suggests, the dispersion of CNTs is achieved on the molecular level by this method. The method consists of dispersing CNTs in copper acetate [$Cu(CH_3COO)_2H_2O$] solution followed by drying the suspension while magnetically stirring it. Good dispersion requires that the CNTs are functionalized by some means. The resultant salt powder has nicely dispersed CNTs in it. The salt is converted to CuO by calcining in air at 300 °C. The CuO/CNT mixture is reduced with H_2 at 250 °C to obtain Cu/CNT powder. Figure 2.12 shows the SEM and TEM micrographs of the Cu/CNT powder. It can be seen that the CNTs are distributed uniformly inside the particle. Fabrication of Cu-CNT composite by SPS of these powders led to extraordinary strengthening. The yield strength of Cu-5 vol.% CNT composite was measured to be 360 MPa, which is 2.3 times that of pure copper [33]. The hardness and sliding wear resistance of Cu-10 vol.% CNT composite was improved by two and three times, respectively, compared to copper [34]. This study highlighted the significance of CNT dispersion on the mechanical properties. In a variation of this method, Cu_2O-CNT particles were prepared by precipitation from suspension of $CuSO_4$-CNT by addition of NaOH. The oxide particles were subsequently reduced to Cu with H_2 at 400 °C. This resulted in Cu powders containing excellent dispersion of CNTs.

There are several other studies on preparation of metal-CNT composite powders that can be discussed in this category. Sn-CNT and $SnSb_{0.5}$-CNT composite powders have been produced by reduction of mixtures of $SnCl_2$ and $SbCl_3$ solutions with

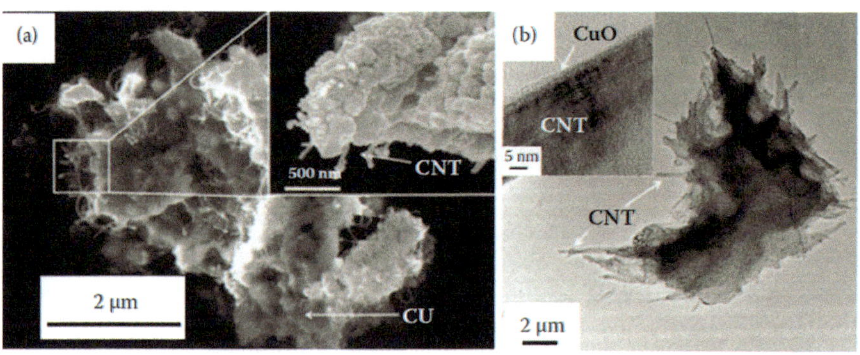

FIGURE 2.12 Micrographs showing (a) SEM and (b) TEM images of Cu-CNT powders prepared by the molecular level mixing method [33]. (Reproduced with permission from Wiley Interscience).

alkaline KBH_4 [35]. Such composites were shown to have increased capacity for lithiation and de-lithiation compared to Sn and $SnSb_{0.5}$ without the CNTs and could serve as anodes for Li-ion battery applications. A similar method of reduction of $CuSO_4$ by alkaline KBH_4 has been carried out to obtain CNTs coated with Cu [36]. CNTs were acid functionalized in order to facilitate dispersion. These Cu-CNT mixtures showed superior catalytic performance in thermal decomposition of ammonium perchlorate due to the large surface area.

Molecular-level mixing is a promising technique for the production of metal powders containing dispersed CNTs. In fact, this process can be used for any metal that forms salts that are soluble in water and decomposes to form oxides on heating. The product would be an oxide containing a good dispersion of CNTs. Metallic oxides can be reduced with hydrogen to yield metal-CNT powders with excellent CNT dispersion. Since there are chances of oxygen impurity due to incomplete reduction by H_2, the use of these powders may or may not be suitable. Kim and co-workers have showed that the remaining oxygen is good for the interfacial stress transfer, which leads to effective strengthening [16]. However, for electrically conductive applications, the presence of oxygen may not be suitable. Hence, the molecular level mixing method must be adopted judiciously.

2.2.6 In Situ Growth on Metal Particles

Apart from mixing CNTs and metallic powders, another efficient way to improve dispersion is by growing CNTs directly on the surface of metallic powders [37]. By depositing catalysts on metal powders, they can be made as substrates for the CNT growth. This method also improves the adherence of CNTs to the metal particles. In this method, $Ni(OH)_2$ is precipitated on Al powders by adding NaOH to a mixture of Al and aqueous $Ni(NO_3)_2$ solution. The $Ni(OH)_2$ is reduced to NiO by calcination and finally to Ni by reduction with H_2. Figure 2.13 shows the TEM images of the Al particle with nanometer-size Ni particles on Al. CNTs were deposited on the Al

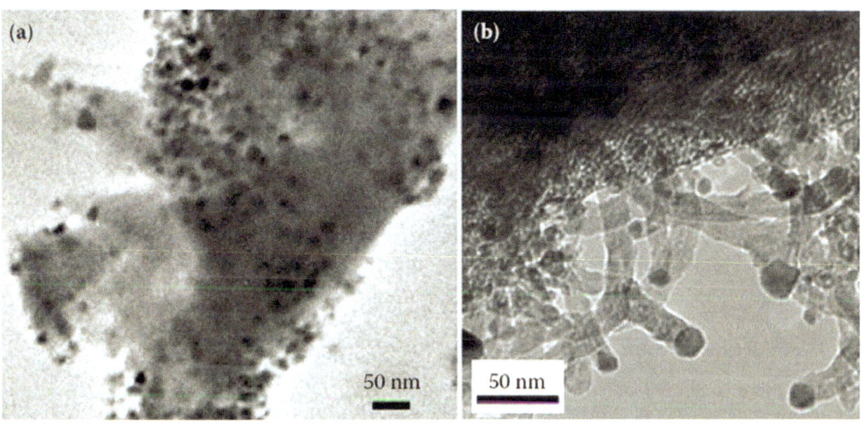

FIGURE 2.13 TEM images showing Si-coated CNTs using cycled vacuum-feeding CVD (the inset image is uncoated CNTs) [14]. (Reproduced with permission from Wiley).

particle by catalytic decomposition of CH_4 gas. Figure 2.13 shows the TEM image of CNTs grown on Al particles, which shows good dispersion as well as bonding between the two.

Si-CNT composites have been probed for use as an active anode layer in lithium ion batteries. CNTs have been grown on Si particles (coated with Ni by electroless plating) by CVD technique [38,39]. CNTs have been reported to form a cage-like structure entangling the Si particles into it and leaving many voids and pores in the composite. These voids are required to accommodate the shape and size change of Si particles during battery cycles, without creating the stress in the structure. Si-coated CNTs have been produced by decomposition of silane (SiH_4) oven CNTs in order to increase thermal stability of CNTs, when they are used as reinforcement in metal and ceramic matrix composites [40]. A continuous, well-covered coating of 10 nm was achieved as shown in the TEM image in Figure 2.14. In another study, Si particles were coated with nanotubes by vapor deposition of tetramethylsilane $(CH_3)_4Si$ [41]. Si-supported Ni catalyst prepared by wet chemical methods was used. The Si-CNT composite was found to have a high H_2 storage capacity. The CVD process has also been used to produce Al-CNT composite powder by growing CNT on Al particle using Ni catalyst [14]. Using this composite particle, Al-5 wt.% CNT composite was prepared by a powder metallurgy technique. Hardness and tensile strength of these composites were found to increase by 200% and 180%, respectively, as compared to the composite made from blended powders.

In addition, CNTs grown by in situ methods may undergo further processing by other methods such as mechanical milling. Yang et al. [42] grow CNTs on Al-0.5 wt.% Co catalysts by CVD. CNTs are grown at 600 °C for 0–60 min (depending on desired CNT fraction) under a flowing mixture of C_2H_2/Ar (20/240 mL min^{-1}). The composite powders are then balled milled in a planetary ball mill at 500 rpm for 90 min using a ball to powder ratio of 10:1. Composite powders with CNT content

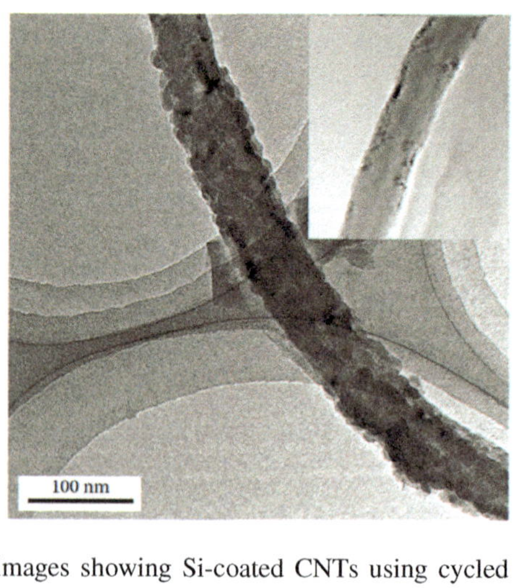

FIGURE 2.14 TEM images showing Si-coated CNTs using cycled vacuum-feeding CVD (the inset image is uncoated CNTs) [40]. (Reproduced with permission from IOP Publishing).

ranging from 1.5–4.5 wt.% were synthesized using this method; reinforcement loading of 4.5 wt.% led to a 241% increase in UTS relative to Al matrix.

2.2.7 Physical Vapor Deposition

These methods are carried out to obtain composite powders, coated CNTs, and one-dimensional nanostructures. These methods are not used to synthesize bulk CNT-MMCs. CNTs have been coated with tungsten through a physical vapor deposition (PVD) method [43]. CNTs were put in an alumina crucible, and a W filament at 8 mm distance was heated to 2200 °C under H_2 flow of 1200 sccm. Gray coating of W was observed on the CNTs.

2.2.8 Spray Drying

Spray drying is another technique that has been used to uniformly disperse CNTs in micron-sized Al-Si powders [48]. Spray drying is a popular technology in the food packaging industry and has gained importance in the thermal spray industry as well [44–46]. Using spray drying, very fine particles can be agglomerated into large ones that help in their transportation through tubes required in thermal spraying. CNTs and metal powder are mixed with water to form aqueous slurry with a little amount of PVA as the binder. The dispersion of the CNTs in the slurry can be increased by subjecting to ultrasonication. This aqueous slurry is then atomized in a column through which hot air is also passed. The hot air could be in the same direction as that of the droplets or in the counter direction. The droplets solidify on their descent and form agglomerates that settle on the bottom of the column. The fine droplets are carried with the gas flow and can be recycled. The agglomerate size, morphology, and porosity are dependent on the slurry content, slurry viscosity, binder type, drying temperature, and process used [44, 45]. Figure 2.15 shows the SEM images of the spray-dried Al-5 wt.% CNT (Al-5 CNT) and Al-10 wt.% CNT (Al-10 CNT) agglomerates at low (a and b) and high magnifications (c and d). The particle size of the starting Al-Si eutectic powder was 2.4 ± 1.2 μm, while that of the Al-5 CNT and Al-10 CNT agglomerates was found to be 57 ± 21 μm and 39 ± 15 μm, respectively. Spray drying resulted in formation of large spherical agglomerates of sizes up to 25 times the constituent particles. Figure 2.15c and 2.15d show the high-magnification SEM image of the outer surface of spray-dried Al-5 CNT and Al-10 CNT agglomerates. CNTs are distributed uniformly on the surface as well as inside of the Al-5 CNT agglomerates. A dense network of CNTs forming a mesh on the outer surface that seems to hold Al alloy particles together was observed in Al-10 CNT agglomerates. However, no mesh formation was observed on the inside of the agglomerate [48]. During spray drying, the CNT-metal slurry is atomized into droplets and the CNTs tend to segregate on the surface of the droplets, owing to their non-wetting property. Shrinkage caused by drying of the droplets brings CNTs closer. Also during drying, the movement of vapor from inside the droplet to outside leads to transport of low-density CNTs to the surface. This leads to surface mesh formation in the Al-10 CNT powder. Mesh formation was not seen in spray dried Al-5 CNT due to the lower concentration of CNTs. CNTs were more or less uniformly distributed

FIGURE 2.15 SEM images of spray-dried agglomerates of (a) Al-5 wt.% CNT and (b) Al-10 wt.% CNT powder, and (c) and (d) the high-magnification images of the surface of (a) and (b), respectively [47]. (Reproduced with permission from Elsevier).

between the particles in the agglomerate. The spherical shape of the particle causes low inter-particle friction and hence leads to excellent flowability, which enables fabrication of bulk Al-CNT composite cylinders up to 5 mm in thickness by plasma spray forming [48]. Spray drying is an industrial-scale process, and this could be the choice for large-scale production of CNT-MMC powders due to economical reasons. Hence, there are multiple advantages of the spray-drying process to make CNT-MMC powder. By employing nano-powders, the dispersion can be further enhanced.

The methods discussed previously are largely utilized for the dispersion of CNTs in the starting powder mixture. The consolidation mechanisms can further alter the quality of CNT dispersion in the CNT-MMC. Although the dispersion of CNTs in the spray-dried powder was very good, the resulting Al-CNT composite microstructure synthesized by plasma spraying consisted of uniformly distributed CNTs in the matrix along with regions of CNT clusters [48]. However, cold spraying of a mixture of the spray-dried Al-CNT and pure aluminum powder resulted in uniform distribution of CNTs in the composite coating [48]. These results indicate that secondary processing of spray-dried powders also plays an important role in determination of CNT distribution. Liquid and semi-solid state processing like casting and thermal spraying may lead to clustering of CNTs during melting as was observed from analysis of single splats of molten spray-dried particles [48]. This will lead to clustering in the final composite. Solid-state deformation processing such as extrusion assists in breaking the CNT clusters and improving CNT dispersion. Chapter 3 discusses consolidation and post-processing techniques in detail.

2.3 QUANTIFICATION OF CARBON NANOTUBE DISPERSION

Many papers in the literature of CNT reinforced composites suggest that the composite reported attained "uniform" or "good" CNT distribution. However, there has been hardly any study on the quantification of dispersion in carbon nanotube composites to describe how good is "good." The majority of the researchers mention uniform CNT dispersion in the composites, which is based on the visual examination of the microstructure. Quantification of CNT dispersion has received the least attention in the area of composites. Quantification of CNT distribution or dispersion will help in comparing various microstructures and the effectiveness of various methods for dispersion of CNTs in composites. Dirichlet tessellation has been used in the quantification as well to study the effect of dispersion in composite materials [49–52]. Most of these methods [49–51] have focused on characterizing composites with particulate inclusion. Seidel and Lagoudas [56] have used Dirichlet tessellation to construct Voronoi polygons around CNT centers and thereafter considered each polygon as a composite and calculated the elastic modulus using the Mori-Tanaka technique. In this manner, they were able to study the effect of clustering in the microstructure on the computed properties. There are only two studies on the quantification of the spatial carbon nanotube/nanofiber distribution, which are discussed here.

Luo and Koo have proposed a dispersion quantification method based on the statistical distribution of horizontal and vertical separation distances between the peripheries of the particles/carbon fibers in a cross-sectional image of the composite [30,31,53,54]. The horizontal and vertical distances between fibers are measured as shown in Figure 2.16a. A lognormal distribution was found to fit the distribution obtained as seen in Figure 2.16b. The lognormal distribution is described by the following equation.

$$f(x)=\left(\frac{1}{xn\sqrt{2\pi}}\right)\exp\left[-\frac{1}{2}\left(\frac{\ln(x)-m}{n}\right)^2\right] \mathit{for}\, x>0, \mathit{and}=0\, \mathit{for}\, x\leq 0 \tag{2.7}$$

Where $m=\ln\frac{\mu^2}{\sqrt{\mu^2+\sigma^2}}$ $\mathit{and}\, n=\sqrt{\ln\left(\frac{\mu^2+\sigma^2}{\mu^2}\right)}$, μ is the mean and σ is the standard deviation. A large value of the standard deviation indicates that there is a large scatter in the data while a smaller standard deviation indicates narrow distribution. Based on this distribution, two parameters $D_{0.1}$ and $D_{0.2}$ were defined, representing the probability that the values lay between $\mu \pm 0.1\ \mu$ and $\mu \pm 0.2\ \mu$ of the values, respectively, μ being the average distance. The larger the values of $D_{0.1}$ and $D_{0.2}$, the better the distribution since it means uniform separation of the filler materials. The results were applied to the polymer-clay composites. The drawback of the model is that it takes into account only the horizontal and vertical separation. Thus, the analysis has some directionality to it.

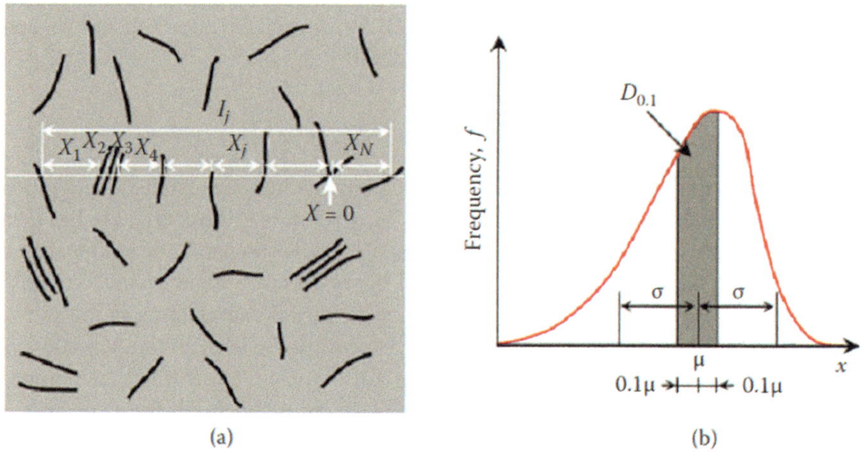

FIGURE 2.16 Figure showing (a) the schematic illustration of measurement of inter-fiber separation, and (b) the statistical distribution of the inter-fiber spacing showing a lognormal distribution [54]. (Reproduced with permission from Elsevier).

Pegel and co-workers [59] made an interesting observation on the dispersion of second phase particles. Figure 2.17a and 2.17c show two model composites having good and poor dispersion, respectively. It is noted that as the size of the reinforcement is increased, as shown in Figure 2.17b and 2.17d, respectively, the area fraction will increase. The variation of the area fraction as a function of the radius of the reinforcement is shown in the Figure 2.18. It is seen that for systems with good dispersion, the convergence to 1 occurs faster as compared to the clustered systems. Using this scheme, an idea on relative level of dispersion can be made. This technique can be very cumbersome and will require processing with advanced computing facilities for the complex images.

Recently, two novel methods to quantify the spatial distribution of carbon nanotubes in nanocomposites have been suggested by our research group [60]. The two parameters area *dispersion parameter* (DP) based on the image analysis and *clustering parameter* (CP) based on the distance between the nearest neighbors (obtained by constructing the Delaunay triangulation of the centers of the nanotubes). To obtain the DP, first a binary image is constructed from an SEM/TEM image of which the dispersion degree is to be quantified. Figure 2.19 shows the SEM image of a fracture surface of Al-0.5 wt.% CNT composite coating prepared by cold spraying and the corresponding binary image showing CNTs. The binary image is now divided into a certain number of parts or cells, for example of maximum dimension 1 μm. The CNT fraction of each cell is subsequently measured. By plotting the CNT content of the cells as a function of the location, one can visualize how the CNT distribution is spatially varying over a length scale of a micron. Figure 2.20a and 2.20b show such plots of CNT spatial distribution. It is noted that for a clustered configuration, as the image is divided into more and more numbers of cells, the number of the division, after which cells with CNT content greater that 90% start showing up, can be used as a measure of the dispersion degree. The larger the number, the better is the

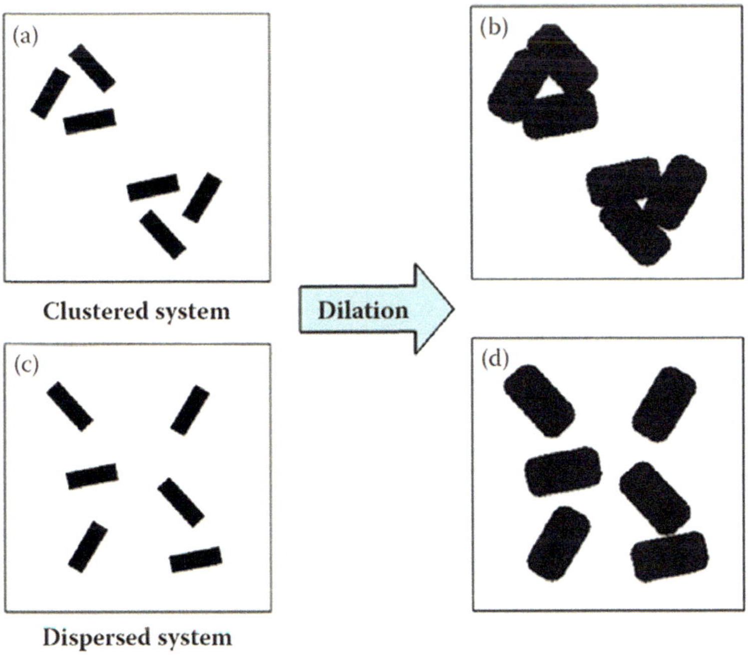

FIGURE 2.17 Schematic illustration of the increase in volume fraction of reinforcement due to increase in reinforcement size (dilation) in the case of (a) and (b) clustered, and (c) and (d) dispersed configuration [55]. (Reproduced with permission from Elsevier).

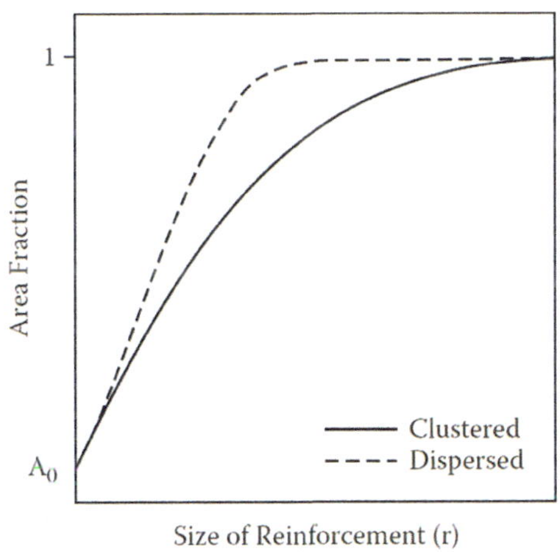

FIGURE 2.18 Graph showing the variation of the area fraction as a function of the dilation [55]. (Reproduced with permission from Elsevier).

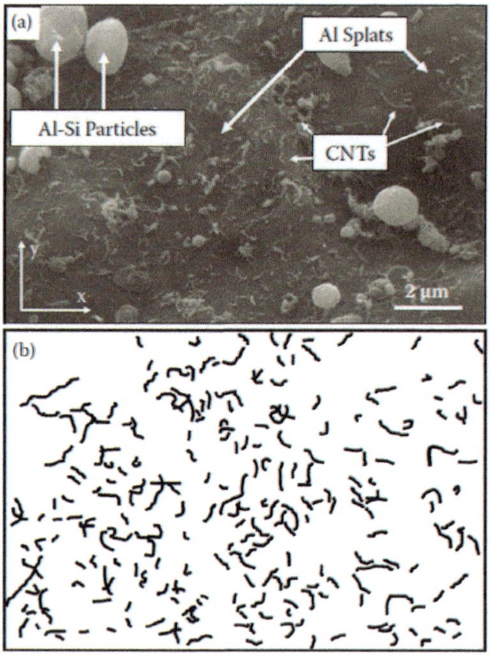

FIGURE 2.19 (a) SEM image of a fracture surface of Al-0.5 wt.% CNT composite showing good CNT distribution, and (b) the corresponding binary image used for computation [56]. (Reproduced with permission from Elsevier).

dispersion. The variation of the maximum CNT fraction recorded for a cell as a function of the number of divisions for the micrograph in Figure 2.19 has been shown in Figure 2.21 along with the number of divisions after which cells with 90% CNT fraction start appearing. The DP is defined by Eq. 2.8.

$$DP = \frac{N(At\,least\,one\,cell\,has\,CNT\,fraction = 0.9)}{N(Cell\ size\ equals\,CNT\ diameter)\times Overall\,CNT\ Fraction} \tag{2.8}$$

The overall CNT fraction was divided in order to make the formula independent of CNT concentration for small ranges of variation. For the SEM image shown in Figure 2.19, DP comes out to be equal to 4489 / (48,400 × 0.094) = 1.07. The higher the value of DP, the better is the dispersion. It was also noted that if the distances between the CNT centers were obtained by Delaunay triangulation and as statistical distribution was plotted, it followed a lognormal curve of Eq. 2.7 as shown in Figure 2.22. The fractional number of inter-tube spacing, which falls below a certain value (five times the CNT diameter of 66 nm = 0.33 μm in this case), can be a measure of the clustering tendency of the CNTs. Thus, the CP is defined as follows.

$$CP = \frac{(Cumulative\ Fraction\ of\ Distance\ Less\ Than\ or\ Equal\ To\,5D_{CNT})}{Overall\,CNT\ Fraction} \tag{2.9}$$

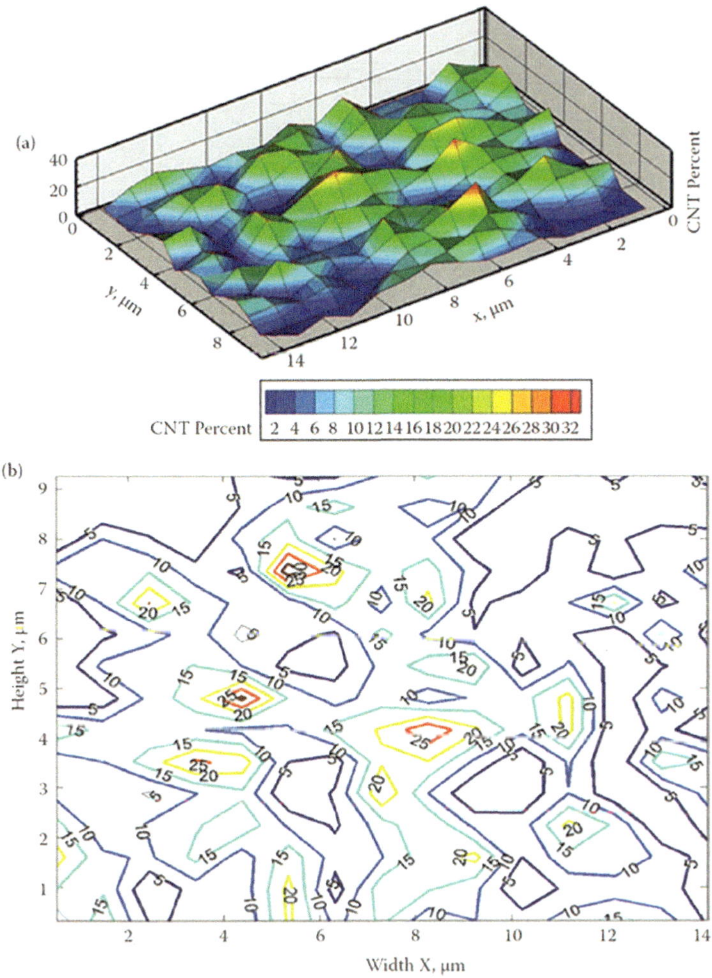

FIGURE 2.20 (a) Surface plot and (b) contour plot of the spatial variation of the CNT distribution on the scale of a micron for the micrograph shown in Figure 2.19a [56]. (Reproduced with permission from Elsevier).

The overall CNT fraction was divided in order to make the formula independent of CNT concentration for small ranges of variation. For the SEM micrograph, the value of CP is equal to 0.13/0.094 = 1.38. The lower the value of CP, the better is the dispersion. These parameters DP and CP can be utilized to compare the dispersion of CNTs for different micrographs arising from different processing techniques. The drawback of this technique is that the CNT distribution binary image is obtained manually, which is time consuming. Another limitation is that several representative images need to be analyzed in order to comment on the dispersion obtained by a given process. Micrographs at different magnifications must be analyzed, and since the CNTs are small, high magnification images are required.

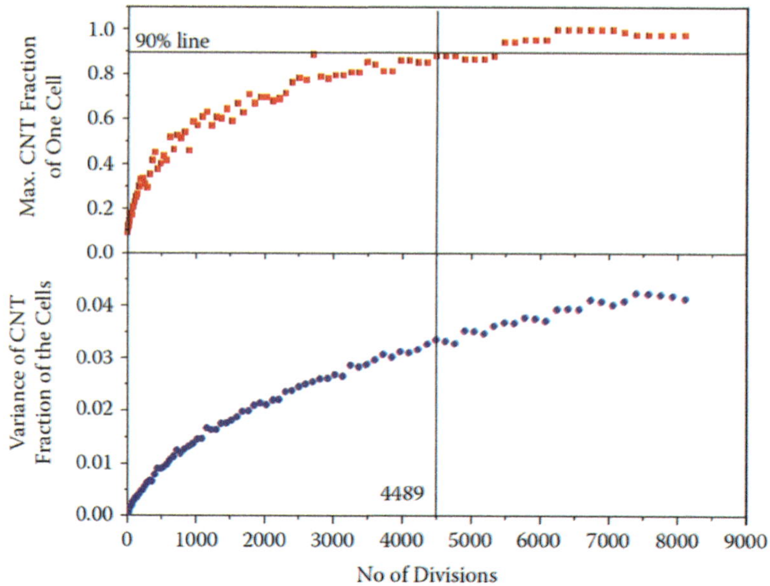

FIGURE 2.21 Graph showing the variation of the maximum CNT fraction recorded for a cell and the variance of the CNT fraction of the cells as a function of the number of divisions

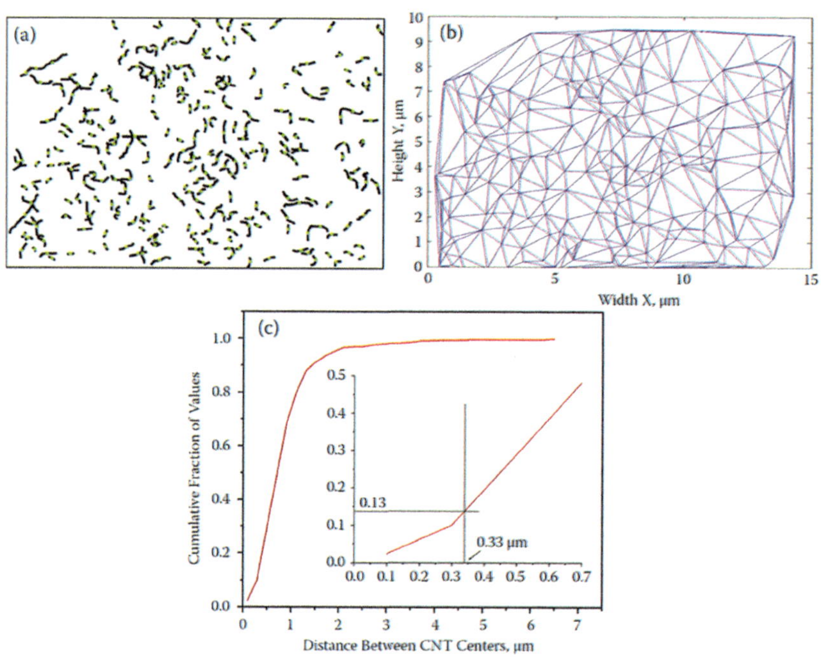

FIGURE 2.22 (a) Binary image showing the points considered along the CNTs for Delaunay triangulation, (b) the Delaunay triangulation of the set of the points, and (c) the variation of the cumulative fraction of the values for different CNT-spacing [56]. The inset in (c) zooms in the low inter-CNT spacing distribution for calculation of clustering parameter. (Reproduced with permission from Elsevier).

2.4 CHAPTER HIGHLIGHTS

CNT dispersion methods and the importance thereof are presented. Several processing methods have been developed to adequately disperse CNTs. These methods have shown some promise, but all of them have limitations. While the quality of dispersion is important, the processes used should also be amenable for bulk production of powders from an economical point of view. NSD leads to good dispersion of CNTs only on the particle surface. Hence, the level of dispersion is dependent on the metal particle size. The molecular-level mixing method may lead to oxide impurities due to incomplete reduction of the powders. These chemical methods are still largely limited to lab scale processes. Perhaps one of the most promising and scalable routes has been flake powder metallurgy for processing routes utilizing pressure, and for routes dependent on particle flowability, ball milling is an excellent choice. Ball milling routines have been modeled and tailored to yield good dispersion, without excessive damage to CNTs. There remains significant scope for CNT dispersion, especially for emerging processing techniques in the additive manufacturing field such as selective laser melting and direct laser deposition. Quantification of CNT dispersion has still not received the attention that is needed to accelerate CNT-MMC reliability and scalability. There are still only a handful of studies on the quantification of spatial distribution of CNTs. Quantification studies based on micrographs have been carried out using the distance between the CNTs and image analysis methods. Parameters like $D_{0.1}$, $D_{0.2}$, and DP have been proposed, which increase with increased quality of dispersion, while the extent of clustering can be represented by CP. There is a need for a method that is robust as well as quick and easy to quantify CNT dispersion from SEM or TEM images. Development of such a standard method will lead to better understanding of the effect of processing on dispersion of CNTs.

REFERENCES

1. Yeh, M-K., Tai, N-H., and J-H. Liu. 2006. Mechanical behavior of phenolic-based composites reinforced with multi-walled carbon nanotubes. *Carbon* 44: 1–9.
2. de Villoria, R. G., and A. Miravete. 2007. Mechanical model to evaluate the effect of the dispersion in nanocomposites. *Acta Mater.* 55: 3025–3031.
3. Yu, G., Cao, A., and C. M. Lieber. 2007. Large-area blown bubble films of aligned nanowires and carbon nanotubes. *Nature Nanotech.* 2: 372–377.
4. Tu, J. P., Yang, Y. Z., Wang, L. Y., Ma, X. C., and X. B. Zhang. 2001. Tribological properties of carbon-nanotube-reinforced copper composites. *Tribology Int.* 10: 225–228.
5. Laha, T., Agarwal, A., McKechnie, T., and S. Seal. 2004. Synthesis and characterization of plasma spray formed carbon nanotube reinforced aluminum composite. *Mater. Sci. Eng. A* 381: 249–258.
6. Feng, Y., Yuan, H. L., and M. Zhang. 2005. Fabrication and properties of silver-matrix composites reinforced by carbon nanotubes. *Mater. Char.* 55: 211–218.
7. Dong, S. R., Tu, J. P., and X. B. Zhang. 2001. An investigation of the sliding wear behavior of Cu-matrix composite reinforced by carbon nanotubes. *Mater. Sci. Eng. A* 313: 83–87.
8. Nai, S. M. L., Wei, J., and M. Gupta. 2006. Lead-free solder reinforced with multiwalled carbon nanotubes. *J. Electr. Mater.* 35: 1518–1522.
9. Nai, S. M. L., Wei, J., and M. Gupta. 2006. Improving the performance of lead-free solder reinforced with multi-walled carbon nanotubes. *Mater. Sci. Eng. A* 423: 166–169.

10. Noguchi, T., Magario, A., Fukuzawa, S., Shimizu, S., Beppu, J., and M. Seki. 2004. Carbon nanotube/aluminum composites with uniform dispersion. *Mater. Trans.* 45: 602–604.
11. Kwon, H., Estili, M., Takagi, K., Miyazaki, T., and A. Kawasaki. 2009. Combination of hot extrusion and spark plasma sintering for producing carbon nanotube reinforced aluminum matrix composites. *Carbon* 47: 570–577.
12. Choi, H. J., Kwon, G. B., Lee, G. Y., and D. H. Bae. 2008. Reinforcement with carbon nanotubes in aluminum matrix composites. *Scripta Mater.* 59: 360–363.
13. Choi, H., Shin, J., Min, B., Park, J., and D. Bae. 2009. Reinforcing effects of carbon nanotubes in structural aluminum matrix nanocomposites. *J. Mater. Res.* 24: 2610–2616.
14. He, C., Zhao, N., Shi, C., Du, X., Li, J., Li, H., and Q. Cui. 2007. An approach to obtaining homogeneously dispersed carbon nanotubes in Al powders for preparing reinforced Al-matrix composites. *Adv. Mater.* 19: 1128–1132.
15. He, C., Zhao, N. Q., Shi, C. S., and S. Z. Song. 2009. Mechanical properties and microstructures of carbon nanotube-reinforced Al matrix composite fabricated by in situ chemical vapor deposition. *J Alloys Compd.* 487: 258–262.
16. Kim, K. T., Cha, S. I., Gemming, T., Eckert, J., and S. H. Hong. 2008. The role of interfacial oxygen atoms in the enhanced mechanical properties of carbon-nanotube-reinforced metal matrix nanocomposites. *Small* 4: 1936–1940.
17. Esawi, A. M. K., and M. A. E. Borady. 2008. Carbon nanotube-reinforced aluminium strips. *Composites Sci. Tech.* 68: 486–492.
18. Esawi, A., and K. Morsi. 2007. Dispersion of carbon nanotubes (CNTs) in aluminum powder. *Composites A* 38: 646–650.
19. Esawi, A. M. K., Morsi, K., Sayed, A., Taher, M., and S. Lanka. 2011. The influence of carbon nanotube (CNT) morphology and diameter on the processing and properties of CNT-reinforced aluminium composites. *Composites: Part A* 42: 234–243.
20. Suryanarayana, C. 1999. Mechanical alloying. In *Non-Equilibrium Processing of Materials*. C. Suryanarayan, Ed. Oxford: Elsevier Science Ltd.
21. Maite, A., Reddy, L., Chen, F., Zhang, L., Schoenung, J. M., Lavernia, E. J., and T. Laha. 2015. Carbon nanotube-reinforced Al alloy-based nanocomposites via spark plasma sintering. *J. Compos. Mater.* 49: 1937–1946.
22. Omidi, M., Khodabandeh, A., Nategh, S., and M. Khakbiz. 2017. Wear mechanism maps of CNT reinforced Al6061 nanocomposites treated by cryomilling and mechanical milling. *Tribology Int.* 110: 151–160.
23. Chen, B., Li, S., Imai, H., Jia, L., Umeda, J., Takahashi, M., and K. Kondoh. 2015. An approach for homogenous carbon nanotube dispersion in Al matrix composites. *Mater. Design* 72: 1–8.
24. Chen, B., Shen, J., Ye, X., Jia, L., Li, S., Umeda, J., Takahashi, M., and K. Kondoh. 2017. Length effect of carbon nanotubes on the strengthening mechanisms in metal matrix composites. *Acta Materialia* 140: 317–325.
25. Liu, Z. Y., Xiao, B. L., Wang, W. G., Ma, Z. Y. 2017. Modeling of carbon nanotube dispersion and strengthening mechanisms in Al matrix composites prepared by high energy ball milling-powder metallurgy method. *Composites: Part A* 94: 189–198.
26. Aiken, B. J. M., Courtney, T. H., and D. R. Maurice. 1991. Reaction rates during mechanical alloying. *Mater. Sci. Eng. A* 147: 229–237.
27. Maurice, D., and T. H. Courtney. 1995. Modeling of mechanical alloying: part III. Applications of computational programs. *Metall. Mater. Trans. A* 94: 2437–2444.
28. Maurice, D., and T. H. Courtney. 1995. Modeling of mechanical alloying: part II. Development of computational modeling programs. *Metall. Mater. Trans. A* 94: 2431–2435.

29. Xu, R., Tan, Z., Xiong, D., Fan, G., Guo, Q., Zhang, J., Su, Y., Li, Z., and D. Zhang. 2017. Balanced strength and ductility in CNT/Al composites achieved by flake powder metallurgy via shift-speed ball milling. *Composites: Part A* 96: 57–66.
30. Jiang, L., Fan, G., Li, Z., Kai, X., Zhang, D., Chen, Z., Humphries, S., Heness, G., and W. Y. Yeung. 2011. An approach to the uniform dispersion of a high volume fraction of carbon nanotubes in aluminum powder. *Carbon* 49: 1965–1971.
31. Jiang, L., Li, Z., Fan, G., Cao, L., and D. Zhang. 2012. The use of flake powder metallurgy to produce carbon nanotube (CNT)/aluminum composites with a homogenous CNT distribution. *Carbon* 50: 1993–1998.
32. Yuuki, J., Kwon, H., Kawasaki, A., Magario, A., Noguchi, T., Beppu, J., and M. Seki. 2007. Fabrication of carbon nanotube reinforced aluminum composite by powder extrusion process. *Mater. Sci. Forum* 534–536: 889–892.
33. Cha, S. I., Kim, K. T., Arshad, S. N., Mo, C. B., and S. H. Hong. 2005. Extraordinary strengthening effect of carbon nanotubes in metal-matrix nanocomposites processed by molecular-level mixing. *Adv. Mater.* 17: 1377–1381.
34. Kim, K. T., Cha, S. I., and S. H. Hong. 2007. Hardness and wear resistance of carbon nanotube reinforced Cu matrix nanocomposites. *Mater. Sci. Eng. A* 449–451: 46–50.
35. Chen, W. X., Lee, J. Y., and Z. Liu. 2003. The nanocomposites of carbon nanotube with Sb and $SnSb_{0.5}$ as Li-ion battery anodes. *Carbon* 41: 959–966.
36. Ping, C., Li, F., Jian, Z., and J. Wei. 2006. Preparation of Cu/CNT composite particles and catalytic performance on thermal decomposition of ammonium perchlorate. *Prop. Explos. Pyrotech.* 31: 452–455.
37. Zhong, R., Cong, H., and P. Hou. 2003. Fabrication of nano-Al based composites reinforced by single-walled carbon nanotubes. *Carbon* 41: 848–851.
38. Shu, J., Li, H., Yang, R., Shi, Y., and X. Huang. 2006. Cage-like carbon nanotubes/Si composite as anode material for lithium ion batteries. *Electrochem. Comm.* 8: 51–54.
39. Kim, T., Mo, Y. H., Nahm, K. S., and S. M. Oh. 2006. Carbon nanotubes (CNTs) as a buffer layer in silicon/CNTs composite electrodes for lithium secondary batteries. *J. Power Sources* 162: 1275–1281.
40. Wang, Y. H., Li, Y. N., Lu, J., Zang, J. B., and H. Huang. 2006. Microstructure and thermal characteristic of Si-coated multi-walled carbon nanotubes. *Nanotechnology* 17: 3817–3821.
41. Ishihara, T., Nakasu, M., Yasuda, I., and H. Matsumoto. 2006. Preparation of Si-carbon nanotube composite by decomposition of tetramethylsilane (TMS) and its hydrogen storage property. *Sci. Tech. Adv. Mater.* 7: 667–671.
42. Yang, X., Zou, T., Shi, C., Liu, E., He, C., and N. Zhao. 2016. Effect of carbon nanotube (CNT) content on the properties of in-situ synthesis CNT reinforced Al composites. *Mater. Sci. Eng. A* 660: 11–18.
43. Zhang, Y., Zhang, Q., Li, Y., Wang, N., and J. Zhu. 2000. Coating of carbon nanotubes with tungsten by physical vapor deposition. *Solid State Comm.* 115: 51–55.
44. Lukasiewicz, S. L. 1989. Spray drying ceramic powders. *J. Amer. Ceram. Soc.* 72: 617–624.
45. Bertrand, G., Roy, P., Filiatre, C., and C. Coddet. 2005. Spray-dried ceramic powders: a quantitative correlation between slurry characteristics and shapes of the granules. *Chem. Eng. Sci.* 60: 95–102.
46. Cao, X. Q., Vassen, R., Schwartz, S., Jungen, W., Tietz, F., and D. Stoever. Spray-drying of ceramics for plasma-spray coating. *J. Eur. Ceram. Soc.* 20: 2433–2439.
47. Bakshi, S. R., Singh, V., Seal, S., and A. Agarwal. 2009. Aluminum composite reinforced with multiwalled carbon nanotubes from plasma spraying of spray dried powders. *Surf. Coat. Tech.* 203: 1544–1554.

48. Bakshi, S. R., Singh, V., Balani, K., McCartney, D. G., Seal, S., and A. Agarwal. 2008. Carbon nanotube reinforced aluminum composite coating via cold spraying. *Surf. Coat. Tech.* 202: 5162–5169.
49. Wray, P. J., Richmond, O., and H. L. Morrison. 1983. Use of the Dirichlet tessellation for characterizing and modeling nonregular dispersions of second-phase particles. *Metallography* 16: 39–58.
50. Tscheschel, A., Lacayo, J., and D. Stoyan. 2005. Statistical characterization of TEM images of silica-filled rubber. *J. Microscopy* 217: 75–82.
51. Ghosh, S., Nowak, Z., and K. Lee. 1997. Quantitative characterization and modeling of composite microstructures by Voronoi cells. *Acta Mater.* 45: 2215–2234.
52. Seidel, G. D., and D. C. Lagoudas. 2006. Micromechanical analysis of the effective elastic properties of carbon nanotube reinforced composites. *Mech. Mater.* 38: 884–907.
53. Luo, Z. P., and J. H. Koo. 2005. Quantifying the dispersion of mixture microstructures. *J. Microscopy* 225: 118–125.
54. Luo, Z. P., and J. H. Koo. 2008. Quantification of the layer dispersion degree in polymer layered silicate nanocomposites by transmission electron microscopy. *Polymer* 49: 1841–1852.
55. Pegel, S., Potschke, P., Villmow, T., Stoyan, D., and G. Heinrich. 2009. Spatial statistics of carbon nanotube polymer composites. *Polymer* 50: 2123–2132.
56. Bakshi, S. R., Batista, R. G., and A. Agarwal. 2009. Quantification of carbon nanotube distribution and property correlation in nanocomposites. *Composites A* 40: 1311–1318.

3 Composite Processing Techniques

The exciting properties of carbon nanotubes were highlighted in previous chapters. Developments in nanoscience have made it possible to measure the mechanical, electrical, and thermal properties of the single carbon nanotube. The main challenge is how to utilize these properties for real-life applications. It is critical that nanoscience is integrated with nanomanufacturing for large-scale integration into devices and components. While the previous chapter discussed incorporation of CNTs into the metal matrix, primarily by powder processing, this chapter focuses on the processing methods to fabricate bulk CNT reinforced metal matrix composites. Processing refers to fabrication of the composite resulting in the integration of the reinforcement (CNTs) in the matrix. This is a critical step because it controls the microstructure, which in turn will determine the properties. The processing methods adopted are subject to the constraints of ensuring minimal damage to CNT structure due to applied stresses or due to reaction with the matrix material at elevated temperature. Depending on the nature of the process, CNTs may be subjected to high temperature and stress or may be in contact with molten metals. This may lead to chemical reactions, which could lead to loss of CNTs through carbide formation. The carbides formed may affect the properties of the composites favorably or adversely. Stress applied during processing or consolidation might damage CNTs or may align them in the matrix. Some processes might not be scalable for the production of bulk composites, while others are restricted for a specific application. Hence, the fabrication technique will have to be selected considering these factors. In this chapter, processing techniques will be described for the fabrication of CNT-MMCs and evaluated based on the factors explained previously. The processes that have been used for CNT-MMC fabrication can be divided according to the classification scheme shown in Figure 3.1. Each of the processes has been described in detail in the following.

3.1 POWDER METALLURGY ROUTES

Powder metallurgy technique is as old as brick making. Animal and human figurines and objects made of fired ceramics and found at sites in Pavlov Hills of Moravia are estimated to be from 28000 BC. In simple terms, it can be described as the fabrication of components from powdered materials. The normal operation would consist of pressing a calculated amount of the powder in a die of required shape. The pressure can be applied in cold (cold compaction) or hot condition (hot compaction). The pressed compact has porosity and is further densified using several processes. The classical consolidation method is sintering, in which the pressed compact is heated in a furnace at temperatures that are below the melting point of the material. Densification

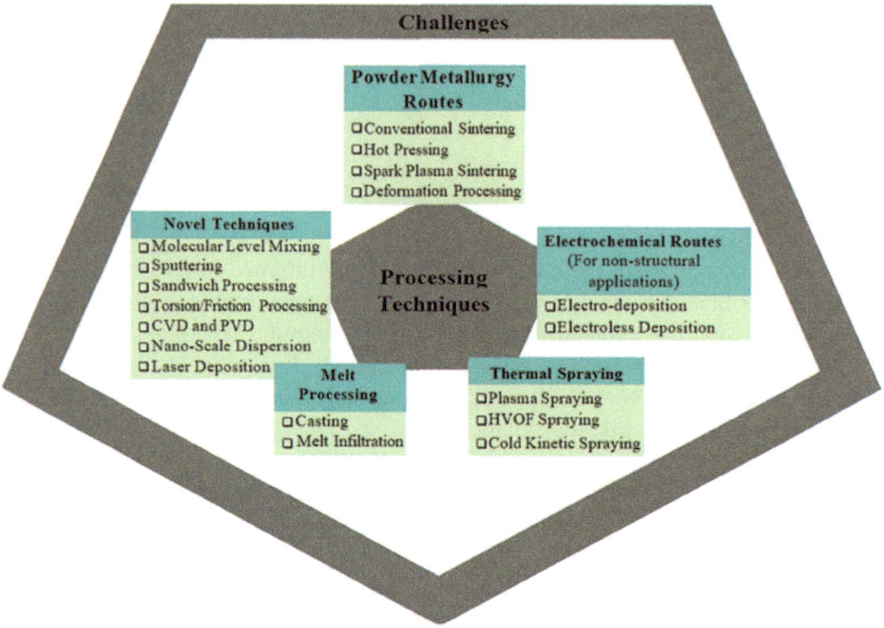

FIGURE 3.1 Classification of the different processes for fabrication of CNT-reinforced MMCs

and change of shape result due to several processes like surface diffusion, grain boundary diffusion, evaporation, and condensation at the inter-particle contacts, which leads to decrease of the inter-particle distances and elimination of pores. There are several advantages of powder metallurgy over other manufacturing processes. The main advantage is that almost any composition can be synthesized, as powder metallurgy is not dictated by thermodynamics and the phase diagram as in the case of ingot metallurgy. There is very small loss of material and intricate shapes can be fabricated using powder processing, with a very small need for machining. The characteristics of the powder are retained in the product because there is no bulk melting. For certain applications like porous bearings or for fabrication of components with refractory materials, this method is indispensable. Sintered compacts can be subjected to further deformation processes like extrusion, rolling, or forging depending on the requirement. Sintering to fully dense compact would require a long time and this could be accompanied by grain growth, which could be deleterious for the application. Sometimes sintering is accompanied by application of pressure in a single direction (hot pressing) or in all directions [hot isostatic pressing (HIP)]. These processes can achieve higher densities at lower temperatures and in less time, leading to lesser grain growth.

Powder metallurgy has been used for the fabrication of composites for a long time. It is easy to disperse the two phases in powder form. The level of mixing is then governed by the size of the components added. Since there is no bulk melting, there is no major rearrangement in the distribution of the phases in the final product.

Powder metallurgy technique has been used extensively for the fabrication of Al-CNT and largely for Cu-CNT composites. It has also been used for a number of other metal-matrix-CNT composites, for example, Mg, Ni, Ti, and Ti-based alloys, various intermetallics, and Ag and Sn-alloys. Powder metallurgy becomes significant because dispersing the CNTs is straightforward using this technique. The quality and level of CNT dispersion attained depends on the powder preparation technique and the size of the starting powders. The smaller the metal powders are, the better is the dispersion of CNTs. However, smaller particles pose a hazard risk and are oxidized easily. Liquid media might be required in order to prevent oxidation and heat generation during processing. In most cases, mechanical alloying has been employed in order to achieve some mechanical bonding as well as for dispersion of CNTs in the powder mixture. In mechanical alloying, metal powders and CNT mixture are fed into a rotating container along with hardened steel/ceramic balls. The repeated impact of the powder mixture between the balls and the ball and container wall leads to deformation, fracture, and welding of metal particles as well as entrapment of CNT within particles. An excellent account of the use of mechanical alloying for the non-equilibrium processing of materials can be found in the text by Suryanarayana [1]. CNT clusters are broken during mechanical mixing and CNTs are attached to metal particles by mechanical interlocking. The ratio of the grinding media (ball) to the powder mixture added, the size of the balls used, the radius of the container, and the rotation speed govern the extent of deformation. Further details of CNT-MMC powder mixing by mechanical alloying is discussed in Chapter 2. The consolidation methods for CNT-MMC processing using powder precursor as the starting material can be classified into four categories. They are (1) conventional sintering, (2) hot pressing, (3) spark plasma sintering, and (4) deformation processing. Each of these processes is described in the following with respect to CNT-MMC processing.

3.1.1 Conventional Sintering

This is the simplest and most conventional method for producing CNT-MMC compacts. The CNTs and metal powders are mixed by a process of mechanical alloying/blending and then compressed to form a green compact, which is then sintered to get the final product. Metallic compacts are subject to oxidation as compared to ceramics and hence the sintering has to be done in an inert atmosphere or under vacuum. This process has been used mainly for fabrication of Al-CNT and Cu-CNT composites. There are few studies for Ag, Mg, and W-Cu composites with CNTs. The dispersion of CNTs in the powder mixture is crucial because compact formation and sintering will not improve the dispersion. Any CNT clusters remaining in the powder mixture will also be present in the final component. CNT clusters are often associated with the porosity and hence poor mechanical properties. Therefore, there is a need to densify the product further using some deformation processes like hot pressing or extrusion.

In some studies, CNTs produced by catalytic decomposition (CVDs) of acetylene over Co nanoparticles on mesoporous silica at 750 °C were purified by refluxing with HNO_3 and HF [2,3]. Subsequently, CNTs were coated with a thin layer of Ni by electroless plating to improve the bonding between the CNTs and Cu matrix.

The CNTs were pre-treated with a sensitizing solution of 0.1 M $SnCl_2$–0.1 M HCl for 30 min followed by activation in a solution of 0.0014 M $PdCl_2$–0.25 M HCl before the electroless plating, which results in a thick (70–90 nm) and irregular coating of Ni on the CNTs. Similar coatings of Ni, Cu, and Ag have been done successfully on SWNTs by electroless plating technique [4]. Composites were prepared by ball milling the coated CNTs with Cu powder followed by isostatic pressing at 600 MPa and sintering at 800 °C for 2 h. The porosity of the composites was found to be low (2.47%) up to 8 vol.% CNTs, which increased to 4.92% for 16 vol.% CNT composite, indicating that the CNTs might be forming clusters at higher concentration [3]. The wear volume and coefficient of friction reduced with CNT content up to 8 vol.% CNT, beyond which it became stable. Dispersion of CNTs in the powder can be increased by direct growth of CNTs on the feedstock powders [5]. For doing this, one needs to deposit catalyst nanoparticles on the surface of metal powders. This was done by precipitating $Ni(OH)_2$ on Al particles, which is followed by drying and calcination at 200 °C for 4 h. This results in formation of NiO-Al, which is reduced to Ni-Al by H_2 reduction at 400 °C for 2 h. This method produced Ni particles 5–20 nm in diameter on the surface of Al particles. CNTs can be grown on this powder by catalytic decomposition of a CH_4-N_2-H_2 mixture at 630 °C. The composite is prepared by pressing the powders into a compact at 600 MPa pressure followed by sintering at 640 °C under vacuum. In order to increase density, the sintered discs are further compacted under a compressive pressure of 2 GPa [5]. Al-CNT composites were also prepared from the ball-milled powders of the same composition. The composite prepared using powders with in situ CVD grown CNTs showed a tensile strength of 398 MPa as compared to 213 MPa for ball-milled powders. This indicates the importance of dispersion of CNTs in starting powder in these processes. Ball milling has been shown to be effective in producing composite powders of Al and CNT with moderate to good dispersion [6,7]. The CNTs were found to penetrate from the surface to the inside of the metal particle as the time for milling increased. However, the powder size reached more than 3 mm due to cold welding of finer particles. Such large powder is usually not suitable for powder metallurgy processing. Ball milling has also been used to prepare composite powders of Mg and CNTs for tapping their properties for hydrogen storage properties [8,9]. Hydrogen is absorbed due to formation of MgH_2. The absorption and desorption kinetics are of importance in the hydrogen storage application. Ball milling reduced the length of the CNTs due to fracture by impact, but they were dispersed nicely in the powder. CNTs did not increase the amount of H_2 intake significantly [9]. Ball milling was also used to prepare CNT-Si composite powders for anode applications in Li-ion batteries [10]. Silver matrix composites reinforced with CNTs have been made by mixing the powders in a mortar and pestle and compacting at 320 MPa pressure followed by sintering at 700 °C for 1 h [11]. In order to enhance adhesive bonding at the surface, the CNTs were treated by acid to roughen the surface through oxidation. Here too, densities achieved by sintering were lower and the composites were pressed again at 400 MPa for further densification. Authors reported to have uniform dispersion of CNT without formation of any cluster up to 8 vol.% CNT content in the composite, after sintering at 700 °C [11]. However, CNTs start agglomerating beyond 10 vol.% CNT addition. This trend was reflected perfectly in the assessment of mechanical property (hardness) of the composites. Interface between CNT and metal matrix has not been analyzed in most of these composites. Similar studies have been carried out to reinforce Sn-Ag-Cu solders with SWNTs. Sintering was carried out at 180 °C, followed by extrusion.

Conventional sintering is one of the earliest techniques utilized to synthesize CNT-MMCs, but it met limited success due to the poor densities of the final structure. CNTs might act as obstacles to the motion of interparticle boundaries and interfaces that lead to the removal of pores during sintering. Most of the studies on sintered CNT-MMCs report poor density, which can be increased only at the expense of matrix grain size. In some cases, repressing was used for the elimination of pores. However, repressing leads to brittleness due to cold working. No mention of chemical reactions between CNT and metal matrix has been made in most of the CNT-MMC work done by conventional sintering.

3.1.2 Hot Pressing

Use of pressure during sintering results in the formation of dense compacts. Hot pressing can be traced back to the 1933 patent by George F. Taylor [12] for an "apparatus for making hard metal compositions," which was an "apparatus for simultaneously pressing and sintering of powdered materials." A typical setup for hot pressing is shown in Figure 3.2a. In hot pressing, powder mixtures or the pressed compact are subjected to high temperature while being pressed in a die. The temperature and pressure result in easy deformation through creep and material transfer and help in achieving high densities. The heating could be done by electrical resistance heating, induction heating, or radiation. The hold time is critical since grain coarsening would occur for extended heating duration. The operation has to be carried out in an inert atmosphere or vacuum for CNT-MMCs to prevent oxidation. Vacuum hot pressing equipment is expensive because of the need of insulation at high temperatures. A hot press can also be used for near-net shape manufacturing of components by low-pressure forging. Because of the small amount of time required for the process, grain

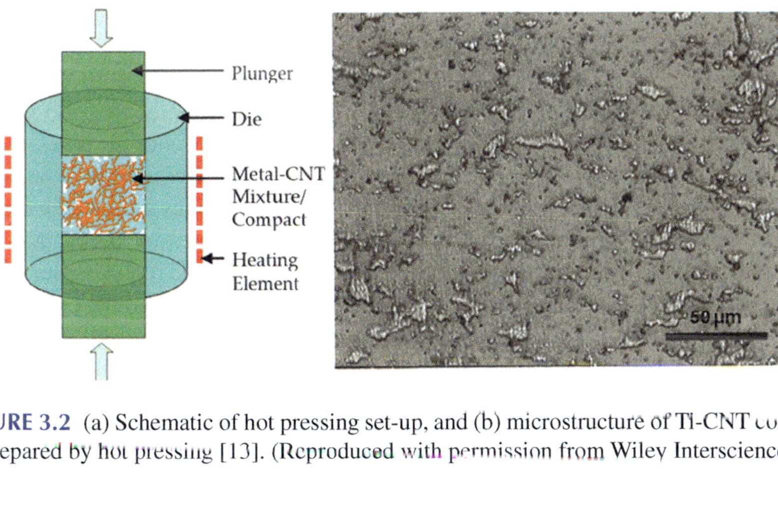

FIGURE 3.2 (a) Schematic of hot pressing set-up, and (b) microstructure of Ti-CNT composite prepared by hot pressing [13]. (Reproduced with permission from Wiley Interscience).

growth can be inhibited. In case of HIP, the pressure is applied uniformly from all directions by using a gaseous or molten salt medium. This leads to uniform densification in all directions and ensures isotropic properties. Hot pressing has been employed for various CNT-MMCs.

Al-CNT composites have been prepared by hot pressing powder mixtures at 520 °C and 25 MPa [14] for more than 30 min. CNT agglomerates were found mainly at the Al grain boundaries, while few single CNTs were found embedded within the matrix. Reaction between Al and CNT leading to compounds with Al:C ratio of 1:1 and 1:2 was observed through EDS. Al-SWNT composites have been prepared by hot pressing of a mixture of nano-Al particles (50 nm in size) and SWNTs at a pressure of 1000 MPa and temperatures between 260 and 480 °C [15]. Ultrasonication was used to disperse the two powders in alcohol, but the composite was found to have CNT agglomerates. No chemical reaction was reported between SWNT and Al. Cu-CNT composites have been prepared by hot pressing ball milled powder mixtures at temperatures of 1100 °C in argon atmosphere in a graphite die at a pressure of 32 MPa [16] for 1 h. Composite with Ni-coated CNT shows better densification and good dispersion of second phase in the matrix, whereas uncoated CNTs segregate at the grain boundaries due to surface tension and poor bonding with the matrix [16]. Mg-CNT composite was prepared by a two-step process of hot pressing in a vacuum at 600 °C at a pressure of 50 MPa for 0.5 h followed by hot isostatic pressing at 600 °C at 180 MPa for 1 h [17]. Ti-20 vol.% CNT composites have been prepared by vacuum hot pressing at 935 °C at a pressure of 30 MPa for 2 h [13]. Figure 3.2b shows the microstructure of Ti-CNT composite showing the presence of CNT clusters. Even though the temperature was high, TiC formation was not observed at the Ti-CNT interface. This study exemplifies the thermal stability of CNTs at high temperature. Vacuum hot pressing has also been used for producing composites of CNTs with novel matrices like iron aluminide (at 1150 °C with 35 MPa for 1 h) [18] and Ti-based bulk metallic glasses (at 450 °C with 1.2 GPa) [19]. Uniform distribution of CNTs throughout the iron aluminide powders has been observed, resulting in improved mechanical properties (hardness, compressive strength, and bend strength). The enhanced mechanical property was also attributed to grain growth inhibition caused by interlocking nanotubes.

Hot pressing has the advantage of producing high-density (>95%) CNT-MMCs. CNTs may provide additional advantages by acting as obstacles to grain growth. The distribution of CNTs in the starting powder is critical because it is retained in the compact. CNT clusters present in the starting powder mixture are expected to be deformed into a flatter morphology under the simultaneous action of temperature and pressure. However, infiltration of the CNT clusters by metal matrix is unlikely due to the solid-state nature of the process. The time required during hot pressing to attain useful densities is approximately 1 h in most studies. This indicates that the reaction between CNT and some metal matrices could be a critical issue. Some of the studies have shown reaction between CNT and metal [14], while others do not [13]. In addition, for the sensitive matrices like metallic glasses which undergo devitrification, hot pressing for long times may not be desirable.

3.1.3 Spark Plasma Sintering

Spark plasma sintering (SPS), also known as electric field assisted sintering technology (FAST), plasma assisted sintering (PAS), and pulsed electric current sintering (PECS), is a variation of hot pressing in which the heat source is a pulsed DC current that is passed through the die or the powders (depending on whether the powder is electrically conducting) during consolidation. A schematic of the SPS setup is shown in Figure 3.3. Spark discharges at the particle interfaces are believed to produce rapid heating, which enhances the sintering rate. The heating rates are quite high compared to hot pressing and can be up to 1000 K s^{-1}. Efficient densification of powder can be achieved in this process through spark impact pressure, joule heating, and electrical field diffusion. This method is, generally, suitable for consolidation of nano powders, without allowing sufficient time for agglomeration or grain growth. Some authors [20] have questioned the existence of sparking, arcing, or plasma. Munir et al. [21] has presented a comprehensive review of the history, mechanisms, and applications of the process. SPS is an attractive process for CNT-MMCs due to its fast nature, which helps in restricting the reactions between the metal matrix and CNTs. Reaction products may often lead to loss of reinforcement or have adverse effects on its mechanical properties.

SPS has been used for the fabrication of Al-CNT and Cu-CNT composites. As with other powder consolidation techniques, the CNT distribution in the starting powders is very crucial. For Cu-CNT composites, CNTs were ultrasonicated in ethanol for separating the clusters. After drying, Cu powders of 100 nm and 200 μm size were mixed with the CNTs to obtain compositions containing 0–15 vol.% of CNT. The powder mixtures were ball milled for 24 h using alumina balls. The resulting mixture was consolidated by SPS at 750 °C for a hold time of 1 min at a pressure of 40 MPa [22]. Dispersion was better in the nano Cu powder compared to the large Cu powder for the obvious reasons of the size. However, final density, after SPS, was

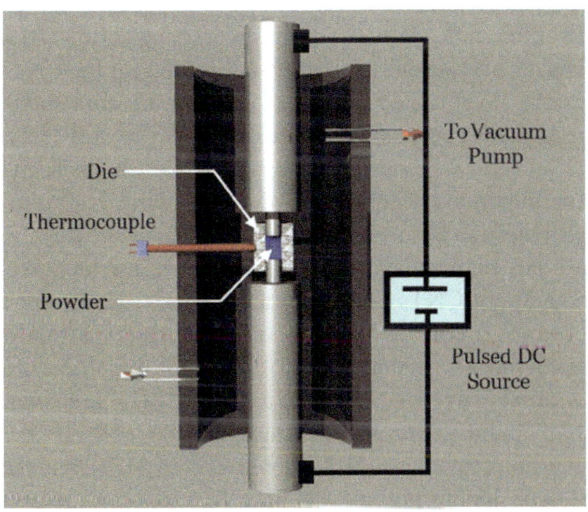

FIGURE 3.3 Schematic of spark plasma sintering set-up.

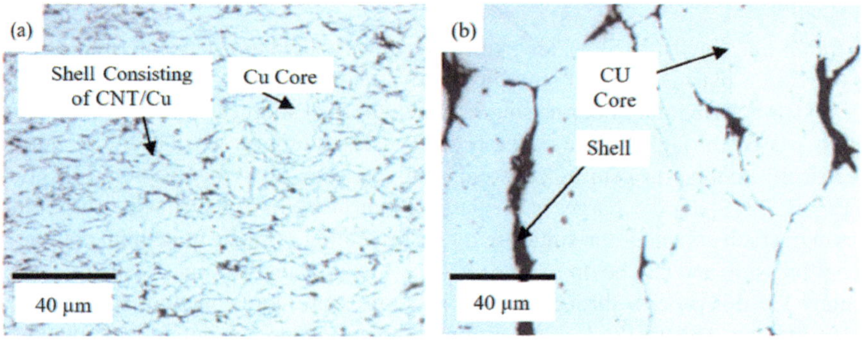

FIGURE 3.4 Microstructures of Cu-CNT composites produced by SPS of (a) nano-Cu powders, and (b) micron-sized Cu powders [22]. (Reproduced with permission from Materials Research Society.)

observed to be unaffected by the powder size or CNT dispersion and was higher as compared to the conventionally sintered composites. The microstructure was core shell type, with the shell region outlining the grain boundary areas containing most of the CNTs. Due to smaller size of the core regions in case of the nano-Cu powder, the dispersion was better as shown in Figure 3.4 resulting in better hardness of nano Cu composite as compared to the larger powder. The CNT/Cu shell structures can be broken down by a cold rolling process up to 50% reduction followed by annealing at 650 °C, which has the benefit of increasing the density and aligning the broken CNT/Cu shells [23]. Further improvement in properties can be achieved by improving the dispersion of CNTs. The same group went on to develop a new method for obtaining powders containing dispersed CNTs, which they termed as molecular level mixing [24]. Extraordinary strengthening was observed in SPS sintered compacts of molecular level mixed powders. The yield strength of Cu-10 vol.% CNT composite was found to be almost three times that of unreinforced composite [25]. Improvement in sliding wear resistance and hardness were also observed for the CNT-MMCs synthesized using molecular-level mixed powders [25]. Similar core shell structures have been reported for Al-CNT composites produced by SPS of powders produced by the nanoscale dispersion (NSD) process, which is described later. The pressure applied was 50 MPa at the maximum temperature of 600 °C with a 20-min hold time. The NSD process distributes the CNTs on the surface of the Al powder. Since starting Al powder was 15 μm in size, the dispersion was not so good in the final microstructure, resulting in segregation of CNTs on the grain boundaries [26]. Post-SPS processing by extrusion lead to the breakdown of these CNT clusters and uniform distribution within the microstructure [26]. A comparative study of the properties of hot pressed and SPS samples of Al-1 wt.% CNT was carried out [27]. It showed that SPS processed samples had slightly improved hardness (53.8 over 49 Hv), improved maximum stress in three-point bend test (201.19 over 113.28 MPa), and reduced wear loss (2.4 over 13.9 g) over the hot pressed samples. SPS has also been employed in the fabrication of CNT reinforced Fe_3Al composite [28]. SPS was carried out at 1000 °C at a pressure of 30 MPa. Addition of CNTs improved the hardness and compressive yield strength. SPS Fe_3Al-CNT composite was found to have better compressive yield strength than HIP Fe_3Al-CNT composite, indicating the effectiveness of SPS over HIP.

Compared to conventional sintering and hot pressing, SPS definitely is a promising method for obtaining high-density CNT-MMC compacts. Since the time of sintering is small, grain growth is inhibited and hence an extremely fine-sized powder mixture can be consolidated. The small sintering time is also favorable in ensuring minimal or no reaction between CNT and metal matrix. The distribution of the CNTs in the final composite is dependent on the distribution of CNTs in the starting powders. Use of smaller powders results in smaller core shell type structures, resulting in better dispersion and properties.

3.1.4 Deformation Processing

As seen in previous sections, distribution of the CNTs is not affected by most of the sintering techniques. Shear force due to deformation is required in order to breakdown CNT shell structures and CNT clusters for improving the dispersion or for aligning them. Deformation of compacts also leads to increase in density. This approach has been used mainly for Cu- and Al-based composites. Hot extrusion, equal channel angular pressing (ECAP), and hot/cold rolling have been employed as deformation techniques. Hot extrusion is the most common process used for deformation processing of CNT-MMCs. In hot extrusion, the sintered or pressed compact is heated to the required temperature and then forced through a die typically conical in shape maintained at the high temperature. Inside the die, the compact undergoes a gradual reduction in area in the conical portion due to the application of shear stresses. The ratio of the area of the material at the entry to the exit is known as the extrusion ratio. ECAP is a technique in which a material is extruded through a die in which the extrusion ratio is 1. However, the material has to pass through a bent region. A lot of shear deformation is induced in the material. There is a considerable grain refinement due to this thermo-mechanical treatment. Hot rolling is believed to improve the properties by improving the density and dispersion of CNTs by breaking CNT clusters. A schematic of the various deformation processes is shown in Figure 3.5.

Kuzumaki et al. made one of the first attempts on synthesis of CNT-MMCs by extrusion of hot pressed compacts of Al-CNT [31]. The powders were ultrasonicated in ethanol followed by drying in vacuum and heating to 600 °C in vacuum and pressing at 100 MPa [31]. The pressed compact was hot extruded at 500 °C with an extrusion ratio of 25:1. However, efficient CNT dispersion could not be achieved. In another case, the ultrasonicated mixture of Al and CNT was subjected to ball milling and then compacted and sintered at 580 °C and hot extruded at 560 °C [32]. A similar approach has been used to produce 2024 Al alloy-CNT composites. The ball-milled powders were first cold isostatically compressed at 300 MPa followed by hot extrusion at 460 °C at an extrusion ratio of 25:1 [33]. A yield strength of 336 MPa was measured for Al-1 wt.% CNT composite in this case [33] as compared to 99 MPa for the Al-2 wt.% composite in the former case [32]. A more comprehensive study on the effect of CNT content on the mechanical properties indicated that tensile strength

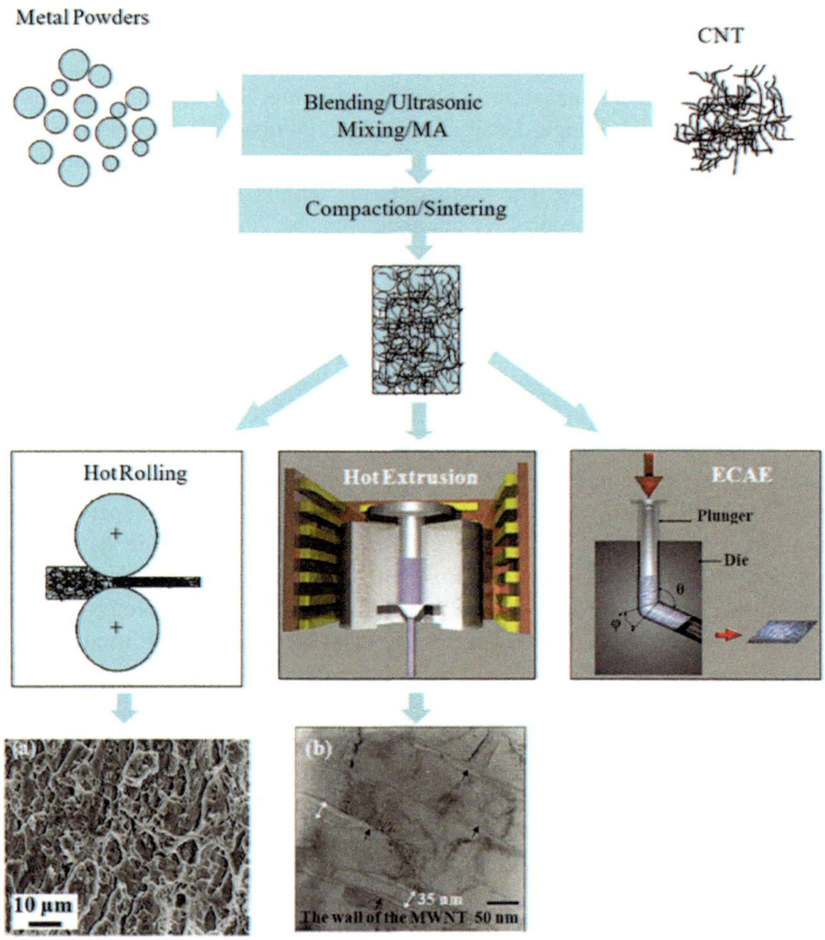

FIGURE 3.5 Schematics of various deformation processes and the typical microstructures resulting from each of the processes. (a) SEM of the fracture surface of an Al-0.5 wt.% CNT composite [29], and (b) TEM image of Al-4 vol.% CNT composite [30].(Microstructures reproduced with permission of Elsevier.)

was greatest for 2024 Al alloy with 1 wt.% CNT (520 MPa). Aluminum composite containing 2 wt.% CNT exhibited lower tensile strength, which was attributed to the presence of micro-voids and poor interfacial strength [34,35]. The 2024 Al-1 wt.% CNT composite was shown to have 11% lower coefficient of thermal expansion compared to the 2024 Al alloy [36]. There was no formation of interfacial product between CNT and Al alloy. It was shown by calorimetric studies that CNT reacts with Al alloy only when the temperature was above 656.3 °C, which is close to the melting point of aluminum [37]. However, Al_4C_3 formation has been detected in X-ray diffraction spectrum of Al-CNT composites prepared by pressure-less sintering of ball-milled powders at 550 °C followed by hot extrusion at 500 °C [38]. A yield strength of 189.2 MPa was reported for Al-1.75 wt.% CNT composite, which was twice that of the value reported earlier [32]. Annealing of hot extruded products can help in improving the strength further. Annealing at 500 °C of Al-2 wt.% CNT composite extruded at 500 °C showed a tensile strength of 345 MPa. One of the advantages of deformation processes is that CNT clusters are broken down and may be aligned in the direction of the shear stress. In Al-4 vol.% CNT composites produced by hot extrusion of ball-milled powders at 470 °C, it was shown that grain refinement and CNT alignment led to the increase in compressive strength by more than 100 MPa [30]. Cold rolling [23] and hot extrusion [26] have been employed to CNT-MMC compacts produced by SPS process. The SPS process produced CNT clusters at the grain boundaries due to the kind of distribution in the starting powders producing the core-shell microstructure. Al-CNT powders were prepared by the NSD process, which is good at dispersing the CNTs on the surface of the particles, but the overall dispersion is determined by the size of the particles used. Al-5 vol.% CNT composite prepared by hot extrusion of the SPS compact at 400 °C had a tensile strength of 194 MPa, which was twice that of pure Al produced in a similar manner [26]. Hot rolling has been used for Al-CNT powders prepared by mechanical alloying [29]. Tensile testing of samples machined out of the strips produced showed a marginal increase in strength for 0.5 wt.% CNT, while the strength reduced for higher loading of CNTs. The decrease in the strength was attributed to presence of CNT clusters at higher loading.

Much work has been carried out on Cu-CNT composites by deformation processing. Compacts prepared by SPS [23] and pressure-less sintering (at 850 °C) [39] have been subjected to cold rolling. Cold rolled Cu-CNT compacts were annealed at 600–650 °C to get rid of the strain hardening. As suggested previously, rolling helps in the densification of the compact as well as breaking and aligning CNT clusters. SPS compacts cold rolled to 50% thickness followed by annealing at 650 °C were found to contain elongated CNT clusters, which acted as reinforcement for the matrix [23]. The yield strength of Cu-10 vol.% CNT composite was found to be ~280 MPa as compared to ~175 MPa for the unreinforced Cu matrix. The wear loss and friction coefficient were found to decrease with increasing CNT content up to 10–15 vol.% CNT. It was concluded that CNTs served as better reinforcements than carbon fiber for improving the wear resistance [39]. Cu-CNT composites have also been processed by ECAP process. Cu-CNT powders prepared by ultrasonication in ethanol were filled in a Cu sheath and subjected to ECAP in cold condition [40–42]. For Cu-5 vol.% CNT composite, the density and the dispersion of the CNTs in the composite improved with increasing number of passes. Repeated ECAP broke the CNT agglomerates to form better dispersion of CNTs in the matrix [40]. ECAP being a severe plastic deformation technique is expected to induce a high amount of deformation to the constituent phases, thus damaging the CNTs. However, the issues of damage or fracture of CNTs during ECAP are yet to be studied and addressed. It is expected that both grain refinement and CNT reinforcement will contribute toward the strengthening.

There have been some efforts in deformation processing of metal-CNT systems as well. Mg-CNT and Mg-SiC-CNT composites have been prepared by hot extrusion of sintered compacts at 350 °C [43,44]. Composites up to 0.3 wt.% CNT were prepared and shown to have good density ($\geq$ 99%) and reduced CTE (~10%), while the mechanical properties improved by a small factor (15% increase in yield strength). Mechanical properties of Mg-based composites were high in presence of SiC (36%

increase in the yield strength) as compared to CNT alone (5% increase) [44]. Sn-Ag-Cu solders reinforced with up to 0.7 wt.% CNTs have been prepared by room temperature extrusion of compacts prepared by sintering the blended mixtures at 175 °C [45,46]. These composites have shown a small decrease (5%) in thermal expansion coefficient and a small increase (3%) in yield strength as compared to unreinforced solder [45,46]. However, Sn-Ag-Cu solders reinforced with 1 wt.% SWNTs produced by hot extrusion showed an 18% increase in tensile strength [47].

It is observed from the previous discussions that deformation processing has been used extensively for fabrication of CNT-MMCs. It has also been found that the mechanical properties are best for these composites CNT-MMCs. The density, dispersion, and bonding between CNT and matrix are improved due to the applied stresses during processing at elevated temperatures. The properties obtained are controlled by the dispersion of CNTs in the powder or the compact stage. Further improvement in properties requires focus on obtaining better powders and compacts for hot deformation processes.

The issues of utmost importance in powder processing of CNT-MMC are dispersion and reinforcement of CNTs in the matrix. Conventional powder processing techniques exhibit a lack of CNT dispersion due to poor mixing, which was carried over to the consolidation stage resulting in CNT clusters in the composite. The solutions to tackle these problems are (1) efficient CNT-metal powder mixing and (2) shorter consolidation time. Thus, ball milling of the initial powder mixture became almost common for all processing routes. The short sintering time during SPS provides an effective option for consolidation of CNT-MMCs. Post-sintering deformation has been found effective to break the CNT clusters in the consolidated structure. Heavy deformation was found to break apart the core/shell type of structures and resulted in dispersion of CNTs that were more homogeneous. However, this process also can lead to damage and fracture of CNTs and formation of some compounds at CNT/metal interface, when the process is carried out at high temperatures. Broken CNTs may lead to deterioration of mechanical properties. At this juncture of development of CNT-MMCs through powder metallurgical routes, SPS and post-sintering still appear to be very promising. However, future research needs to be directed toward solving the current problems of these two techniques. Moreover, detailed studies on important CNT-metal matrix composite systems should be taken up to prepare industry-acceptable process maps.

3.2 MELT PROCESSING

Melt processing is the most primitive form of fabrication of metallic components that can be traced back to 2500 BC, when people in the Indus Valley used to make beautiful bronze statues, a famous one being that of the dancing girl. The processes that fall under this category involve molten metals and alloys as the starting material and hence the temperature is higher than other techniques. Consequently, there are new challenges to be considered for CNT-MMCs, as follows:

1. How is the second phase (CNT) in the molten metal phase dispersed?
2. Can the CNTs retain their structure at high temperatures?

3. How is the chemical stability of CNTs ensured since the molten metal is expected to have high reactivity?

Melt processing can be divided into two types: casting and infiltration methods. The advantage of melt processing methods is that bulk components can be easily manufactured. Thus, these methods would ideally by suitable for large-scale manufacturing of CNT-MMCs. The challenges associated with casting and infiltration methods as applied to CNT-MMC fabrication are discussed in the following sections.

3.2.1 Casting

Casting is a method of manufacturing bulk components from the molten metal. Molten metals are poured into molds of desired shape, and the solid components are obtained after solidification. The challenge here is to disperse the low density CNTs in the melt. One has to design a method for mixing the CNT reinforcement in the molten matrix. Mixing of CNTs will be easy if CNTs wet the molten metal. The contact angle between a liquid (*L*) and substrate (*S*) as shown in Figure 3.6 is obtained from the force balance as

$$cos\theta = \frac{\gamma_{SV} - \gamma_{LS}}{\gamma_{LV}} \quad (3.1)$$

Where *L* denotes the molten metal, *S* denotes the surface of CNT, and γ denotes the surface tension. Wetting occurs when the contact angle θ is less than 90°. It has been found that metals exhibiting good wetting like Cs and Se also fill the CNTs by capillary action [48]. If the CNTs do not wet the liquid, then surface tension forces will tend to bring the CNTs together leading to formation of clusters. It has been found that high surface tension metals (e.g., Te, Pb, Hg, Ga, Al) do not wet the CNTs. Among these, Al has the highest surface tension (~800 mN m^{-1}) and hence poor wetting with CNTs. Low surface tension metals (S, Cs, Rb, 97) have surface tension ranging from 61–97 mN m^{-1} and exhibit good wetting with CNTs [48]. It is observed that when the surface tension of the molten metal (γ_{LV}) is below 200 mN m^{-1}, then

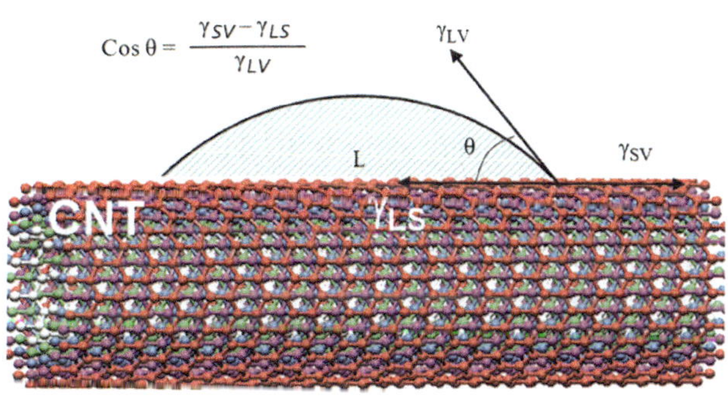

FIGURE 3.6 Definition of a contact angle between a liquid droplet and a carbon nanotube.

wetting occurs. It is also observed that most of the metals do not wet the CNTs. Hence, some treatments are necessary to improve wetting.

A melt processing route is favorable for fabricating composites from those metals that have a low melting point, such as Mg, Al, and some bulk metallic glass matrices. Casting has been used to synthesize Mg-CNT and bulk metallic glass (BMG)-CNT composites. Mg was melted at 700 °C followed by the addition of CNTs to the melt by stirring [49]. After that, the melt was cast into ingots. The CNTs were coated with Ni by electroless deposition before mixing into Mg for better wetting with the matrix. The tensile strength and ductility of Mg-CNT composites increased by 150% and 30%, respectively, for a 0.67 wt.% CNT composite but mechanical properties deteriorated at higher CNT concentration due to cluster formation [49]. To improve the dispersion, a disintegrated melt deposition (DMD) method has been used [50,51]. The molten Mg-CNT pool at 750 °C is stirred at 450 rpm with a mild steel impeller, which is coated with ceramic to avoid contamination of the melt. The melt is allowed to pass through an orifice of 25 mm diameter and is disintegrated by two argon jets. The disintegrated melt is deposited on a metallic mold. The ingot is then extruded at 350 °C to get the Mg-CNT composite rods. Composites with up to 2 wt.% CNT were prepared. Simultaneous improvement of strength and ductility was found due to CNT reinforcement. Fatigue studies indicated that the number of cycles to failure reduced with increasing CNT content, which was attributed to the presence of voids at the CNT/matrix interfaces [51]. To improve the dispersion, CNTs were first dispersed on the AZ91 Mg chips by using a block copolymer [52]. The block copolymer Disperbyk-2150 was dissolved in ethanol and then CNTs were dispersed by ultrasonication followed by stirring. Stirring was continued after addition of Mg chips until the ethanol was evaporated. The chips were melted in an argon atmosphere at 650 °C and the melt was stirred at 370 rpm before being cast into ingots. Grain sizes were found to be similar in the cast alloys with and without CNT. Hence, the improvement in the compressive properties could be attributed to reinforcement by CNTs alone. For making Zr-based BMG-CNT composites [53,54], $Zr_{52.5}Cu_{17.9}Ni_{14.6}Al_{10}Ti_5$ BMG ingot was first prepared by arc-melting a mixture of high purity Zr, Al, Ni, Cu, and Ti under a Ti-gettered purified argon atmosphere. The ingots were crushed into a fine powder with a particle size of approximately 200 μm and mechanically mixed with the CNTs. The mixture was compressed to form a cylinder and melted in a quartz tube rapidly by induction under a high purity Ar atmosphere and then cast into a copper mold to produce rods with a diameter of 5 mm. Rapid melting and solidification ensures that the CNTs do not become segregated. In this way, composites were prepared by adding up to as high as 10 vol.% CNT. CNT clustering was observed as well as an increase in the crystallinity of the BMG due to formation of a ZrC phase. Increase in the hardness and acoustic wave absorption properties were observed for the CNT reinforced composites [54]. Rapid solidification techniques have the advantage that there is no time for CNTs to rearrange or segregate forming clusters. $Fe_{82}P_{18}$ metallic glass-CNT composite ribbons 40 μm in thickness have been prepared by melt spinning [55]. The CNTs and the melt were mixed and poured drop by drop on a rotating copper wheel. The molten droplets were transformed into glassy ribbons upon coming in contact with the wheel rotating at more than 2500 rpm. Retention of undamaged CNTs and the amorphous nature of the composite were observed [55].

Casting can be used to fabricate bulk components of CNT-reinforced metal matrix composites but one has to be careful about the dispersion of the CNTs. CNTs tend to segregate and form clusters due to their non-wetting nature to most molten metals. Due to the increased reactivity of the matrix material in the liquid form, the reaction between CNTs and the matrix can also be deleterious. Thus, a rapid melting and solidification processes with improved wetting conditions would lead to better retention and dispersion of CNTs in the CNT-MMC.

3.2.2 Melt Infiltration

Melt infiltration is one of the widely used techniques for metal matrix composite fabrication. This method produces composites with a high loading of reinforcements. In this method, a porous preform of the reinforcement is prepared first by powder metallurgy techniques. The reinforcement could also be in the form of woven cloths from fibers. The preform is then infiltrated with molten alloy for fabricating the composite. This technique has a higher chance to have uniform distribution of CNTs, but at the same time proper filling up of the pores, to make a good, dense, and composite structure, becomes a critical step. A typical set-up for the pressure-less infiltration is shown in Figure 3.7a. The metal or alloy ingot is put on top of the preform and the temperature is gradually increased. The ingot melts and infiltrates the preform. Aluminum and magnesium alloys have low viscosity and can easily infiltrate porous performs and fiber prepregs. The melt infiltrates the CNT preforms due to gravity. Sometimes, pressure is also required to aid the infiltration process. The pressure-induced infiltration of porous compacts is governed by Darcy's law, given by [57]

$$h^2 = \frac{2k_p t}{\mu(1-V_p)}(P-P_0) \tag{3.2}$$

where h is the infiltrated depth of the metal into the porous CNT compact, k_p is the intrinsic permeability of the compact, μ is the viscosity of melt, t is the infiltration time, V_p is the particulate volume fraction, and P is the applied pressure. P_0 is the

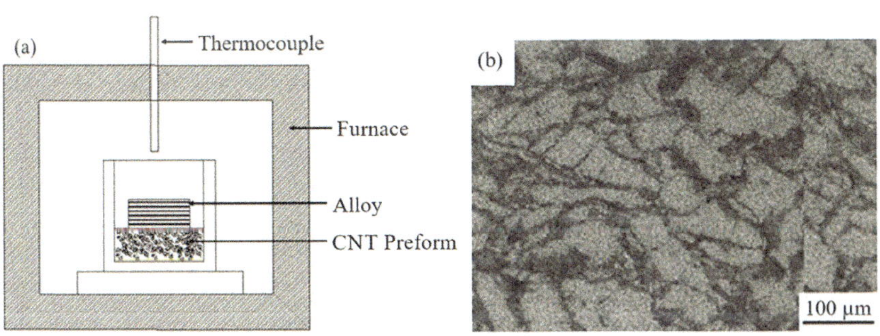

FIGURE 3.7 (a) Schematic showing melt infiltration technique, and (b) microstructure of Al-10 vol.% CNT composite prepared by pressure less infiltration technique [56]. (Microstructures reproduced with permission from Elsevier.)

threshold pressure for infiltration to occur, is governed by capillary forces, and is given by [58]

$$P_0 = \frac{6\lambda}{D}\gamma_{LV}cos\theta\left(\frac{V_p}{1-V_p}\right) \tag{3.3}$$

where λ is a parameter dependent on the particulate shape, γ_{LV} is the liquid vapor surface tension of the infiltrating liquid, and θ is the contact angle. From Eqs. 3.2 and 3.3, it can be seen that the lower the viscosity and surface tension of the infiltrating liquid, the greater the infiltration depth. It is also seen that the smaller the contact angle, that is, the better the wetting, higher is the threshold pressure for infiltration. This is because the liquid would wet the preform and force would be required to force the liquid through the pores.

To prepare a preform, CNTs were grown on Al_2O_3 fibers (Saffil) by a CVD process [59]. The resultant CNT coated Al_2O_3 fibers were infiltrated with magnesium by gas pressure infiltration. Mg-CNT composite prepared in this manner had an enhanced shear modulus. Metal-CNT preforms have also been made by mechanical milling [56]. CNTs and Al powder mixture was ball milled at 300 rpm for 7 h and the resultant powder was compacted to make the preform. This preform was infiltrated spontaneously by LY12 Al alloy at 800 °C in N_2 atmosphere. The distribution of CNTs was limited by its dispersion in the preform. The back scattered scanning electron micrograph in Figure 3.7b showed a core-shell type microstructure. Composites with up to 20 vol.% CNT were prepared. Wear rate of the melt infiltrated Al-CNT composites was found to decrease with the increasing CNT content, while friction coefficient also reduced up to 15 vol.% and then became constant [56]. Hardness was found to increase up to 10 vol.% CNT content and became constant thereafter. CNT preforms have also been prepared by mixing with an organic binder (50 wt.%) followed by sintering of the compact at 500–600 °C [60]. Molten Al-Mg and Al-Si and Mg-Al alloys infiltrated the compact by application of pressure. The calculated and measured threshold pressures were found to be close in value.

Melt infiltration is an economical method of producing composites with high CNT loading. The quality of CNT dispersion in the composite is dependent on the CNT dispersion in the preform. However, the infiltrating alloy may be devoid of reinforcement since there is very little transfer of CNTs from the preform to the melt. A combination of methods described in casting and infiltration could lead to superior composites and needs to be explored.

It should be observed that melt processing techniques have been used to a lesser extent as compared to powder metallurgy techniques for the synthesis of CNT-MMCs. The challenge associated with melt processing techniques is the same, that is, how to disperse the CNTs in the melt pool. The problem of dispersion is more critical in melt processing methods because of poor wettability of the CNTs with most molten metals. Thus, rapid solidification techniques have the advantage of not allowing much time for CNT segregation to occur. In addition, the melt pool has to be stirred continuously. Pressure infiltration of CNT compact is a promising method. Due to the high temperatures and the increased activity of the molten metal, formation of interfacial compounds due to reactions is inevitable. A coating can be applied on the CNT surface to improve the wetting as well as to inhibit the chemical reaction. Research should be directed toward preparing better CNT compacts and designing shorter processing times.

3.2.3 Induction Melting

Induction melting utilizes induction coils to propel a magnetic flux through the material to be processed, causing it to heat up and subsequently melt. Akin to joule heating thru the use of flowing electric current, induction melting generates heat internally due to the rapidly oscillating magnetic fields. Mansoor et al. [61] proposed that this electromagnetic stirring action would aid the dispersion of CNTs in molten aluminum. The processing steps utilized to fabricate Al-CNT composites by induction melting are shown schematically in Figure 3.8. Aluminum bars are first introduced into an aluminum oxide crucible placed in the center of the induction coils. After partial melting of the aluminum at 660 °C, the appropriate amounts of CNTs along with a low melting temperature flux metal are introduced; the flux serves to consume the surface oxide layer on the melt, reduce the effective melting temperature of Al, and enhance reactivity with CNTs. The temperature is then increased to 760 °C and

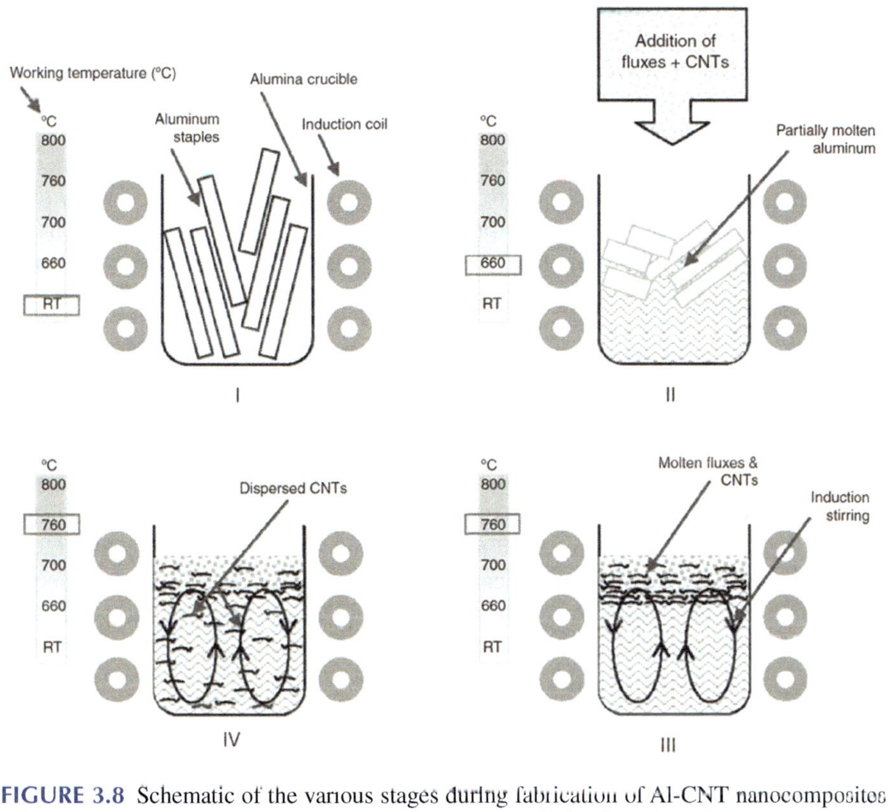

FIGURE 3.8 Schematic of the various stages during fabrication of Al-CNT nanocomposites using air induction furnace [61]. (Reproduced with permission from Elsevier.)

CNTs are dispersed throughout the Al matrix via the stirring action of induction heating. This is done for 10 min, after which the melt is poured into a preheated (150 °C) copper mold and held at 540 °C for 1 h in argon atmosphere. Samples were then cold rolled to obtain Al-0.1 vol.% CNT and Al-0.2 vol.% CNT composites. Both composite compositions exhibit enhanced ductility and strength over pure Al, with the Al-0.2 vol.% CNT composite attaining a 77% increase in yield strength, 52% increase in tensile strength, 45% increase in hardness, and 44% increase in ductility.

3.3 THERMAL SPRAYING

Thermal spray is an industrial scale processing technique that can be used to produce coatings or free standing structures. In thermal spray techniques, material to be sprayed is fed into a heat source in the form of fine powder or wire, where it is converted into a molten or semi-molten state that is accelerated by a carrier gas and made to impinge on a substrate. The molten/semimolten particles strike the substrate and form splats. Accumulation of splats layer by layer results in formation of a coating. Thermal spray is a nearly 100-year-old process. Quoting from the Calendar of Patents Records published in the journal Nature, "The modern metal-spraying process for coating iron and steel is largely due to the Swiss chemical engineer, Dr. M. U. Schoop, whose first patent was applied for in Germany on April 27, 1909. The English patent was granted the following year" [62]. The technology of thermal spraying has vastly improved in the last three decades due to much research, resulting in newer applications. Thermal spray coatings are used in a variety of industries like automobile, aerospace, chemical, and heavy machinery for protection from wear, corrosion, and high temperature environments. Based on the source of heat, the techniques can be classified as wire arc, flame spray, plasma spray, high velocity oxy-fuel spray, and detonation gun spray process. In certain spray processes, there is no thermal energy input like cold spraying. A comparison of the various processes based on the gas temperatures involved and the particle velocities attained is shown in Figure 3.9.

Researchers at Florida International University pioneered thermal spray processes for synthesis of metal-CNT [63–67] and ceramic-CNT composite [68–72] coatings. A schematic of thermal-sprayed CNT composite coating microstructure is shown in Figure 3.10. The unit cell of the microstructure is a splat. If CNTs can be distributed uniformly within the single splat, then the entire composite coating structure, which is built layer by layer, will also have an overall uniform CNT distribution. Thus, thick coatings and bulk near-net shape structures of CNT-MMC for real-world engineering applications can be fabricated by thermal spraying. The use of thermal spray techniques for synthesizing CNT-MMCs is discussed in the following subsections.

3.3.1 Plasma Spraying

In plasma spraying, the heat source is an arc struck between a tungsten cathode and a copper anode. Argon gas passed through the arc is ionized to form a plasma, which releases a lot of heat as the ions and electron recombine. The temperature of the

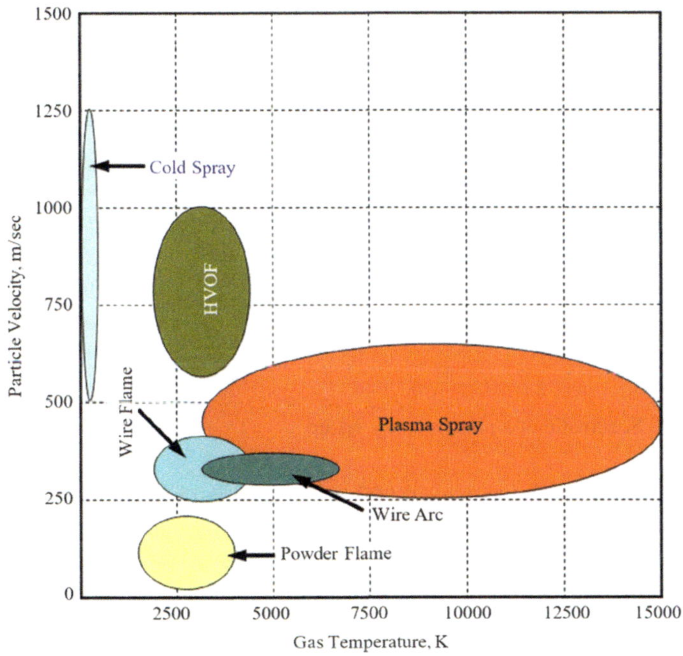

FIGURE 3.9 Comparison of various thermal spray processes in terms of the gas temperatures and particle velocities involved.

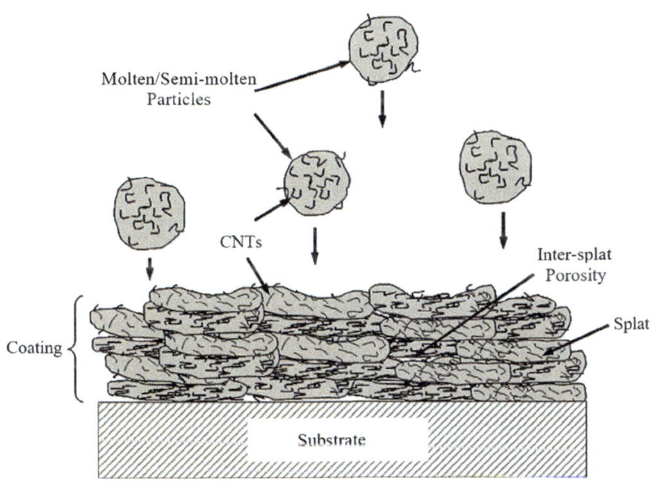

FIGURE 3.10 Schematic showing the uniform distribution of CNT in powder particles and single spat leading to uniform distribution of CNT in the entire composite coating structure fabricated by thermal spraying.

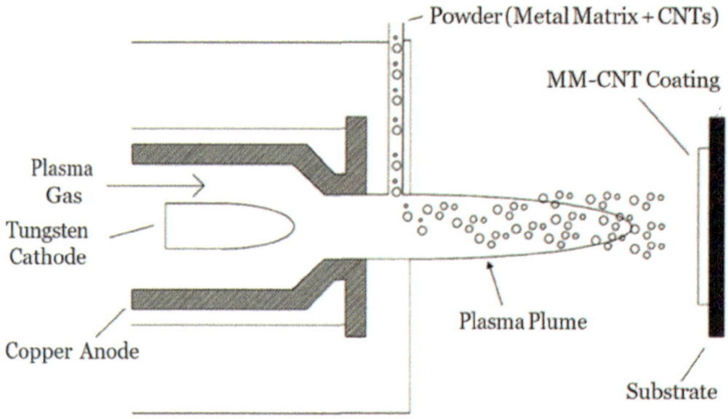

FIGURE 3.11 Schematic of a plasma spray set-up.

resulting plasma plume is close to 10,000 K and can melt virtually all known materials. The powder material to be deposited as a coating is fed into the plume with the help of a carrier gas (usually argon). Hence, consistent flowability of powder particles is important for depositing uniform and dense structure/coating. Generally, particles that are spherical in nature and 20–70 μm in size would flow easily. Smaller particles are difficult to flow due to inter-particle friction and they will clog the feed tubes. Larger particles (>100 μm) might be heavy enough to be carried by the powder. Therefore, a tradeoff has to be made. A schematic of a plasma spraying set-up is shown in Figure 3.11.

Plasma spraying was first explored for the fabrication of Al-Si alloys reinforced with CNTs by our research group [63–65]. Al-Si powder of hypereutectic composition (23 wt.% Si) was blended with 10 wt.% CNT for 48 h by ball milling. Blending was not an efficient process for dispersing CNTs within the mixture and CNT clusters were observed as shown in the SEM image in Figure 3.12a. The mixed powders were plasma sprayed on to a cryogenically cooled, rotating 6061 aluminum mandrel.

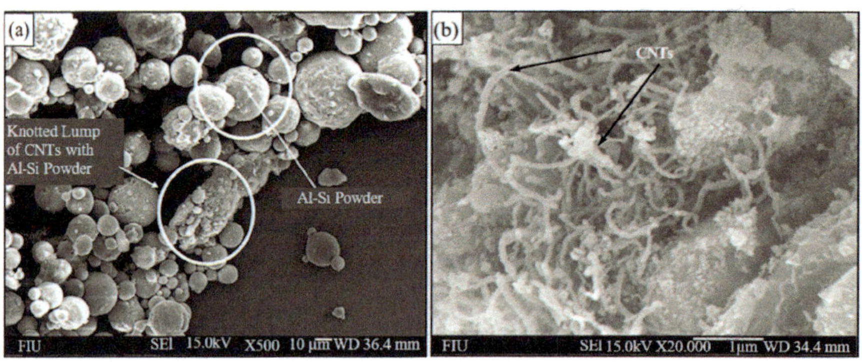

FIGURE 3.12 SEM images showing (a) the blended mixture of Al-23 wt.% Si powder with 10 wt.% CNTs [63], and (b) the fracture surface of the plasma sprayed composite showing retention of CNTs after plasma spraying. (Reproduced with permission from Elsevier and TMS).

FIGURE 3.13 Picture of the freestanding cylinder prepared by (a) plasma spray forming, and (b) HVOF process [73]. (Image reproduced with permission from Taylor & Francis).

CNTs were found to be retained in the plasma sprayed composites as shown in the SEM image of the fracture surface in Figure 3.12b. This was the first study to prove that CNTs can withstand the harsh temperature environment of the plasma [63]. Free standing structures 62 mm in diameter, 100 mm in length, and 0.3 mm in thickness were produced as shown in Figure 3.13a [73]. X-ray diffraction spectrum of the sprayed sample indicated absence of formation of carbides due to reaction between CNTs and the matrix. Thermodynamic/kinetic analysis along with TEM micrographs indicated the formation of an SiC layer 5 nm in thickness on the CNTs [74]. X-ray diffraction technique could not detect carbide formation due to low content. The phenomena occurring at the CNT/Al interface is discussed in detail in Chapter 7. Sintering of the plasma sprayed Al-CNT composites at 400 °C for up to 72 h led to an increase in the size of the primary silicon (from 0.8–2.0 to 1.1–2.7 μm), an increase in the primary silicon content (from 14.1 ± 1.9 to 17.7 ± 1.5 vol.%), and a decrease in pore content (from 6.7 ± 0.4 to 3.2 ± 0.3 vol.%) but did not affect the interfacial layer thickness of the carbide [75]. Sintering did not produce any noticeable damage to the CNT structure. Uniaxial tensile tests showed an increase in the elastic modulus by 78% for 10 wt.% CNT reinforcement [76]. Due to the poor flowability of the blended Al-CNT powders, only a 0.3-mm-thick cylinder could be sprayed. Clogging of the feed tubes prevented spraying for longer times to deposit thicker structure. In order to improve the flowability and the dispersion of CNTs in the powder, a spray-drying process was adopted [67]. Al-Si particles of eutectic composition (12 wt.% Si) were added with 5 wt.% CNTs (denoted as Al 5 CNT) and 10 wt.% CNTs (denoted as Al-10 CNT) and an aqueous suspension was made with a little bit of

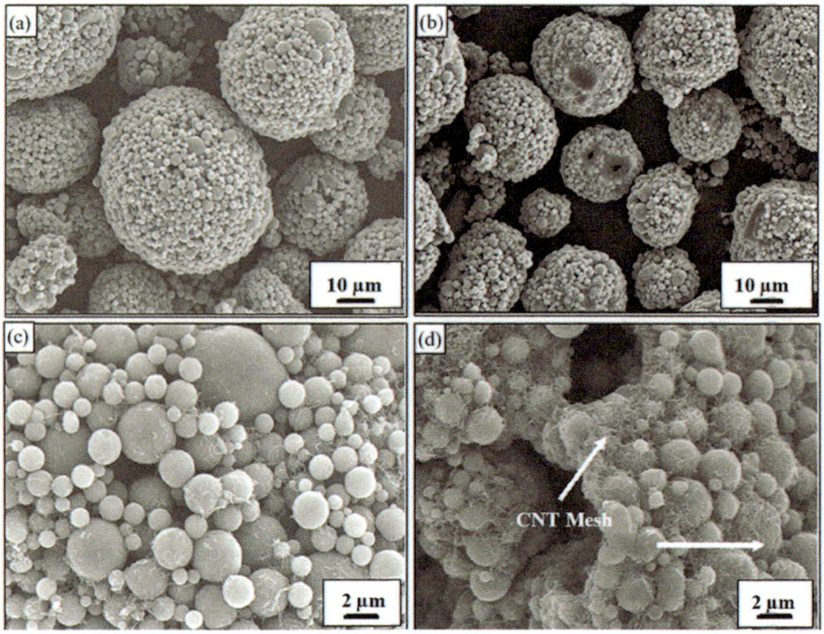

FIGURE 3.14 SEM images of spray-dried agglomerates of (a) Al-5 CNT, (b) Al-10 CNT, and (c) and (d) the corresponding high magnification images [67]. (Reproduced with permission from Elsevier).

polyvinyl alcohol (PVA) as a binder. The slurry was atomized in a chamber where hot air was allowed to pass to dry the droplets formed by atomization. The droplets solidified to form agglomerates of fine particles with very nice distribution of CNTs in the powder. Figure 3.14a–d shows the SEM images of the spray-dried Al-5 CNT and Al-10 CNT powder showing uniform distribution of CNTs. It can be seen that the agglomerates are quite spherical and the CNTs are distributed uniformly in the agglomerates. These powders had excellent flowability, and cylinders as thick as 5 mm and length over 100 mm could be prepared as shown in Figure 3.15a. It can be seen that as compared to the composite prepared from blended powders (Figure 3.12a), the outer surface of the composite made from the spray-dried powders is smoother and uniform. The microstructure of the coatings was found to be two-phase of a matrix containing dispersed CNTs and CNT clusters as shown in Figure 3.15b and 3.15c. The matrix has a good distribution of CNTs in the intersplat region and within the splats, especially for the Al-5 CNT coating. This shows the benefit of spray drying over blending. The clusters are a result of the non-wetting nature of CNTs to molten Al-Si alloy, which causes them to segregate when the powders melt to form droplets. CNTs were also found to be uniformly distributed in the intersplat regions. A detailed thermodynamic and kinetic analysis of the reaction between Al-Si alloys and CNTs to predict interfacial products in these composites is discussed in Chapter 7 [77]. Mechanical properties showed improvement as measured by nanoindentation of the matrix portions and is discussed in Section 4.7.1.1.

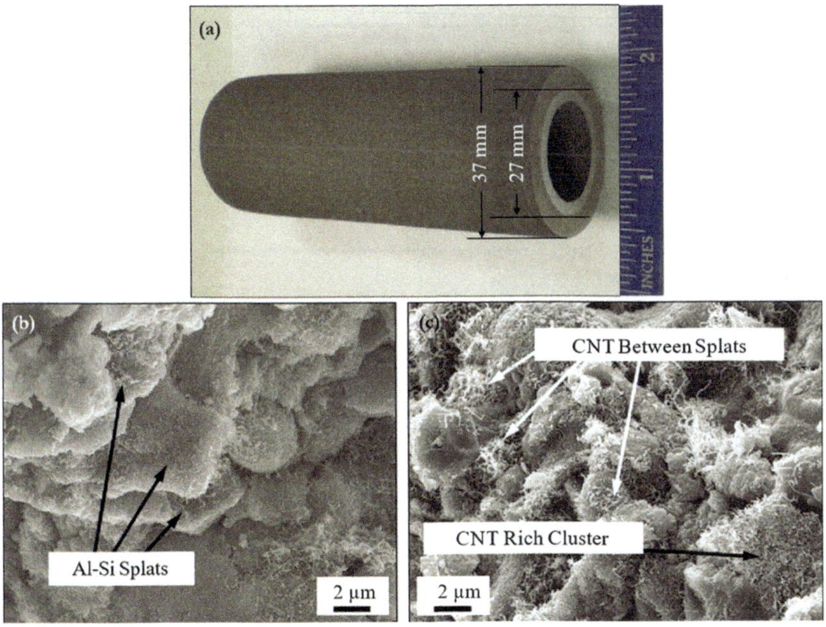

FIGURE 3.15 (a) Picture of Al-Si cylinder reinforced with 10 wt.% CNT prepared by plasma spraying of spray-dried powders, and (b) and (c) SEM images of the fracture surfaces of Al-5 CNT and Al-10 CNT coatings, respectively [67]. (Image reproduced with permission from Elsevier).

Plasma spraying can also fabricate near-net shaped CNT-MMC structures with complex shapes and configurations. The advantage of such processing is that they need minimal machining after spraying. Near-net shape processing through plasma spraying involves simultaneous melting of composite powder through plasma and accelerating the molten particles for spray deposition on a rotating mandrel or substrate. Upon rapid cooling, the desired shape is formed, which is released from the mandrel surface to get the freestanding composite structure. Figure 3.16a shows the schematic of plasma spraying of a near-net shape structure. The mandrel used for such fabrication needs to have the negative shape of the desired structure [78]. Mandrels, used for fabricating CNT-MMC structures, are broadly of two different types. The first category is the sacrificial mandrels, which can be made using the 3-D Z-printing technique (Z Corporation, Burlington, MA) and can be etched away after spraying to release the composite structure. The second category consists of the non-sacrificial or permanent mandrel, which is made out of metal or graphite. Agarwal's group has studied the fabrication of Al-CNT near-net structures using both sacrificial and non-sacrificial mandrels. Figure 3.16b–e shows the pictures of both the sacrificial mandrels and corresponding near-net shaped Al-CNT composite structures fabricated through plasma spraying.

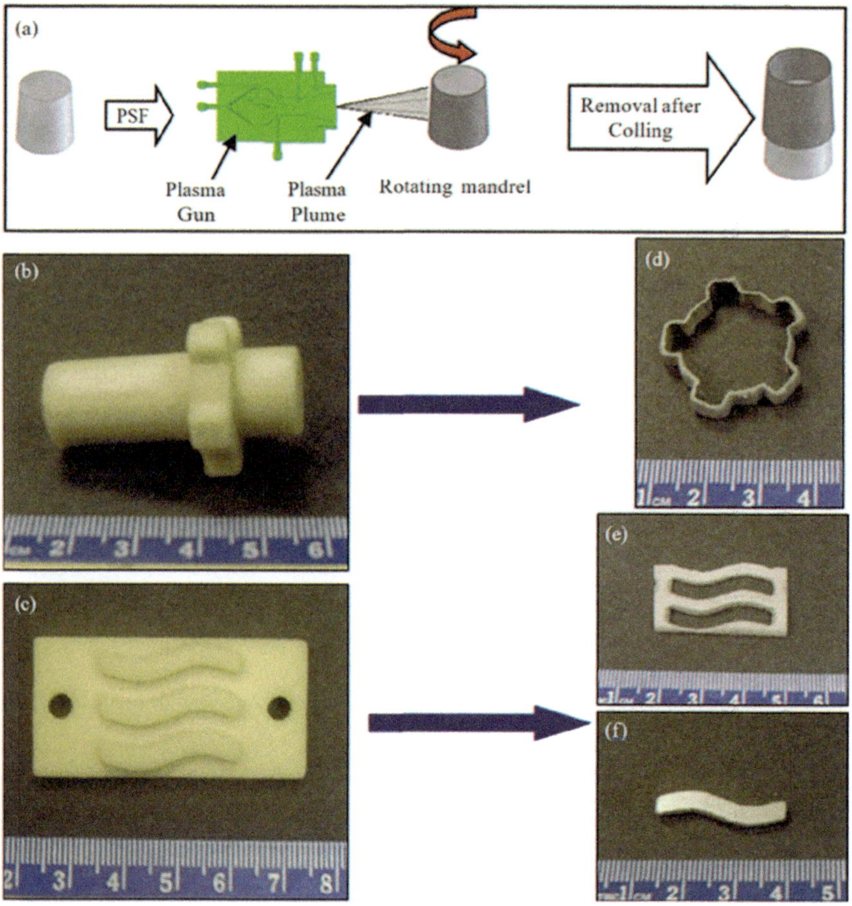

FIGURE 3.16 (a) Schematic showing plasma spraying of a near-net shape structure, (b) and (c) sacrificial mandrels of different shapes, and (d) and (f) plasma sprayed near-net shape metallic structures fabricated using the corresponding mandrels.

3.3.2 High Velocity Oxy-Fuel Spraying

High velocity oxy-fuel (HVOF) spraying is a relatively new thermal spray technique introduced in 1988 (Sulzer METCO Diamond Jet HVOF). Figure 3.17 shows a schematic of an HVOF torch. Here the heat source is high-pressure combustion of fuel (kerosene, propylene) in oxygen. Flame temperatures reach up to 3500 °C and the particle velocities are quite high (Mach 4–5) compared to plasma spraying (subsonic to Mach 3). Consequently, the coatings obtained have better bond strengths and very high densities. High velocity oxy-fuel has also been pursued for fabricating bulk CNT reinforced Al-23 wt.% Si hypereutectic alloy composites [64,65]. The free-standing cylinder prepared by HVOF is shown in Figure 3.13b [73]. Figure 3.18 shows the SEM image of the polished cross-section of the composites Al-10 CNT prepared by plasma spray forming (PSF) and HVOF. The microstructures indicated

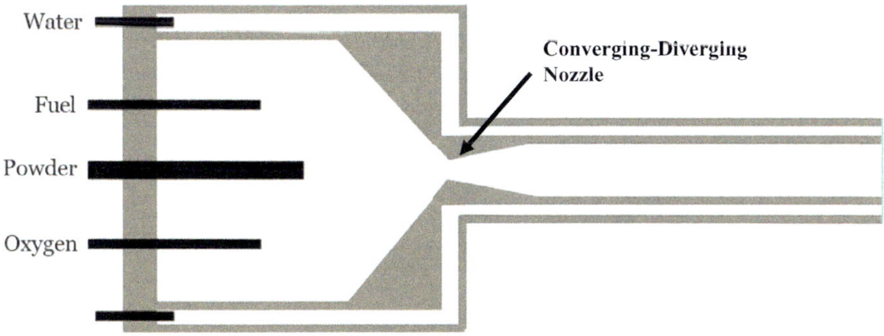

FIGURE 3.17 Schematic of a typical HVOF spray torch.

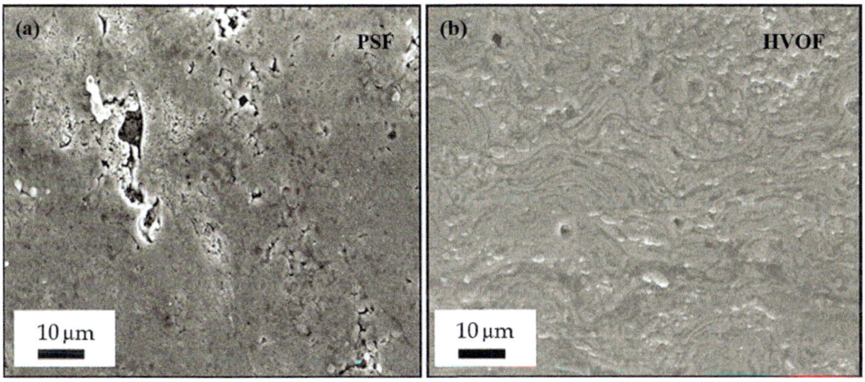

FIGURE 3.18 SEM images of the polished and etched cross-section of Al-23 wt.% Si composite reinforced with 10 wt.% CNTs prepared by (a) plasma spraying, and (b) HVOF.

dense packing of splats with porosity (3.2 ± 0.3%), which was less than that of the plasma sprayed composite (6.7 ± 0.4%). The primary silicon size was found to be lower in HVOF composite (10–30 nm) compared to plasma sprayed coating (40–90 nm), which could be due to the rapid heat dissipation and high impact fracture in HVOF composite. The HVOF structure shows 49% improvement in elastic modulus and 17% improvement in micro-hardness values as compared to plasma sprayed composite coating of the same composition of Al-10 wt.% CNT [65]. HVOF offers a great potential in synthesizing CNT-MMCs as it produces composites with high density, improved CNT distribution due to high velocity impact, and hence improved mechanical properties.

3.3.3 Cold Spraying

Cold spraying is a relatively new technique developed in the developed in mid-1980s at the Institute for Theoretical and Applied Mechanics of the Siberian Division of the Russian Academy of Sciences in Novosibirsk [79]. In this method, powder particles

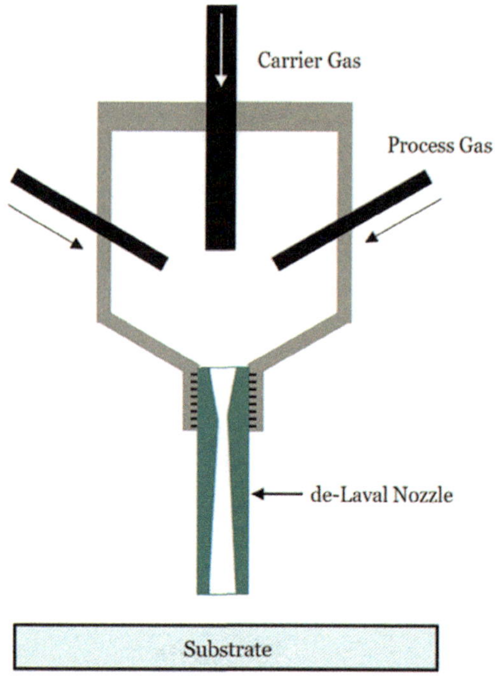

FIGURE 3.19 Schematic of cold spray setup.

are accelerated to very high velocities (Mach 2–5) and made to impact on a substrate. A high-pressure gas (up to 6.5 MPa pressure difference) flowing through a de-Laval type of nozzle is used for accelerating the particles. Typical gasses used are helium, nitrogen, and air. A schematic of the cold spraying set-up is shown in Figure 3.19. Plastic deformation of the particles due to impact leads to formation of splats. Adiabatic shear instabilities occur at particle-particle and particle-substrate interfaces due to local melting; this leads to curvature generation [80]. Thus, mechanical interlocking is promoted. In addition, it is believed that the oxide layers are broken leading to chemically clean surfaces for atomic diffusion bonding to occur. The cold spray process has several advantages, as there is no oxidation and phase transformation involved due to the low temperature of the process.

In collaboration with University of Nottingham (UK), the Plasma Forming Lab at Florida International University synthesized Al-CNT composite coatings by cold spray technique [66]. Al-CNT composites were prepared from blended powders of pure Al and spray-dried agglomerates containing eutectic Al-Si and CNTs as shown in the schematic in Figure 3.20. The spray-dried agglomerates containing CNTs were cold sprayed but did not form a deposit. This was because the energy of impact was utilized in disintegration of the spray-dried agglomerates. In addition, the presence of Si in eutectic Al alloy leads to bouncing off of the particles because Si has poor plastic deformation characteristics. Hence, Al-CNT agglomerates were blended with pure Al powder. The pure Al powder served as the binder to trap the Al-CNT agglomerates because of the higher plastic deformation of pure Al particles.

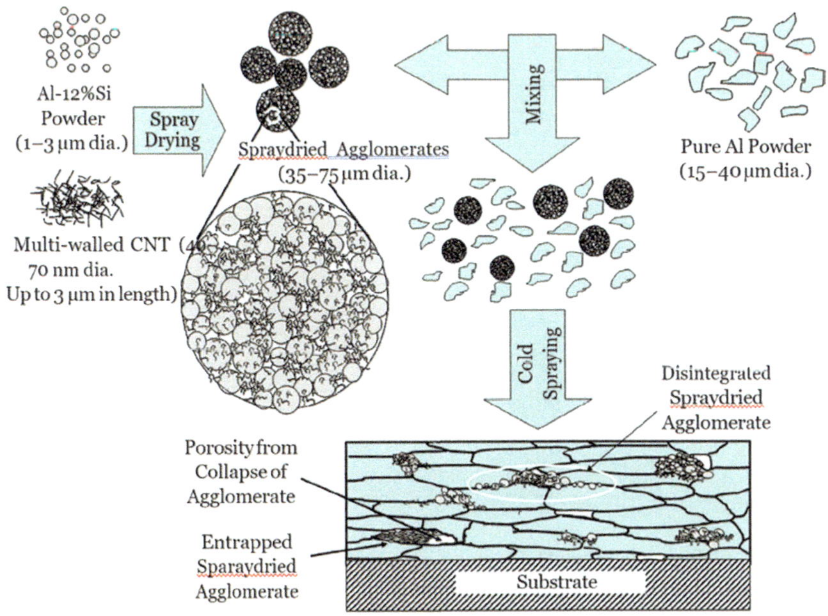

FIGURE 3.20 Schematic showing the methodology adopted for synthesis of Al-CNT coatings using cold spraying [66]. (Reproduced with permission from Elsevier).

Compositions up to 1 wt.% CNT were prepared. Figure 3.21a shows the optical micrograph of Al-0.5 wt.% CNT composite showing the dense microstructure (>98%) and trapped agglomerates [66]. Cold sprayed Al-CNT composite coatings were 500 μm thick. CNTs were shortened in length due to fracture by impact and shearing between particles. A uniform distribution of CNTs was observed on the fracture surface as shown in Figure 3.21b. Very interesting phenomena on deformation and fracture of CNTs were observed under high impact conditions during cold

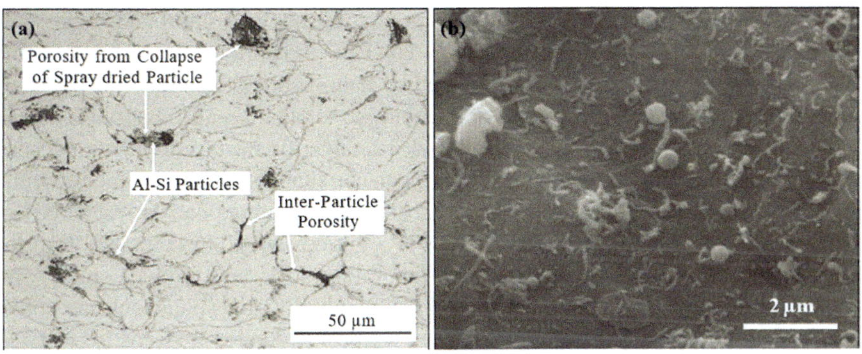

FIGURE 3.21 (a) Optical micrograph of polished and etched cross-section showing the microstructure [66], and (b) SEM image of the fracture surface of cold sprayed Al-0.5 wt % CNT coating. (Reproduced with permission from Elsevier).

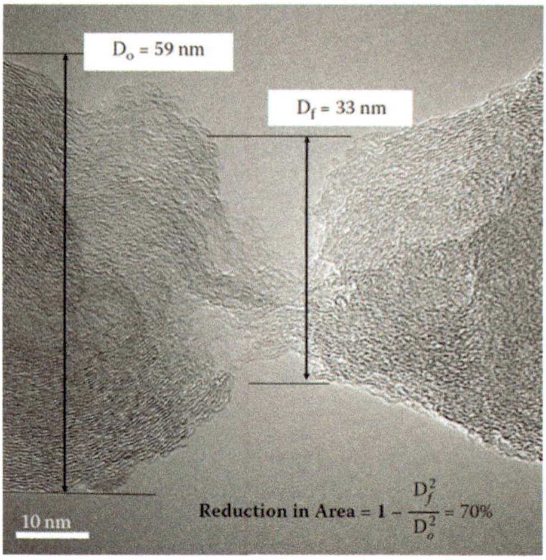

FIGURE 3.22 High-resolution TEM image of fractured CNT in cold sprayed Al-CNT composite showing necking and cup and cone type of failure [81]. (Reproduced with permission from Elsevier).

spraying. Necking and cup and cone type of failure, which is typical of mild carbon steels, were observed in a CNT due to impact as shown in Figure 3.22 [81]. Formation of ripples and kinks in the CNT structure was also observed, which could be attributed to the axial shock wave generated during impact deformation. These ripples were observed to be different from ones observed during bending. Due to shearing action between the Al-Si particles and the CNTs, peeling off of the graphene layers of the CNTs. The cold spray method is an advantageous method for the synthesis of oxidation-sensitive materials like high pure metals, for example, Ti. The bonding between CNTs and the matrix may not be very good because no heat is provided in this technique and there may not be any chemical interaction between the matrix and the CNT. Cold sprayed Al-CNT composite coating showed improved nano-scale wear resistance. Addition of a mere 1 wt.% CNT in cold sprayed Al coating showed a 40% increase in wear resistance measured by nano-scratch technique [82].

Thermal spraying is a very promising technique to obtain real-life applications of CNT-MMCs. It can be used to produce thick coatings for various applications, such as improving the wear resistance, to bulk near-net shape structures. Since thermal spray techniques are already scaled up, CNT-MMC coatings can be prepared for real-life application in a short period. This is a big advantage in comparison to the powder metallurgy techniques, which are laboratory scale and produce small samples. By ensuring the uniform dispersion of CNTs in individual splats, one can obtain a uniform distribution of CNTs in complex shaped structures and coatings. Thus, novel powder processing techniques must be adopted to ensure better dispersion of CNTs in starting powder. Surface treatment or coating of CNTs is another area to look at in order to improve wettability and prevent clustering of CNTs upon melting of the metallic powders. HVOF technique offers the advantage of very high velocity impact along with high temperature, which might result in disintegration of any CNT clusters and result in dense coatings.

3.4 ELECTROCHEMICAL ROUTES

Electrochemical and electroless deposition techniques are utilized to synthesize CNT-MMC coatings and thin films. The composite structures produced through different deposition techniques are restricted in thickness (<200 μm) [83]. It is not feasible to synthesize freestanding CNT-MMC of larger thickness for load-bearing structural applications through an electrochemical processing route. CNT-MMC coatings, made by the electrode-position techniques, can be divided broadly into three categories based on the fabrication method, as well as the morphology of the composite.

1. Coating or thin film synthesized by co-deposition of CNTs and metal-ions from the electrochemical or chemical bath on a substrate [3,83–121]. Most of the research on CNT-MMCs by electrodeposition has been carried out in this category.
2. Metal is deposited on a uniform, aligned array of CNTs or CNT film [100, 122–124]. A uniform and homogeneous distribution of CNT is achievable in the metal matrix through this process.
3. A single CNT is coated by the electrodeposition of metallic particles on it, also known as one dimensional (1-D) composite [4,92,125–128]. This is primarily used in different types of nano-sensors, electrodes, inter-connects, and magnetic recorder heads in computer applications and also as the precursor powder for larger CNT-MMC structures, processed through a powder metallurgy route [92].

The first category of electrodeposited CNT-MMC films and coatings is the ones fabricated through co-deposition by both electrochemical and electroless processes. Synthesis of CNT-MMC through co-deposition route involves a similar approach for electrodeposition [83–97,99–104,121] and electroless deposition route [3,98,105–120]. CNTs are thoroughly suspended in the electrolytic bath and simultaneously deposited on the substrate along with the metal ions, thus producing a composite layer. The process is explained schematically in Figure 3.23.

As discussed for other processing techniques in this chapter, the main challenge for electrochemical techniques is also the dispersion of CNTs in the metal matrix. The natural tendency of agglomeration of CNTs, due to their high surface energy, hinders uniform dispersion and good suspension in the electrolytic bath. This ultimately results in inhomogeneous distribution of CNTs in the metal matrix coating. The most commonly used technique to improve CNT dispersion in the electrolytic bath includes agitation by means of mechanical stirrer, ultrasonication, and magnetic forces. Ball milling of CNTs prior to mixing in the bath also contributes to improving the CNT dispersion. It helps in making the CNTs shorter, reducing the tendency of agglomeration, and thus resulting in better dispersion [84,85,87,108,109,113,118]. Surface treatment of CNTs can also help in increasing the wettability to keep CNTs

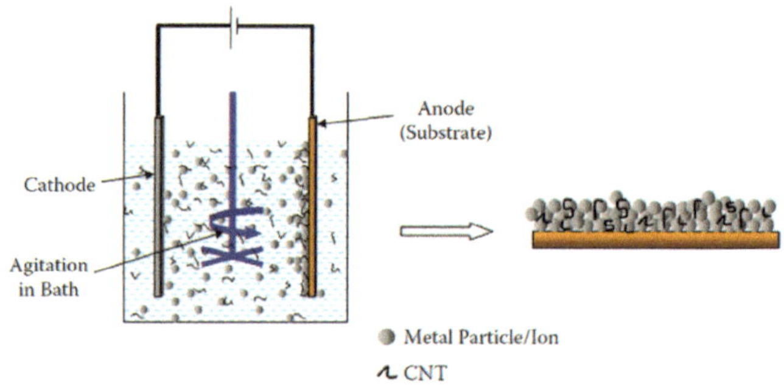

FIGURE 3.23 Schematic showing the setup for electrochemical deposition of metal-CNT coatings.

in the suspension and to modify their surface charge to reduce tendency of agglomeration. These surface treatments include acid treatment and sensitization [118], acid cleaning [88,92], adding surfactants [88,92,105,112–114,117], and pre-coating the CNTs with the metal [105,109,114]. Modifying the bath by adding reagents (adding polyacrylic acid for Ni-CNT electro-deposition [91,97,126]) is another means to achieve better dispersion of CNTs in the bath.

For electrodeposition and electroless deposition techniques, the CNT content in the composite coating is not directly proportional to the amount of CNT added to the bath. It has been observed for both processing techniques that there is a critical concentration of CNT in the bath, which is related to the deposition efficiency and hence CNT content in the composite coating. Electrodeposition and electroless deposition techniques being fundamentally very different methods, the optimum concentration for CNTs in the bath for best co-deposition is different for both techniques. The CNT content in the composite is also directly related to the agglomeration tendency of the CNTs in the bath. The CNT content in the bath being higher results in their agglomeration and hence creates a decrease in the amount of dispersed and suspended CNTs available for co-deposition on substrate. For the electrochemical technique, the optimum CNT content achieved in the composite structure is 14 vol.% with a bath concentration of 1.9 g l^{-1} [85]. In the case of the electroless deposition technique, an optimum CNT content of 13 vol.% was achieved for 1.1 g l^{-1} concentration [117]. The final concentration of CNTs in the deposited structure and coating is a combined effect of CNT concentration in the bath and process parameters of the respective techniques. Hence, dependence on multi-variables makes the analysis of CNT content more complicated. Research is ongoing to isolate the individual and cumulative effect of all the parameters on the CNT content in the composite coating and thin films deposited by electrochemical techniques.

The processing parameters that govern the nature and morphology of the deposited structure are also different in both electrochemical and electroless deposition techniques. Current density is an important factor in the case of electrodeposition. An increase in the current density helps in the deposition of CNTs due to increase in the electrostatic force between CNT and the substrate. However, at the same time, the high electrical conductivity of CNTs causes them to be engulfed by metallic particles vary rapidly, creating very large particles to be deposited on the surface. Hence, a critical current density is desired for an optimum CNT content, high deposition rate, smoothness, and uniformity of the CNT-MMC deposited coating and thin films [85,97,121].

Apart from the current density, the type of power source (AC or DC) also affects the composite structure fabricated by the electrodeposition process [84,94]. Pulse current provides more nucleation sites for metal ions to deposit on the CNT surface. In addition, during the reverse cycle of pulse current, metal is selectively dissolved as it forms the top layer of the deposit, engulfing the CNTs. Thus, the CNT content of the deposit increases and results in better mechanical properties, namely hardness and wear resistance for the composite coatings and thin films. The protruding parts of the deposited coating dissolve more during the reverse cycle, thus making the surface smoother for further deposition. This helps in uniform deposition and decreased porosity in the deposited composite coating. Increase in the pulse frequency and reverse ratio has also been reported to increase CNT content and smoothness of deposition [94], but more research is needed for the concluding evidence. Since electroless deposition is a thermo-chemical process, bath temperature and pH are two other important factors, which determine the coating composition and morphology. These two parameters vary with respect to the composition of the bath and the target composition of the deposit [98,107,109–112,116–120]. A thorough study of these parameters on CNT-MMC morphology, composition, and properties is yet to be carried out. Most of the studies carried out for this first type of composite coating and film are on Ni-CNT system, with a few others on Cu [92], Zn [93], and Cr [101] based systems. Figure 3.24 shows uniformly distributed and embedded CNTs in an electrodeposited Ni matrix fabricated through a co-deposition route.

The second type of CNT-MMC synthesized through the electrodeposition route is by depositing the metal in the voids between the CNT array and network. A CNT network on a non-metallic substrate (silicon, alumina, or glassy carbon) is prepared by two different techniques: (1) growing aligned CNTs on a substrate (Si or porous alumina) [100,122,123] and (2) dropping CNT solution on a substrate surface followed by drying (glassy carbon electrode) [123]. Studies have been carried out on copper- [122–124] and nickel-based [100] composites fabricated by electro-deposition of pre-arrayed CNTs. Prior distribution of CNTs on the surface ensures their uniform distribution in the composite. However, these types of composite structures are restricted in thickness, with a maximum reported value of ~50 μm [97], which is much lower than the maximum thickness (~180 μm [81]) that could be obtained through a co-deposition route.

The third category of the CNT-MMC is metal-coated freestanding CNTs, which can be synthesized by both electrodeposition [125,126] and electroless deposition techniques [4,92,127,128]. Such one-dimensional CNT-MMCs can be obtained by co-deposition of metal and CNT from the bath on a substrate. Subsequently, the coated CNTs are separated from the substrate by ultrasonication. The coating deposited on the CNT surface is not very uniform and often results in the formation of metal particles [126]. A better way of controlling the size and maintaining the uniformity of metallic coating on CNT through the electrodeposition route is by coating the vertically aligned array of CNTs inside a porous template substrate (e.g., porous alumina template), which can be leached away further [125].

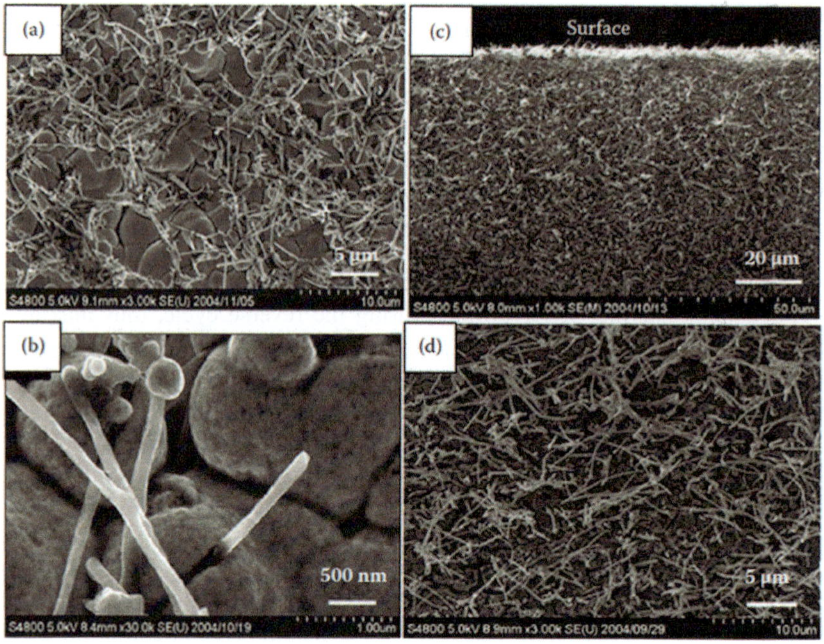

FIGURE 3.24 SEM images of (a) and (b) surface, and (c) and (d) cross-section for Ni-0.7 mass% CNT electrodeposited composite film of 100 μm thickness [91]. (Reproduced with permission from The Electrochemical Society).

Electroless deposition of metal particles on CNTs is performed by suspending the CNTs in an electroless-plating bath. The reducing agent present in the bath reduces the metallic salts to release metal, which coats the CNTs. CNTs are functionalized prior to this process to create nucleation sites on the CNT surface to aid metal deposition. Better surface treatment produces uniform metallic coating on CNTs [4,127,128]. Reports are also available on such one-dimensional coated CNT composites for metal particles like nickel, copper, silver, and cobalt with the coating thickness varying between 20 and 600 nm. Figure 3.25 presents SEM micrographs of cobalt-coated CNTs fabricated through electroless deposition.

Electrodeposition and electroless deposition techniques are very effective in producing thin CNT-MMCs and coatings with good dispersion of CNTs. Metal ions either co-deposit with CNTs or nucleate on CNT surfaces, producing a strong bonding. These factors are very important with respect to the properties exhibited by the composites and their performance in service condition. Deposition techniques are most suitable for producing one-dimensional CNT-MMCs. However, there are two major limitations of electrochemical processing routes. First, these techniques are mostly developed for Ni and Ni-alloys, with a few investigations on Cu and Co. These techniques require more research to optimize the process for other metals including Cu and Co. Second, these processing methods can be applied mainly for synthesizing the CNT-MMC in the form of thin coating or a film less than 200 μm.

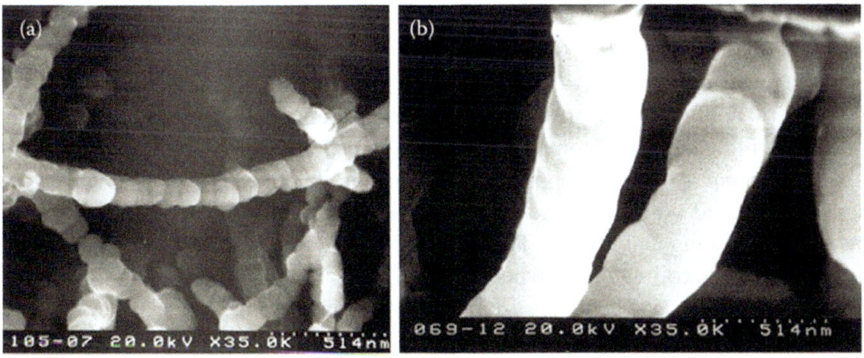

FIGURE 3.25 SEM micrographs of Co-coated carbon nanotube after (a) 30 min and (b) 35 min of electroless deposition [127]. (Reproduced with permission from Elsevier).

With the current knowledge, it is almost impossible to produce thick and freestanding CNT-MMCs for structural applications by electrochemical processing.

3.5 NOVEL TECHNIQUES

It can be concluded that conventional processing techniques for CNT-MMCs synthesis have been applied with partial success. However, due to the unique problems associated with the fabrication of CNT composites, several new strategies have to be designed. Some of these novel routes fall in the category of well-established techniques but are modified according to the suitability for processing metal-matrix composites, while others are based on completely novel concepts. This section is a compendium of the scattered novel processing methods developed to overcome the main challenges in fabricating CNT-MMCs.

3.5.1 Sputtering

Sputtering techniques can be used to deposit metal over CNTs and produce composites with small dimensions (of the length scale of the CNT). This could also be used for production of one-dimensional (1-D) nanostructures and aligned CNT composites. Magnetron sputtering has been used to deposit Al on CNTs, which were then subjected to several annealing treatments to study the interfacial reaction between Al and CNT [129]. Sputter-deposited Au on SWNT bundles show self-organization into evenly spaced clusters [130]. Au, Ag, and Cu were found to form an array of nano crystals of ~10 Å on the surface of CNTs, whereas Ti, Zr, and Mo formed nanowires at the grooves of the SWNT bundles. This difference in morphology has been explained in terms of interactions between carbon and respective metal atoms. Particle formation in Au, Ag, and Cu indicates a weak interaction of those metals with C, whereas strong interaction of C with Ti, Mo, and Zr helps them in forming elongated islands [130]. This method is quite effective in fabricating such one-dimensional nanostructures. It is not suitable for fabrication of large structural applications. However, it may be a useful tool for coating CNTs for improvement of wettability with the metal matrices.

3.5.2 Sandwich Processing

Sandwich processing involves dispersing CNTs between several thin metallic films followed by application of pressure to cold-weld them. This method has been used for fabricating Cu-SWNT composites [131]. Well-dispersed SWNT suspensions in acetone were sprayed onto copper films (10 μm in thickness). Twenty layers of Cu were sandwiched with 19 layers of SWNT and the assembly was cold rolled followed by annealing at 1050 °C. This was followed by cold rolling again to smaller thicknesses. Cold rolling has also been adopted for fabrication of Al-CNT composite. Figure 3.26a shows a schematic of fabrication of a roll-bonded Al-CNT composite,

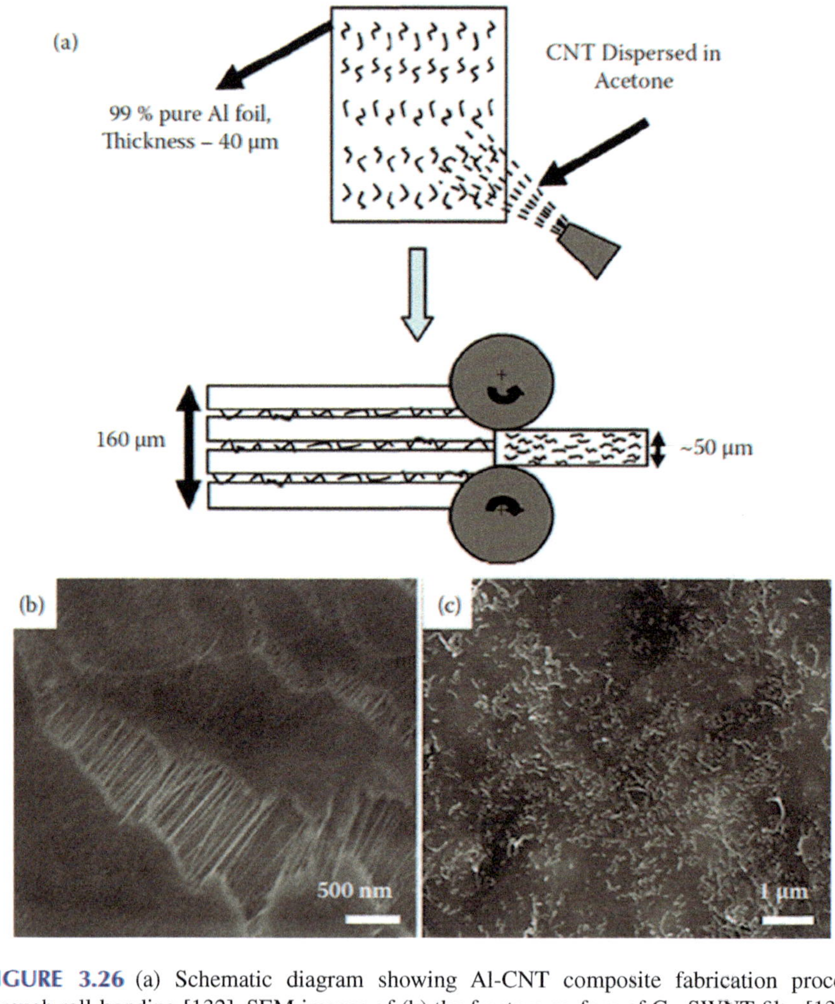

FIGURE 3.26 (a) Schematic diagram showing Al-CNT composite fabrication process through roll-bonding [132]. SEM images of (b) the fracture surface of Cu-SWNT film [131], and (c) delaminated surface of a Al-2 vol.% CNT film [132] prepared by sandwich processing showing good CNT dispersion that can be achieved. (Reproduced with permission from IOP Publishing and Elsevier).

processed by spray dispersion of CNTs on Al-foil and then rolling multiple layers of such foils together [132]. Good bonding between CNT and Cu was observed, which was evident from the SEM image of a fracture surface as shown in Figure 3.26b. Fracture strength of the films in the rolled and annealed condition increased from ~175 MPa to ~220 MPa due to reinforcement by SWNTs (26% increase). A similar processing route was used for production of Al-CNT composites [132]. Good dispersion of CNTs was observed on the delaminated surface as shown in the SEM image in Figure 3.26c for an Al-2 vol.% CNT composite. An increase in the elastic modulus by 59% was observed for Al-2 vol.% CNT composite, while the tensile strength increased by 250% for the Al-9.5 vol.% CNT composite [132]. Improvement in the mechanical properties was also attributed to an increase in dislocation density due to strain hardening. Explosive shock wave consolidation of a mixture of Al powder and CNTs has been carried out to produce composites with 2 and 5 wt.% CNT content [133]. CNTs are observed to cluster at grain boundaries and triple points of the matrix. The tendency of agglomeration increases with increasing CNT content as well. This has resulted in deterioration of mechanical properties in the composite. Sandwich processing by itself cannot synthesize large structures but has a good potential in dispersing CNTs. With thin foils, one can get a good dispersion. The composite foils can then be used for further solid-state processing, which will retain the good dispersion of the foils.

3.5.3 High Pressure Torsion (HPT)

High pressure torsion (HPT) utilizes high torsional forces to consolidate material via severe plastic deformation, as shown schematically in Figure 3.27a. HPT has been used to consolidate ultrasonicated mixture of Al and SWNTs [134]. A torsion force of 2.5 GPa at a rotation speed of 1 rpm was applied to produce Al-5 wt.% SWNT composite having 98% theoretical density. A decrease in the grain size by 80% was observed as shown in the TEM image in Figure 3.27b. This was attributed to the presence of SWNTs in the matrix causing constrained movement of dislocation toward grain boundary and subsequent annihilation. Addition of CNTs can improve the property of the welded joint prepared by friction stir welding (FSW). The challenge of uniformly dispersing the CNTs in the weld region remains unsolved. High-pressure torsion may be a good method for synthesizing composites with smaller grain sizes. In addition, the pressure applied will lead infiltration of any CNT clusters present and may lead to improved properties.

3.5.4 Friction Stir Processing

In this method, frictional forces are been applied to weld CNT and metal together to form CNT-MMCs. Figure 3.28a shows a schematic of the processes. Friction stir processing (FSP), which is a variation of friction stir welding, has been used for fabrication of Mg-CNT [135] and Al-CNT composites [136,137]. CNTs were put on a groove made in an Mg alloy block and the FSP tool rotating at 1500 rpm was rastered on the groove at various speeds to mix the CNTs in the matrix and produce an Mg CNT composite on the surface. The grain size was found to decrease by the

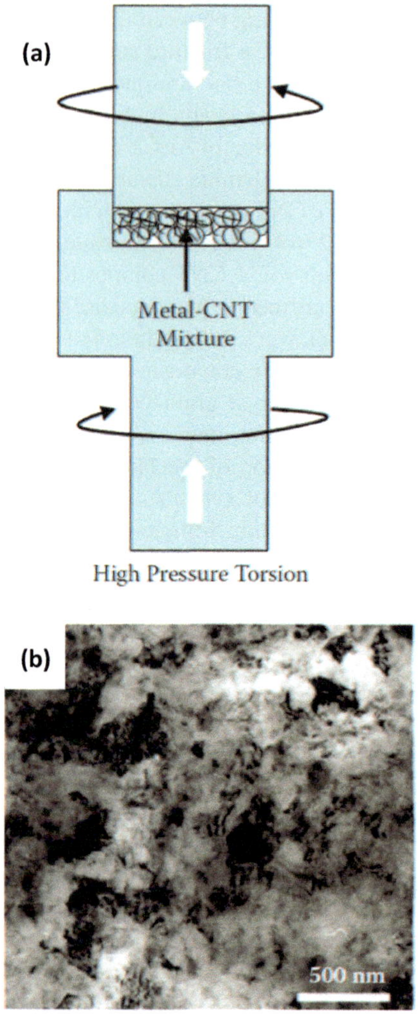

FIGURE 3.27 (a) Schematic of high pressure torsion (HPT) setup, and (b) TEM image showing small grain structure of Al-5 vol.% SWNT composite prepared by HPT method [134]. (Microstructures reproduced with permission from Elsevier).

addition of nanotubes. CNTs were found to be distributed as elongated clusters in the direction of tool movement as shown in the optical image of the top surface in Figure 3.28c [135]. The hardness was found to be higher for CNT containing composite. Al-CNT composites were prepared in the same manner at rotation speeds between 1500 and 2500 rpm by Lim et al. [136]. Lamellar microstructure was observed in the stirred region. CNTs were observed to be retained, although they seemed to be shortened and damaged. These results indicated that stronger welds can be obtained by incorporating CNTs in the weld pool. Liu et al. [137] investigated the effect of varying amounts of MWCNT in Al, ranging from 1.6 to 6.0 vol.%. Holes of 3.5 mm depth

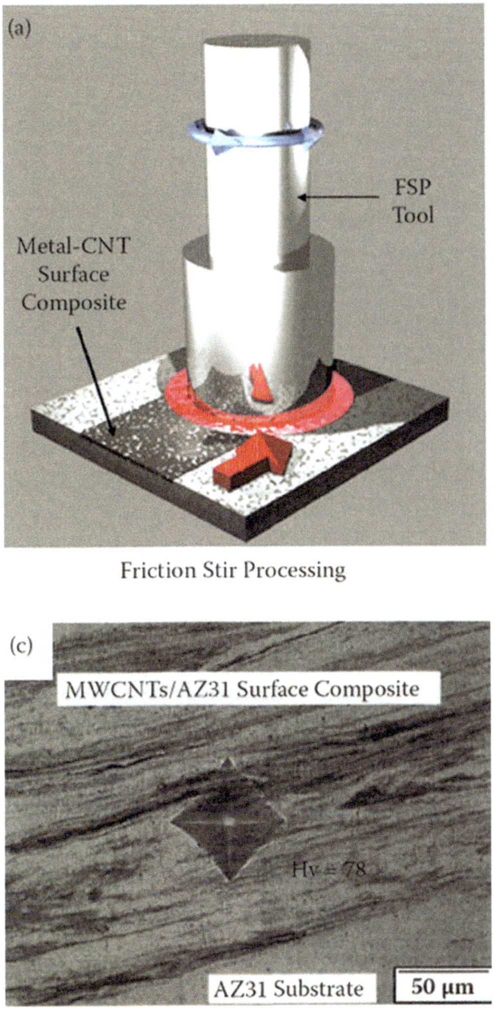

FIGURE 3.28 (a) Schematic of friction stir processing (FSP) setup, and (b) optical image of the top surface of the AZ31-CNT composite layer [135]. (Microstructures reproduced with permission from Elsevier).

and diameters ranging from 0–10 mm were drilled in an A1016 aluminum alloy. The holes are filled with various amounts of MWCNT to obtain desired composition and then processed via FSP using a rotational speed of 950 rpm and forward velocity speed of 30 mm min^{-1}. Five passes on each hole were conducted to promote good dispersion of CNTs throughout the matrix. Mechanical properties of the FSP composites are shown in Figure 3.29, it can be seen that microhardness is uniform throughout the FSP processed portion of the material. Al-6 vol.% MWCNT exhibits the highest microhardness (65 Hv) and ultimate tensile strength (190.2 MPa), which are over two times higher than the control FSP processed Al. Strength enhancement

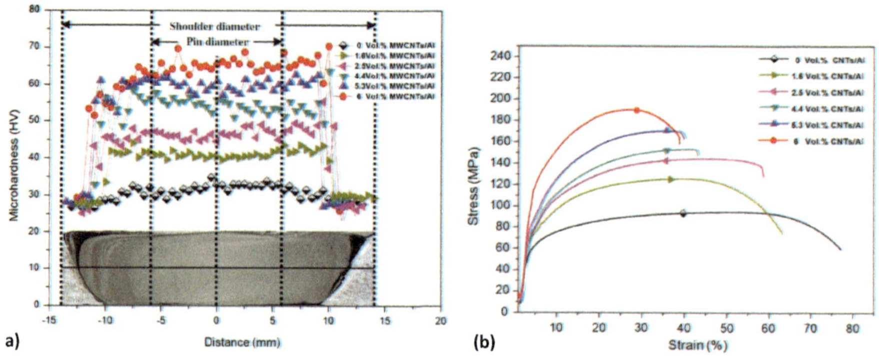

FIGURE 3.29 Mechanical properties of FSP Al-CNT (a) microhardness variations of MWCNTs/Al composite specimens containing different MWCNT content, and (b) stress–strain curves of FSP MWCNT/Al composites [137]. (Reproduced with permission from Elsevier).

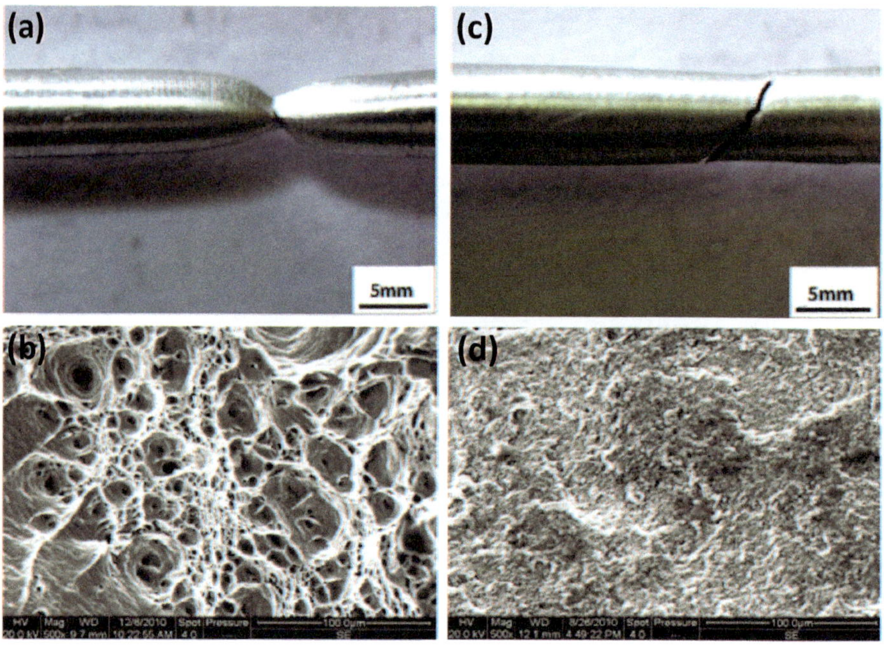

FIGURE 3.30 (a) Fractured tensile specimen of FSP Al, (b) SEM of fracture surface of FSP Al, (c) fractured tensile specimen of FSP 6 vol.% MWCNT/Al, and (d) SEM of fracture surface of FSP 6 vol.% MWCNT/Al [137]. (Reproduced with permission from Elsevier).

was attributed to both CNT reinforcement and grain refinement. However, the fracture behavior reveals that the increased strength comes at the expense of ductility (Figure 3.30). The Al-6 vol.% MWCNT sample exhibits a brittle fracture response with failure strains almost half that of FSP Al.

3.5.5 Layer Deposition

Lasers could be used as a sintering tool for application of coatings and sintering of powder mixtures. Laser deposition techniques have been used to synthesize Ni-CNT composite after roller mixing of the CNTs and Ni powders [138]. Ni-graphite and Ni-CNT composites with a cylindrical geometry of diameter ~10 mm and height ~10 mm were prepared. Although the process incurs very high temperature, CNTs have been shown to survive in the composite. However, an increase in defect density and graphitization of CNTs was observed, which is quite reasonable considering the high processing temperature. Laser engineered net shaping (LENS) has the advantage of rapid processing, but the coupling of the laser wavelength with different metals presents a challenge.

3.6 CONCLUSION

It can be concluded from this chapter that the main challenge during processing of CNT-MMCs is to achieve homogeneous dispersion and good reinforcement of CNTs in the matrix. Processing methods involve different types of starting materials such as powders, molten metals, salt solution, and thin films and could be processed by liquid or solid state routes. Electrochemical deposition and sandwich processing techniques produce good CNT dispersion. However, the main limitation for those techniques is that they can produce thin coatings/films only and not bulk freestanding composite structures. Casting techniques are suitable for only low melting point materials or amorphous structures although uniform dispersion of CNTs in the melt pool is still a problem due to the non-wetting nature of CNTs to most metals. High pressure is required in order to infiltrate CNT clusters formed in the process. Powder metallurgy is, by far, the most popular and feasible route for preparing bulk CNT-MMCs. Spark plasma sintering and hot extrusion have been very successful in achieving high densities. Thermal spraying offers the unique advantage of being an industrially scaled-up process and hence is closer to application-oriented results. Coatings as well as bulk near-net shape structures of CNT-MMC can be directly processed using this technique.

3.7 CHAPTER HIGHLIGHTS

Rapid developments in nanoscience and nanotechnology have pushed the limits of measurements to the level of nano-Newtons and nanometers. Atomic force microscopy (AFM), in situ SEM, and in situ TEM testing have made it able to see the extraordinary properties of nanomaterials as small as a SWNT. Making use of the mechanical properties of CNTs in real-life applications, which are often in the macroscale, requires efficient integration of CNTs in the bulk structure, which is the theme of this chapter; that is, processing. In this chapter, the processing techniques adopted by different researchers have been summarized and critically analyzed for their advantages and limitations. It is observed that a uniform integration of CNTs in the structure is a very big challenge. The processing methods have been classified into several types based on the starting material and processing states involved.

By far the most widely used method has been powder metallurgy (PM) because of easy dispersion and relative low temperature solid-state processing of many PM techniques. Application of pressure and deformation results in dense composites with efficient reinforcement. Processes using molten metals have the advantage of fabrication of bulk components but are limited by the non-wettability of CNTs to most metals, leading to formation of CNT clusters. Thin coatings with uniform dispersion can be prepared by electrodeposition or sandwich rolling techniques and can be used as precursors for bulk component fabrication. Thermal spraying techniques can directly form bulk coatings and bulk CNT-MMC structures. By controlling CNT dispersion in a single splat, bulk components with uniform CNT distribution can be engineered. Again, it is limited by dispersion of CNTs in the starting powder and non-wetting property of CNTs. Thus, improvement of CNT wettability is a critical requirement in many processing techniques. Sputtering techniques can be utilized for surface functionalization of the CNTs to improve their wetting characteristics. There is still significant scope for innovation in development of processes to overcome the challenges with CNT dispersion, especially for composites with high (~10 vol.%) CNT concentration.

REFERENCES

1. C. Suryanarayana. 1999. Mechanical alloying. In *Non-Equilibrium Processing of Materials*. C. Suryanarayan, Ed. Oxford: Elsevier Science Ltd.
2. Tu, J. P., Yang, Y. Z., Wang, L. Y., Ma, X. C., and X. B. Zhang. 2001. Tribological properties of carbon-nanotube-reinforced copper composites. *Tribology Int.* 10: 225–228.
3. Chen, W. X., Tu, J. P., Wang, L. Y., Gan, H. Y., Xu, Z. D., and X. B. Zhang. 2003. Tribological application of carbon nanotubes in a metal-based composite coating and composites. *Carbon* 41: 215–222.
4. Zeng, Q., Luna, J., Bayazitoglu, Y., Wilson, K., Imam, M. A., and E. V. Barrera. 2007. Metal coated functionalized single-walled carbon nanotubes for composites application. *Mater. Sci. Forum* 561–565: 655–658.
5. He, C., Zhao, N., Shi, C., Du, X., Li, J., Li, H., and Q. Cui. 2007. An approach to obtaining homogeneously dispersed carbon nanotubes in Al powders for preparing reinforced Al-matrix composites. *Adv. Mater.* 19: 1128–1132.
6. Esawi, A., and K. Morsi. 2007. Dispersion of carbon nanotubes (CNTs) in aluminum powder. *Composites A* 38: 646–650.
7. Morsi, K., and A. Esawi. 2007. Effect of mechanical alloying time and carbon nanotube (CNT) content on the evolution of aluminum (Al)–CNT composite powders. *J. Mater. Sci.* 42: 4954–4959.
8. Chen, D., Chen, L., Liu, S., Ma, C. X., Chen, D. M., and L. B. Wang. 2004. Microstructure and hydrogen storage property of Mg/MWNTs composites. *J. Alloys Compd.* 372: 231–237.
9. Huang, Z. G., Guo, Z. P., Calka, A., Wexler, D., and H. K. Liu. 2007. Effects of carbon black, graphite and carbon nanotube additives on hydrogen storage properties of magnesium. *J. Alloys Compd.* 427: 94–100.
10. Eom, J. Y., Park, J. W., Kwon, H. S., and S. Rajendran. 2006. Electrochemical insertion of lithium into multiwalled carbon nanotube/silicon composites produced by ball milling. *J. Electrochem. Soc.* 153: A1678–A1684.
11. Feng, Y., Yuan, H. L., and M. Zhang. 2005. Fabrication and properties of silver-matrix composites reinforced by carbon nanotubes. *Mater. Char.* 55: 211–218.

12. G. F. Taylor. 1933. An application for making hard metal compositions. U.S. Patent 1896854.
13. Kuzumaki, T., Ujiie, O., Ichinose, H., and K. Ito. 2000. Mechanical characteristics and preparation of carbon nanotube fiber-reinforced Ti composite. *Adv. Eng. Mater.* 2: 416–418.
14. Xu, C. L., Wei, B. Q., Ma, R. Z., Liang, J., Ma, X. K., and D. H. Wu. 1999. Fabrication of aluminum–carbon nanotube composites and their electrical properties. *Carbon* 37: 855–858.
15. Zhong, R., Cong, H., and P. Hou. 2003. Fabrication of nano-Al based composites reinforced by single-walled carbon nanotubes. *Carbon* 41: 848–851.
16. Chen, X-H., Li, W-H., Chen, C-S., Xu, L-S., Yang, Z., and J. Hu. 2005. Preparation and properties of Cu matrix composite reinforced by carbon nanotubes. *Trans. Nonferrous Met. Soc. China* 15: 314–318.
17. Carreño-Morelli, E., Yang, J., Couteau, E., Hernadi, K., Seo, J. W., Bonjour, C., Forró, L., and R. Schaller. 2004. Carbon nanotube/magnesium composites. *Phys. Stat. Sol.* 201: R53–R55.
18. Pang, L. X., Sun, K. N., Ren, S., Sun, C., and J. Q. Bi. 2007. Microstructure, hardness, and bending strength of carbon nanotube – iron aluminide composites. *J. Composite Mater.* 41: 2025–2031.
19. Hsu, C-F., and P-Y. Lee. 2007. Effect of carbon-nanotube addition on thermal stability of ti-based metallic glass composites. *Mater. Sci. Forum* 534–536: 865–868.
20. Hulbert, D. M., Anders, A., Andersson, J., Lavernia, E. J., and A. K. Mukherjee. 2009. A discussion on the absence of plasma in spark plasma sintering. *Scripta Mater.* 60: 835–838.
21. Munir, Z. A., Anselmi-Tamburini, U., and M. Ohyanagi. 2006. The effect of electric field and pressure on the synthesis and consolidation of materials: a review of the spark plasma sintering method. *J. Mater. Sci.* 41: 763–777.
22. Kim, K. T., Lee, K. H., Cha, S. I., Mo, C. B., and S. H. Hong. 2004. Characterization of carbon nanotubes/Cu nanocomposites processed by using nano-sized Cu powders. *Mater. Res. Soc. Symp. Proc.* 821: P3.25.1–P3.25.6.
23. Kim, K. T., Cha, S. I., Hong, S. H., and S. H. Hong. 2006. Microstructures and tensile behavior of carbon nanotube reinforced Cu matrix nanocomposites. *Mater. Sci. Eng. A* 430: 27–33.
24. Cha, S. I., Kim, K. T., Arshad, S. N., Mo, C. B., and S. H. Hong. 2005. Extraordinary strengthening effect of carbon nanotubes in metal-matrix nanocomposites processed by molecular-level mixing. *Adv. Mater.* 17: 1377–1381.
25. Kim, K. T., Cha, S. I., and S. H. Hong. 2007. Hardness and wear resistance of carbon nanotube reinforced Cu matrix nanocomposites. *Mater. Sci. Eng. A* 449–451: 46–50.
26. Kwon, H., Estili, M., Takagi, K., Miyazaki, T., and A. Kawasaki. 2009. Combination of hot extrusion and spark plasma sintering for producing carbon nanotube reinforced aluminum matrix composites. *Carbon* 47: 570–577.
27. Kim, I-Y., Lee, J-H., Lee, G-S., Baik, S-H., Kim, Y-J., and Y-Z. Lee. 2009. Friction and wear characteristics of the carbon nanotube–aluminum composites with different manufacturing conditions. *Wear* 267: 593–598.
28. Pang, L-X., Sun, K-N., Ren, S., Sun, C., Fan, R-H., and Z-H. Lu. 2007. Fabrication and microstructure of Fe_3Al matrix composite reinforced by carbon nanotube. *Mater. Sci. Eng. A* 447: 146–149.
29. Esawi, A. M. K., and M. A. E. Borady. 2008. Carbon nanotube-reinforced aluminium strips. *Composites Sci. Tech.* 68: 486–492.
30. Choi, H. J., Kwon, G. B., Lee, G. Y., and D. H. Bae. 2008. Reinforcement with carbon nanotubes in aluminum matrix composites. *Scripta Mater.* 59: 360–363.
31. Kuzumaki, T., Miyazawa, K., Ichinose, H., and K. Ito. 1998. Processing of carbon nanotube reinforced aluminum composite. *J. Mater. Res.* 13: 2445–2449.

32. George, R., Kashyap, K. T., Rahul, R., and S. Yamdagni. 2005. Strengthening in carbon nanotube/aluminium (CNT/Al) composites. *Scripta Mater.* 53: 1159–1163.
33. Deng, C. F., Zhang, X. X., Wang, D. Z., Lin, Q., and A. B. Li. 2007. Preparation and characterization of carbon nanotubes/aluminum matrix composites. *Mater. Lett.* 61: 1725–1728.
34. Deng, C. F., Wang, D. Z., Zhang, X. X., and A. B. Li. 2007. Processing and properties of carbon nanotubes reinforced aluminum composites. *Mater. Sci. Eng. A* 444: 138–145.
35. Deng, C. F., Zhang, X. X., Ma, Y. X., and D. Z. Wang. 2007. Fabrication of aluminum matrix composite reinforced with carbon nanotubes. *Rare Metals* 26: 450–455.
36. Deng, C. F., Ma, Y. X., Xhang, P., Zhang, X. X., and D. Z. Wang. 2008. Thermal expansion behaviors of aluminum composite reinforced with carbon nanotubes. *Mater. Lett.* 62: 2301–2303.
37. Deng, C. F., Zhang, X. X., and D. Z. Wang. 2007. Chemical stability of carbon nanotubes in the 2024Al matrix. *Mater. Lett.* 61: 904–907.
38. Pérez-Bustamante, R., Gómez-Esparza, C. D., Estrada-Guel, I., Miki-Yoshida, M., Licea-Jiménez, L., Pérez-García, S. A., and R. Martínez-Sánchez. 2009. Microstructural and mechanical characterization of Al–MWCNT composites produced by mechanical milling. *Mater. Sci. Eng. A* 502: 159–163.
39. Dong, S. R., Tu, J. P., and X. B. Zhang. 2001. An investigation of the sliding wear behavior of Cu-matrix composite reinforced by carbon nanotubes. *Mater. Sci. Eng. A* 313: 83–87.
40. Quang, P., Jeong, Y. G., Hong, S. H., and H. S. Kim. 2006. Equal channel angular pressing of carbon nanotube reinforced metal matrix nanocomposites. *Key Eng. Mater.* 326–328: 325–328.
41. Quang, P., Jeong, Y. G., Yoon, S. C., Hong, S. H., and H. S. Kim. 2007. Consolidation of 1 vol.% carbon nanotube reinforced metal matrix nanocomposites via equal channel angular pressing. *J. Mater. Proc. Tech.* 187–188: 318–320.
42. Quang, P., Jeong, Y. G., Yoon, S. C., Hong, S. I., and H. S. Kim. 2007. Carbon nanotube reinforced metal matrix nanocomposites via equal channel angular pressing. *Mater. Sci. Forum* 534–536: 245–248.
43. Goh, C. S., Wei, J., Lee, L. C., and M. Gupta. 2006. Development of novel carbon nanotube reinforced magnesium nanocomposites using the powder metallurgy technique. *Nanotechnology* 17: 7–12.
44. Thakur, S. K., Gan, T. K., and M. Gupta. 2007. Development and characterization of magnesium composites containing nano-sized silicon carbide and carbon nanotubes as hybrid reinforcements. *J. Mater. Sci.* 42: 10040–10046.
45. Nai, S. M. L., Wei, J., and M. Gupta. 2006. Lead-free solder reinforced with multiwalled carbon nanotubes. *J. Electronic Mater.* 35: 1518–1522.
46. Nai, S. M. L., Wei, J., and M. Gupta. 2006. Improving the performance of lead-free solder reinforced with multi-walled carbon nanotubes. *Mater. Sci. Eng. A* 423: 166–169.
47. Kumar, K. M., Kripesh, V., and A. A. O. Tay. 2008. Single-wall carbon nanotube (SWCNT) functionalized Sn–Ag–Cu lead-free composite solders. *J. Alloys Compd.* 450: 229–237.
48. Dujardin, E., Ebbesen, T. W., Hiura, H., and K. Tanigaki. 1994. Capillarity and wetting of carbon nanotubes. *Science* 265: 1850–1852.
49. Li, S., Song, S., Yu, T., Chen, H., Zhang, Y., and H. Chen. 2005. Microstructure and fracture surfaces of carbon nanotubes/magnesium matrix composite. *Mater. Sci. Forum* 488–489: 893–896.
50. Goh, C. S., Wei, J., Lee, L. C., and M. Gupta. 2006. Simultaneous enhancement in strength and ductility by reinforcing magnesium with carbon nanotubes. *Mater. Sci. Eng. A* 423: 153–156.
51. Goh, C. S., Wei, J., Lee, L. C., and M. Gupta. 2008. Ductility improvement and fatigue studies in Mg-CNT nanocomposites. *Composites Sci. Tech.* 68: 1432–1439.

52. Li, Q., Viereckl, A., Rottmair, C. A., and R. F. Singer. 2009. Improved processing of carbon nanotube/magnesium alloy composites. *Composites Sci. Tech.* 69: 1193–1199.
53. Bian, Z., Pan, M. X., Zhang, Y., and W. H. Wang. 2002. Carbon-nanotube-reinforced $Zr_{52.5}Cu_{17.9}Ni_{14.6}Al_{10}Ti_5$ bulk metallic glass composites. *Appl. Phys. Lett.* 81: 4739–4741.
54. Bian, Z., Wang, R. J., Wang, W. H., Zhang, T., and A. Inoue. 2004. Carbon-nanotube-reinforced Zr-based bulk metallic glass composites and their properties. *Adv. Funct. Mater.* 14: 55–63.
55. Li, Y. B., Ya, Q., Wei, B. Q., Liang, J., and D. H. Wu. 1998. Processing of a carbon nanotubes-Fe82P18 metallic glass composite. *J. Mater. Sci. Lett.* 17: 607–609.
56. Zhou, S-M., Zhang, X-B., Ding, Z-P., Min, C-Y., Xu, G-L., and W-M. Zhu. 2007. Fabrication and tribological properties of carbon nanotubes reinforced Al composites prepared by pressure less infiltration technique. *Composites A* 38: 301–306.
57. Garcia-Cordovilla, C., Louis, E., and J. Narciso. 1999. Pressure infiltration of packed ceramic particulates by liquid metals. *Acta Mater.* 18: 4461–4479.
58. Molina, J. M., Saravanan, R. A., Arpon, R., Garcia-Cordovilla, C., Louis, E., and J. Narciso. 2002. Pressure infiltration of liquid aluminium into packed SiC particulate with a bimodal size distribution. *Acta Mater.* 50: 247–257.
59. Yang, J., and R. Schaller. 2004. Mechanical spectroscopy of Mg reinforced with Al_2O_3 short fibers and C nanotubes. *Mater. Sci. Eng. A* 370: 512–515.
60. Uozumi, H., Kobayashi, K., Nakanishi, K., Matsunaga, T., Shinozaki, K., Sakamoto, H., Tsukada, T., Masuda, C., and M. Yoshida. 2008. Fabrication process of carbon nanotube/light metal matrix composites by squeeze casting. *Mater. Sci. Eng. A* 495: 282–287.
61. Mansoor, M., Shahid, M. 2016. Carbon nanotube-reinforced aluminum composite produced by induction melting. *Journal of Applied Research and Technology* 14. 215–224.
62. Calendar of Patent Records. 1929. *Nature* 123: 655–655.
63. Laha, T., Agarwal, A., McKechnie, T., and S. Seal. 2004. Synthesis and characterization of plasma spray formed carbon nanotube reinforced aluminum composite. *Mater. Sci. Eng. A* 381: 249–258.
64. Laha, T., Agarwal, A., and T. McKechnie. 2005. Comparative evaluation of plasma and high velocity oxy-fuel spray formed carbon nanotube reinforced Al-based composite. In *Surface Engineering in Materials Science-III*. A. Agarwal, N. B. Dahotre, S. Seal, J. J. Moore, and C. Blue, Eds. Warrendale, PA: The Minerals, Metals and Materials Society.
65. Laha, T., Liu, Y., and A. Agarwal. 2007. Carbon nanotube reinforced aluminum nanocomposite via plasma and high velocity oxy-fuel spray forming. *J. Nanosci. Nanotech.* 7: 515–524.
66. Bakshi, S. R., Singh, V., Balani, K., McCartney, D. G., Seal, S., and A. Agarwal. 2008. Carbon nanotube reinforced aluminum composite coating via cold spraying. *Surf. Coat. Tech.* 202: 5162–5169.
67. Bakshi, S. R., Singh, V., Seal, S., and A. Agarwal. 2009. Aluminum composite reinforced with multiwalled carbon nanotubes from plasma spraying of spray dried powders. *Surf. Coat. Tech.* 203: 1544–1554.
68. Balani, K., Anderson, R., Laha, T., Andara, M., Tercero, J., Crumpler, E., and A. Agarwal. 2007. Plasma-sprayed carbon nanotube reinforced hydroxyapatite coatings and their interaction with human osteoblasts in vitro. *Biomaterials* 28: 618–624.
69. Balani, K., Bakshi, S. R., Chen, Y., Laha, T., and A. Agarwal, 2007. Role of powder treatment and CNT dispersion in the fracture toughening of plasma-sprayed aluminum oxide–carbon nanotube ceramic nanocomposite. *J. Nanosci. Nanotech.* 7: 3553–3562.
70. Balani, K., Zhang, T., Karakoti, A., Li, W. Z., Seal, S., and A. Agarwal. 2008. In situ carbon nanotube reinforcements in a plasma-sprayed aluminum oxide nano-composite coating. *Acta Mater.* 56: 571–579.
71. Keshri, A. K., Balani, K., Bakshi, S. R., Singh, V., Laha, T., Seal, S., and A. Agarwal. 2009. Structural transformations in carbon nanotubes during thermal spray processing. *Surf. Coat. Tech.* 203: 2193–2201.

72. Keshri, A. K., Huang, J., Singh, V., Choi, W., Seal, S., and A. Agarwal. 2009. Synthesis of aluminum oxide coating with carbon nanotube reinforcement produced by chemical vapor deposition for improved fracture and wear resistance. *Carbon* 48: 431–442.
73. Bakshi, S. R., Lahiri, D., and A. Agarwal. 2010. Carbon nanotube reinforced metal matrix composites—a review. *Int. Mater. Rev.* 55(1): 41–64.
74. Laha, T., Kuchibhatla, S., Seal, S., Li, W., and A. Agarwal. 2007. Interfacial phenomena in thermally sprayed multiwalled carbon nanotube reinforced aluminum nanocomposite. *Acta Mater.* 55: 1059–1066.
75. Laha, T., and A. Agarwal. 2008. Effect of sintering on thermally sprayed carbon nanotube reinforced aluminum nanocomposite. *Mater. Sci. Eng. A* 480: 323–332.
76. Laha, T., Chen, Y., Lahiri, D., and A. Agarwal. 2009. Tensile properties of carbon nanotube reinforced aluminum nanocomposite fabricated by plasma spray forming. *Composites A* 40: 589–594.
77. Bakshi, S. R., Keshri, A. K., Singh, V., Seal, S., and A. Agarwal. 2009. Interface in carbon nanotube reinforced aluminum silicon composites: thermodynamic analysis and experimental verification. *J. Alloys Compd.* 481: 207–213.
78. Patel, R. R., Keshri, A. K., Dulikravich, G. S., and A. Agarwal. 2010. An experimental and computational methodology for near net shape fabrication of thin walled ceramic structures by plasma spray forming. *Journal of Materials Processing Technology* 210: 1260–1269.
79. Alkimov, A. P., Papyrin, A. N., Kosarev, V. F., Nesterovich, N. I., and M. M. Shushpanov. 1994. Gas-dynamic spraying method for applying a coating. U.S. Patent 5302414.
80. Balani, K., Agarwal, A., S. Seal, and J. Karthikeyan. 2005. Transmission electron microscopy of cold sprayed 1100 aluminum coating. *Scripta Mater.* 53: 845–850.
81. Bakshi, S. R., Singh, V., McCartney, D. G., Seal, S., and A. Agarwal. 2008. Deformation and damage mechanisms of multiwalled carbon nanotubes under high-velocity impact. *Scripta Mater.* 59: 499–502.
82. Chen, Y., Bakshi, S. R., and A. Agarwal. 2010. Correlation between nanoindentation and nanoscratch properties of carbon nanotube reinforced aluminum composite coatings. *Surf. Coat. Technol.* 204: 2709–2715.
83. Tan, J., Yu, T., Xu, B., and Q. Yao. 2006. Microstructure and wear resistance of nickel–carbon nanotube composite coating from brush plating technique. *Tribology Lett.* 21(2): 107–111.
84. Arai, S., Fujimori, A., Murai, M., and M. Endo. 2008. Excellent solid lubrication of electrodeposited nickel-multiwalled carbon nanotube composite films. *Mater. Lett.* 62: 3545–3548.
85. Chen, X. H., Cheng, F. Q., Li, S. L., Zhou, L. P., and D. Y. Li. 2002. Electrodeposited nickel composites containing carbon nanotubes. *Surf. Coat. Technol.* 155: 274–278.
86. Chen, X. H., Peng, J. C., Li, X. Q., Deng, F. M., Wang, J. X., and W. Z. Li. 2001. Tribological behavior of carbon nanotubes—reinforced nickel matrix composite coatings. *J. Mater. Sci Lett.* 20: 2057–2060.
87. Tu, J. P., Zhu, L. P., Chen, W. X., Zhao, X. B., Liu, F., and X. B. Zhang, 2004. Preparation of Ni-CNT composite coatings on aluminum substrate and its friction and wear behavior. *Trans. Nonferrous Met. Soc. China.* 14(5): 880–884.
88. Shi, Y. L., Yang, Z., Li, M. K., Xu, H., and H. L. Li. 2004. Electroplated synthesis of Ni–P–UFD, Ni–P-CNTs, and Ni–P–UFD–CNTs composite coatings as hydrogen evolution electrodes. *Mater. Chem. Phys.* 87: 154–161.
89. Shi, Y. L., Yang, Z., Xu, H., Li, M. K., and H. L. Li, 2004. Preparation of electroplated Ni-P-ultrafine diamond, Ni-P-carbon nanotubes composite coatings and their corrosion properties. *J. Mater. Sci.* 39: 5809–5815.

90. Gang, A. B., Xang, L., and L. Hong. 2005. Preparation of carbon nanotube composite material with metal matrix by electroplating. *Trans. Nonferrous Met. Soc. China.* 15(5): 1045–1048.
91. Arai, S., Endo, M., Sato, T., and A. Koide. 2006. Fabrication of nickel-multiwalled carbon nanotube composite films with excellent thermal conductivity by an electrodeposition technique. *Electrochem. Solid-State Lett.* 9(8): C131–C133.
92. Chen, X. H., Li, W. H., Chen, C. S., Xu, L. S., Yang, Z., and J. Hu. 2005. Preparation and properties of Cu matrix composite reinforced by carbon nanotubes. *Trans. Nenferrous Met. Soc. China.* 15(2): 314–318.
93. Praveen, B. M., Venkatesha, T. V., Naik, Y. A., and K. Prashantha. 2007. Corrosion studies of carbon nanotubes–Zn composite coating. *Surf. Coat. Tech.* 201: 5836–5842.
94. Guo, C., Zuo, Y., Zhao, X., Zhao, J., and J. Xiong. 2007. The effects of pulse reverse parameters on the properties of Ni–carbon nanotubes composite coatings. *Surf. Coat. Tech.* 201: 9491–9496.
95. Shi, L., Sun, C. F., Gao, P., Zhou, F., and W. M. Liu. 2006. Electrodeposition and characterization of Ni–Co–carbon nanotubes composite coatings. *Surf. Coat. Tech.* 200: 4870–4875.
96. Dai, P. Q., Xu, W. C., and Q. Y. Huang. 2008. Mechanical properties and micro-structure of nanocrystalline nickel-carbon nanotube composites produced by electrodeposition. *Mater. Sci. Eng. A.* 483–484: 172–174.
97. Arai, S., Saito, T., and M. Endo. 2007. Low-internal stress nickel multiwalled carbon nanotube composite electrodeposited from a sulfamate bath. *J. Electrochem. Soc.* 154(1): D530–D533.
98. Chao, T. Y., Shen, G. R., and Y. T. Cheng. 2006. Comparative study of Ni–P-diamond and Ni–P-CNT nanocomposite films. *J. Electrochem. Soc.* 153(1): 98–104.
99. Sun, Y., Sun, J., Liu, M., and Q. Chen. 2007. Mechanical strength of carbon nanotube–nickel nanocomposites. *Nanotechnology.* 18: 505704.
100. Popp, A., Engstler, J., and J. J. Schneider. 2009. Porous carbon nanotube-reinforced metals and ceramics via a double templating approach. *Carbon.* 47: 3208–3214.
101. Liu, B., Zeng, Z., and Y. Lin. 2009. Mechanical properties of hard Cr–MWNT composite coatings. *Surf. Coat. Tech.* 203: 3610–3613.
102. Deng, M., Ding, G., Wang, Y., Wu, H., Yao, Y., and L. Zhu. 2009. Fabrication of Ni-matrix carbon nanotube field emitters using composite electroplating and micromachining. *Carbon.* 47: 3466–3471.
103. Yang, Y. L., Wang, Y. D., Ren, Y., He, C. S., Deng, J. N., Nan, J., Chen, J. G., and L. Zuo. 2008. Single-walled carbon nanotube-reinforced copper composite coatings prepared by electrodeposition under ultrasonic field. *Mater. Lett.* 62: 47–50.
104. Jeon, Y. S., Byunb, J. Y., and T. S. Oha. 2008. Electrodeposition and mechanical properties of Ni–carbon nanotube nanocomposite coatings. *J. Phys. Chem. Sol.* 69: 1391–1394.
105. Deng, F. M., Chen, X. H., Chen, W. X., and W. Z. Li. 2004. Electroless plating Ni-P matrix composite coating reinforced by carbon nanotubes. *Trans. Nonferrous Met. Soc. China.* 14(4): 681–685.
106. Wang, F., Arai, S., and M. Endo. 2005. The preparation of multi-walled carbon nanotubes with a Ni–P coating by an electroless deposition process. *Carbon*, 43: 1716–1721.
107. Shen, G. R., Cheng, Y. T., and L. T. Tsai. 2005. Synthesis and characterization of Ni–P-CNT's nanocomposite film for MEMS applications. *IEEE Trans. Nanotech.* 4(5): 539–547.
108. Wang, L. Y., Tu, J. P., Chen, W. X., Wang, Y. C., Liu, X. K., Olk, C., Cheng, D. H., and X. B. Zhang. 2003. Friction and wear behavior of electroless Ni-based CNT composite coatings. *Wear.* 254: 1289–1293.

109. Chen, W. X., Tu, J. P., Xu, Z. D., Chen, W. L., Zhang, X. B., and D. H. Cheng. 2003. Tribological properties of Ni–P-multi-walled carbon nanotubes electroless composite coating. *Mater. Lett.* 57: 1256–1260.
110. Chen, X., Zhang, G., Chen, C., Zhou, L., Li, S., and X. Li. 2003. Carbon nanotube composite deposits with high hardness and high wear resistance. *Adv. Eng. Mater.* 5(7): 514–518.
111. Yang, Z., Xu, H., Li, M. K., Shi, Y. L., Huang, Y., and H. L. Li. 2004. Preparation and properties of Ni/P/single-walled carbon nanotubes composite coatings by means of electroless plating. *Thin Solid Films.* 466: 86–91.
112. Yang, Z., Xu, H., Shi, Y. L., Li, M. K., Huang, Y., and H. L. Li. 2005. The fabrication and corrosion behavior of electroless Ni–P-carbon nanotube composite coatings. *Mater. Res. Bull.* 40: 1001–1009.
113. Chen, C. S., Chen, X. H., Yang, Z., Li, W. H., Xu, L. S., and B. Yi. 2006. Effect of multi-walled carbon nanotubes as reinforced fibres on tribological behaviour of Ni–P electroless coatings. *Diamond Relat. Mater.* 15: 151–156.
114. Zhao, G., and F. Deng. 2005. Electroless plating of Ni–P-CNTs composite coating. *Key Eng. Mater.* 280–283: 1445–1448.
115. Zhao, G., Zhang, H., Wang, Z., and Y. C. Zhang. 2007. Corrosion resistance properties of carbon nanotube reinforced electroless composite coating. *Key Eng. Mater.* 336–338: 2142–2145.
116. Wu, Y., Ren, R., Wang, F., Yuan, Z., Wang, T., and X. Hu. 2008. Preparation and characterization of Ni–Cu–P/CNTs quaternary electroless composite coating. *Mater. Res. Bull.* 43(12): 3425–3432.
117. Chen, X. H., Chen, C. S., Xiao, H. N., Liu, H. B., Zhou, L. P., Li, S. L., and G. Zhang. 2006. Dry friction and wear characteristics of nickel/carbon nanotube electroless composite deposits. *Tribo. Inter.* 39: 22–28.
118. Xue, R. J., and Y. C. Wu. 2007. Mechanism and microstructure of electroless Ni-Fe-P plating on CNTs. *J. China Univ. Min. Tech.* 17(3): 424–427.
119. Liu, Y. M., Sung, Y., Cheng Chen, Y. C., Lin, C. T., Chou, Y. H., and M. D. Ger. 2007. A method to fabricate field emitters using electroless co-deposited composite of MWNTs and nickel. *Electrochem. Sol. State Lett.* 10(9): 101–104.
120. Li, Z. H., Wang, X. Q., Wang, M., Wang, F. F., and H. L. Ge. 2006. Preparation and tribological properties of the carbon nanotubes–Ni–P composite coating. *Tribo. Inter.* 39: 953–957.
121. Guo, C., Zuo, Y., Zhao, X., Zhao, J., and J. Xiong. 2008. The effects of electrodeposition current density on properties of Ni–CNTs composite coatings. *Surf. Coat. Tech.* 202: 3246–3250.
122. Ngo, Q., Cruden, B. A., Cassell, A. M., Walker, M. D., Ye, Q., Koehne, J. E., Meyyappan, M., Li, J., and C. Y. Yang. 2004. Thermal conductivity of carbon nanotube composite films. *Mat. Res. Soc. Symp. Proc.*, 812: 3.18.1–3.18.6.
123. Kang, X., Mai, Z., Zou, X., Cai, P., and J. Mo. 2007. A sensitive nanoenzymatic glucose sensor in alkaline madia with copper nanocluster/multimalled carbon nanotube-modified glassy carbon electrode. *Anal. Biochem.* 363: 143–150.
124. Chai, Y., Zhang, K., Zhang, M., Chan, P. C. H., and M. M. F. Yuen. 2007. Carbon nanotube/copper composites for via filling and thermal management. *Electronic Components and Technology Conference*, 1224–1229.
125. Xu, Q., Zhang, L., and J. Zhu. 2003. Controlled growth of composite nanowires based on coating Ni on carbon nanotubes by electrochemical deposition method. *J. Phys. Chem. B.* 107(33): 8294–8296.
126. Arai, S., Endo, M., and N. Kaneko. 2004. Ni-deposited multi-walled carbon nanotubes by electrodeposition. *Carbon.* 42: 641–644.

127. Chen, X., Jintong Xia, J., Peng, J., Li, W., and S. Xie. 2000. Carbon-nanotube metal-matrix composites prepared by electroless plating. *Comp. Sci. Tech.* 60: 301–306.
128. Dingsheng, Y., and L. Fingliang. 2006. Electroless deposition of Cu on multi-walled carbon nanotubes. *Rare Metals.* 25(3): 237–240.
129. Ci, L., Ryu, Z., Jin-Phillipp, N. Y., and M. Ruhle. 2006. Investigation of the interfacial reaction between multi-walled carbon nanotubes and aluminum. *Acta Mater.* 54: 5367–5375.
130. Huang, W., Chen, H., and J-M. Zuo. 2006. One-dimensional self-assembly of metallic nanostructures on single-walled carbon-nanotube bundles. *Small* 2: 1418–1421.
131. Li, Y-H., Housten, W., Zhao, Y., and Y. Q. Zhu. 2007. Cu/single-walled carbon nanotube laminate composites fabricated by cold rolling and annealing. *Nanotechnology* 18: 205607.
132. Lahiri, D., Bakshi, S. R., Keshri, A. K., Liu, Y., and A. Agarwal. 2009. Dual strengthening mechanisms induced by carbon nanotubes in roll bonded aluminum composites. *Mater. Sci. Eng. A* 523: 263–270.
133. Salas, W., Alba-Baena, N. G., and L. E. Murr. 2007. Explosive shock-wave consolidation of aluminum powder/carbon nanotube aggregate mixtures: optical and electron metallography. *Met. Mater. Trans. A* 38: 2928–2935.
134. Tokunaga, T., Kaneko, K., and Z. Horita. 2008. Production of aluminum-matrix carbon nanotube composite using high pressure torsion. *Mater. Sci. Eng. A* 490: 300–304.
135. Morisada, Y., Fujii, H., Nagaoka, T., and M. Fukusumi. 2006. MWCNTs/AZ31 surface composites fabricated by friction stir processing. *Mater. Sci. Eng. A* 419: 344–348.
136. Lim, D. K., Shibayanagi, T., and A. P. Gerlich. 2009. Synthesis of multi-walled CNT reinforced aluminum alloy composite via friction stir processing. *Mater. Sci. Eng. A* 507: 194–199.
137. Liu, Q., Ke, L., Liu, F., Huang, C., Xing, L. 2013. Microstructure and mechanical property of multi-walled carbon nanotubes reinforced aluminum matrix composites fabricated by friction stir processing. *Materials and Design* 45: 343–348.
138. Hwang, J. Y., Neira, A., Scharf, T. W., Tiley, J., and R. Banerjee. 2008. Laser-deposited carbon nanotube reinforced nickel matrix composites. *Scripta Mater.* 59: 487–490.

4 Characterization Techniques for CNT-MMC Materials

The objective of this chapter is to introduce those characterization techniques that are critical for evaluating the properties of CNT-reinforced metal matrix composites. The chapter is intended to inform the reader about the nuances and limitations while characterizing CNT-reinforced metal matrix composites. One of the focuses is to understand the microstructural features such as CNT distribution in the matrix, interfacial reaction between CNT and the matrix, CNT retention and damage, and grain size changes in the matrix due to CNT addition. The second objective is to evaluate the mechanical, thermal, electrical, and corrosion properties of MM-CNT composite and understand the correlation between the results and microstructure of the composites.

4.1 X-RAY DIFFRACTION

X-ray diffraction (XRD) is a very important tool in materials science. It is a versatile technique for the identification of phases present in a sample. Much information on CNT-MMCs can be gathered by this technique. The first application is in determining the phase composition after processing. It gives immediate information of the effect of the processing technique on the changes taking place in the phases present in the matrix. For example, in the case of bulk-metallic-glass-CNT composites, it shows whether the matrix is still crystalline [1]. The formation of carbides due to reaction between CNT and the matrix could adversely alter the mechanical properties. If the vol.% of the carbide is high enough (>5%), then one can observe peaks corresponding to the phase in the XRD pattern [1–3]. XRD study of randomly oriented CNTs will result in a large peak at $2\theta = 26°$ corresponding to the graphite (002) plane spacing = 0.34 nm. Several small intensity peaks corresponding to (100) plane at $2\theta = 42.4°$ and (110) planes at $2\theta = 77.7°$ may also be present indicating CNT presence. Qualitative information on the alignment of the CNTs can be made from the relative intensities of the different peaks [4]. In the case of composites, the mass fraction of CNTs is generally low and only the (002) peak is discernible. XRD as such is not a confirmative technique for the presence of carbon nanotubes because the peaks are of graphite.

XRD has been used to study the reaction between metals and carbon species and qualitatively compare the reactivity of different carbon species. Using a small scan rate of $0.05° s^{-1}$, the changes in the intensity of the (111) peak of Co (at $2\theta = 44.24°$) was examined before and after annealing with the carbon species at 1000 °C for 10 h [5].

The intensity of the (111) peak reduces as the Co is consumed to form CO_2C. It was found that the reactivity of activated carbon > multi-walled CNT > SWNT > layered graphite. Another use of XRD is in determining the crystallite/grain size from the broadening of the peaks using the simplified Scherrer formula

$$\beta = \sqrt{\beta_s^2 - \beta_i^2} = \frac{0.9\lambda}{d\cos(\theta)} \quad (4.1)$$

Where β is the corrected full width at half maximum (FWHM) for a peak at 2θ (in radians), β_s is the FWHM for the sample (in radians), β_i is the FHWM due to instrumental broadening (in radians), λ is the wavelength, θ is the corresponding Bragg's angle, and d is the average grain size of the material.

This method was applied to determine the grain size of Al in Al-CNT composites [6]. Depending on the degree of dispersion, CNTs play a complex role in altering the grain size of the metal matrix. Due to high thermal conductivity of dispersed CNTs, the metal matrix near isolated CNTs experiences a higher cooling rate, which results in a fine grain size. On the contrary, CNT clusters have very low thermal conductivity and could lead to grain growth near CNT clusters [7,8]. CNT clusters also provide the pinning effect and resist grain growth. Hence, all these factors need to be kept in mind while computing matrix grain size from the Scherrer formula. Information can also be obtained from the peak broadening by fitting pseudo-Voigt function to the curve and analyzing the Cauchy and Gauss contributions to the broadening. Using such an analysis, the contribution due to the crystallite size and micro-strain in the matrix can be determined. Subsequently, dislocation density in the matrix can be computed [9]. The increase in the dislocation density was predicted and experimentally observed due to addition of CNTs in roll bonded Al-CNT composites by this method [9]. Peak broadening method is also applied to assess the crystallite size, lattice strain, and dislocation density of Al-CNT composites, synthesized from milled and ultrasonicated powders, consolidated through equi-channel angular pressed [10]. Residual stresses in the composite can also be obtained from the shifts in the XRD peaks of the matrix. However, it has not been applied to CNT-MMC yet.

4.2 RAMAN SPECTROSCOPY

Raman spectrum originates due to interaction of radiation with the vibrational modes of a molecule. When a coherent light radiation is incident on a sample, a part of the incident beam may be used up to excite a characteristic vibration or a vibration may die out giving its energy to the incident photon. If the frequency of a vibrational mode of a molecule and the incident beam are υ_m and υ_0, respectively, the resulting new photon will have a frequency of either $\upsilon_0 - \upsilon_m$ (Stokes Raman Scattering) or $\upsilon_0 + \upsilon_m$ (Anti-Stokes Raman Scattering), respectively. This is known as the Raman Effect. It is a very weak phenomenon with an intensity of scattered radiation approximately 10^{-5} to 10^{-7} times that of the incident beam. Hence, lasers are employed as the incident light source owing to their high intensities and coherent nature. Some of the common radiations that have been used are the red light (wavelength $\lambda = 633$ nm) from an He-Ne laser, red light ($\lambda = 785$ nm) from a Ti-sapphire laser and green light

(λ = 514.5 nm) from an argon ion laser. The reflected beams with lower energy have a lower wave number (reciprocal of λ) and the difference, known as Raman shift (expressed in cm^{-1}), corresponds to the characteristic vibration frequency of the molecule. The dipole moment induced in a molecule must be affected by the molecular vibration for the Raman Effect to occur and hence only those species for which the polarizability changes with vibration are Raman active. An excellent introduction to Raman spectroscopy can be found in the textbook by Ferraro et al. [11].

Raman spectroscopy is one of the most widely used characterization techniques for carbon nanotube and reinforced composites. Most of the research studies on CNT composites use Raman spectroscopy for CNT characterization. This is because Raman spectroscopy provides the most conclusive signature for the presence of CNTs in the matrix. In addition, this is non-destructive test with almost no sample preparation. Raman spectrum of a natural graphite single crystal shows a single peak at 1575 cm^{-1} corresponding to the E2g vibration mode [12]. The Raman spectrum of different types of carbon nanotubes and species has been well documented in the literature [13–15]. CNTs show a peak at 1575 cm^{-1}, which is called the G-peak. Disorder in the CNTs and presence of sp^3 defects leads to an A1g radial breathing mode peak at 1348 cm^{-1}, which is known as the D-peak. A second-order peak of the D-peak is observed at 2691 cm^{-1} and is called the G'-peak. SWNTs show an additional unique peak at ~180 cm^{-1} for vibrations corresponding to the radial expansion and compression. This peak is known as radial breathing mode (RBM). Figure 4.1 shows the Raman spectra of an SWNT. The Raman shift of the RBM peak changes with the diameter of the small nanotubes with diameter <2 nm and is given by the relation

$$\nu\left(cm^{-1}\right) = \frac{248}{d\left(nm\right)} \tag{4.2}$$

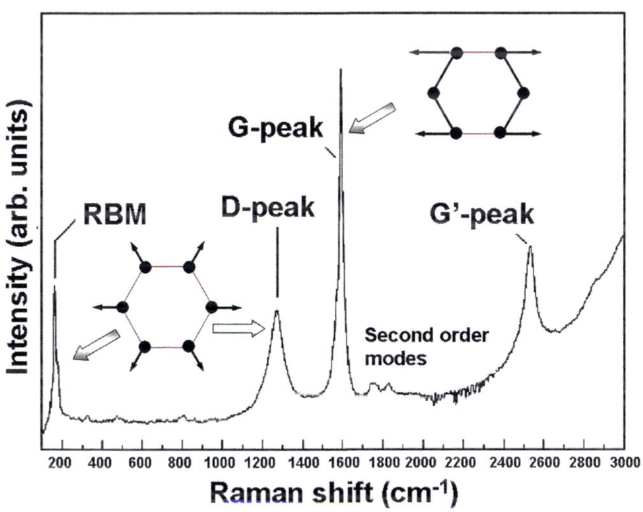

FIGURE 4.1 Raman spectra for an SWNT showing the RBM, G-peak, D-peak, and G'-peak [15]. (Reproduced with permission from Elsevier).

where d is the diameter of the SWNT in nanometers. The G'-peak, which is the overtone of the D peak, has been observed to shift significantly when the nanotube composites are stressed [16] and can be used to predict the stress state of the CNTs in the composite. The position of the G' band has been utilized to calculate the strain in the nanotubes and by calculating the corresponding thermal stresses, the approximate elastic modulus of CNTs has been estimated [17]. Additionally, the shift in G peak in Al-CNT composite has been correlated to differential compressive stress field acting upon CNT, due to thermal mismatch with matrix, as a function of different processing conditions, like sintering, hot working, and cooling down. Micro-Raman spectroscopy enables gathering information from very small areas (<10 mμ diameter) and can be used to obtain specific information like the infiltration of CNT clusters by metal. From the low intensity of the Raman peak for Si obtained from the CNT cluster region, it was concluded that the CNT cluster was poorly infiltrated with Al-Si alloy in plasma sprayed coatings [3]. Important information pertaining to the structure of the CNTs can be obtained from the ratio of the intensities of the D- and G-peak (I_D/I_G). An increase in the I_D/I_G ratio after CNT-MMC synthesis means an increase in the defect density and hence implies some sort of damage incurred to CNTs while processing. Similarly, a decrease in the I_D/I_G ratio may indicate graphitization by annealing effect produced during processing. By studying the changes in the I_D/I_G ratio, one can make qualitative conclusions on whether changes in the CNT structure during various stages of processing. Mechanical milling of Al-CNT powder mixture followed by hot rolling leads to an increase in I_D/I_G ratio and a shift in the G-peak from 572.7 to 1596.3 cm^{-1}. This has been attributed to the increase in the inter-atomic distance of carbon atoms during the milling procedure, which may cause a decrease in the bonding potentials and result in decline of the Raman vibration frequency [18]. Upshift in G band is also noted for CNTs coated with electrodeposited Al. This shift in G band is attributed to the electron transfer from CNT to Al^{3+}, in order to reduce the latter and to form a strong interface [19].

4.3 SCANNING ELECTRON MICROSCOPY WITH ENERGY DISPERSIVE SPECTROSCOPY

SEM is a very important imaging tool for materials scientists. From imaging nanopowders and CNTs to the fracture surface of the composites, it can yield a lot of information. Particle sizes of powders and the length and diameter distribution of multiwalled CNTs can be estimated from secondary electron images. For SWNTs, one must be careful that there is no charging, in which case the image may be blurred and the diameter will appear larger than the actual value. All samples need to be sputter-coated with gold in order to avoid charging. Upon very high magnification (>70,000×), one can see the discontinuous thin layer of gold deposited on the CNTs as shown in Figure 4.2. SEM is helpful in studying dispersion of CNTs in the powders and in CNT clusters in the composite. SEM investigations of fracture surface and wear tracks of MM-CNT composite provide important information on the mechanism of failures and the role of CNTs in strengthening. CNT pullouts and crack bridging and deflection by CNTs are readily visible in SEM images of fracture

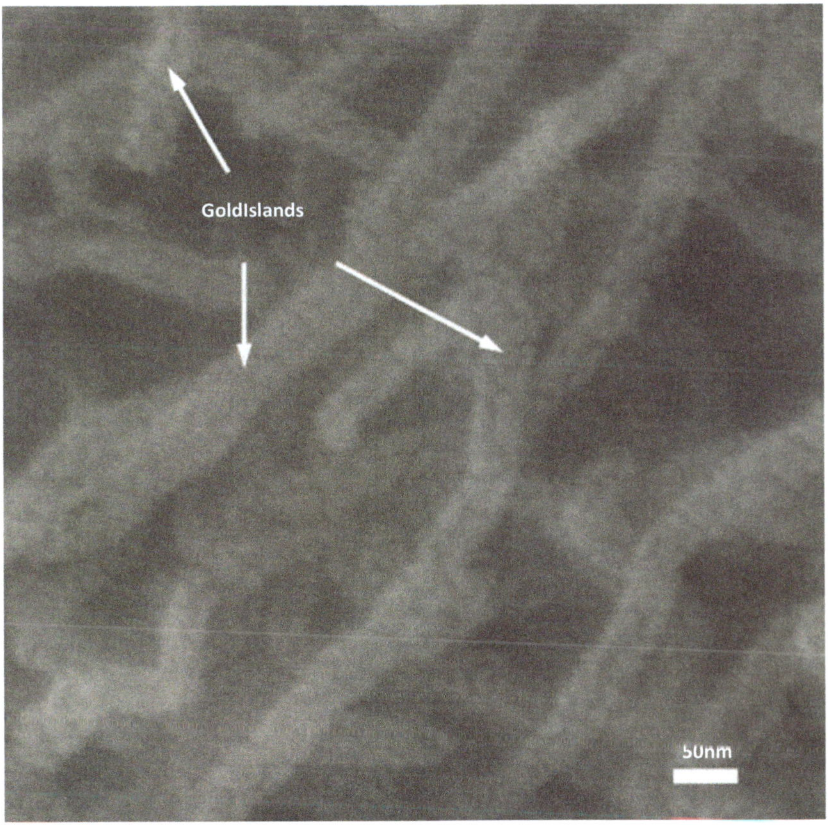

FIGURE 4.2 High-magnification SEM image of multi-walled carbon nanotubes showing a thin discontinuous gold layer deposited due to sputter coating.

surface and show information of the strengthening mechanisms involved. Using special fixtures, tensile test of single CNT has also been carried out in a SEM that allows in situ observations during the test [20]. A SEM with a field emission electron gun (FE-SEM) is recommended for studying CNT-MMC. FE-SEM has a better resolution (almost twice that of conventional SEM with a thermionic emission gun) and higher brightness and reduced noise level in the micrographs. This is very advantageous while observing details at high magnifications (>20,000×).

Energy dispersive X-ray analysis (EDS) helps in identifying the elemental composition of the phases formed during processing. With windowless or ultra-thin window detectors, elements up to boron could be detected with good enough sensitivity. In the case of Al-CNT composites, phases with composition of Al:C ratio of 1:1 and 1:2 were observed [21]. Similar conclusions about the formation of metal carbides can be made. X-ray mapping can be carried out to get information on the distribution of various elements in the microstructure. Oxidation is a problem during materials processing. A quantitative idea about the distribution of oxygen can also be obtained from elemental maps. Elemental maps and EDS are also helpful in characterizing the worn surface after sliding wear tests to ascertain the mechanism of wear in CNT-MMC. The presence of oxygen on the wear surface indicates an oxidative wear mechanism.

4.4 HIGH-RESOLUTION TRANSMISSION ELECTRON MICROSCOPY

One of the reasons for the late discovery of carbon nanotubes was the lack of high-resolution imaging facilities. Although CNTs may have been produced in the arc deposition experiment by Bacon [22], their structure was not determined due to lack of imaging facilities at the nanometer level. Conventional TEM operating at 200 keV accelerating voltage may be good enough for information on the matrix properties but is not adequate to get a complete picture of the phenomena occurring in CNTs and the CNT/matrix interfaces. A TEM operating at 300 keV provides very nice lattice images that are useful for in-depth studies. Through the high-resolution TEM (HRTEM) of multiwalled CNTs, Iijima could count the number of walls and could easily make out that they consisted of concentric tubules made of graphene sheets instead of being scroll-like as was proposed by Bacon [23]. HRTEM images can give a lot of information on CNT composites like wetting between the metal matrix and the CNTs, interfacial compound formation due to reaction between CNTs and the matrix, orientation relationship between carbide and the CNT and carbide and matrix, which will determine the load transfer properties of the composites. Using HRTEM images, one can see the walls of the CNT in projection and make conclusions on the crystallinity and quality of the nanotubes and the defects present in the nanotubes.

By taking a Fourier transform and refining the diffraction pattern and then performing an inverse Fourier transform, the lattice images of the CNT walls can be refined to determine the inter-tube spacing accurately. In fact, TEM is the only method to estimate the number of walls and the inner diameter of the CNTs. TEM enables understanding the fracture mechanisms of CNTs [24] as shown in Figure 4.3. A symmetric tip of a fractured CNT in a cold-sprayed Al-CNT composite as shown in Figure 4.3a indicates that the CNT was fractured in tension or due to impact, while an asymmetric tip as shown in Figure 4.3b indicates a shear fracture [24]. Based on these observations, the mechanism of fracture process during cold spraying can be deduced as shown in Figure 4.3c and 4.3d. It is also to be remembered that, while observing under TEM, electron irradiation causes local temperatures to increase and could lead to transformations in the structure and formation of caps and closure of CNTs [25]. TEM is an excellent tool to study interfacial phenomena and has been used to probe the interfacial reaction between aluminum and CNTs [26–31]. Aluminum carbide was found to form due to reaction between Al and Al-12 wt.% Si alloys with CNT while SiC was observed in the case of an Al-23 wt.% Si alloy matrix. Aluminum carbide nanorods were also found to form as a result of interfacial reaction in Al-CNT composites [32,33]. In case of Mg-Al alloy (AZ series) reinforced with CNT, ternary carbides were identified at the interface [34,35]. On the contrary, Cu-CNT composite reveals clean interface [36], indicating the absence of any reaction product, which is supported thermodynamically also. Selective area electron diffraction (SAED) patterns from the specific areas can be used to examine the phases present in the small regions. Tensile test of individual CNTs can be carried

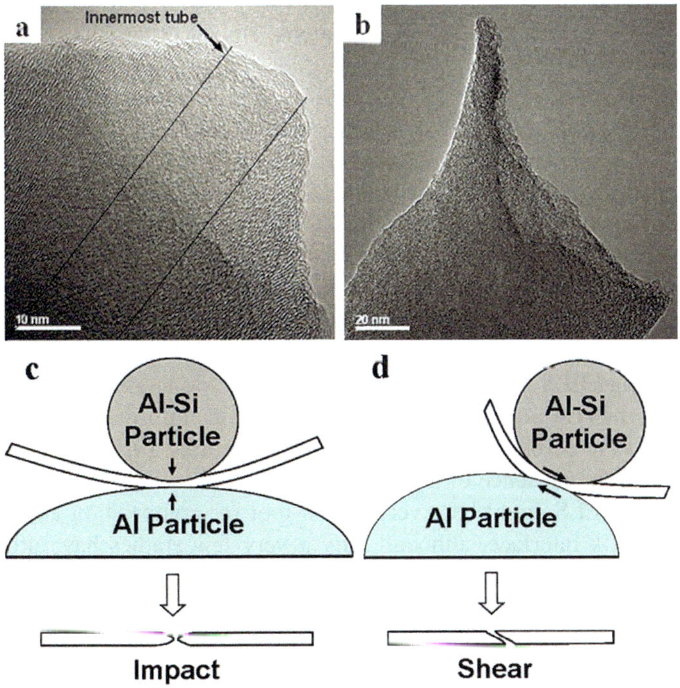

FIGURE 4.3 TEM images showing (a) tip of a CNT broken due to impact, and (b) tip of a CNT broken due to shearing. Schematic showing the two fracture mechanisms for CNT during cold spraying by (c) impact and (d) shear [24]. (Reproduced with permission from Elsevier).

out inside a TEM while observing the phenomena occurring in situ [37]. This gives a lot of information on the mechanism of the fracture of the CNTs like the "sword in sheath" mechanism [20] or the "breakage-sliding-fusion" mechanism [38]. This method can be extended to measure properties of nanorods and CNTs coated with metals.

Scanning transmission electron microscopy (STEM) can also be carried out to generate images with compositional or Z-contrast. The use of a high angle annular dark field (HAADF) detector makes it possible to collect only the electrons scattered at large angles (also known as Rutherford scattering), which depends on the atomic number of the element scattering the electrons. Thus, it gives rise to Z-contrast. STEM images can be used to study filling of CNTs, infiltration of CNT clusters by metal matrix, and segregation of elements due to preferred reaction with CNTs. However, STEM has been not utilized extensively to study CNT-MMCs.

4.5 ELECTRON ENERGY LOSS SPECTROSCOPY

In this spectroscopic technique, the loss of the electron energy is measured as it passes through the thin film specimen in a TEM and the intensity of the electrons is plotted as a function of energy loss. The energy loss is due to the interaction of electrons with atoms leading to ionization of electrons from inner shells. The lower loss spectra (<50 eV) is due to plasmon excitations. By selecting the area using TEM bright field image, one can obtain spectroscopic information from the very small areas of a few square nanometers. Also by using the information from specific regions and energy-filtered TEM (EFTEM), an image can be constructed that is similar to the elemental X-ray map images generated by EDS but at the scale of nanometers. Elemental analysis can be performed along a line as small as the diameter of CNTs and carbon onions. Electron energy loss spectroscopy (EELS) is particularly useful in detection of low atomic number elements like carbon. The characteristic peak at 284.5 eV corresponds to the transitions to the unoccupied π^* states. EELS technique has been used to show the increase in sp^2 carbons over sp^3 carbon atoms on heat treatment by annealing. This technique is useful for studying CNTs filled with metals. It was shown using EELS that the filling of CNTs by Cr was facilitated by formation of sulfides rather than carbides [39]. EELS can also provide information on the transformation of CNTs to carbon onions and diamond structures. EELS study of Fe-filled CNTs indicated the presence of a semi-coherent interface and bonding between Fe and CNT [40]. EELS could be a very useful tool for composition analysis across CNT-metal matrix interface, although only a very few studies have utilized it. A comprehensive report on the application of EELS studies for carbon nanotubes has been provided by Stockli [41].

4.6 X-RAY PHOTOELECTRON SPECTROSCOPY

X-ray photoelectron spectroscopy (XPS) is based on the photoelectric effect. When an incident photon of high-energy radiation (e.g., X-ray or UV) is incident on an atom, it may be absorbed and an electron may be ejected from the inner shell. The phenomenon is known as the photoelectric effect. The kinetic energy of the emitted electron *Ek* is given by

$$E_k = E_i - E_b \tag{4.3}$$

where E_i and E_b are the energies of the incident radiation and the binding energy of the electron, respectively. By employing a monochromatic source of known energy like Mg Kα (1254 eV) or Al Kα (1487 eV) X-rays and measuring the energies of the emitted photoelectrons, one can find out the binding energies of the electrons. The binding energies of the electrons depend on the environment of the atoms and its oxidation state. By comparison with the standard binding energies, conclusion can be drawn on the bonding between atoms. Since photoelectric effect occurs from the surface, the technique is helpful in analyzing the chemical composition on the surface. XPS has been used to prove the formation of SiC in CNTs coated with Si by the CVD process at 1000 °C [28]. Binding energies corresponding to Si-C and Si-O bonds were observed. In a study on wetting of Al on CNT surface, XPS analysis was used to reveal the presence of 40% Al-C bonds, along with signature of Al-O bonds and Al metal [19]. XPS is also an important tool for characterizing the wear surfaces of MM-CNT composites. XPS analysis of the wear track of Al-CNT composite revealed that higher Al_4C_3 content with higher CNTs reinforcement (>5 wt.%) leads to poor wear resistance [42]. The identification of intermediate phases and compounds forming during wear of MM-CNT composites can also be studied with XPS.

4.7 MECHANICAL PROPERTIES EVALUATION

The most important potential application of CNT reinforced metal matrix composites is as the load bearing structural materials. Mechanical properties like Young's modulus, tensile strength, and strain to failure are dependent on several factors like the CNT distribution, type of CNT, defects in CNT, and processing technique used to synthesize the CNT-MMCs.

For large structures, flaws like cracks and CNT clusters also affect the mechanical properties. For small-size applications like in MEMS, flaws may not be present and the mechanical properties may be superior to those predicted by bulk mechanical tests. The local properties of the composite would depend on the local CNT content and will have an effect on the global property. Thus, there could be a correlation between the properties measured at different length scales. The processing techniques developed may not be amenable for the synthesis of bulk samples for large-scale mechanical testing. Thus, there is an inherent need to study the properties of CNT-MMCs at multiple length scales. This section summarizes the various testing techniques that are available and commonly utilized for the mechanical property evaluation of CNT-MMCs.

4.7.1 Nanoscale Mechanical Testing

Nanoscale mechanical testing of CNT-MMCs can be performed using a relatively novel technique of nano-indentation. Development of novel transducers with load and depth resolution on the order of 1 nN and 1 nm, respectively, has enabled accurate measurement of the properties from small regions/samples (tens of microns). Depending on the capability of the equipment, nanoscale mechanical testing can be performed in multiple modes including quasistatic indentation, nano-dynamic modulus analysis (DMA), modulus mapping, and nanoscratch techniques. These techniques are very important in studying the mechanical properties of CNT-MMCs when the microstructural features are very small like CNT reinforcement, CNT clusters, and reaction products.

4.7.1.1 Nano-Indentation

Nano-indentation is a depth sensing indentation test from which the elastic modulus (E) and the hardness (H) of a material are obtained. In nano-indentation, the load and depth of penetration are measured with high precision as the diamond indenter penetrates the sample. The load-depth curve can be recorded during both loading and unloading stages of the indentation. The slope of the unloading curve at maximum load is governed by the modulus of elasticity of the material. Information about the elastic and plastic work and yield strength can also be obtained from the loading–unloading curve obtained from nano-indentaton. The indentation depth is of the order of tens and hundreds of nanometers. The advantage of this technique is that it can be carried out on small samples and features as small as a human red blood cell (of a few microns in size). Therefore, it has become a popular tool when the samples produced are in small quantities with fine features. The load and displacement are measured as a diamond indenter is indented into the surface of the sample. The load depth data is obtained with very high accuracy and resolution during both the loading and unloading part of the test and can be plotted as shown in Figure 4.4b. With advancement in instrumentation and sensing capabilities, it is now possible to have a load resolution of the order of nano-Newtons and depth resolution in nanometers. Commonly used tips during nano-indentation are the Berkovich tip, conical/cono-spherical tip, and the cube corner tip [43]. The Berkovich tip is in the form of a three-sided pyramid with an included half angle of 65.3° and has been extensively used for nano-indentation. The hardness and elastic modulus are calculated using the most commonly used Oliver and Pharr method [44] using the following equations:

$$E_r = \frac{\sqrt{\pi}}{2}\frac{S}{\sqrt{A}} \tag{4.4}$$

$$E_r = \frac{1-\nu_s^2}{E_s} + \frac{1-\nu_i^2}{E_i} \tag{4.5}$$

where E and ν stand for the elastic modulus and Poisson's ratio and the subscripts r, s, and i are for reduced modulus, that of the specimen and the indenter, respectively, and A is the projected area of the indentation which depends on the true contact depth (h_c) by the geometric relation $A = 25.4h_c$. The E_i and ν_i values correspond to that of diamond being equal to 1141 GPa and 0.07, respectively. $S = dP/dh$ is the contact stiffness of the initial portion of the unloading curve, which is obtained by fitting the unloading portion of the curve to the equation $P = B(h-h_f)^m$, where h is the depth, h_f is the final or residual depth, and B and m are constants. The true contact depth at maxi mum load is given by $h_c = h\text{max}-\varepsilon,\ _S$, where $h$max is the maximum depth of the indent, P_{max} is the maximum load used, and ε is a factor that depends on the geometry of the indenter and is evaluated based on the value of m. The value of ε is 1 for flat punch indenter, 0.72 for conical indenter, and 0.75 for paraboloids of revolution. In practice, a standard aluminum and quartz sample having known Er values is used to calibrate the contact area as a function of the contact depth and the area function and thus formed can be used readily. This helps in knowing the true contact areas at very low depths where the radius of curvature of the tips becomes significant.

Quasistatic nano-indentation has been used to determine the elastic modulus of CNT-MMCs. Usually a scatter is obtained in the data due to the localized nature of the test. The values obtained correspond to the local micro-structure and CNT content. In the case of cold-sprayed Al-0.5 wt.% CNT and Al-1.0 wt.% CNT, the E values were found to be between 40 and 120 GPa with a mean around 69 GPa (*E* value for pure Al) [24]. Some locations with higher local CNT content were found to have an elastic modulus as high as 229 GPa. In the case of the Al-CNT composites produced by plasma spraying, nano-indentation was carried out on the matrix portion of the two-phase microstructure (the other phase being the CNT clusters). Considerable increase in the yield strength, elastic modulus, and hardness was observed by the addition of CNTs [3] at the nano/microscale level. Figure 4.4a shows the scanning

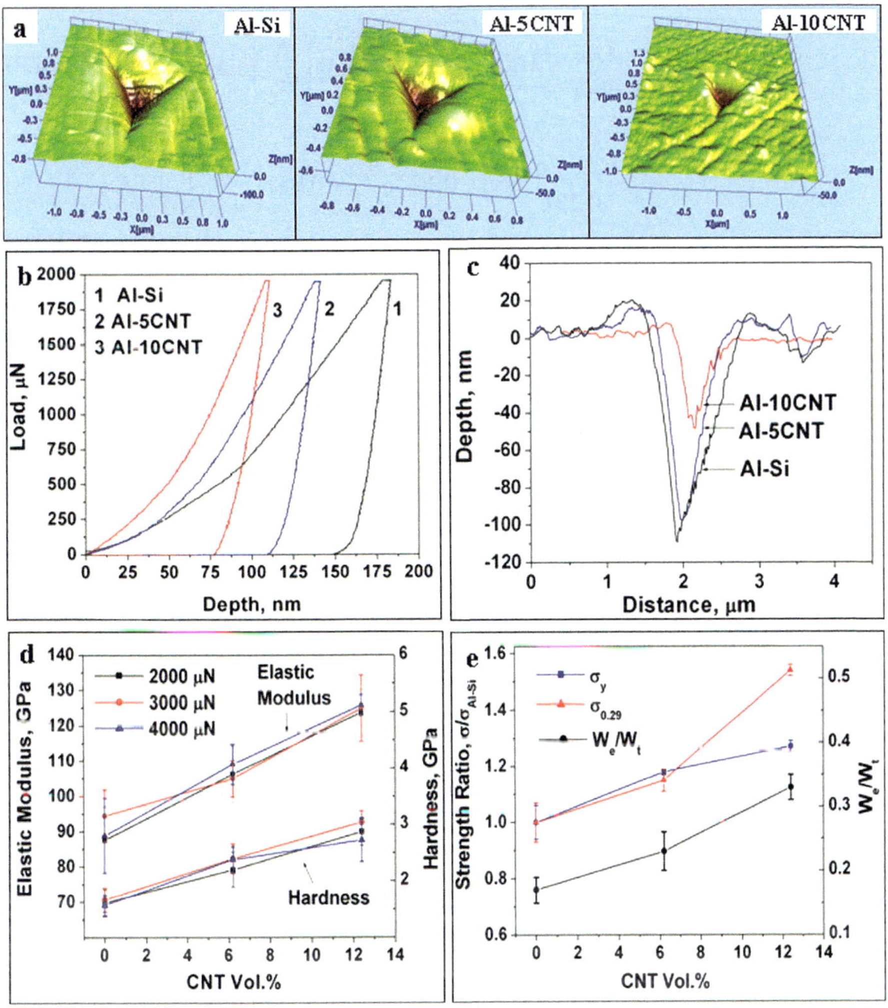

FIGURE 4.4 (a) SPM images of the indent on Al-Si and Al-CNT (5 wt.% and 10 wt.%) coatings obtained with 2000 μN load, (b) representative load displacement curve from nanoindentation, (c) line profiles along a median of the indentation obtained with 2000 μN load showing pile-up surface, (d) variation of elastic modulus and hardness of the coatings with CNT vol.%, and (e) calculated yield strength ratio of the coatings and the ratio of elastic work to total work [3]. (Reproduced with permission from Elsevier).

probe microscope (SPM) images of the indents produced in the Al-CNT composites and the corresponding load-displacement curves (Figure 4.4b). It is observed that, with addition of CNTs, the residual depth of the indent decreases implying an increase in the hardness (Figure 4.4c). Much information can be generated using the load-depth curves from the following equations [45,46]:

$$\frac{W_e}{W_t} = 1 - \frac{h_r}{h_{max}} \tag{4.6}$$

$$\frac{P_{max}}{h_{max}^2} = M_1\sigma_{0.29}\left[1+\frac{\sigma_y}{\sigma_{0.29}}\right]\left\{\ln\left(\frac{E}{\sigma_y}\right)+M_2\right\} \tag{4.7}$$

$$\frac{\sigma_{0.29}-\sigma_y}{0.29E} = 1-0.142\frac{h_r}{h_{max}}-0.957\left(\frac{h_r}{h_{max}}\right)^2 \tag{4.8}$$

where W_e and W_t refer to the elastic work and total work done during indentation, h_{max} and h_r are the maximum depth and residual depth, P_{max} is the maximum load, σ_y is the yield strength, $\sigma_{0.29}$ is the stress at strain of 0.29, E is the elastic modulus, and $M_1 = 6.618$ and $M_2 = -0.875$ are constants for the Berkovich-type indenter.

The elastic modulus and hardness and the strength properties calculated for the aluminum composites with and without CNT are shown in Figure 4.4d and 4.4e, respectively [3]. An increase of 45% and 80% in hardness, 19% and 39% in elastic modulus, 18 and 27% in σ_y, and 35% and 94% in W_e/W_t was observed for 5 and 10 wt.% CNT composites, respectively, as calculated from the nano-indentation curves [3].

Nano-indentation techniques have a great potential in the study of the mechanical behavior of CNT-MMCs. Since the response can be generated from sub-micron size regions, several important studies can be performed. Mechanical properties can also be directly obtained from regions like CNT clusters, CNT-metal interfaces, and reaction products like carbides formed using nano-indentation techniques. However, the main limitation is the inability to locate visually these fine microstructural features for placing an indent. Most of the nano-indenters are coupled with an optical microscope and cannot resolve fine microstructural features such as CNT cluster, CNT/matrix interface, and reaction product that is a few nanometers thick. New techniques like in situ nano-indentation inside SEM [47] and TEM [48] are being developed which can provide a wealth of information and understanding on the deformation behavior of CNT-MMCs. There is a lot of scope in studying the nanoscale mechanical behavior and the role of CNT reinforcement in the metal matrix composites.

4.7.1.2 Nano Dynamic Modulus Analysis

The nano dynamic modulus analysis (nano DMA) technique involves use of an alternating load in addition to a static load applied at different frequencies. The material response can be expressed by the following equation.

$$\Phi 0 \sigma\iota\nu\left(\omega\tau\right) = \mu\xi + \mathrm{X}\xi + \kappa\xi \tag{4.9}$$

where the displacement response at the same frequency of oscillation is

$$\xi = \Xi\sigma\iota\nu\left(\omega\tau - \phi\right) \tag{4.10}$$

F_0 is the maximum force, m is the mass of center plate, C and k are combined damping and stiffness, respectively, X is the amplitude of displacement oscillation, ω is angular frequency, and ϕ is phase shift of displacement. Dynamic modulus is a complex term comprising a real part known as the storage modulus and an imaginary part known as the loss modulus. The storage modulus (E'), loss modulus (E''), and tan delta can be calculated using the following equations.

$$E' = \frac{k_s\sqrt{\pi}}{2\sqrt{A_c}} \tag{4.11}$$

$$E'' = \frac{\omega C_s\sqrt{\pi}}{2\sqrt{A_c}} \tag{4.12}$$

$$\tan delta = \frac{\omega C_s}{k_s} \tag{4.13}$$

where k_s and C_s are the stiffness and compliance of a sample deduced by subtracting the instrument's stiffness and compliance, respectively; A_c is the contact area, which is dependent on the contact depth. Contact depth of the indenter is described through tip area function during instrument calibration. The loss modulus is related to the damping properties of the composite and is reflected in the tan delta value. Nano DMA is a useful technique for viscoelastic materials like plastic and biological materials [49]. Nano DMA can also provide a lot of information in damping characteristics of -MMCs, especially when carried out in CNT-rich regions like CNT clusters. This technique has not been employed much due to the elastic nature of most of the CNT-MMCs. However, for very high CNT concentration composites and areas like CNT clusters, information on the improvement of the damping capacity can be obtained [50].

4.7.1.3 Modulus Mapping

Modulus mapping refers to the plot of the spatial variation of the elastic modulus of a given composite as measured by dynamic mechanical testing using a nano-indenter. The dynamic mechanical properties are measured while the tip is rastering the surface, in order to generate a scanning probe image by application of dynamic force with the help of a lock-in amplifier. The frequency of the dynamic load is kept at 200 Hz, which is large enough not to interfere with the imaging. The phase and amplitude of the dynamic signal is measured by the lock-in amplifier and used to calculate the storage and loss modulus of the material at a given pixel by the following relations [51]:

$$X_0 = \frac{F_0}{\sqrt{\left(k - m\omega^2\right) + \left[\left(C_i - C_s\right)\omega\right]^2}} \tag{4.14}$$

$$\varphi = \tan^{-1}\frac{\left(C_i + C_s\right)\omega}{k - m\omega^2} \tag{4.15}$$

where X_0 and ϕ refer to the amplitude and phase difference between the ac force and the displacement, F_0 is the force amplitude, ω is the frequency, m is the indenter mass, C_i is the damping coefficient of the air gap between the plates of the capacitive transducer, C_s is the damping coefficient of the sample, k is combined stiffness given by $K_i + K_s$ where K_s is the contact stiffness and K_i is the spring constant of the springs holding the indenter. C_i and K_i are obtained by tests made in air. For viscoelastic

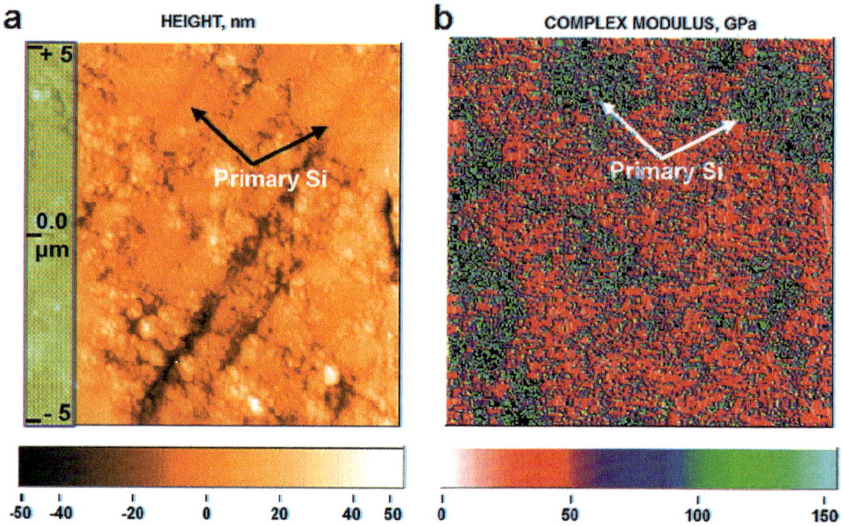

FIGURE 4.5 (a) SPM image of 10 wt.% CNT reinforced Al-23 wt.% Si alloy composite, and (b) the corresponding modulus map image [52]. (Reproduced with permission from Elsevier).

materials, the storage component of the stiffness is given by K_s and the loss component is given by ωC_s. The elastic modulus is related to the contact stiffness by

$$k_s = 2Eh\sqrt{\frac{24.5}{\pi}} \tag{4.16}$$

where E^* refers to the reduced modulus and h is the penetration depth.

This force modulation technique provides the elastic modulus at each image pixel resulting in 65,536 values per image, which can be used to prepare the modulus map over the selected microstructure of interest. This allows direct correlation of the elastic modulus variation with the microstructural features. Figure 4.5 shows the SPM image of 10 wt.% CNT reinforced Al-23 wt.% Si alloy and its corresponding modulus map [52]. It is seen that the modulus map correlates well with the microstructure. This very powerful technique has not received much attention. This technique can give information on microstructural features like reaction products at the CNT-metal interfaces and the present of regions of varying CNT content. The values obtained from modulus maps can also be used for studying the damping capacity of the composite and its variation in and around the CNT-rich regions [50].

4.7.1.4 Nanoscratch

Nanoscratch tests can provide information about the tribological properties (wear resistance and coefficient of friction) of CNT-MMC at the nanoscale level. Nanoscratch involves scratching the sample surface using a diamond-tip indenter at low loads on the order of hundreds of micro-Newtons to a few milli-Newtons. Usually an axisymmetrical indenter like the conical indenter is preferred for the test. The test can be carried out under ramp loading conditions or constant load conditions during the scratch. The lateral force required for scratching is measured along the scratch length. The coefficient of friction is obtained as the ratio of the instantaneous lateral force to the normal force. SPM image in conjunction with SEM can reveal much information on the fundamental wear mechanisms operating and the effect of CNT addition on the recovery properties.

Bakshi et al. have developed a method for evaluation of wear resistance in terms of wear volume loss from the nanoscratch test [53]. There are two depths associated with the nanoscratch test. First is the instantaneous depth (h_{inst}), which is recorded by the indenter during scratching. The second is the true depth of the scratch after elastic recovery processes have taken place (h_{true}), which can be measured by SPM image of the scratch. Consequently, there are two volumes associated with scratch, namely the contact volume (V_C) and true volume (V_T), respectively. Figure 4.6a shows the difference between the h_{inst} and h_{true}, which is a measure of the elastic recovery property of the composite. The area of the cross-section of the triangular grove made by scratching is dependent on the angles ϕ and θ as shown in the line profile in Figure 4.6b. The volume of the scratch can be calculated by the general formula:

$$V = \frac{1}{2}\left[\tan(\phi) + \tan(\theta - \phi)\right] \int_{-1/2}^{+1/2} h^2 dx = C \int_{-1/2}^{+1/2} h^2 dx \qquad (4.17)$$

Using the values of ϕ and θ as calculated from SPM images, the true wear volumes can be estimated. When a Berkovich tip is used, the orientation angle of the tip, α, with respect to the scratch direction as shown in Figure 4.6c, has an effect on the angles ϕ and θ during scratching so that the orientation factor $C = f(\alpha)$ and can be read directly from Figure 4.6d for calculation of contact wear volumes. By using SEM of wear tracks and by noting the depth of the scratches during and after scratching, it was observed that the CNT containing aluminum coating had increased elastic recovery after scratch [53]. CNT-rich regions were shown to have slightly lower coefficient of friction and increased recovery. A significant amount of information on wear at nanoscale level can be obtained from the analysis of nanoscratch data. Correlations between nanoscratch and macro-wear can be studied using this technique [54]. For applications like MEMS and thin coatings, information generated from nanoscratch can be directly applicable to the wear performance of the devices.

4.7.2 Macroscale/Bulk Mechanical Testing

The mechanical properties of bulk CNT-MMC are of great interest for their application as structural components. The mechanical properties depend on the macro features such as the strength of the matrix containing dispersed CNTs, presence of CNT clusters, and presence of pores. The different test methods for mechanical property evaluation of MM-CNT composites at large scale can be broadly divided into two categories: (1) tensile/compressive test and (2) tribological tests. Both test methods and the associated constraints are described in the following subsections.

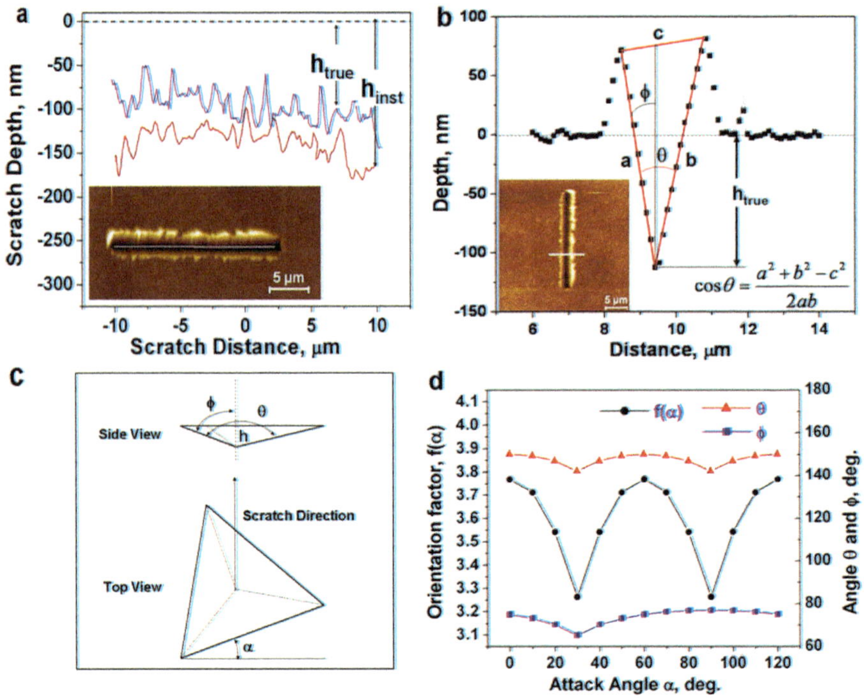

FIGURE 4.6 (a) SPM image and the depth profile along the scratch length showing the difference between h_{inst} and h_{true}, (b) SPM image of scratch and the depth profile across the scratch showing angle ϕ and θ used in volume calculations, (c) geometrically accurate diagram of the top view and side view of a Berkovich tip, and (d) the plot of the influence of orientation angle α on angles ϕ and θ and orientation factor C [53]. (Reproduced with permission from Elsevier).

4.7.2.1 Tensile/Compression Test

This section describes mechanical testing of bulk CNT-MMC samples using conventional/standard testing methods to obtain tensile and compressive strength properties. However, standardization of mechanical testing techniques and fabrication of large tensile samples of CNT-MMC have been difficult due to the difficulties associated with the processing of large test samples as discussed in Chapter 2. There are several reasons for use of non-standard samples for the mechanical property evaluation. In some cases, it is just impossible to fabricate samples with the standard geometries due to the nature of the process. For example, by processes like thermal spraying and electrodeposition, it is difficult to prepare freestanding samples with round geometries. Some processes cannot produce bulk samples like pressing and sintering and require additional processes like extrusion. Some processes are not meant for fabrication of large structural members like friction stir processing and high-pressure torsion. The discrepancy in the use of standard sample sizes results in a large scatter in data on the mechanical properties. The scatter in the mechanical properties is discussed in detail in Chapter 6. The bulk mechanical properties are highly dependent on the level of dispersion of CNTs, which is governed by the process used. In some cases, the grain sizes are affected by addition of CNTs, which results in additional strengthening. Flat and round samples have been employed for tensile tests. With flat samples, care must be taken in machining the curved edges so that the stress intensity is minimum; otherwise, it will lead to failure outside the gage length leading to wrong results. In the case of round samples produced by extrusion, similar care needs to be taken. Figure 4.7 shows an innovative method to fabricate tensile test specimens of Al-CNT composite using wire EDM from thin-walled cylinders fabricated by plasma spray forming [55]. The convex and concave stubs ensure that the load is applied axially along the specimen and there is no torsion generated. Usually a scaled-down sample size similar to standard samples is a good bet for obtaining good results. Compression tests have been carried out on cylindrical and cube specimens. Usually the ratio of the height to the diameter or edge is 1–2. Fatigue studies have not been carried out on CNT-MMCs but are important to look at. Fatigue tests can be carried out using the tensile testing equipment or a rotating beam test. The number of cycles to failure at a given stress is plotted, which would give information on the existence of a fatigue limit (load below which fatigue will not occur) for the samples.

4.7.2.2 Tribological Property Evaluation

It is important to study the tribological properties of CNT MMCs to evaluate the effect of CNTs on the improvement in wear and friction properties. CNTs play a dual role toward improvement of the wear performance in CNT-MMCs. CNTs not only strengthen the metal matrix but also reduce the coefficient of friction. Formation of graphitic layers due to shearing action during wear may produce a lubrication effect reducing further wear volume loss. Tribological properties of CNT-MMCs are evaluated using conventional techniques, like ball-on-disc [56,57], pin-on-disc [58–60],

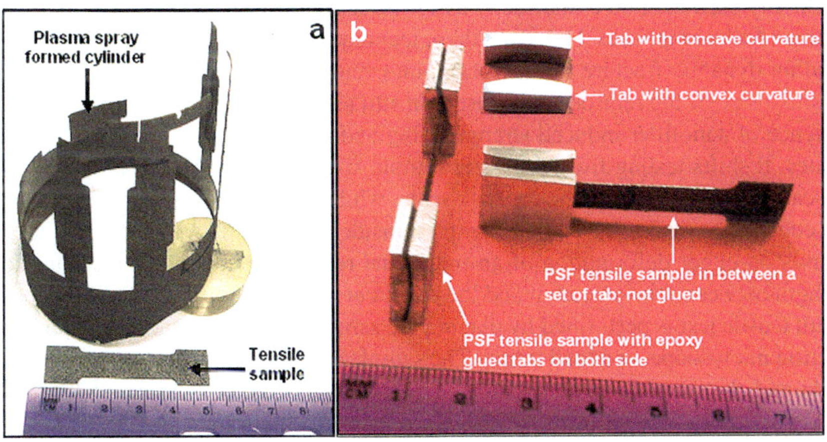

FIGURE 4.7 Photographs of (a) tensile samples machined out from the spray-formed cylindrical Al-CNT nanocomposite structures, and (b) attachment of aluminum tabs on the grip sections of tensile samples for facilitating the griping during the testing [55]. (Reproduced with permission from Elsevier).

and ring-on-block techniques [61–63]. A ball-on-disc test involves application of a constant load on a ball (made of hard steel, tungsten carbide, or aluminum oxide) placed on the CNT-MMC sample, while the sample is rotated at a selected speed. The tests can be carried out in air or in liquid medium. Wear phenomena is highly dependent on the type of mating surfaces and comparisons must be made only for the same mating surfaces and geometry. It is important to measure the wear of CNT-MMCs in terms of volume loss instead of weight loss. Weight loss does not provide a real sense of the structural degradation of composite materials due to difference in the density between the CNT and metal matrix. The volume of the wear track can be measured accurately using a contact (rastering sharp tip) or non-contact type (laser/optical-based) profilometer. Non-contact profilometers are preferred as they are not restricted by the correlation between tip size and wear track dimensions. The wear data is plotted as the volume of the material removed as a function of the sliding distance. The slope of the plot is a measure of the wear rate and indicates the improvement in the wear properties. The observation of wear track at high magnification can provide much information on the wear mechanism in CNT-MMCs [58–62]. Formation of oxides, graphitic layers, gauging by hard reaction products, micro cracking, and chip formation can be observed. SEM observation of morphology and size of wear debris also provides insight about the mechanism of wear, that is, abrasion or chipping, etc. Specific consideration is required for conducting wear tests on CNT-MMCs. However, care should be taken while handling the wear debris, which might contain loose CNTs, as the cytotoxic effect of CNTs is still being debated [64]. The effect of CNTs on tribological performance of CNT-MMCs is discussed in detail in Section 8.6.

4.7.3 In Situ Mechanical Testing

Post fracture surface observation using high-resolution electron microscopy is a commonly used method to indirectly asses the strengthening and deformation mechanism in bulk composites. However, these techniques are not ideally suited for understanding dynamic fracture and deformation events in nanocomposites, considering the small size of the reinforcement used. In situ testing is great tool to provide direct evidence of nanofiller-induced strengthening and deformation mechanisms in composites. It shifts testing from the basic sample sizes to in situ load frame. This way one can measure the properties at bulk scale, while performing real-time imaging on the constituent scale, by testing samples inside SEM/TEM/AFM. In situ testing focuses on either deformation of an individual reinforcement and/or the interfacial strength between reinforcement and the matrix at the scale of the reinforcement. The in situ tensile test is performed by placing the samples inside the vacuum chamber of electron microscope, in a load frame. The samples are loaded in displacement control mode and the imaging is done during the entire test duration. In situ SEM/TEM/AFM experiments have been intensively applied to investigate failure of CNT in polymer matrix composites (PMCs) [65,66] and ceramic matrix composites (CMCs) [67,68]. However, reports on metal matrix composites are limited [69–72]. All the studies reported so far have performed tensile testing inside SEM to look into the evidence of CNT strengthening in Al matrix. The in situ tensile test is performed by operating the tensile stage with the CNT-Al sample inside the SEM chamber. The tensile specimen is machined from the fabricated composite into a flat dumbbell shape [69–72]. The sample is placed into the two clamping heads of the tensile unit and then loaded at a given strain rate [69–72]. Imaging is done during the entire test duration. The tensile test can be manually paused during the test to capture high-quality SEM micrographs. In situ testing can concentrate on the deformation mechanism of an individual entity of the reinforcement phase or the interfacial strength between reinforcement and the matrix at the scale of the reinforcement or both simultaneously. In case of CNT-MMC matrix elongation, crack initiation as well as propagation, reinforcement pull out or fracture are observed in real time. Such observations are enormously helpful in validating the predicted mechanisms of strengthening in CNT reinforced metal matrix composites. The in situ studies reported so far have used tensile testing inside SEM to provide the evidence of CNT strengthening Al matrix [69–72]. Pull-out of several CNTs from the matrix and the deformation provides the direct evidence of load transfer/sharing as dominant strengthening mechanism in CNT-MMC composites.

4.8 THERMAL PROPERTIES

Thermo-physical properties like specific heat capacity, thermal conductivity, and coefficient of thermal expansion are very important in the selection of materials. They are critical for applications involving thermal management in electronic packaging. Specific heat capacities can be measured using a differential scanning calorimeter (DSC). The sample is enclosed in an aluminum pan and heated along with an empty reference pan at a given heating rate ~5 °C min^{-1}. The heat input to the cells is maintained such that the temperature difference between the sample and the reference is zero during the heating event. Reactions occurring in the sample require a lower (in the case of exothermic) or higher (in the case of endothermic) input of heat equal to the enthalpy of the reaction occurring, which is recorded with change in temperature. By subtracting the heating curve for an empty pan, the heat input for the sample can be calculated. Specific heat of unknown samples can be measured by comparing the heat required for a given temperature program with that required for a standard sample of known specific heat (e.g., sapphire). The samples should be in the form of powder or chips so that the thermal contact between the sample and the pan is good and without any thermal resistance. DSC has been used to show that Al reacts with CNTs to form Al_4C_3 at temperatures above 656 °C and the activation energy was calculated [73]. Such studies can be conducted on other metal matrices to study the onset of carbide formation in CNT-MMCs. DSCs are available with good sensitivity up to 700 °C. High temperature specific heat capacity can be measured using a drop calorimeter.

Thermal conductivity is widely measured using the laser flash technique. In this technique, a disc-shaped sample is irradiated with a short duration laser pulse on the front side. The temperature increase on the rear side is monitored using an infrared sensor. Depending on the thermal diffusivity of the sample, the heat diffuses throughout the sample over a period of time until the whole body has attained an equilibrium temperature. The time required for the temperature of the rear side to reach half of the maximum increase in temperature ($t_{1/2}$) can be related to the thermal diffusivity (α) and thickness of the sample (l) by Parker's formula [74]:

$$\alpha = 0.1388 \frac{l^2}{t_{1/2}} \tag{4.18}$$

The thermal conductivity (k) is then calculated from the definition of thermal diffusivity using the formula $k = \alpha\rho C_p$, where ρ is the geometric/bulk density and C_p is the specific heat capacity of the material. The specific heat capacity of the composite can be taken as the mole fraction weighted average of the specific heats of the matrix and the CNTs. Data for specific heat capacity of SWNT [75] and multiwall CNT [76] are available only up to low temperatures (<400 °C) and is expected to vary with the type of CNT. The specific heat capacity of the composite can be measured accurately using the DSC [7]. Care must be taken to ensure good heat transfer efficiency between the DSC pan and the sample; otherwise, the values obtained may be in error and will lead to error in the calculation of thermal conductivity. Some authors have assumed the specific heat capacity of CNT equal to graphite, which is a reasonable estimate at temperatures above room temperature, but will result in errors if used at temperatures below 0 °C [77]. Thermal conductivity of CNT-MMCs has not been focused on much except in a few studies [7,8]. An increase in thermal conductivity was noted in Al matrix with incorporation of CNT, which was explained in terms of enhanced phonon transport [78]. There is a large scope for study in this area and in the development of high thermal conductivity materials.

CTE is an important property for many applications. Large CTE is undesirable because it leads to generation of thermal stresses and dimensional changes. In addition, larger CTE leads to early failure during thermal cycling due to larger strains. CTE is measured using a dilatometer. The sample is in the form of a thin rod that is heated at a pre-determined rate and the length of the sample is determined as a function of the temperature from which the CTE is calculated. Addition of SWNTs has proved to be beneficial by reducing the thermal expansion coefficient of Al-composite [79]. Even multiwalled CNTs were also found effective in significantly reducing CTE of Al matrix, while the experimental data were in good agreement with the theoretical predictions by Schapery's model [80]. Mg-CNT also shows decrease in CTE with increasing content of CNT [81].

4.9 ELECTRICAL PROPERTIES

Electrical properties of CNT-reinforced materials are of great interest as CNTs possess high electrical conductivity. Carbon nanotubes have been added to polymers and ceramics to make them electrically conductive. However, in the case of conductive metals and alloys, CNTs are added primarily to improve the mechanical strength without affecting the electrical conductivity significantly [82]. With uniform sample geometries like wires or bars, the electrical conductivity can be measured by simply applying a voltage and measuring the current using a simple setup. However, for planar samples, a four-probe instrument is used. The probes are usually made of tungsten carbide and are sharp for precise contact. For better contact, the probe tips on the sample may be sputter-coated with gold. The probes are arranged in an array or at the corners of a square. In the case of an array, the outer probes are connected

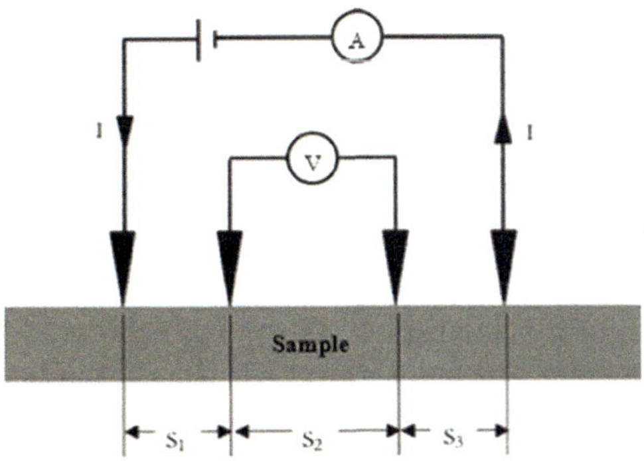

FIGURE 4.8 Schematic of the four-probe test for measuring electrical conductivity of flat specimens.

to the powder source, the inner ones are connected to a voltmeter, and then the current and voltage are measured. For a square geometry, opposite ends are connected to a voltmeter and a battery. For the array geometry as shown in Figure 4.8, the conductivity σ is estimated from the following equation:

$$\sigma = \frac{I}{2\pi V}\left[\frac{1}{S_1}+\frac{1}{S_2}-\frac{1}{(S_1+S_2)}-\frac{1}{(S_2+S_3)}\right] \tag{4.19}$$

where I and V are the current and voltage recorded and S_1, S_2, and S_3 are the distances between the probes as shown in Figure 4.8. The electrical resistivity of Al-CNT composites has been found to increase slightly with CNT content [21]. This is because the additional interfaces introduced and the aluminum carbide formed serve as scattering sites for electrons. Similarly, in the case of Ag-CNT composites, the electrical resistivity was found to increase slightly up to approximately 15 vol.% CNT [83]. Cu composites reinforced by SWNTs prepared by electrodeposition have been found to have better mechanical properties while maintaining similar electrical conductivity to OHFC Cu [84]. However, formation of CNT network in Al matrix, by tailoring the processing route, have offered improved electrical conductivity in the composite structure [85].

4.10 ELECTROCHEMICAL PROPERTIES

Electrochemical or corrosion properties of CNT-MMCs are important aspects but have not been studied extensively. It is expected that CNT-MMCs with improved mechanical properties should also have comparable or better corrosion properties in comparison with the metal matrix without reinforcement. Corrosion properties for CNT-MMCs are usually investigated using conventional techniques like potentiodynamic polarization tests and electrochemical impedance measurement.

The potentiodynamic polarization test can be carried out in acidic or salt solution using a potentiostat. The potentiodynamic plots elucidate the properties of the passive film and the corrosion rate. The corrosion rate in penetration units (mils per year, i.e., mpy) can be estimated from the corrosion current density by the following equation:

$$Corrosion\ rate(mpy) = \frac{i_{corr} X \varepsilon}{\rho X 96500} X393.7\left({}^{mils}\!/\!_{cm} \right) \quad (4.20)$$

where i_{Corr} is corrosion current density (A cm^{-2}), ε is equivalent weight of the alloy/composite, and ρ is density. The corrosion current density can be calculated from the potentiodynamic polarization test. Electrochemical impedance spectroscopy technique can also be utilized to predict the corrosion lifetime of an MM-CNT composite. A potentiostat with electrochemical impedance spectroscopy software can be used to generate and analyze the data. The impedance is expressed in terms of a constant (Z_0) and a phase shift (f) between potential and current, as follows.

$$Z = \frac{E(t)}{I(t)} = \frac{E_0 \cos(\omega t)}{I_0 \cos(\omega t - \varphi)} = Z_0 \frac{\cos(\omega t)}{\cos(\omega t - \varphi)} \quad (4.21)$$

The time for failure (t_{fail}) can be estimated applying the following equation.

$$t_{fail} = \theta \left[\ln \frac{Z_0 - Z_m}{Z_{fail} - Z_m} \right] \quad (4.22)$$

where Zm is the limiting impedance, θ is the characteristic decay times for the MM-CNT composite under consideration, and Z_{fail} is the impedance value during failure of the composite.

Apart from these techniques, measurement of weight loss or assessment of porosity after immersion in the corrosive medium for periodic intervals has also been adapted to quantify the corrosion behavior of CNT-MMCs [83,86]. The corrosion resistance in CNT-MMCs is mostly correlated with the chemical inertness of CNTs, which helps in forming a passive film on the composite surface. Wider plateau and lower current in the polarization curve [84] and more positive corrosion potential [73,74] denotes higher corrosion resistance for CNT-MMC in potentiodynamic polarization tests. On the other hand, larger semicircular loops in Nyquist plots for the impedance measurement indicate smaller impedance range and higher resistance to corrosion for CNT-MMC coating [83,86,87]. The current density vs. potential plot indicates higher pitting corrosion potential for CNT-MMCs [74]. Resistance to pitting or localized corrosion has been justified in terms of uniform distribution of CNT in matrix, which assists in micro-cell (between metal-CNT) formation throughout the matrix. As a result, uniform corrosion is favored over localized pitting [83]. However, no remark has been included on the contribution of such micron-sized cells on total corrosion resistance of the composites. At the same time, no account is found on the presence of CNT clusters in metal matrix that can make macro-sized cells with surrounding metals and aggravate the corrosion. The effect of CNT clustering on corrosion resistance could be assessed by calculation of corrosion potential and rate

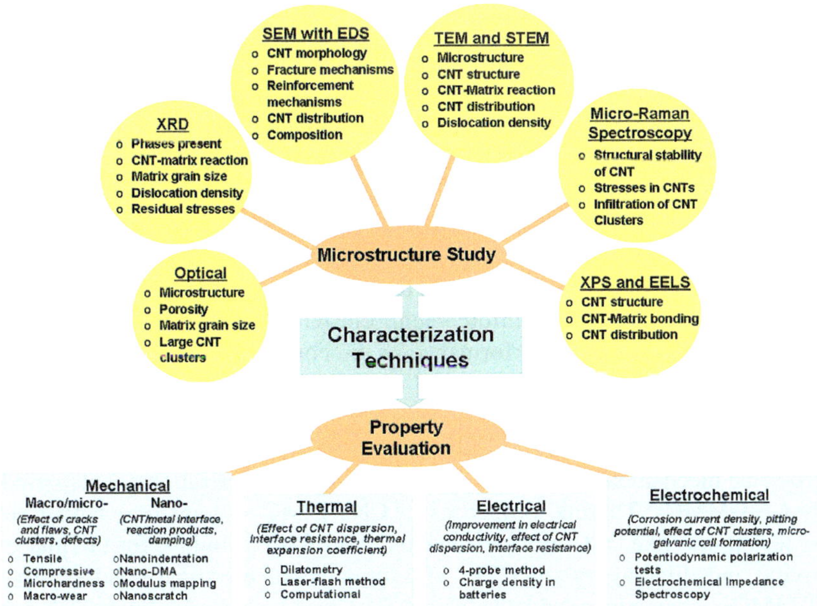

FIGURE 4.9 Summary of the characterization techniques available for studying MM-CNT composites.

as a function of fraction and size of CNT-clustering in metal matrix. A detailed account of the corrosion properties of CNT-metal matrix composites is provided in Chapter 8. The various tools available for characterizing CNT-MMCs and their unique features with respect to carbon nanotubes are summarized in Figure 4.9. With the help of suitable characterization, processing parameters can be modified to synthesize CNT-MMCs with improved CNT dispersion and mechanical, thermal, and electrochemical properties.

4.11 CHAPTER HIGHLIGHTS

In this chapter, the characterization techniques have been discussed in a manner to make the reader aware of the nuances and limitations while characterizing CNT-reinforced metal matrix composites. These characterization techniques can be divided broadly into two types. The first set of techniques provides information on the microstructural features in CNT-MMCs, whereas the second type of technique provides data on the mechanical, thermal, and corrosion properties of the composite. The different types of information about CNT-MMC systems that can possibly be obtained using each of these techniques have been summarized. The XRD technique can detect the metal-CNT reaction product formation, CNT orientation in matrix, effect of CNT reinforcement on grain size, and dislocation density in matrix. Raman spectroscopy is used as the most conclusive evidence about the presence of CNT in the matrix. SEM helps in finding out qualitative information about distribution of CNT and the nature of their interfacial bonding with metal matrix. HRTEM provides much information about CNT-MMCs, for example, wetting between the metal matrix and the CNTs, interfacial compound formation due to reaction between CNTs and the matrix, orientation relationship between carbide (reaction product) with CNT and matrix, and defects introduced in CNTs during processing. Information about filling of CNT with metal and transformation of CNT to any other carbon nano-structure could be obtained using EELS. The reaction products of carbon and metal are detected using XPS. Nano-indentation and nano-scratch techniques can identity and discriminate the effect of microstructural features, for example, CNT reinforcement, clusters, or reaction products on the mechanical and tribological performance of the composite structure. Nano-DMA and modulus mapping are very powerful characterization techniques to study the effect of CNT addition on damping property of the composite, as well as the spatial distribution of the elastic modulus in the matrix. Conventional techniques of macro-scale mechanical and tribological property characterization are also used to study CNT-MMCs. Thermal properties of CNT-MMC systems are characterized using techniques like DSC for heat capacity, laser flash technique for thermal conductivity, and dilatometry for CTE. Measurement of electrical conductivity on CNT-MMCs is mostly conducted using four-probe methods. Potentiodynamic polarization tests and electrochemical impedance measurement, being the two most common methods for corrosion property characterization, are popular for CNT-MMCs CNT-MMCs as well. Conventional techniques used for microstructural and property characterization need attention to account accurately the effect of CNT reinforcement in metal matrix. Such considerations, if any, have been mentioned in the respective sections of this chapter.

REFERENCES

1. Bian, Z., Pan, M. X., Zhang, Y., and W. H. Wang. 2002. Carbon-nanotube-reinforced $Zr_{52.5}Cu_{17.9}Ni_{14.6}Al_{10}Ti_5$ bulk metallic glass composites. *Appl. Phys. Lett.* 81: 4739–4741.
2. Pérez-Bustamante, R., Gómez-Esparza, C. D., Estrada-Guel, I., Miki-Yoshida, M., Licea-Jiménez, L., Pérez-García, S. A., and R. Martínez-Sánchez. 2009. Microstructural and mechanical characterization of Al–MWCNT composites produced by mechanical milling. *Mater. Sci. Eng. A* 502: 159–163.
3. Bakshi, S. R., Singh, V., Seal, S., and A. Agarwal, 2009. Aluminum composite reinforced with multiwalled carbon nanotubes from plasma spraying of spray dried powders. *Surf. Coat. Tech.* 203: 1544–1554.
4. Cao, A., Xu, C., Liang, J., Wu, D., and B. Wei. 2001. X-ray characterization on the alignment degree of carbon nanotubes. *Chem. Phys. Lett.* 344: 13–17.
5. Zhong, Z., Liu, B., Sun, L., Ding, J., Lin, J., and K. L. Tan. 2002. Dispersing and coating of transition metals Co, Fe and Ni on carbon materials. *Chem. Phys. Lett.* 362: 135–143.
6. Choi, H. J., Kwon, G. B., Lee, G. Y., and D. H. Bae. 2008. Reinforcement with carbon nanotubes in aluminum matrix composites. *Scripta Mater.* 59: 360–363.
7. Chu, K., Wu, Q., Jia, C., Liang, X., Nie, J., Tian, W., Gai, G., and H. Guo. 2010. Fabrication and effective thermal conductivity of multi-walled carbon nanotubes reinforced Cu matrix composites for heat sink applications. *Composites Sci. Tech.* 70: 298–304.

8. Shi, X. L., Yang, H., Shao, G. Q., Duan, Z. L., Yan, L., Xiong, Z., and P. Sun. 2007. Fabrication and properties of W–Cu alloy reinforced by multi-walled carbon nanotubes. *Mater. Sci. Eng. A* 457: 18–23.
9. Lahiri, D., Bakshi, S. R., Keshri, A. K., and A. Agarwal. 2009. Dual strengthening mechanisms induced by carbon nanotubes in roll bonded aluminum composites. *Mater. Sci. Eng. A* 523: 263–270.
10. Zare, H., Jahedi, M., Toroghinejad, M. R., Meratian, M., and M. Knezevic. 2016. Compressive, shear, and fracture behavior of CNT reinforced Al matrix composites manufactured by severe plastic deformation, *Mater. Des.* 106: 112–119.
11. Ferraro, J. R., Nakamoto, K., and C. W. Brown. 1994. *Introductory Raman Spectroscopy*. San Diego, CA: Academic Press.
12. Tuinstra, F., and J. L. Koenig. 1970. Raman spectrum of graphite. *J. Chem. Phys.* 53: 1126–1130.
13. Eklund, P. C., Holden, J. M., and R. A. Jishi. 1995. Vibrational modes of carbon nano tubes; spectroscopy and theory. *Carbon* 33: 959–972.
14. Rao, A. M., Richter, E., Bandow, S., Chase, B., Eklund, P. C., Williams, K. A., Fang, S., Subbaswamy, K. R., Menon, M., Thess, A., Smalley, R. E., Dresselhaus, G., and M. S. Dresselhaus. 1997. Diameter-selective raman scattering from vibrational modes in carbon nanotubes. *Science* 275: 187–191.
15. Belin, T., and F. Epron. 2005. Characterization methods of carbon nanotubes: a review. *Mater. Sci. Eng. B* 119: 105–118.
16. Cooper, C. A., Young, R. J., and M. Halsall. 2001. Investigation into the deformation of carbon nanotubes and their composites through the use of Raman spectroscopy. *Composites A* 32: 401–411.
17. Lourie, O., and H. D. Wagner. 1998. Evaluation of Young's modulus of carbon nanotubes by micro-Raman spectroscopy. *J. Mater. Res.* 13: 2418–2422.
18. Choi, H., Shin, J., Min, B., Park, J., and D. Bae. 2009. Reinforcing effects of carbon nanotubes in structural aluminum matrix nanocomposites. *J. Mater. Res.* 24: 2610–2616.
19. So, K. P., Lee, I. H., Duong, D. L., Kim, T. H., Lim, S. C., An, K. H., and Y. H. Lee. 2011. Improving the wettability of aluminum on carbon nanotubes, *Acta Mater.* 59: 3313–3320.
20. Yu, M. F., Lourie, O., Dyer, M. J., Moloni, K., Kelly, T. F., and R. S. Ruoff. 2000. Strength and breaking mechanism of multiwalled carbon nanotubes under tensile load. *Science* 287: 636–640.
21. Xu, C. L., Wei, B. Q., Ma, R. Z., Liang, J., Ma, X. K., and D. H. Wu. 1999. Fabrication of aluminum–carbon nanotube composites and their electrical properties. *Carbon* 37: 855–858.
22. Bacon, R. 1960. Growth, structure, and properties of graphite whiskers. *J. Appl. Phys.* 31: 283–290.
23. IIjima, S. 1991. Helical microtubules of graphitic carbon. *Nature* 354: 56–58.
24. Bakshi, S. R., Singh, V., Balani, K., McCartney, D. G., Seal, S., and A. Agarwal. 2008. Carbon nanotube reinforced aluminum composite coating via cold spraying. *Surf. Coat. Tech.* 202: 5162–5169.
25. Krasheninnikov, A. V., and F. Banhart. 2007. Engineering of nanostructured carbon materials with electron or ion beams. *Nature Mater.* 6: 723–733.
26. Ci, L., Ryu, Z., Jin-Phillipp, N. Y., and M. Rühle. 2006. Investigation of the interfacial reaction between multi-walled carbon nanotubes and aluminum. *Acta Mater.* 54: 5367–5375.
27. Kwon, H., Estili, M., Takagi, K., Miyazaki, T., and A. Kawasaki. 2009. Combination of hot extrusion and spark plasma sintering for producing carbon nanotube reinforced aluminum matrix composites. *Carbon* 47: 570–577.

28. Laha, T., Kuchibhatla, S., Seal, S., Li, W., and A. Agarwal. 2007. Interfacial phenomena in thermally sprayed multiwalled carbon nanotube reinforced aluminum nanocomposite. *Acta Mater.* 55: 1059–1066.
29. Bakshi, S. R., Keshri, A. K., Singh, V., Seal, S., and A. Agarwal. 2009. Interface in carbon nanotube reinforced aluminum silicon composites: thermodynamic analysis and experimental verification. *J. Alloys Compd.* 481: 207–213.
30. Lipecka, J., Andrzejczuk, M., Lewandowska, M., Janczak-Rusch, J., and K. J. Kurzydlowski. 2011. Evaluation of thermal stability of ultrafine grained aluminium matrix composites reinforced with carbon nanotubes. *Compos Sci. tech.* 71: 1881–1885.
31. Liu, X., Li, J., Liu, E., Li, Q., He, C., Shi, C., and N. Zhao. 2018. Effectively reinforced load transfer and fracture elongation by forming Al4C3 for in-situ synthesizing carbon nanotube reinforced Al matrix composites. *Mater. Sci. Eng. A.* 718: 182–189.
32. Eswai, A. M. K., Morsi, K., Sayed, A., Taher, M., and S. Lanka. 2011. The influence of carbon nanotube (CNT) morphology and diameter on the processing and properties of CNT-reinforced aluminium composites. *Composites A.* 42: 234–243.
33. Guo, B., Chen, B., Zhang, X., Cen, X., Wang, X., Song, M., Ni, S., Yi, J., Shen, T., and Y. Du. 2018. Exploring the size effects of Al4C3 on the mechanical properties and thermal behaviors of Al-based composites reinforced by SiC and carbon nanotubes, *Carbon.* 138: 135–224.
34. Fukuda, H., Kondoh, K., Umeda, J., and B. Fugentsu. 2011. Interfacial analysis between Mg matrix and carbon nanotubes in Mg–6 wt.% Al alloy matrix composites reinforced with carbon nanotubes. *Compos. Svi. Tech.* 71: 705–709.
35. Liang, J., Li, H., Tian, W., Li, X., Chao, X., and J. Wei. 2017. Fabrication and mechanical properties of CNTs/Mg composites prepared by combining friction stir processing and ultrasonic assisted extrusion. *J. Alloys Compd.* 728: 282–288.
36. Cho, S., Kikuchi, K., Miyazaki, T., Takagi, K., Kawasaki, A., and T. Tsukuda. 2010. Multiwalled carbon nanotubes as a contributing reinforcement phase for the improvement of thermal conductivity in copper matrix composites, *Scrita Mater.* 63: 375–378.
37. Peng, B., Locascio, M., Zapol, P., Shuyou, L., Mielke, S. L., Schatz, G. C., and H. D. Espinosa. 2008. Measurements of near-ultimate strength for multiwalled carbon nanotubes and irradiation-induced crosslinking improvements. *Nature Nanotech.* 3: 626–631.
38. Hwang, G. L., and K. C. Hwang. 2001. Breakage, fusion, and healing of carbon nanotubes. *Nano Lett.* 1: 435–438.
39. Demoncy, N., Stephan, O., Brun, N., Colliex, C., Loiseau, A., and H. Pascard. 1998. Filling of carbon nanotubes with metals by the arc discharge method: the key role of sulphur. *Eur. Phys. J. B-Cond. Matter Complex Sys.* 4: 147–157.
40. Jin-Phillipp, N. Y., and M. Rühle. 2004. Carbon nanotube/metal interface studied by cross-sectional transmission electron microscopy. *Phys. Rev. B* 70: 245421.
41. Stockli, T. 2003. Electron energy loss spectroscopy of carbon nanotubes and onions. In *Electron Microscopy of Nanotubes.* Z. L. Wang and C. Hui, Eds. Norwell, MA: Kluwer Academic Press.
42. Kim, I-Y., Lee, J-H., Lee, G-S., Baik, S-H., Kim, Y-J., and Y-Z. Lee. 2009. Friction and wear characteristics of the carbon nanotube–aluminum composites with different manufacturing conditions. *Wear* 267: 593–598.
43. http://www.hysitron.com/page_attachments/0000/0636/Hysitron_Probe_Selection_Guide.pdf.
44. Oliver, W. C., and G. M. Pharr. 1992. An improved technique for determining hardness and elastic modulus using load and displacement sensing indentation experiments. *J. Mater. Res.* 7: 1564–1583.
45. Giannakopoulos, A. E., and S. Suresh. 1999. Determination of elastoplastic properties by instrumented sharp indentation. *Scripta Mater.* 40: 1191–1198.

46. Chen, Y., Laha, T., Balani, K., and A. Agarwal. 2008. Nanomechanical properties of hafnium nitride coating. *Scripta Mater.* 58: 1121–1124.
47. Rzepiejewska-Malyska, K. A., Buerki, G., Michler, J., Major, R. C., Cyrankowski, E., Asif, S. A. S., and O. L. Warren. 2008. In situ mechanical observations during nanoindentation inside a high-resolution scanning electron microscope. *J. Mater. Res.* 23: 1973–1979.
48. http://www.hysitron.com/products/pi-series-tem-picoindenter/.
49. Balani, K., Brito, F. C., Kos, L., and A. Agarwal. 2009. Melanocyte pigmentation stiffens murine cardiac tricuspid valve leaflet. *J. Roy. Soc. Interface.* doi:10.1098/rsif.2009.0174.
50. Balani, K., and A. Agarwal. 2008. Damping behavior of carbon nanotube reinforced aluminum oxide coatings by nanomechanical dynamic modulus map-ping. *J. Appl. Phys.* 104: 063517.
51. Asif, S. A. S., Wahl, K. J., and R. J. Colton. 1999. Nanoindentation and contact stiffness measurement using force modulation with a capacitive load-displacement transducer. *Rev. Sci. Instr.* 70: 2408–2413.
52. Laha, T., and A. Agarwal. 2008. Effect of sintering on thermally sprayed carbon nanotube reinforced aluminum nanocomposite. *Mater. Sci. Eng. A* 480: 323–332.
53. Bakshi, S. R., Lahiri, D., Patel, R. R., and A. Agarwal. 2010. Nanoscratch behavior of carbon nanotube reinforced aluminum coatings. *Thin Solid Films* 518: 1703–1711.
54. Balani, K., Harimkar, S. P., Keshri, A., Chen, Y., Dahotre, N. B., and A. Agarwal. 2008. Multiscale wear of plasma-sprayed carbon-nanotube-reinforced aluminum oxide nanocomposite coating. *Acta Mater.* 56: 5984–5994.
55. Laha, T., Chen, Y., Lahiri, D., and A. Agarwal. 2009. Tensile properties of carbon nanotube reinforced aluminum nanocomposite fabricated by plasma spray forming. *Composites: A* 40: 589–594.
56. Chen, X. H., Peng, J. C., Li, X. Q., Deng, F. M., Wang, J. X., and W. Z. Li. 2001. Tribological behavior of carbon nanotubes—reinforced nickel matrix composite coatings. *J. Mater. Sci. Lett.* 20: 2057–2060.
57. Liu, B., Zeng, Z., and Y. Lin. 2009. Mechanical properties of hard Cr–MWNT composite coatings. *Surf. Coat. Technol.* 203: 3610–3613.
58. Tu, J. P., Zhu, L. P., Chen, W. X., Zhao, X. B., Liu, F., and X. B. Zhang. 2004. Preparation of Ni-CNT composite coatings on aluminum substrate and its friction and wear behavior. *Trans. Nonferrous Met. Soc. China* 14(5): 880–884.
59. Tu, J. P., Yang, Y. Z., Wang, L. Y., Ma, X. C., and X. B. Zhang. 2001. Tribological properties of carbon-nanotube-reinforced copper composites. *Tribol. Lett.* 10(4): 225–228.
60. Zhou, S. M., Zhang, X. B., Ding, Z. P., Min, C. V., Xu, G. L., and W. M. Zhu. 2007. Fabrication and tribological properties of carbon nanotubes reinforced Al composites prepared by pressureless infiltration technique. *Composites A* 38: 301–306.
61. Chen, W. X., Tu, J. P., Wang, L. Y., Gan, H. Y., Xu, Z. D., and X. B. Zhang. 2003. Tribological application of carbon nanotubes in a metal-based composite coating and composites. *Carbon.* 41: 215–222.
62. Dong, S. R., Tu, J. P., and X. B. Zhang. 2001. An investigation of the sliding wear behavior of Cu-matrix composite reinforced by carbon nanotubes. *Mater. Sci. Eng.* 313: 83–87.
63. Chen, X., Zhang, G., Chen, C., Zhou, L., Li, S., and X. Li. 2003. Carbon nanotube composite deposits with high hardness and high wear resistance. *Adv. Eng. Mater.* 5(7): 514–518.
64. Fiorito, S. 2008. *Carbon Nanotube: Angels or Demons?* Singapore: Pan Stanford Publishing.
65. Wagner, H. D., Lourie, O., Feldman, Y., and R. Tenne. 1998. Stress-induced fragmentation of multiwall carbon nanotubes in a polymer matrix. *Appl. Phys. Lett.* 72: 188.

66. Deng, F., Ogasawara, T., and N. Takeda. 2007. Tensile properties at different temperature and observation of micro deformation of carbon nanotubes-poly(ether ether ketone) composites. *Compos. Sci. Technol.* 67: 2959–2964.
67. Yamamoto, G., Shirasu, K., Hashida, T., Takagi, T., Suk, J-W., An, J., Richard, R. P., and R. S. Ruoff. 2011. Nanotube fracture during the failure of carbon nanotube/alumina composites. *Carbon.* 49: 3709–3716.
68. Xia, Z., Riester, L., Curtin, W. A., Li, H., Sheldon, B. W., Liang, J., Chang, B., and J. M. Xu. 2004. Direct observation of toughening mechanisms in carbon nanotube ceramic matrix composites. *Acta Mater.* 52: 931–944.
69. Boesl, B., Lahiri, D., Behdad, S., and A. Agarwal. 2014. Direct observation of carbon nanotube induced strengthening in aluminum composite via in situ tensile tests. *Carbon.* 69: 79–85.
70. Chen, B., Li, S., Imai, H., Jia, L., Umeda, J., Takahashi, M., and K. Kondoh. 2015. Load transfer strengthening in carbon nanotubes reinforced metal matrix composites via in-situ tensile tests. *Compos. Sci. Technol.* 113: 1–8.
71. Chen, B., Shen, J., Ye, X., Imai, H., Jia, L., Umeda, J., Takahashi, M., and K. Kondoh, 2017. Solid-state interfacial reaction and load transfer efficiency in carbon nanotubes (CNTs)-reinforced aluminum matrix composites. *Carbon* 114: 198–208.
72. Chen, B., Kondoh, K., Umeda, J., Li, S., Jia, L., and J. Li. 2019. Interfacial in-situ Al2O3 nanoparticles enhance load transfer in carbon nanotube (CNT)-reinforced aluminum matrix composites. *J. Alloys Compd.* 789: 25–29.
73. Deng, C. F., Zhang, X. X., Wang, D. Z., and Y. X. Ma. 2007. Calorimetric study of carbon nanotubes and aluminum. *Mater. Lett.* 61: 3221–3223.
74. Parker, W. J., Jenkins, R. J., Butler, C. P., and G. L. Abbott. 1961. Flash method of determining thermal diffusivity, heat capacity, and thermal conductivity. *J. Appl. Phys.* 32: 1679–1684.
75. Hone, J., Batlogg, B., Benes, Z., Johnson, A. T., and J. E. Fischer. 2000. Quantized phonon spectrum of single-wall carbon nanotubes. *Science* 289: 1730–1733.
76. Yi, W., Lu, L., Zhang, D. L., Pan, Z. W., and S. S. Xie. 1999. Linear specific heat of carbon nanotubes. *Phys. Rev. B* 59: R9015–R9018.
77. Bakshi, S. R., Balani, K., and A. Agarwal. 2008. Thermal conductivity of plasma-sprayed aluminum oxide-multiwalled carbon nanotube composites. *J. Am. Ceram. Soc.* 91: 942–947.
78. Wu, J., Zhang, H., Zhang, Y., and X. Wang. 2012. Mechanical and thermal properties of carbon nanotube/aluminum composites consolidated by spark plasma sintering, *Mater. Des.* 41: 344–348.
79. Tang, Y., Cong, H., Zhong, R., and H-M. Cheng. 2004. Thermal expansion of a composite of single-walled carbon nanotubes and nanocrystalline aluminum. *Carbon* 42: 3260–3262.
80. Liu, Z. Y., Xiao, B. L., Wang, W. G., and Z. Y. Ma. 2012. Elevated temperature tensile properties and thermal expansion of CNT/2009Al composites, *Compos. Sci. Tech.* 72: 1826–1833.
81. Goh, C. S., Wei, J., Lee, L. C., and M. Gupta. 2006. Development of novel carbon nanotube reinforced magnesium nanocomposites using the powder metallurgy technique. *Nanotechnology.* 17: 7–12.
82. Feng, Y., Yuan, H. L., and M. Zhang. 2005. Fabrication and properties of silver-matrix composites reinforced by carbon nanotubes. *Mater. Char.* 55: 211–218.
83. Chen, X. H., Chen, C. S., Xiao, H. N., Cheng, F. Q., Zhang, G., and G. J. Yi. 2005. Corrosion behavior of carbon nanotubes–Ni composite coating. *Surf. Coat. Technol.* 191: 351–356.
84. Yang, Y. L., Wang, Y. D., Ren, Y., He, C. S., Deng, J. N., Nan, J., Chen, J. G., and L. Zuo. 2008. Single-walled carbon nanotube-reinforced copper composite coatings prepared by electrodeposition under ultrasonic field. *Mater. Lett.* 62: 47–50.

85. Shin, S. E., Choi, H. J., and D. H. Bae. 2012. Electrical and thermal conductivities of aluminum-based composites containing multi-walled carbon nanotubes. *J. Composites Mater.* 47: 2249–2256.
86. Praveen, B. M., Venkatesha, T. V., Naik, Y. A., and K. Prashatha. 2007. Corrosion studies of carbon nanotubes–Zn composite coating. *Surf. Coat. Technol.* 201: 5836–5842.
87. Yang, Z., Xu, H., Shi, Y. L., Li, M. K., Huang, Y., and H. L. Li. 2005. The fabrication and corrosion behavior of electroless Ni–P-carbon nanotube composite coatings. *Mater. Res. Bull.* 40: 1001–1009.

5 Carbon Nanotube Reinforced Metal Matrix Composite (CNT-MMC) Systems

Several metal matrices have been reinforced with carbon nanotubes in order to study the feasibility of composite fabrication and their properties for potential applications. Figure 1.5 in Chapter 1 showed the pie chart of the percentage of publications on various metal matrices. Based on the number of total articles published in journals, it is observed that Ni-CNT composites have been researched the most for non-structural applications, whereas Cu-CNT and Al-CNT composites have received major attention for structural applications. Other matrices like Mg, Ti, and Si have also received some attention. Attempts have also been made to incorporate CNTs in novel materials like metallic glasses. The aim of this chapter is to provide the reader an idea of the various attempts made by researchers in the area of CNT-MMCs. In this chapter, we look into specific metal-CNT systems and summarize the majority of the work carried out to date. This work has been summarized in the form of exhaustive tables for each metal-CNT system. These tables list important information for each metal-CNT system such as: (1) composite processing technique, (2) CNT dispersion method used, (3) CNT content and quality of dispersion, (4) reaction at the CNT/matrix interface, if any and (5) material properties. Since the largest intended application of the CNT-MMCs is for structural purposes, most of the researchers have reported the mechanical properties of these composites. Hence, mechanical properties (elastic modulus, hardness, yield/tensile strength, strain to failure) of CNT-MMCs are also tabulated to highlight the role of CNT addition. It was discussed in Chapter 4 that mechanical testing and properties are highly dependent on the processing methods, which often dictates the shape and dimensions of the test samples made out of CNT-MMCs. The samples utilized for the mechanical property evaluation are often non-standard in size. The sample size used to evaluate the mechanical properties of the metal matrices with and without CNT reinforcements is also outlined in the table. The table includes a separate column that outlines the improvement in other properties, namely electrical, thermal, tribological, etc. A detailed discussion on the improvement of other properties such as wear, and thermal and electrical properties, is provided in Chapter 8. A short discussion for each CNT-MMC system is also included to highlight the most significant studies related to the mechanical properties. This chapter also provides the reader a clear understanding of the CNT-MMCs with the scope for further research.

5.1 ALUMINUM-CARBON NANOTUBE SYSTEM

Aluminum and its alloys, being the most abundantly used non-ferrous structural materials, were the first choice for reinforcement with carbon nanotubes. Table 5.1 is a compendium of the research work carried out on Al-CNT composites. The first publication on metal-CNT system was on Al-CNT in 1998 [1]. The powder metallurgy route has been used extensively due to ease of dispersing the CNTs within the aluminum matrix. A secondary consolidation by deformation processing is advantageous in obtaining higher densities and improved CNT distribution. Hot extrusion has been used extensively because it results in highly dense composites. In addition, hot extrusion has been shown to break CNT clusters and even align them along the extrusion direction. Several methods have been used for dispersing CNTs in aluminum powder such as dispersing in liquid medium by ultrasonic mixing, blending, mechanical milling, spray drying, NSD, and synthesizing CNTs on powder by CVD. Ball milling is found to result in moderate to good dispersion with poor to excellent mechanical properties.

Most of the studies (>60%) have used ball milling for the powder preparation. Dispersion of the CNTs and the presence of porosity are two major factors that affect the mechanical properties of the CNT-MMC. CNT clusters are very detrimental to the properties because they act as notches and areas of stress concentrations. Processes using inefficient methods for dispersion like mixing by stirring in alcohol [1], blending by roll mixing [2], etc. result in formation of CNT clusters in the final product. A decrease in the tensile strength by 9% was observed in a 10 vol.% CNT composite in which stirring was used to disperse the CNTs in the powder [1]. Further annealing the composites resulted in softening of the sample without CNTs, while the ones containing CNTs showed negligible softening. CNT reinforcement in an Al alloy (AA 7075) matrix has also been found to improve the impact and flexural strength by a significant 90% and 125%, respectively [3]. Fatigue strength is also enhanced by incorporating CNT into Al alloy (Al2024) matrix, which is reported to be due to crack bridging by CNTs, during failure [4].

Agrawal's group has studied synthesis of Al-CNT composites by thermal spray processes, namely plasma spraying [2,5], HVOF [6,7], and cold spraying [8]. Various aspects like interfacial phenomena, quantification of CNT dispersion, and mechanical properties have been studied at length. Al-CNT composite prepared by plasma spraying of *blended* powders also showed very small increase in the tensile strength (4%) for 12.5 vol.% CNT addition [9]. Tensile tests on plasma spray formed bulk tensile samples using *spray-dried* powder and indicated that there is a 23% and 17% decrease in elastic modulus, 24% and 25% decrease in fracture strength, and 29% and 45% decrease in fracture strain by addition of 5 and 10 wt.% CNTs [10]. Improper milling also leads to poor dispersion and has been shown to reduce the strength by 52% for just 2.4 vol.% CNT addition [11]. Deterioration in the hardness in shock wave consolidated Al-CNT composite with up to 5 vol.% CNT addition has been observed [12]. All of these studies showed the presence of CNT clusters in the microstructure, which indicates that they are detrimental to the properties of the composites.

TABLE 5.1
Table of the Summary of the Work Carried Out in an Al-CNT System

CNT Content	Composite Processing Technique	Tensile Test Sample Size	Mechanical Properties (UTS, YS, E, Strain to Failure, Hardness)		CNT Dispersion and Interface	Other Properties	Ref.
			With CNT	Without CNT			
5 and 10 vol.%	Stirred in ethanol (0.5 h, 300 rpm), dried in vacuum and preheated (600°C, 1.5 h, vacuum), compacted (100 MPa), hot extruded (500°C, extrusion ratio—25)	Diameter.: 3 mm Gauge length: 15 mm	σ_{TS}: at 873K ~ 80 MPa (for 0–100 h of annealing)	σ_{TS}: (0–100 h) 90–40 MPa (for 0–100 decreasing with annealing time)	Non-homogeneous distribution of CNT in matrix is reported	—	[1]
1, 4, and 10 wt.%	Hand ground (30 min), hot pressed (525°C, 25 MPa)	—	—	—	Agglomerate of CNT at grain boundary is reported	Electrical resistivity drops at low temperature	[97]
5wt.% SWNT	Ultrasonicated in alcohol (30 min), pressed (1.5 GPa), hot pressed (260–480°C, 1 GPa, 30 min)	–	H: 2.89 GPa	H: 1.6 GPa	–	–	[99]
10 wt.%	Blended and mixed, ball milled, plasma sprayed (28 kV)	–	H: 146 VHN	85 VHN	Dangling CNT and entrapment of CNT in successive splats is found	–	[2]
0.8 and 1.5 vol.%	Precursor with natural rubber, roll milled, compressed in mold, heated (800°C, 1 h, nitrogen atmosphere)	Compression test, dimensions not mentioned	Al-1.6 vol.% CNT $\sigma_{YS(T)}$: 225 MPa	$\sigma_{YS(T)} < 50$ MPa	Good dispersion and good adhesion of CNT in matrix is reported	–	[14]

(Continued)

TABLE 5.1 ***(Continued)***

Table of the Summary of the Work Carried Out in an Al-CNT System

CNT Content	Composite Processing Technique	Tensile Test Sample Size	Mechanical Properties (UTS, YS, E, Strain to Failure, Hardness)		CNT Dispersion and Interface	Other Properties	Ref.
			With CNT	Without CNT			
0–20 vol.% SWNT	Mixed, compacted (1.5 GPa), room temperature, consolidated at high temperature and pressure (380°C, 1 GPa, 30 min, vacuum)	–	–	–	Homogeneous dispersion of CNT in matrix is reported	Coefficient of thermal expansion reduces with SWNT addition	[100]
10 wt.%	Blended and mixed, ball milling (48 h), high velocity oxy-fuel and plasma sprayed (28 kV)	–	H (PSF): 1.57 GPa H (HVOF): 1.72 GPa	H:1.0 GPa (conventionally casted)	CNT are found coated with Al-Si resulting into good bonding	–	[6]
0.5–2 vol.% CNT 1–2 vol.% SWNT	Mixed by ball milling (5 min, 200 rpm), compacted (120 kN), sintered (580°C, 45 min), and hot extruded (560°C)	–	Al: 2 vol.% CNT $\sigma_{YS(T)}$: 99MPa σ_{TS}: 150 MPa Al: 2 vol.% SWNT $\sigma_{YS(T)}$: 98.7 MPa σ_{TS}: 181 MPa	–	–	–	[101]
–	Al deposited by magnetic sputtering on vertically aligned film of CNT	–	–	–	Al_4C_3 formation at interfaces is found	–	[91]
10 wt.%	Blended and mixed by ball milling (48 h), HVOF (high velocity oxy fuel forming) and plasma sprayed (28 kV)	–	–	–	Formation of SiC at interface is found	–	[7]

5 wt.%	CVD growth of CNT on Al powder (using Ni catalyst), pressed (600 MPa), sintered (640°C, 2 h), re-pressed (2 GPa)	–	H: 0.65 GPa σ_{TS}: 398 MPa	H: 0.15 GPa σ_{TS}: 140 MPa	Homogeneous dispersion and good interfacial bonding of CNT with matrix is reported	–	[13]
2 wt.%	Ball milling (1–48 h, 200 rpm, argon atmosphere)	–	–	–	Good distribution after 48 h of milling, CNT embedded in deformed Al particle	–	[102]
0–20 vol.%	Al-Mg-CNT-mixed by ball milling (7 h, 300 rpm, argon atmosphere), pressed to preform, pressureless infiltration of Al, (800°C, 5 h, nitrogen atmosphere)	–	cx gvioAl: 15 vol.% CNT H(Brinell): 175	H(Brinell): 106	Fully dispersed and embedded CNTs in matrix are found	Decrease in coefficient of friction and wear loss with CNT addition	[103]
1 wt.%	Mixed and ultrasonicated, cold isostatically pressed (300 MPa, 5 min), hot extruded in Al case (460°C, extrusion ratio—25)	–	E: 102.2 GPa σ_{TS}: 521.7 MPa %ε: 17.9 MPa H: 136 MPa	E 72.3 GPa σ_{TS}: 384.5 MPa %ε: 18.8 H: 104	Homogeneous distribution, good bonding with Al matrix, short pull out of CNTs are found, resulting in better elongation	–	[104]
2 and 5 vol.%	Layers of pure Al, and Al-CNT, green compacted, shock wave consolidated (7 GPa initial pressure)	D638 type: V ASTM Gauge length: 9 mm	Al-2 vol. % CNT H: 39 HRE* Al-5% vol. CNT H: 33 HRE* σ_{TS}: 20 MPa, %ε: 2 (*Rockwell hardness in E scale)	E: 40 HRE* $\sigma_{YS(T)}$: 120 MPa %ε: 6.5	Non homogeneous dispersion, CNT agglomerates, weak bonding with matrix is reported	–	[12]
5 wt.%	Mixed and ultrasonicated (30 min), dried (120°C, vacuum), cold isostatic pressed	–	–	–	CNT react with Al to form Al_4C_3 above T_m of Al (656.3°C)	–	[90]

(Continued)

TABLE 5.1 (*Continued*)
Table of the Summary of the Work Carried Out in an Al-CNT System

CNT Content	Composite Processing Technique	Tensile Test Sample Size	Mechanical Properties (UTS, YS, E, Strain to Failure, Hardness)		CNT Dispersion and Interface	Other Properties	Ref.
			With CNT	Without CNT			
0 – 2 wt.%	Mixed and ultrasonicated, cold isostatically pressed (300 MPa, 5 min), hot extruded in Al case (460°C, extrusion ratio = 25)	Not mentioned	Al-1 wt.% CNT σ_{TS}: 520 MPa E: 102 GPa %ε: 19%	σ_{TS}: 385 MPa E: 72 GPa %ε: 20	Reported homogeneous dispersion of CNTs with good bonding, bridging across crack, short pullouts, form Al_4C_3 phase at interface (~656.3°C)	–	[105]
2–5 wt.%	Mechanically alloyed using ball milling	–	–	–	After 48 h of milling, CNT gets embedded in plastically deformed Al particles	–	[106]
1 wt.%	Mixed and ultrasonicated, cold isostatically pressed (300 MPa, 5 min), hot extruded in Al case (460°C, extrusion ratio = 25)	Damping specimen, gauge side 1 × 7 × 38 mm	Damping capacity storage modulus: 82.3 GPa (at 400°C) 98 GPa (at room temperature)	Damping capacity storage modulus: 71 GPa (at room temperature)	Reported homogeneous dispersion of CNTs with good bonding, bridging across crack, short pullouts, form Al_4C_3 phase at interface (~656.3°C)	–	[107]
1 wt.%	Mixed and ultrasonicated, cold isostatically pressed (300 MPa, 5 min), hot extruded in Al case (460°C, extrusion ratio = 25)	Dog bone shape sample Gauge length: 15 mm	H: 136 MPa $\sigma_{YS(T)}$: 336 MPa σ_{TS}: 474 MPa E : 88 GPa	HV : 104 MPa $\sigma_{YS(T)}$: 289 MPa σ_{TS}: 384 MPa E: 71 GPa	Reported homogeneous dispersion of CNTs with good bonding, bridging across crack, short pullouts, and formation of Al_4C_3 phase at interface (~656.3°C)	–	[108]

5 vol.%	Mixed and ultrasonicated, cold isostatically pressed (300 MPa, 5 min), hot extruded in Al case (460°C, extrusion ratio = 25)	–	–	–	CNT transforms to nano Al_4C_3 needle shaped particle, mainly found at grain boundaries of Al particle	–	[109]
1 wt.%	Mixed and ultrasonicated, cold isostatically pressed (300 MPa. 5 min), hot extruded in Al case (460°C, extrusion ratio = 25)	–	–	–	CNTs found homogeneously embedded in matrix, short pullout length	Coefficient of thermal expansion decreases with CNT addition	[110]
0.5, 1.0, and 2.0 wt.%	Mixed by blending (300 rpm), encapsulated, hot rolled (50% reduction per pass), vacuum sintered (300°C, 3 h), air sintered (550°C, 45 min)	Mechanical properties were measured with resonance frequency	Al-0.5 wt.% CNT $\sigma_{YS(T)}$: 100 MPa σ_{TS}: 150 MPa E: 60 GPa	$\sigma_{YS(T)}$: 70 MPa σ_{TS}: 130 MPa E: 50 GPa	CNT dispersion worsens with increase in CNT amount and CNT clusters are formed	–	[11]
–	Blended and ball milled, compacted, sintered, hot extruded (560°C)	–	–	–	CNTs are pinned at subgrain boundaries	E of CNTs has been calculated by assessing bending of CNTs at subgrain boundaries	[111]
4 vol.%	Mixed and ball-milled (6/12 h), hot extruded (470°C, extrusion ratio = 15)	Rectangular 2:1 length to width ratio	σ_{TS}: 400 MPa E: 104.19 GPa	σ_{TS}: 350 MPa E: 70.05 GPa	CNTs are found well dispersed, deeply embedded, aligned along the extrusion direction	–	[112]
5 wt.% SWNT	Mixed and ultrasonicated (5 min), high pressure torsion to form composite disks (2.5 GPa, 1 rpm, 30 turns)	Dog bone shape sample Length: 1mm Width: 1mm Thickness: 0.5mm	H: 76 VHN σ_{TS}: ~215 MPa	Hardness: H: 43 VHN σ_{TS}: ~150 MPa	SWNTs are found sitting at grain boundaries	–	[113]

(*Continued*)

TABLE 5.1 (*Continued*)
Table of the Summary of the Work Carried Out in an Al-CNT System

CNT Content	Composite Processing Technique	Tensile Test Sample Size	Mechanical Properties (UTS, YS, E, Strain to Failure, Hardness)		CNT Dispersion and Interface	Other Properties	Ref.
			With CNT	Without CNT			
–	CNT preform fabricated by sintering (2500°C, 20 min, argon atmosphere), infiltration of liquid metal in preform by squeeze casting	NA	NA	NA	Claimed good wetting, i.e., good reinforcement and good dispersion	–	[114]
0.5 wt.%	Al-CNT composite powder synthesized by spray drying, cold sprayed (pressure difference = 2.9 MPa)	–	–	–	–	–	[115]
0.5, 1 wt.%,	Al-CNT composite powder synthesized by spray drying, cold sprayed (pressure difference – 2.9 MPa)	–	Al-0.5 wt.% CNT E-68.6 GPa Al-1 wt.%CNT E-68.1 GPa	–	Uniform dispersion of CNT in Al matrix is claimed	–	[8]
–	Friction stir welded (1500 – 2500 rpm)	–	H – 213 HVN	H: 140HVN	Good reinforcement but not completely uniform distribution is reported	–	[116]
5, 10 wt.%	Al-CNT composite powder synthesized by spray drying, plasma sprayed (22 kW)	–	Al-10 wt.% CNT E: 125 GPa H (nano): 2.89 GPa H (micro): 2.10 GPa	E: 90 GPa H (nano): 1.61 GPa H (micro): 0.87 GPa	Two phase microstructure – matrix having good distribution of CNT and CNT-rich clusters; aluminium carbide (Al_4C_3) forms at CNT-AL interface	–	[5]

0 6 vol.%	Al-CNT composite powder synthesized by spray drying, cold sprayed (pressure difference: 2.9 MPa)	–	–	–	CNTs are reported uniformly dispersed in Al-Si matrix – the dispersion is quantified in this study	–	[117]
5, 10 wt.%	Al-CNT composite powder synthesized by spray drying, plasma sprayed (22 kW)	–	–	–	Study on interfacial reaction: formation of Al_4C_3 at interface with low Si content and SiC at higher Si content	–	[118]
5 vol.%	Composite powder prepared by NSD technique using natural rubber as precursor, SPS (600°C, 50 MPa, 20 min), hot extruded (400°C, 500 kN, extrusion ratio – 20)	Diameter: 3 mm Gauge length ~ 15 mm	σ_{TS}: 194 MPa	σ_{TS}: 85 MPa	Homogeneous and good dispersion of CNT in Al matrix, CNTs are oriented in matrix Al-carbide formation at Al-CNT interface	–	[15]
10 wt.%	Blended ball milled (48 h), plasma sprayed (28 kW)	Gauge length: 26 mm Width: 8 mm Thickness: 0.635 mm	E: 120.4 GPa σ_{TS}: 83.1 GPa Strain to failure: 8.8×10^{-4}	E: 57.5 GPa σ_{TS}: 79.8 GPa Strain to failure: 19.2×10^{-4}	Some degree of CNT clustering and inhomogeneous distribution in matrix Thin layer of SiC formed at CNT-Al interface	–	[9]
1.5, 3, 4.5, 6 vol.%	Ball milled (500 rpm, argon atmosphere), hot rolled (480°C, 12% reduction)	Gauge length: 12.5 mm Thickness: 1.5 mm 3-Point Bend Test Length: 15 mm Width: 3.75 mm Thickness: 1.88 mm	Al: 4.5 vol.% CNT E: 110.05 GPa $\sigma_{YS(T)}$: 610 MPa K_{IC} : 60.79 MPA. $mm^{1/2}$	E: 70.063 GPa $\sigma_{YS(T)}$: 262 MPa K_{IC} : 33.22 MPA. $mm^{1/2}$	CNTs reported uniformly dispersed and embedded in Al matrix and aligned along rolling direction	–	[16]

(Continued)

TABLE 5.1 (*Continued*)
Table of the Summary of the Work Carried Out in an Al-CNT System

CNT Content	Composite Processing Technique	Tensile Test Sample Size	Mechanical Properties (UTS, YS, E, Strain to Failure, Hardness)		CNT Dispersion and Interface	Other Properties	Ref.
			With CNT	Without CNT			
0–2 wt.%	Mechanically milled (5 h, argon atmosphere), pressureless sintered (550°C, 3 h, vacuum), hot extruded (500°C , extrusion ratio – 16)	Diameter: 10 mm Gauge length: 30 mm	Al: 2 wt.% CNT $\sigma_{YS(T)}$: 189.2 MPa σ_{TS}: 243 MPa H – 73 VHN	$\sigma_{YS(T)}$: 105 MPa σ_{TS}: 159 MPa H: 49.2 VHN	Good dispersion reinforcement of CNT is reported in matrix with Al-carbide formation at Al-CNT interface	–	[19]
0–2 wt.%	Mechanically milled (5 h, argon atmosphere), pressureless sintered (550°C, 3 h, vacuum), hot extruded (500°C , extrusion ratio – 16)	–	–	–	Reported uniform dispersion of CNT in Al matrix and formation an amorphous interface that causes better adhesion between Al and CNT	–	[119]
2 wt.%	Ball milled (3/6 h, 200 rpm, argon atmosphere), compacted (475 MPa), hot extruded (500°C, extrusion ratio – 4)	Diameter: 4 mm Length: 65 mm Gauge length: 20 mm	Ball milled: 3 h σ_{TS}: 345 MPa %ε: 5.7 Ball milled: 5 h σ_{TS}: 348 MPa %ε: 7.9	Ball milled: 3 h σ_{TS}: 284.5 MPa %ε: 8.6 Ball milled: 5 h σ_{TS}: 348.5 MPa %ε: 8.4	Good dispersion and alignment of CNT is reported in matrix	–	[120]
0–5 wt.%	Acid treatment of CNTs, ultrasonicated with Al (20 min), SPS (600 °C, 50 MPa, 10 min, vacuum)	–	Al-5 wt.% CNT H: 55 VHN	H: 45 VHN	Al-carbide formed at interface with 5 wt.% CNT	Coefficient of friction decreases and wear decreases with 1 wt% CNT With 5 wt.% CNT, properties become poor	[121]

0–9.5 vol.%	CNTs suspended in acetone, sprayed on Al foil, sandwiched cold rolled (70% reduction), intermediate annealed (250°C, 1 h)	Length: 50 mm Gauge length: 15 mm Width: 4 mm Thickness – 50 µm	Al-9.5 vol.% CNT σ_{TS}: 97 MPa Al-2 vol.% CNT E: 75 GPa	σ_{TS}: 28 MPa E: 47 GPa	Good dispersion and reinforcement of CNT in Al matrix is reported at 2 vol.% CNT, agglomeration of CNT in matrix occurs at higher CNT content	–	[122]
5 vol.%	Ball milled (6 h), hot rolled (480°C, 12% reduction)	Gauge length: 12.5 mm Width: 6 mm Thickness: 1 mm	$\sigma_{YS(T)}$: 520 MPa	$\sigma_{YS(T)}$: 400 MPa	–	–	[123]
2 wt.%	Mechanically alloyed using ball milling (12/24/48/72 h, 200 rpm, argon atmosphere)	–	–	–	Uniform dispersion of CNT is reported in ball-milled composite powder	–	[124]
0–6.5 wt.%	CNT grown on Al powder by CVD, pressed (600 MPa), sintered (540°C , vacuum, 3 h), further pressed (2 GPa), annealed (850°C, 2 h)	Gauge length: 20 mm Width: 5.5 mm Thickness: NA	Al-5 wt.% CNT E: 95.4 GPa σ_{TS}: 398 MPa H: 0.65 GPa	E: 71.1 GPa σ_{TS}: 140 MPa H: 0.15 GPa	Homogeneous dispersion of CNT is reported in matrix along with formation of Al_4C_3 at Al-CNT interface observed	–	[125]
0–2 wt.%	Mechanical mixing of powders (2h, 200 rpm), uniaxial cold compacted (2 ton/cm²), sintered (580°C, 90 min), cold extruded (extrusion ratio = 2.25)	Diameter: 7 mm Gauge length: 10 mm	Al: 2 wt.% CNT $\sigma_{YS(T)}$: 176 MPa σ_{TS}: 184 MPa H: 74 VHN	$\sigma_{YS(T)}$: 91 MPa σ_{TS}: 98 MPa H: 69 VHN	Uniform dispersion and preferential alignment of CNTs in the matrix and formation of Al_4C_3 at Al-CNT interface is reported	–	[126]
0–2 wt.%	Mixing of powder in roller mill, spark plasma sintered (500°C, 20 min, vacuum), hot extruded (500°C, extrusion ratio = 9)	Diameter: Gauge length: 28 mm	Al-0.5 wt.% CNT $\sigma_{YS(T)}$: 96 MPa σ_{TS}: 174 MPa H: 50 VHN	$\sigma_{YS(T)}$: 66 MPa σ_{TS}: 153 MPa H: 45 VHN	CNTs are found distributed at Al grain boundaries and tend to form clusters at higher CNT contents	–	[127]

(Continued)

TABLE 5.1 (*Continued*)
Table of the Summary of the Work Carried Out in an Al-CNT System

CNT Content	Composite Processing Technique	Tensile Test Sample Size	Mechanical Properties (UTS, YS, E, Strain to Failure, Hardness)		CNT Dispersion and Interface	Other Properties	Ref.
			With CNT	Without CNT			
2 vol.%	WC layer deposited on CNTs, melted with Al and casted, aging treatment performed (natural and artificial at 185–195°C)	–	Al-CNT H: 40 MPa 2024Al alloy – CNT H: 120 MPa	H: 15MPa 2024Al alloy H: 102 MPa	Homogeneous distribution of CNT in matrix is found	–	[128]
1 vol.%	Composite powder prepared through NSD method, SPS (480/500/560/600°C, 50 MPa, 20 min), hot extruded (400°C, 500 kN, extrusion ratio = 20)	Diameter: 3 mm (ICS 59.100.01)	σ_{TS}: 207 MPa %ε: 21.5	σ_{TS}: 52 MPa %ε: 19.5	Formation of Al_4C_3 at Al-CNT interface is observed	–	[18]
5, 10 wt.%	Al-CNT composite powder synthesized by spray drying, plasma sprayed (22 kW)	–	–	–	Uniform distribution of CNT as well as presence of clusters in matrix are reported	Wear resistance increases with CNT content, but no effect is found on coefficient of friction	[129]
2.5 wt.%	Mixing by ball milling (90 min, argon atmosphere), spark plasma extrude (433°C, extrusion ratio = 16)	Compression test sample Disk shape Length/ Diameter = 1.5	σ_{CS}: 415 MPa H: 99 VHN	σ_{CS}: 377 MPa H: 74 VHN	–	–	[130]

Note: — Data not available or not applicable; E – elastic modulus; H – hardness; σ_{TS} – tensile strength; σ_{CS} – compressive strength; $\sigma_{YS(T)}$ – yield strength in tension; %ε – percentage elongation; K_{IC} – fracture toughness.

Growing CNTs on Al powder surface by CVD method leads to better bonding and dispersion. Composites prepared from CVD Al-CNT powder by sintering followed by repressing lead to an increase in tensile strength by 184% and hardness by 333% in a 5 vol.% CNT composite as compared to unreinforced material [13]. An increase in the compressive yield strength by 350% has been reported for samples prepared by NSD that resulted in better dispersion [14]. Composites prepared by hot extrusion of compacts consolidated by SPS of NSD Al-CNT powders showed good dispersion due to breakup of CNT clusters. The composite containing 5 vol.% CNT was found to have a tensile strength 128% higher than the unreinforced material [15]. Ball milling has been used extensively to disperse the powders in the Al powder. Depending on the degree of dispersion, different studies reported different degrees of strengthening. Aluminum 4.5 vol.% CNT composite prepared by hot rolling of ball milled powders was shown to have a tensile yield strength of 620 MPa and fracture toughness of 61 MPa $mm^{1/2}$, which are, respectively, 15 and 7 times more than that for aluminum [16]. Plasma sprayed aluminum composite coatings made by blended powder have been shown to improve the hardness by 72%, elastic modulus by 78%, marginal improvement in tensile strength, and 46% decrease in ductility with 10 wt.% CNT addition [13]. Sintering at 673 K for 72 h of the plasma sprayed Al-10 wt.% CNT coating has been reported to further increase the elastic modulus of the composite coating by 80%, which has been attributed to reduction in porosity and residual stress [17]. Al-12 vol.% CNT composite produced by plasma spraying of spray-dried powders shows 40% increase in the elastic modulus [5]. CNT addition results in an increase in the elastic recovery [5]. Al-1 vol.% CNT prepared by hot extrusion of SPS compacts have displayed tensile strengths up to four times (198 MPa) of aluminum (52 MPa) [18]. Strengthening has been observed irrespective of the formation of Al_4C_3 [15,19]. Significant strengthening has been achieved in samples produced by hot extrusion technique because the technique can produce high densities and can lead to breakdown of CNT clusters [15]. These results show that homogeneous distribution of CNTs, strong bonding with the matrix, and high density are the key factors to control the mechanical properties of the aluminum-CNT composites.

5.2 COPPER-CARBON NANOTUBE SYSTEM

Much work has been devoted to developing copper-CNT composites. These composites are excellent candidates for thermal management applications due to the high conductivity of Cu (~400 W m^{-1} K^{-1}) as well as of CNTs (~3000 W m^{-1} K^{-1}). Most of the researchers have utilized the powder metallurgy technique. A few studies are on developing sensors where Cu particles are deposited on the CNTs or they are electrochemically deposited. Most of the studies have used ball milling to disperse CNTs in Cu powder. Ni-coated CNTs synthesized by electro-deposition or electroless deposition are better reinforcement because they lead to better bonding between the CNT and the Cu matrix. The molecular level mixing (MLM) method was carried out in a Cu-CNT system and excellent dispersion was obtained in the powder. In this method, CNTs are dispersed in a Cu-salt solution followed by drying, calcinations, and H2 reduction to get the powders. Consolidation has been carried out by pressing and sintering, rolling, equal channel angular pressing, SPS, sandwich processing, and high-pressure torsion. Laser Engineered Near-net shaping has also been used effectively for synthesizing macro-scale composites from Cu-CNT composite powders with up to 10 vol.% of reinforcement content [20]. Micro-scale composites for thermal management have been prepared by electrochemically depositing Cu on aligned CNT arrays. Copper-CNT composites have also been prepared by electrodeposition and electroless deposition of Cu on CNTs. They have also been prepared by mixing in mineral oil and making paste for sensor applications. Formation of carbides or any interfacial products has not been reported in any of the studies mentioned previously. Table 5.2 shows the various aspects of the work carried out on Cu-CNT composites.

Reports on Cu-CNT systems are focused equally on the improvement in mechanical and electrical properties. Mechanical properties of Cu-CNT composites clearly show the effects of processing techniques on their improvement. Conventional powder metallurgy techniques, comprising compaction and sintering, help increase the hardness up to 20% with 15 vol.% CNT addition [21]. An electroless coating of the CNTs with Ni improves their bonding with the Cu matrix and increases the hardness by ~80 to 100% for even 9–12 vol.% CNT addition [22–24]. SPS improves the hardness by 79% with 10 vol.% CNT addition [25]. Further deformation by rolling leads to improvement on dispersion and alignment of CNT clusters and improved the hardness by 207% and the elastic modulus up to 95% [26]. Molecular-level mixing leads to excellent dispersion and elimination of CNT clusters in SPS composites [27]. This causes an extraordinary strengthening of the composite with a 200% increase in the yield strength and 70% increase in the elastic modulus. Molecular level mixing, followed by cold pressing and vacuum sintering, has been found to increase the flexural strength of the composites also by 31% [28]. The strengthening was explained by improved load transfer to the CNTs due to the presence of Cu and O atoms at the CNT interface that helped in bonding as well as dispersion. Samples prepared by shock wave consolidation of molecular level mixed powders showed increased hardness (51%) than that predicted from the Hall–Petch relation [29]. A similar increase was observed in 1 wt.% CNT composite prepared by high pressure torsion of ball milled powders [30]. Out of the total 640 MPa increase in the hardness, 280 MPa was attributed to the grain refinement and the balance of 360 MPa was attributed to CNT addition. For coatings deposited by electrodeposition technique, a 36.4% increase in hardness was reported for a 10 vol.% SWNT composite, which might be good improvement considering the inherent porous nature of electrodeposited coatings [31]. Cu-CNT composite, processed by cold rolling of sandwiched layers of metal and CNT, showed an 8% increase in tensile strength and a 12.8% increase in the elastic modulus [32]. Cu-CNT composites prepared by SPS of electroless Cu-plated CNTs showed an excellent increase in elastic modulus by 100% and yield strength by 183% for a 15 vol.% CNT addition [33]. However, further increase in CNT content to 20 vol.% led to lowering of the yield strength compared to 15 vol.% CNT (~96%), although the elastic modulus increased slightly (112%). These results again indicate the significance of dispersion and bonding between CNTs and the

TABLE 5.2
Table of the Summary of the Work Carried Out in a Cu-CNT System

CNT Content	Composite Processing Technique	Tensile Test Sample Size	Mechanical Properties (UTS, YS, E, Strain to Failure, Hardness)		CNT Dispersion and Interface	Other Properties	Ref.
			With CNT	Without CNT			
0–25 vol.%	Cu powder and CNT mixed by ball milling (30 min), isostatically pressed (350 MPa, 5 min), isothermally sintered(850°C, 2 h, vacuum), cold rolled, annealed (600°C, 3 h)	–	Cu-15 vol.% CNT H – 118 VHN	H – 98 VHN	Homogeneous distribution of CNT is reported	Coefficient of friction decreases with CNT addition, wear loss decreases up to 12 vol.% CNT and then slightly increases	[21]
4–16 vol.%	Nickel-coated Cu powder and CNT, ball milled (30 min), isostatically pressed (100°C, 600 MPa, 10 min), isothermally sintered (800°C, 2 h)	–	Cu – 12 vol.% CNT H – 21.5 HRB* (* Rockwell hardness with B scale)	H – 10.2 HRB	–	Coefficient of friction decreases with increasing CNT content. Wear volume decreases with increase in CNT content up to 12 vol.%	[22]
4, 8, 12, and 16 vol.%	Nickel-coated Cu powder and CNT, ball milled (30 min), isostatically pressed (100°C, 600 MPa, 10 min), isothermally sintered (800°C, 2 h)	–	Cu – 12 vol.% CNT H – 21.5 HRB* (* Rockwell hardness with B scale)	H – 10.2 HRB	–	Coefficient of friction decreases with increasing CNT content. Wear volume decreases with increase in CNT content up to 12 vol.%	[23]
SWNT	Cu nanoparticles are electrodeposited on SWNTs	–	–	–	–	Glucose detecting sensitivity increases four times	[131]

(Continued)

TABLE 5.2 (*Continued*)
Table of the Summary of the Work Carried Out in a Cu-CNT System

CNT Content	Composite Processing Technique	Tensile Test Sample Size	Mechanical Properties (UTS, YS, E, Strain to Failure, Hardness)		CNT Dispersion and Interface	Other Properties	Ref.
			With CNT	Without CNT			
0–15 vol.%	Ball milled(24 h), SPS (750°C, 40 MPa, 1 min)	–	Cu – 10 vol.% CNT H ~ 100 MPA	H ~56 MPa	Better dispersion of CNT in composites with Cu-nano powders is reported	–	[132]
–	CNTs grown in arrays and in between places were filled with Cu by electrodeposition	–	–	–	–	Cu-CNT films show lower thermal resistance than only CNTs	[133]
–	Cu-CNT-mineral oil hand mixed and put in fused silica tube with a Cu wire for maintaining electrical contact	–	–	–	–	Better sensitivity of microchips with CNT for carbohydrate detection	[134]
5 and 10 vol.%	CNT and Cu ion are suspended in a solvent, dried, calcination, reduction, performed for composite powder (molecular level mixing), pre-compacted (10 MPa), SPS (550°C, 50 MPa, 1 min, vacuum)	–	Cu – 10 vol.% CNT $\sigma_{YS(C)}$ ~ 455 MPa	$\sigma_{YS(C)}$ ~150Mpa E ~80 GPa	Homogeneous distribution of CNT is reported with high interfacial strength	—	[27]
0–5.25 vol.%	CNTs coated with Ni by electroless deposition, mixed with Ni powder, ball milled (30 min), hot pressed (1100°C, 32 MPa, 1 h)	—	Cu – 2.25 vol.% CNT: H ~280 VHN	H ~155 VHN	CNTs are found to get agglomerated at higher concentrations	Coefficient of friction reduces drastically for 2.25 vol.% CNT and then reduction is lowered, wear loss is minimum at 2.25 vol.% CNT	[24]

–	Dispersion of CNT and Cu microparticles in mineral oil, used as electrode	–	–	–	–	Excellent performance with detection limits in micro-molar levels for non-electroactive amino acids for composite electrode	[135]
–	Cu nanoparticle added to SWNTs in Nafion solution and sonicated, solution is dropped on polished electrode and dried	–	–	–	Good dispersion of CNT is observed in dried film forming network	Cu-SWNT Composite gives most synergistic signal effect	[136]
5 vol.%	Cu powder, CNT-mixed, ultrasonicated (1 h), dried, equichannel angular pressed through 8 passes	–	Hardness increases with number of passes, i.e., amount of deformation H ~ 115 VHN (after 8 passes)	–	Reduction in agglomerates observed along with improved distribution; improves upon deformation	–	[137]
5 and 10 vol.%	Spray-dried Cu powder and CNT, ball milled (24 h, 150 rpm), precompacted (10 MPa), SPS (700°C, 50 MPa, 1 min, vacuum), rolled (50% reduction), annealed (650°C, 3 h)	ASTM-E8M Dogbone-shaped sample Gauge length – 9 mm Width – 2.5 mm	Cu-10 vol.% CNT E – 137 GPa $\sigma_{YS(T)}$ – 197 MPa σ_{TS} – 281 MPa H– 1.75 GPa	E – 70 GPa $\sigma_{YS(T)}$ –135MPa σ_{TS} – 175 MPa H – 0.57 GPa	CNT-rich and CNT-free regions are observed distinctly in matrix	–	[26]
30 and 55 vol.% SWNT	Thick network of SWNT, prepared by suspension of CNT dropped on substrate and dried, Cu is electrochemically deposited on CNT network	–	–	–	–	Electrical conductivity and thermal expansion coefficient of composite is same as Cu	[138]

(Continued)

TABLE 5.2 (*Continued*)
Table of the Summary of the Work Carried Out in a Cu-CNT System

CNT Content	Composite Processing Technique	Tensile Test Sample Size	Mechanical Properties (UTS, YS, E, Strain to Failure, Hardness)		CNT Dispersion and Interface	Other Properties	Ref.
			With CNT	Without CNT			
–	Cu-CNT composite particle, precipitated from dispersion of $CuSO_4$ and CNT in distilled water by adding NaOH and KBH_4	–	–	–	CNTs are found covered by Cu particles	Catalytic activity increases in presence of CNT	[139]
–	Cu – electroless deposited on CNT from $CuSO_4$ solution	–	–	–	Cu particles deposited on surface and inside of CNTs, Cu particle on surface of CNTs are found uniform in size and distribution	Composite possesses fine electron conductivity	[140]
SWNT	Cu – argon sputtered on SWNT bundles at high vacuum	–	–	–	Formation of 1D array of nanoclusters prefereably at the groove of CNTs in the bundle is reported	–	[141]
–	Cu nanoclusters – electrochemically deposited on CNT (on electrode)	–	–	–	Some aggregation of nanoclusters is observed	High sensitivity, good reproducibility, and fast response	[142]
–	CVD-grown aligned CNT – copper deposited by electrochemical plating	–	–	–	Cu fillings are found occupying the voids between CNTs forming compact channels	Composite shows better electrical and thermal conductivity than only aligned CNTs	[143]

1 vol.%	Cu powder and CNT, mixed, ultrasonicated (1 h), filled in a sheathe, cold ECAP through 8 passes	–	Hardness increases with number of passes, i.e., amount of deformation H ~ 115 VHN (after 8 passes)		Reduction of agglomeration and better dispersion with greater number of ECAP passes is reported	–	[144]
0–10 vol.%	CNT and Cu ion is ssuspended in a solvent, dried, calcination, reduction, performed for composite powder (molecular level mixing), SPS (550°C, 50 MPa, 1 min, vacuum)	–	Cu – 10 vol.% CNT H – 1.1 GPa	H – 0.8 GPa	Homogeneous dispersion of CNT is found with good inter-facial bonding (CNTs embedded in Cu powders)	Wear loss reduces with addition of CNT	[25]
1 vol.%	Cu powder and CNT – mixed, ultrasonicated (1 h), dried, filled in a sheathe, ECAP for 8 passes	–	Hardness increases with number of passes, i.e., amount of deformation H ~ 115 VHN (after 8 passes)	–	Homogeneous distribution of CNT in matrix is obsrved	–	[145]
SWNT	CVD grown SWNT films (19), sandwiched with Cu films (20), cold rolled, annealed (1050°C, 10 h), cold rolled	Length – 4.9 –5.2 mm Width – .8 – .09 mm Thickness –0.025 mm	σ_{TS} –361 MPa E– 132 GPa	σ_{TS} – 334 MPa E – 117 GPa	Good interfacial adhesion reported	–	[32]
–	CNT-dispersed, ultrasonicated, $CuSO_4$ added, NaOH added, dried, reduced (molecular level mixing)	–	–	–	CNTs are found homogeneously implanted in the Cu spheres	–	[146]

(Continued)

TABLE 5.2 (*Continued*)
Table of the Summary of the Work Carried Out in a Cu-CNT System

CNT Content	Composite Processing Technique	Tensile Test Sample Size	Mechanical Properties (UTS, YS, E, Strain to Failure, Hardness)		CNT Dispersion and Interface	Other Properties	Ref.
			With CNT	Without CNT			
–	CNT and Cu nano powder, mixed in mineral to prepare a paste electrode	–	–	–	–	Detection sensitivity is sufficient with reasonable repeatability and operational stability	[147]
–	CNT and Cu micro powder, mixed in mineral to prepare a paste electrode	–	–	–	–	Highly sensitive and fast detector	[148]
7–10 vol.% SWNT	CNT dispersed in $CuSO_4$ electrolyte, electrochemically deposited under ultrasonic field	–	Cu-10 vol.% CNT H – 1.61 GPa	H – 1.18 GPa	Interface is found wettable and in good adhesion	Electrical conductivity is comparable to pure Cu for low to high temperatures	[31]
5 and 10 vol.%	Composite powder, prepared by oxidation–reduction process (molecular level mixing), compacted, SPS (550°C , 50 MPa, 1 min, vacuum)	Not mentioned	$\sigma_{YS(C)}$ – 455MPa E (NI) – 138 GPa	$\sigma_{YS(C)}$ – 150 MPa E (NI) – 100 GPa	Uniform dispersion and good reinforcement of CNT with matrix is reported	—	[149]
1 wt.%	Ball milled (5 h), isostatically compacted (500 MPa), consolidation by high pressure torsion (6 GPa, 5 revolutions)	—	H – 3.5 GPa	H – 2.8 GPa	Homogeneous and good dispersion of CNT in matrix is reported	—	[30]

10 vol.%	Composite powder, prepared by oxidation–reduction process (molecular level mixing), shock wave consolidated using propellant gun system fixture	—	H – 1.19 GPa	H – 0.80 GPa	Homogeneous and good dispersion of CNT in matrix is reported	—	[29]
1 wt.%	Ball milled (5 h, argon atmosphere), isostatically compacted, consolidation by high pressure torsion (6 GPa)	Micro-pillar compression testing Diameter – 5 µm Length – 11µm	$\sigma_{YS(C)}$ – 1125 MPa	$\sigma_{YS(C)}$ – 738 MPa	Homogeneous distribution of CNT in matrix is found	—	[150]
0–20 vol.%	Composite powder prepared by electroless deposition of Cu on CNT, SPS (550°C, 50 MPa, 1 min)	Gaige length – 9 mm Width – 2.5 mm	$\sigma_{YS(T)}$ – 350 MPa E – 105.9 GPa H – 1.4 GPa	$\sigma_{YS(T)}$ – 120 MPa E – 51.6 GPa H – 0.7 GPa	Uniform distribution of CNT in matrix is found	Electrical conductivity decreases with increasing CNT content	[33]
0.5 vol.%	Mechanically mixed, hot pressed (600°C, 45 MPa, 30 min)	–	–	–	Uniform distribution of CNT in Cu-Ni matrix is reported	No improvement is recorded in electrical and thermal conductivity, initial coefficient of Friction decreased for Cu, reinforced with Ni-coated CNTs	[151]
—	Pulse reversed electrochemical deposited	Gauge length – 4 mm Width – 0.4 mm Thickness – 0.04 mm	For CNT diameter of 1.5 – 3 nm σ_{TS} – 670 MPa	σ_{TS} – 230 MPa	Uniform dispersion of CNT in Cu matrix is reported	–	[152]

(Continued)

TABLE 5.2 (*Continued*)
Table of the Summary of the Work Carried Out in a Cu-CNT System

CNT Content	Composite Processing Technique	Tensile Test Sample Size	Mechanical Properties (UTS, YS, E, Strain to Failure, Hardness)		CNT Dispersion and Interface	Other Properties	Ref.
			With CNT	Without CNT			
5, 10, and 15 vol.%	Dry impact blended of Cu and CNT powder (40 min, 5000 rpm), SPS (600°C , 50 MPa, 5 min)	–	–	–	Distribution of CNT is uniform up to 10 vol.% and at 15 vol% clustering starts	No improvement in thermal conductivity with 5 and 10 vol.% CNT addition; decrease in the same with 15 vol.% CNT – due to interface resistance, CNT clustering porosity	[34]

Note: E – elastic modulus; H – hardness; σ_{TS} – tensile strength; $\sigma_{YS(T)}$ – yield strength in tension; $\sigma_{YS(C)}$ – yield stress in compression; — Data not available or not applicable.

matrix in Cu-CNT composites. Cu-CNT composites prepared by SPS of dry-impact-blended powders showed similar thermal conductivity as copper up to 15 vol.% reinforcement and a decrease thereafter [34].

5.3 NICKEL-CARBON NANOTUBE SYSTEM

In terms of the number of publications, Ni-CNT composites and coatings of thickness less than 200 μm have received the maximum attention but they are restricted for non-structural applications. Electro- and electroless-deposition are the most researched methods for the deposition of Ni-CNT and Ni-P-CNT coatings. This method is also utilized to deposit Ni coating on CNTs because Ni results in better wetting and bonding with other metal matrices. Pulsed reverse electrodeposition has also been used since it results in coatings with lower porosity and nano-crystalline structure. The CNTs are added to the electrolytic bath and their dispersion is maintained by providing agitation and adding a dispersant. Ni has also been co-deposited with Co by electrodeposition. The idea is to replace Ni plating with Ni-CNT plating, which could have an increased lifetime due to higher wear resistance. Hence, most of the studies report the hardness, and tribological and electrochemical properties of Ni-CNT coatings. Very few studies have been carried out using other consolidation processes like SPS and LENS [35] for synthesizing Ni-CNT composites. Formation of nickel carbide has not been reported in any of the studies previously mentioned. Table 5.3 summarizes the efforts in synthesis of Ni-CNT coatings.

The hardness of the electroless Ni composite coating has been found to improve by 44% with the addition of 2 vol.% CNT [36]. Another study shows that the hardness increased by only 11% with 12 vol.% CNT addition, which was due to the agglomeration of CNTs in the bath at higher concentration [37]. Other studies have reported an increase in hardness of 45% with 15.3 vol.% CNT addition [38] and 39% increase in hardness with 11 vol.% CNT addition [39]. Extraordinary improvement of more than 300% increase in hardness and elastic modulus of electroless Ni-CNT composite coating for MEMS application has been reported [40]. This was attributed to the acid functionalization treatment of the CNTs, which helped in obtaining a good dispersion of the CNTs in the bath and the coating. Freestanding Ni-CNT composite structures have also been fabricated using the electrochemical co-deposition technique. A large increase in the tensile strength by 233% for SWNT and 167% for CNT was reported [41]. However, the CNT content of the coatings was not determined. Another study has reported a 26.9% increase in tensile strength, although the processing technique was exactly the same as in Ref. [42]. These studies indicate that the dispersion of the CNTs in the bath is critical. Ultrasonic agitation is better than mechanical stirring during deposition. Pulsed reverse deposition also assists in agitation and improved CNT dispersion. The deposition parameters like current density need to be optimized to control the rate of metal deposition and the CNT dispersion obtained. CNT reinforcement has been found to increase the coefficient of thermal expansion for bulk scale Ni-CNT structure, up to 12%, due to mismatch in the expansion of matrix and reinforcement [43].

TABLE 5.3
Table of the Summary of the Work Carried Out in Ni-CNT System

CNT Content	Composite Processing Technique	Tensile Test Sample Size	Mechanical Properties (UTS, YS, E, Strain to Failure, Hardness)		CNT Dispersion and Interface	Other Properties	Ref.
			With CNT	Without CNT			
–	CNTs ball milled and ultrasonically dispersed in the electrolytic bath, electrodeposited from Ni-sulfate bath	–	–	–	Good dispersion of CNTs at both grain boundary and grain body is reported	Friction coefficient lower for CNT composite Wear volume is lower for CNT composite	[153]
–	CNTs ball milled and ultrasonically dispersed in the electrolytic bath, electrodeposited from Ni-sulfate bath	–	–	–	Good dispersion of CNTs is found	Effect of deposition parameters studied	[154]
–	Electroless deposited from Ni-sulfate bath	–	–	–	–	Lower friction coefficient and higher wear resistance with CNT addition	[23]
–	Electroless deposited from Ni-sulfate bath	–	H – 946 VHN	H – 562 VHN	Good dispersion of CNTs is observed	Lower friction coefficient and higher wear resistance with CNT addition	[155]
–	Ni deposited electrochemically from Ni-sulfate bath on vertically aligned	–	–	–	–	–	[156]
12 vol.%	nanotubes Electroless co-deposited	–	H – 520 VHN	H – 467 VHN	–	Lower friction coefficient and higher wear resistance with CNT addition	[37]

5.1, 10, 11.2, and 12.5 vol.%	CNTs ball milled, electrodeposited from Ni-sulfate bath	–	Ni – 11.2 vol.% CNT H – 1524 VHN	H – 1095 VHN	Well dispersed and embedded CNTs are found in matrix	Lower coefficient of friction and higher wear resistance with CNT addition up to 11.2 vol%, with further increase in CNT wear resistance decreases	[39]
–	CNTs dispersed in Ni-sulfate bath with polyacrylic acid, electrodeposited	–	–	–	Ni was not found uniformly deposited on CNTs	–	[157]
–	CNTs ultrasonically dispersed, electro-deposited	–	–	–	–	Shows improved catalytic activity as a hydrogen evolving electrode	[158]
1, 2.5, and 5 wt.%	SWNTs heat treated to cut shorter, added to surfactant solution, and ultrasonically dispersed, electroless plated	–	Ni – 5 wt.% CNT H – 1204 HVN	H ~ 875 VHN	–	Shows lower friction coefficient, higher wear resistance, and higher corrosion resistance	[159]
0–15.3 wt.%	CNTs ball milled, electroplated along with Ni	–	Ni – 15.3 wt.% CNT H – 822 VHN	H – 564 VHN	CNTs are found well dispersed and embedded in Ni matrix	Shows improved wear resistance at lower load with CNT addition but at higher load wear resistance decreases with CNT	[38]
0.1–0.52 wt.%	CNTs dispersed Ni-sulfate and sodium hypophosphite bath, electroless deposited	–	Ni – 0.52 wt.% CNT H ~ 890 MPa	H ~ 630 MPa	Some clusters are found along with CNTs, which are well embedded in the matrix	Coefficient of friction and wear loss decreases with CNT addition	[36]
–	Ni-P electroless deposited on CNT	–	–	–	Ni-P is found forming uniform layer on CNTs	–	[160]
11.79, 16.4, and 28.2 vol.%	CNTs dispersed in solution and sonicated, Ni-P plating solution added, electroless deposited	–	Ni – 28.2 vol.% CNT E: 665.9 GPa H: 28.9 GPa	E: 165.1 GPa H: 6.7 GPa	–	Electrical resistivity increases linearly with CNT content, resonance frequency is 3.9 times (28% CNT) of pure Ni	[40]

(Continued)

TABLE 5.3 (*Continued*)
Table of the Summary of the Work Carried Out in Ni-CNT System

CNT Content	Composite Processing Technique	Tensile Test Sample Size	Mechanical Properties (UTS, YS, E, Strain to Failure, Hardness)		CNT Dispersion and Interface	Other Properties	Ref.
			With CNT	Without CNT			
0.15–0.52 wt.%	CNTs dispersed in a bath containing Ni-sulfate and Na-hypophosphate, electroless deposited	–	Ni – 0.52 wt.% CNT H – 880 HVN at	H – 610 HVN	Clusters of CNTs are found deposited at nodes of Ni-P, CNTs are also found embedded deeply in the matrix while part of them are protruded out	Friction coefficient and volume of wear decreases with CNT addition	[161]
8.48 wt.%	CNTs dispersed in a bath containing Ni-sulfate and Na-hypophosphate, electroless deposited, heat treated	–	H – 1200 HVN	H – 900 HVN	–	Corrosion resistance of composite is Better	[162]
Ni-CNT coating	CNTs dispersed in a Ni-sulfate bath, electrodeposited	–	–	–	CNTs are found well dispersed and embedded in matrix	Corrosion resistance improves significantly with CNT addition	[163]
–	CNTs dispersed in electrolyte bath, ultrasonicated, Cu and Ni plates used as electrodes, Ni ions get deposited on CNTs dispersed in the bath	–	–	–	Coating is found to become more uniform with more time, i.e., deposition	–	[164]
–	SPS (pressure – 600 N, electric current – 400 A)	–	–	–	Good dispersion of CNTs in matrix is observed	–	[165]

–	CNTs ultrasonically dispersed in Ni bath, electroless deposited	–	H – 946 MPa	H – 562 MPa	CNTs are found dispersed and embedded in the matrix	Coefficient of friction and wear rate reduces with CNT addition	[166]
–	CNTs dispersed in $NiSO_4$ bath, electrodeposited	–	–	–	CNTs are found dispersed in the matrix homogeneously	Thermal conductivity increases with CNT addition	[167]
–	CNTs dispersed in Ni-P bath, electroless plated	–	E – 665.9 GPa H – 28.9 GPa	E – 165.1 GPa H – 6.7 GPa	–	Resonance frequency increases 4 times	[168]
–	CNTs dispersed in Ni-Co bath – electrochemically co-deposited	—	E – 236 GPa H – 5.87 GPa	E – 202 GPa H – 4.41 GPa	CNTs are found uniformly dispersed in matrix	Wear resistance increases with CNT addition	[169]
–	Electroless deposited from Ni-sulfate bath	–	H – 865 VHN	H – 562 VHN	CNTs are found well dispersed in the matrix and embedded deeply	Coefficient of friction and wear loss decreases with CNT addition	[170]
–	Electroless deposited from Ni-sulfate bath	–	–	–	CNTs are found dispersed in the matrix and embedded deeply	Coefficient of friction and wear loss decreases with CNT addition	[171]
–	CNTs ball milled with Ni sulfate, brush plated	–	H – 605 VHN	H – 479 VHN	–	Increase in wear resistance with CNT addition	[172]
–	Electroless deposited from Ni-sulfate bath	–	–	–	Good dispersion of CNT is observed	Field emission properties are improved	[173]
–	Electroless plating of Ni-Fe-P on CNTs from an Ni-sulfate, Fe-sulfate, and sodium hypophosphite bath	–	–	–	–	Processing conditions optimized for getting continuous, unique, and smooth coating	[174]
–	CNT dispersed in coating bath, electrodeposited	–	–	–	Good dispersion of CNT is observed	Change in bath composition reduces internal stress of the composite fi formed	[174]

(*Continued*)

TABLE 5.3 (*Continued*)
Table of the Summary of the Work Carried Out in Ni-CNT System

CNT Content	Composite Processing Technique	Tensile Test Sample Size	Mechanical Properties (UTS, YS, E, Strain to Failure, Hardness)		CNT Dispersion and Interface	Other Properties	Ref.
			With CNT	Without CNT			
22.5 vol.%	Electrodeposited from sulfate bath	–	Ni – 14.5 vol.% CNT Fracture stress ~780 MPa	Fracture stress ~ 605 MPa	Agglomeration of CNT at higher concentration is reported	–	[175]
–	Solution is ultrasonicated for good dispersion and electrochemically co-deposited	Dog bone shape: Length – 4 mm Width – 400 µm Thickness – not mentioned	CNT σ_{TS} ~ 1600 MPa SWNT σ_{TS} ~ 2000MPa	σ_{TS} ~ 600 MPa	Well dispersed and embedded CNTs in matrix are reported	–	[41]
–	CNT dispersed in Ni-sulfate bath by ultrasonication and electrochemically co-deposited	–	Microhardness is correlated with process parameter (reverse pulse ration)	-	–	Corrosion property is correlated with process parameter (reverse pulse ration)	[176]
—	Ultrasonically dispersed CNT added to	Gauge length – 20 mm	σ_{TS} –1475 MPa H – 645 VHN % ε – 2.09	σ_{TS} – 1162 MPa H – 572VHN % ε – 2.39	Homogeneous dispersion of CNTs in matrix is reported	Coefficient of friction and wear loss decreases with addition of CNT	[42]

—	CNT dispersed in Ni-sulfate bath by ultrasonication and electrochemically co-deposited	Dog bone shape sample 10 × 20 × 2 mm	Hardness first increases with current density and then decreases % ε – 4.99	–	–	Carbon content, cracking resistance, adherence, and corrosion resistance first increase with current density (8 A/dm^2) and then decrease	[177]
0.1 and 0.5 wt.%	Ultrasonically dispersed CNT added to Ni-sulfate and Ni-chloride bath, electrochemically co-deposited	–	–	–	CNTs are found homogeneously distributed and tightly incorporated in the matrix	Coefficient of friction reduces with CNT addition	[178]
10 vol.%	Powders mixed in roller mixer, melt pool created on substrate by laser, the powder mixture is added through inert gas	–	–	–	Agglomerates of CNTs are found along with wetting at the surface of CNT but no reaction	–	[179]
–	Ni-CNT on Cr/Cu micromachined conducting layer is electroplated, at room temperature to etch away Ni and leave behind protruded CNT tips	–	–	–	CNTs are found dispersed in electroplated layer	Shows good field emission properties, e.g., high current density, low turn on field, good stability	[180]
–	CNTs are grown on porous alumina template, alumina removed, Ni is electrochemically deposited in the grooves between CNTs	–	–	–	–	–	[181]

(Continued)

TABLE 5.3 (*Continued*)
Table of the Summary of the Work Carried Out in Ni-CNT System

CNT Content	Composite Processing Technique	Tensile Test Sample Size	Mechanical Properties (UTS, YS, E, Strain to Failure, Hardness)		CNT Dispersion and Interface	Other Properties	Ref.
			With CNT	Without CNT			
–	Electroless co-deposited on glass substrate	–	–	–	–	Better field emission property with CNT addition in terms of lower turn on and threshold electric field, but the film cracks and performance degrades with time due to high emission current density	[182]
10 vol.%	Powders mixed by mixing roller, coating made on steel substrate using LENS process	–	–	–	Uniform distribution of CNT in Ni matrix and absence of any interfacial product is reported	Coefficient of friction decreases with CNT addition to Ni due to low interfacial shear strength graphitic film formation	[183]
–	Electroless co-deposited on Cu substrate	–	E – 124 GPa	E – 102 GPa	–	–	[184]
0.7–1.2 mass%	Electrochemically co-deposited on stainless steel substrate	–	Before heat treatment H ~ 600 HVN After heat treatment H ~ 1100 VHN	Before heat treatment H ~ 500 HVN After heat treatment H ~ 950 VHN	Uniform distribution of CNT in matrix is reported	Coefficient of friction decreases with CNT addition	[185]

Note: E – elastic modulus; H – hardness; σ_{TS} – tensile strength; σ_{CS} – compressive strength; $\sigma_{YS(T)}$ – yield strength in tension; $\sigma_{YS(C)}$ – yield stress in compression; %ε – percentage elongation, — Data not available or not applicable.

5.4 MAGNESIUM-CARBON NANOTUBE SYSTEM

Magnesium and its alloys are important candidates for structural applications in the automobile and aerospace industry due to their low density and good castability. They also find applications in the electronic industry such as in cell phones and laptop casings. The number of reports on Mg-CNT composite is fewer as compared to those in Al, Cu, and Ni-CNT composites. Mg-CNT composites are being developed largely for improved mechanical properties for structural applications. Some researchers have also focused on the effect of CNTs on the hydrogen storage abilities of Mg for fuel cell applications. Addition of CNT to Mg has been done mainly by powder metallurgical and casting route. Most of the studies on metal-CNT composites by the casting method are on Mg-CNT systems. A DMD method has been developed for the deposition of Mg-CNT composites [44]. The molten metal is mixed with CNTs and stirred by a stirrer that is coated with ceramic to avoid contamination. The molten mixture is then allowed to pass through a nozzle and the stream is atomized with two argon jets. The droplets of the composite are then deposited on a mould. Such a technique has shown to improve the strength and ductility simultaneously. The ductility improvement was found to be the result of the high activity of the basal slip system and the initiation of prismatic slip. Powder metallurgy methods of hot pressing and hot isostatic pressing have been attempted as well. None of these studies has reported the formation of magnesium carbide or any reaction products between CNT and Mg. Table 5.4 summarizes the work carried out on Mg-CNT composites.

An increase in the tensile strength by 200% was observed in Mg-0.55 vol.% CNT composite prepared through the melting and casting route [45]. The hardness increased by 90% for Mg-CNT composites prepared by the friction stir welding route [46]. The processing route has a profound effect on the properties of the composite. For composites prepared by the melting and casting route, a 15% increase in yield strength on 1 vol.% CNT addition was obtained while the same amount of improvement was obtained with only 0.25 vol.% CNT addition when sintering followed by extrusion route was used [44,47]. Investigations have also revealed that low amount of CNT reinforcement (<1 vol.%) in Mg alloy can offer an impressive improvement in failure strain as high as 150%, without impairing strength [48–50]. However, the fatigue property deteriorated due to 1.3 wt.% CNT addition. Poor fatigue properties are attributed to the surface porosity and presence of CNT clusters. Similarly, corrosion resistance of Al containing Mg alloy (AZ series) is also found to deteriorate with CNT reinforcement, which is claimed to be due to formation of galvanic cells between CNT and matrix [51]. An account of the hydrogen storage properties of Mg-CNT composite is provided in Chapter 8. There is a large scope for development of lightweight Mg-CNT composites.

TABLE 5.4
Table of the Summary of the Work Carried Out in an Mg-CNT System

CNT Content	Composite Processing Technique	Tensile Test Sample Size	Mechanical Properties (UTS,YS, E, Strain to Failure, Hardness)		CNT Dispersion and Interface	Other Properties	Ref.
			With CNT	Without CNT			
5 and 20 wt.%	Ground, mixed, ball milled (hydrogen atmosphere)	–	–	–	–	Hydrogen storage increases for 5 wt.% CNT but decreases for 20 wt.% CNT decreases for 20 wt.% CNT	[186]
1 vol.%	CNTs were grown on Al_2O_3 preform by CVD, infiltration of liquid Mg in preform by gas pressure	Sample for torsion test – 40 mm × 4 mm × 1 mm	Shear modulus of Mg-Al_2O_3 CNT is 20% greater	Shear modulus of Mg-Al_2O_3 is 20% lower	CNTs are found well embedded in the Matrix.	—	[187]
2 wt.%	Ball milled, compacted, hot isostatically pressed (600°C, 1800 Bar)	4mm × 1 mm × 50 mm	E-38.6 GPa $\sigma_{YS(T)}$ – 89 MPa σ_{TS} – 140 MPa %ε ~ 3	E – 35.3 GPa	Uniform dispersion and good interface bonding of CNT with matrix is observed	—	[188]
0.67 and 1 wt.%	Electroless Ni plating on CNT – Mg melted (700°C), CNT added, stirred, and casted	—	σ_{TS} is doubled for Mg – 0.67 wt.% CNT 150% increase with plated CNT for Mg – 1 wt.% CNT σ_{TS} decreases for both plated and unplated CNTs	σ_{TS} is half of 0.67% CNT Composite	Non-uniform dispersion predicted at higher CNT Content. Interfacial bonding is reported better for Ni-plated CNTs	—	[189]

0.3, 1.3, 1.6, and 2 wt.%	Mg and CNT slurry, stirred at 750°C, released and disintegrated with argon jet, deposited on mold	ASTM E8M-01, round specimen Diameter – 5 mm Gauge length – 25 mm	Mg – 1.3 wt.% CNT $\sigma_{YS(T)}$ – 140 MPa σ_{TS} – 210 MPa %ε – 13.5	$\sigma_{YS(T)}$ – 126 MPa σ_{TS} – 192 MPa %ε – 8	Good dispersion of CNT in matrix is claimed	–	[44]
–	CNT filled in a grove on the alloy plate, friction stir processed (1500 rpm)	–	H – 78 HVN	H – 55 HVN	Dispersion is found dependent on traveling speed of the columnar probe	–	[45]
0.06, 0.18 and 0.3 wt.%	Mixed in V-blender (10 h, 50 rpm), compacted (728 MPa), sintered (630°C, 2 h, argon atmosphere), hot extruded (350°C, extrusion ratio – 20.25)	ASTM test method – (E8M-01)	Al – 0.3 wt.% CNT $\sigma_{YS(T)}$ – 146 MPa σ_{TS} – 210 MPa	$\sigma_{YS(T)}$ – 127 MPa σ_{TS} – 205 MPa	–	Coefficient of thermal expansion reduces with increase in CNT content	[47]
5 wt.%	Reactive mechanical alloying, through vibrational ball milling in H_2 atmosphere (24 h)	–	–	–	–	Hydrogen storage property remains similar	[190]
0.3, 0.5, 0.7, and 1%	Blended (1 h, 200 rpm), compacted (50 tons), microwave sintered (640°C, 25 min), hot extruded (350°C, extrusion ratio – 25)	Cylindrical specimen Diameter 5 mm Gauge length 25 mm	Mg – 1 wt.% CNT H – 43 HVN $\sigma_{YS(T)}$ – 117 MPa σ_{TS} – 154 MPa	H – 41 HVN $\sigma_{YS(T)}$ – 112 MPa σ_{TS} – 155 MPa	Uniformly distributed CNTs in matrix are reported	Coefficient of thermal expansion reduces with addition of CNT content	[191]
0.3, 1.3, 1.6, and 2 wt.%	Ingots produced by disintegrated melt deposition, hot extruded (350°C, extrusion ratio – 20.25)	Fatigue test specimens Cylindrical Diameter 6 mm Gauge length 30 mm	Number of cycle to fatigue failure is lower in Mg-CNT than in Mg	–	–	CNT composites are more prone to fatigue failure	[192]

(Continued)

TABLE 5.4 (*Continued*)
Table of the Summary of the Work Carried Out in an Mg-CNT System

CNT Content	Composite Processing Technique	Tensile Test Sample Size	Mechanical Properties (UTS,YS, E, Strain to Failure, Hardness)		CNT Dispersion and Interface	Other Properties	Ref.
			With CNT	Without CNT			
–	CNT sintered into preform (2500°C, 20 min, argon atmosphere), preform infiltrated by liquid metal through squeeze casting	–	–	–	Good wetting and reinforcement of CNT in matrix is observed	–	[100]
1 vol.% – Al core	Mg alloy (AZ31) and CNT (in Al foil) cast together (750°C, argon atmosphere), Co-extruded (hot) along with Al core placed inside Mg-CNT billet (350°C, extrusion ratio – 20.25)	Diameter – 5 mm Gauge length – 25 mm	In Tension E – 61 GPa $\sigma_{YS(T)}$ – 169 MPa σ_{TS} – 296 MPa %ε – 12.2 In Compression E – 126 GPa σ_{CS} – 491 MPa %ε – 27.9 H – 84 VHN	In Tension E – 44 GPa $\sigma_{YS(T)}$ – 172 MPa σ_{TS} – 263 MPa %ε – 10.4 In Compression E – 93 GPa σ_{CS} – 486 MPa %ε – 19.7 H – 63 VHN	CNTs have been reported to be reasonably uniformly dispersed in Mg-alloy matrix	–	[193]
2 wt.%	Turbula mixed, hot pressed (600°C, 50 MPa, 30 min)	–	–	–	Good dispersion reported in CNT in matrix Defect sites are reported at CNT-matrix interface	Hydrogen storage capacity increases by 25 times with addition of CNT in Mg-Ni system	[194]
0.1wt.%	Mg chips coated with CNT by ultra-sonication with polymer, melted and casted in mold through melt stirring, same as NSD	Diameter – 5 mm Length – 7 mm	σ_{CS} – 412 MPa $\sigma_{YS(C)}$ – 272 MPa %ε – 24.4	σ_{CS} – 344 MPa $\sigma_{YS(C)}$ – 248 MPa %ε – 18	Homogeneous dispersion of CNT in Mg matrix is found	–	[45]

Note: E – elastic modulus; H – hardness; σ_{TS} – tensile strength; σ_{CS} – compressive strength; $\sigma_{YS(T)}$ – yield strength in tension; $\sigma_{YS(C)}$ – yield stress in compression; %ε – percentage elongation; — Data not available or not applicable.

5.5 CNT-MMC FOAMS

Metal foams are solid materials having a cellular structure made by filling metals with pores (or pore formers) such that 95% of the volume consist of gas-filled voids. These metallic cellular materials have light weight and high damping capacity. However, low stiffness and strength limit their potential to energy absorption applications [52,53]. In order to strengthen these metallic foams, different strategies have emerged, which include strengthening by adding alloying elements (e.g., Mg, Zn, Ni) to stimulate the formation of intermetallic phases or incorporation of micro/nano-sized reinforcement [54]. Both open- as well as closed-cell metal foams have been reinforced with different kinds of micro/nano-sized reinforcements to improve mechanical properties of their metallic matrix. Reinforcing metal foams provides high strength and stiffness to the matrix of the cell edges and walls. The open-cell metallic foams are widely used as shock absorbers, heat exchangers, electrodes, filters exploiting their high specific surface area and high thermal and electrical conductivities [54,55]. Meanwhile, the closed-cell metal foams have been used in different structural engineering applications that require lightweight structures with high specific strength and stiffness, high damping capacity, and/or high impact resistance [52,53]. Ceramic fibers and ceramic nanoparticles are the most commonly used reinforcement for metallic foams [54]. Recently, exploring carbon-based nanomaterials as reinforcing agents for metal alloys has been a great area of interest for researchers [55,56]. Till now, research has been carried out on Al [57–69] and Mg [70] due to their low density and high specific strength, with the main focus on Al. High aspect ratio and properties, like high stiffness, strength, electrical and thermal conductivities make CNTs the most preferred reinforcement for fabricating composite materials for structural and functional applications. However, it is very difficult to successfully incorporate CNTs into the metal-matrices due to their high tendency to agglomerate, and poor wettability of carbon by molten metal (due to a large difference of surface tension) and the formation of interfacial reaction products. Various processing strategies have been employed to overcome these problems, but with limited success [70].

Durate et al. published the first two articles in the year 2015 in this field [58,59]. A novel approach was used to fabricate CNT reinforced Al foam that combines colloidal processing and conventional powder metallurgy technique. A schematic for the same is reproduced in Figure 5.1. It consisted of three steps (i) preparation of spherical granules; (ii) preparation of foamable-precursor; and (iii) fabrication of reinforced Al-alloy foams. Three main starting materials, used in this study, were Al alloy as matrix, COOH functionalized CNTs as reinforcement, and TiH_2 as blowing agent. The technique used was effective in incorporating and dispersing the CNTs easily into the micro-sized Al-alloy and TiH_2 powders, as seen from Figure 5.2a–e. The fractured surface images of foam clearly show uniformly distributed CNTs throughout the matrix (Figure 5.2f–h). CNTs were seen stretched and aligned perpendicular to the direction of hot compaction, retaining their structure (Figure 5.2g and 5.2h). Highly spherical grains with diameters <1200 μm were obtained. The increase in Vickers micro-hardness of Al-foams reinforced with 0.5 wt.% CNTs was in the range of 55–125%, measured on the cell walls of the foam [8]. The high

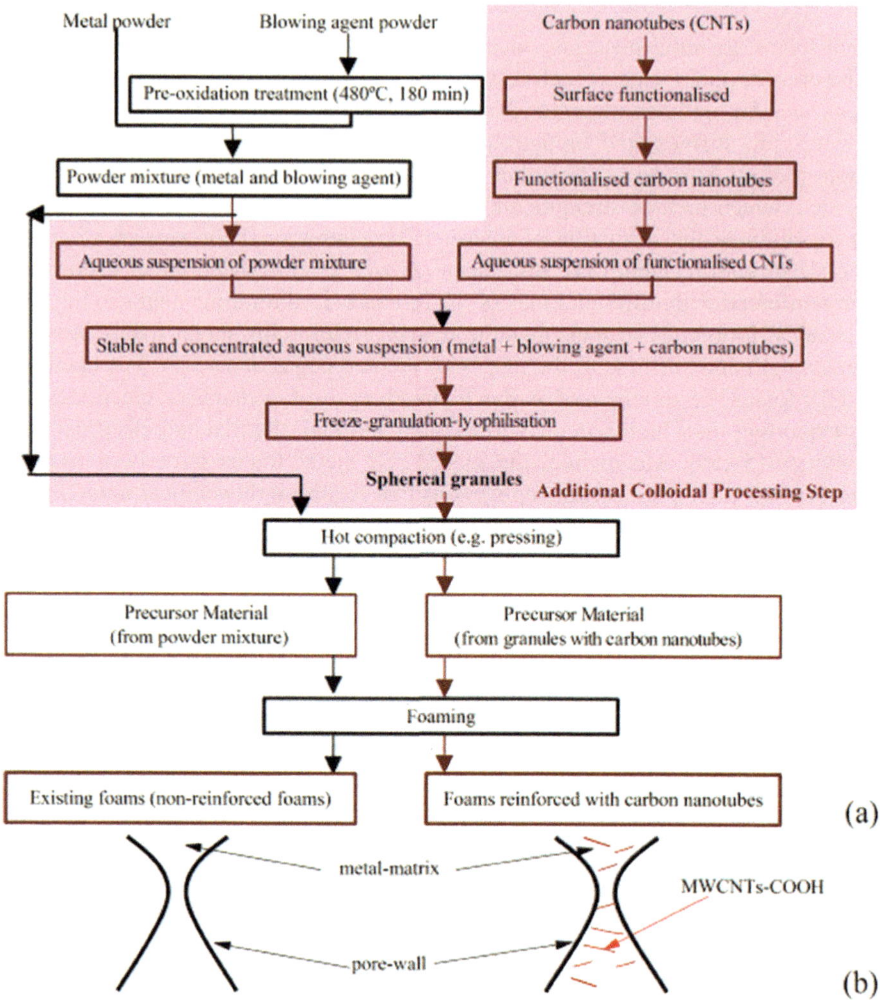

FIGURE 5.1 A schematic showing a comparison between (a) conventional powder metallurgy method for fabricating non reinforced closed cell metal foam and (b) a combination of colloidal processing and conventional powder metallurgy method for fabricating CNT reinforced Al foam [59].

variation in hardness value is attributed to lower CNT content, which results in higher hardness in the vicinity of CNT.

In another study, CNT reinforced closed-cell Al foam was fabricated using modified melt foaming method (Figure 5.3) for reinforcement contents of 0, 0.2, 0.5, 0.8, and 1.0% wt.% [60]. Ball milling parameters, that is, CNTs to aluminum powder ratio (CAR), ball to powder ratio (BPR), milling rate, and milling time were optimized to achieve effective dispersion and less structural damage. Based on degree of dispersion and I_D/I_G ratio, the optimal parameters of ball-milling were determined to be CAR-1 : 20, BPR-10 : 1, milling rate of 200 rpm, and milling time of 5 h. SEM

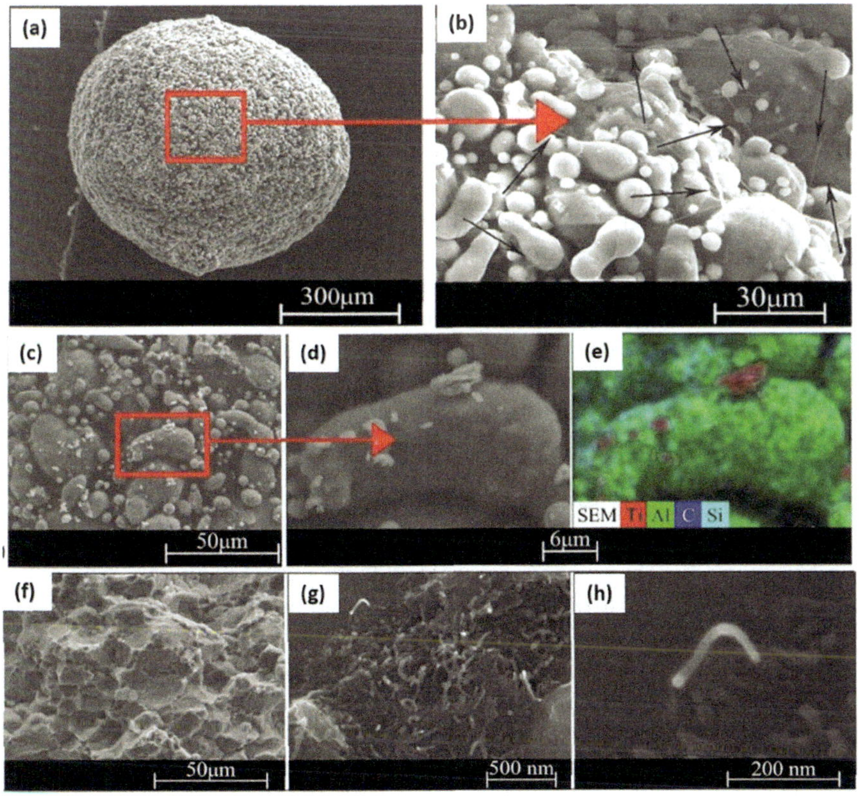

FIGURE 5.2 SEM micrograph (a) and (b) of spherical granules of Al-alloy and TiH_2 powders containing 0.5 wt.% of CNTs at different magnifications, (c) and (d) showing surface details under high magnification, (e) EDX elemental mapping and (f)–(h) fracture surface morphology of precursor material prepared from granule at different magnifications [59].

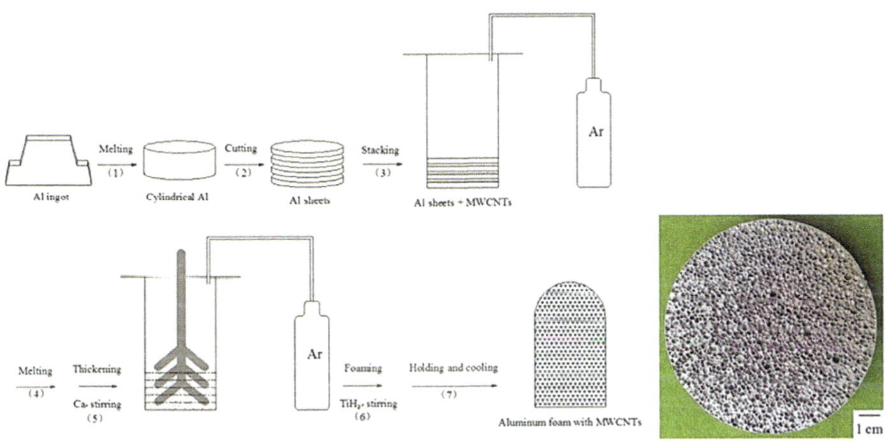

FIGURE 5.3 Schematic of CNT reinforced closed cell Al foam, fabricated using modified melt foaming method [60].

images confirmed the presence of CNTs inside the foams, in three different forms: (i) totally embedded in the cell wall, (ii) partially embedded in the cell wall, and (iii) totally exposed on cell wall surface. The presence of carbide was confirmed by XRD analysis for 1 wt.% CNT-Al composite, although the Al_4C_3 peaks were weak. The presence of defects and amorphous carbon layer on CNT surface improved the interfacial bonding leading to effective load transfer increasing the compressive yield strength by 42% and structural stiffness by 94%, along with energy absorption capacity with 0.5 wt.% CNT. This was the highest improvement achieved, beyond which the properties reduced due to agglomeration. Second phase strengthening due to the presence of carbide was claimed as they tend to bind and hinder dislocation. However, the claim was made just on the basis of XRD peaks which were very weak.

Wang et al. proposed an approach to obtain in situ grown carbon nanotube reinforced aluminum foams [61]. The method used a combination of in situ CVD, ball milling, and space-holder technique (Figure 5.4). CNTs were grown on Al powder by CVD, followed by ball milling. The ball milled powder was then mixed with carbamide particles and cold pressed into cylindrical samples under a pressure of 500 MPa for 5 m. Sintering the sample at 650 °C for 3 h helped achieve CNT-Al foam with 55% porosity. The compressive yield strength increases by 25% with only 2 wt.% CNT, as compared to only Al foam [10]. This study was followed by other studies by the same group [62–67].

The effect of CNT content (0, 2, 2.5, 3, and 3.5 wt.%) and milling time (30, 60, 90, and 120 min) on compressive properties of CNT-Al foam were studied [62]. Strong interfacial bonding between CNTs and Al was achieved with 90 min ball-milling, without any structural damage of CNTs. Post milling, Al powder probably gets cold-welded and forms large spherical particles with weak bonding. The CNT content influenced the compressive and energy absorption properties of the composite foams. Addition of 3 wt.% CNT improves the yield strength and energy absorption capacity of composite foams by 1.3 and 3.6 times, respectively. This is the highest improvement achieved and the properties deteriorated with higher CNT contents, due to agglomeration of CNTs [62]. Further, CNT-Al foams with CNT content (2, 2.5, and 3 wt.%) as well as porosity content (50, 60, and 70%) offered higher damping capacity than Al foams. The loss factor value at 25–200 °C were found to be 2.71 times higher for 3 wt.% CNT-Al composite foams [62]. When the temperature is increased to 390 ^{0}C, the value of loss factor was 1.77 times higher than pure Al foam [63]. The slippage between the inter-tubes and interlayer of CNTs, as well as the interfacial slip of CNT-Al contributes toward the increased energy dissipation

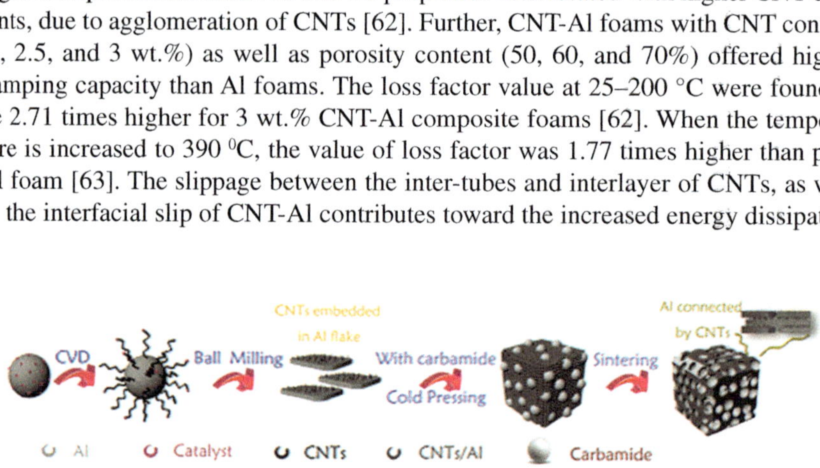

FIGURE 5.4 Schematic of approach to obtain in situ grown carbon nanotube reinforced aluminum foams [61].

ability of CNT-Al composite foams. Also foams with higher porosity possess a superior damping capacity [63]. The dynamic compressive properties of CNT-Al foam were investigated with different CNT content (0, 2, 2.5, and 3 wt.%) and at strain rate varying from 0.001 to 2130 s^{-1} [64]. The compressive strength and energy absorption capacity of the CNT-Al composite foams increase with increase in CNT content and strain rate. The peak stress increases by ~86% at 2130 s^{-1} as compared to 0.001 s^{-1} for 3.0 wt.% CNT-Al foams. The increase in stress bearing ability with higher CNT content is due to the higher number of CNT available for efficient load sharing. As the strain rate increases from 0.001 to 800 s^{-1}, the macro deformation behavior of the pure Al foam changes from the local failure to layers collapsing. Whereas, for CNT-Al foams shear deformation is the macro failure mechanism both under quasi-static as well as dynamic conditions. Further, the compression–compression fatigue performance of CNT-Al foams was investigated [65]. The strain (Ɛ) vs no. of cycle (N) graph for foams is composed of three stages, the fatigue strain accumulates very rapidly in stage III for CNT-Al foam as compared with Al foams. The fatigue strength for CNT-Al foam with CNT contents of 2, 2.5 and 3.0 wt.%, increases by 6, 44, and 102%, respectively. The deformation behavior for Al foam under fatigue loading is layer-by-layer, while the main failure modes of CNT-Al foams are the brittle fracture, as well as the collapse of pores with significant shear deformation bands [65].

These CNT-Al foams are also being used to strengthen or improve the compressive behavior of the thin-walled tubes that are being used as primary energy absorbing structures in various structural applications [66]. The CNT-Al foams are inserted into 6061 Al alloy empty tubes. The crushing behavior and energy absorption property of these tubes were evaluated using uniaxial compressive tests. Effective interaction between foams and pipe was observed (Figure 5.5). The highest stress and energy absorption capacity observed with 2 wt.% CNT-Al foam filled tubes are ~3.5 and 3.7 times higher, respectively, with very small increase in weight, as confirmed from the deformation images in Figure 5.5 [66]. The CNT-foam, filled inside the tube, effectively supports the tube due to the strong interaction between them while the CNTs efficiently shares load [66].

The technique used by Wang and group was further modified to achieve better dispersion [16]. Instead of using a combination of in situ CVD and space holder technique, CNT-Al foam was fabricated using in situ CVD and powder metallurgy

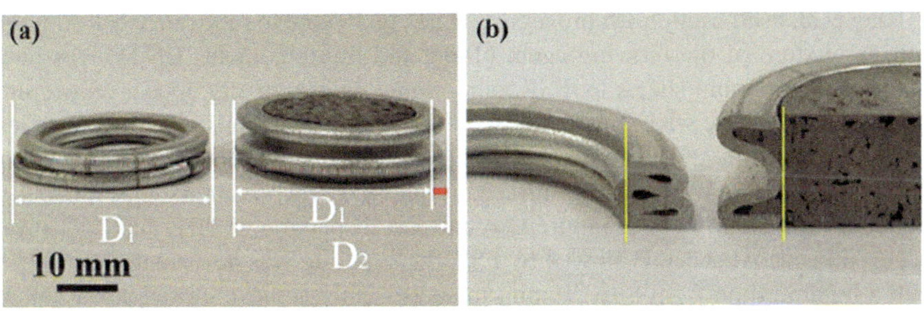

FIGURE 5.5 Digital images of (a) final deformed and (b) cross section of tubes and foam filled tubes [66].

method [67]. AlSi7 matrix was successfully fabricated using this method, with uniform distribution of 1.5 wt.% CNTs and 60% porosity. The CNTs provided more nucleation sites for decomposition of hydrogen (TiH_2 foaming agent) and resist merging of bubbles generated during foaming process, thus resulting in reduced pore size. The cell structure achieved was quite uniform, which is due to the stabilization of CNTs during sintering. As a result, the improvement in compressive stress and the energy absorption capacity is increased by ~95% and ~94%, respectively [67].

Stir casting is also used for fabrication of CNT-Al foams. Al-Cenosphere (ceramic microballoons)-SWCNTs closed-cell foams of varying relative densities were synthesized through stir casting technique using TiH_2 (0.6 wt.% of alloy) as a foaming agent [68]. The CNTs content was varied for 0, 0.1, 0.2, 0.3, and 0.4 wt.%. The content of cenospherewas kept at 4.25 wt.%/15 vol.%. The role of the cenosphere is to provide reinforcement to the cell wall induce microporosity that serves to decrease weight. The cell size and wall thickness varies with CNT content and also with the foaming temperature. With increase in CNT content, the nucleating points for solidification increase. As a result, the viscosity increases, leading to reduction in cell size and increase in cell wall thickness. On the contrary, the cell size increases and cell wall thickness decreases with increase in foaming temperature, due to the decrease in viscosity and surface tension of the melt. The foam is observed to have porosities at multiscale range, (i) the macro porosities due to presence of dissociated gases produced from the TiH_2 entrapped in the melt and (ii) micro-porosities within the cell wall due to the cenosphere particles. The compressive strength and modulus increases with increase in CNT content with highest improvement of ~47% and ~34%, respectively with 0.3 wt.% CNT, above which it reduces slightly due to agglomeration. The presence of uniformly distributed CNTs, having high surface area increase the interfacial bonding between cenosphere and matrix thus leading to efficient load transfer. Also, the large difference in coefficient of thermal expansion between CNTs and matrix increases dislocation density within the matrix in a random direction. This facilitates dislocation–dislocation interaction and its pile-up. Above 0.3 wt.% content, CNTs start to agglomerate, reducing dislocation generation and stress concentration [68]. As observed from the quasi-static compression deformation mechanism, the deformation of foams begins near the loading surface and moves toward the rear surface, layer by layer. Also, it is observed that hybrid foams with same relative density but different CNT contents possess similar kind of deformation behavior.

Pang et al. used friction stir processing (FSP) technique to fabricate CNT-Al foam [69]. A mixture of the blowing agent (TiH_2) and reinforcement (CNTs) was integrated into aluminum sheets by FSP route. Some holes/porosity appear on the surface of the Al sheet, when processed with FSP conditions consisting of a rotation speed of 1000 rpm and welding speed of 100 mm min^{-1}. Increase in the welding speed to 300 mm min^{-1} made the surface morphology smoother and denser, due to a more uniform frictional heat input. Also uniform dispersion of TiH_2 and CNTs was achieved inside Al sheet. Followed by FSP, the structure was heat treated at 680 °C with a holding time from 8 to 20 min, to successfully achieve closed-cell CNT-Al composite foams [69]. The porosity and pore diameter of closed-cell CNTs-Al composite foams increases with the higher holding time. Specifically, when the holding time is 15 min at 680 °C, the porosity is uniformly distributed and spherical, due to the sufficient expansion of the precursor. Increasing holding time further led to large pores and increased cracking within pores. The compressive yield stress of CNT-Al composite foams was 3–4 times higher than that of pure Al foam [69]. Similarly, in another study, Al foams reinforced with silicon carbide (0, 2, 4, and 6 wt.%) and carbon nanotubes (0, 0.2, 0.5, 0.8, and 1 wt.%), fabricated by using powder metallurgy process (ball milling, compaction followed by extrusion), performed better than only Al foam [19]. The foams with relative density of up to 94% were fabricated. A hardness of 0.11 GPa was obtained in Al, which increased to 0.77 GPa with 6 wt.% SiC and 1 wt.% CNT nanofiller addition [70].

Only one study has been reported so far on CNT reinforced Mg foams. AZ61 was added to CNT-ethanol-surfactant mixture and ultrasonicated at 60°C to disperse CNT in Mg alloy and evaporated the ethanol. The mixture is than ball milled at 300 rpm for 60 min. The technique used for mixing was very effective in dispersing and incorporating CNTs into the matrix without, loss of structural integrity, as confirmed through SEM and raman spectroscopy. Spherical carbamide granules were used to control the porosity distribution, shape, and size in the foam. The pores are spherical, undistorted, and evenly distributed throughout the matrix, with pore sizes varying in the range of 0.2–0.9 mm. When the porosity was ~49%, the pores were closed and isolated, having minimal interconnectivity. Increasing porosity (62%) led to nearly fully interconnected pores [71].

The samples were rapidly consolidated in 20 min, through microwave sintering. The SiC/graphite-lined kiln acts as an external susceptor for initially heating the samples externally, until the CNTs incorporated into the matrix absorb heat and act as internal susceptor to initiate volumetric heating. This results in rapid temperature rise in the samples, which reduces the formation of brittle interfacial products. The compression test results showed improvements in compressive strength and energy absorption by ~93% and 153%, respectively, with 3 wt.% CNTs and 49% porosity level [71]. Almost similar improvement was reported in foam samples with 62% porosity levels. The major mechanisms responsible for the failure of the foam samples are the initiation of cracks and their propagation along the loading direction, as well as a mixed mode of shear and brittle failure of the pore walls and matrix. Therefore, this approach successfully fabricates CNT-Al foam with reduced processing time and cost, along with release of toxic gases [71].

In short, CNT-Al foam has been successfully fabricated using various techniques having various CNT and porosity content, uniformly distributed throughout the matrix with and without formation of interfacial products. CNT content along with processing parameters like temperature, time, etc. are seen to influence the quasi static and dynamic compressive properties, energy absorption capacity, fatigue, as well as damping capacity of metallic foams.

5.6 CNT-METAL NANOLAMINATES

In recent years, various approaches have been adopted for fabricating CNT reinforced MMCs [32,72–75]. These approaches have focused on balancing both strength and toughness, by tailoring the orientation and/or distribution of reinforcements in the matrix, rather than changing the constituents. It has been widely accepted that among various architectures, laminated structures can effectively inhibit the penetration of dislocations through the high density of interfaces introduced by nanoscale matrix layers. In addition, laminated structure is also quite helpful for deflecting crack and resisting its propagation, which makes a material tough. Techniques, such as rolling [76], selective dip coating [77], electrodeposition [75,78], flake powder metallurgy [79] via slurry blending [74,80,81], shift speed ball milling [82], and in situ approach [83,84], have been used by different research groups to fabricate CNT-Al [74,80,82,83] and CNT-Cu [75–79,81] matrix laminated composites, with optimized strength and ductility.

Initially, simple and effective methods like electrodeposition were used to fabricate Cu/SWCNT/Cu sandwich-type laminated nanocomposites [78]. Directly grown SWCNT films, with continuous reticulate structure, was used as a template. The estimated Young's modulus of the SWCNT films is in the range of 860–960 GPa. The resulting Cu/SWCNTs/Cu laminated nanocomposites exhibited improvement of ~120% and ~197% in strength and Young's modulus, respectively, over Cu. Improvement in properties is attributed to effective load-transfer capacity of reticulated architecture of SWCNT films and the strong interfacial interaction between the SWCNTs and Cu matrix [78]. In another effort, CNTs were electrodeposited on Cu foil as template, followed by hot pressing of CNT-Cu foil laminates and hot rolling of the laminated composite in sequence [75]. Similarly, Zhu et al. [76] fabricated CNT-Cu laminated composites by repeated cold rolling of CNT films, sandwiched between Cu thin foils, followed by annealing. However, Kim et al. obtained a similar laminate microstructure by selective dip-coating and electroplating [77]. Selective dip-coating was performed at room temperature by immersing the photoresist patterned substrate vertically into the suspension of CNTs and then withdrawing it at a velocity of 3 mm min^{-1}. Cu was electroplated in a sulfuric acid bath at room temperature. Selective dip-coating and electroplating were carried out alternately to form the laminated structures until desired thickness was achieved. Improvement in both strength and ductility were realized by these laminate structures. However, these methods can hardly reduce the laminate thickness to submicron or nanoscale.

Yang et al. found that the in situ chemical vapor deposition (CVD) processed CNT-Al composite powders could also be flattened into CNT-Al flakes by ball milling, and cold-welded into particles with a 90 min milling time [83,84]. It was found that the cold welding favored better inter-flake bonding and thus led to the best combination of strength and ductility in the as-fabricated bulk composites. In situ approach [83,84], along with other methods, like spray pyrolysis [85] or conventional powder metallurgy (PM) [83,84], has been proposed to facilitate dispersion of CNTs in metal matrix for laminated composite with small milling time. However, it is very difficult to achieve uniform dispersion, while retaining the structural integrity of CNTs.

Flake powder metallurgy has been explored to overcome the drawback of conventional powder metallurgy [74,80]. Spherical Al particles were processed by ball milling into 2-D nanoflakes (Figure 5.6) and then modified using polyvinyl alcohol (PVA) hydrosol to make their surface properties and geometries more compatible. This resulted in a uniform and individual distribution of CNTs in Al powders simply by direct slurry blending. The blend was further consolidated by hot extrusion, with

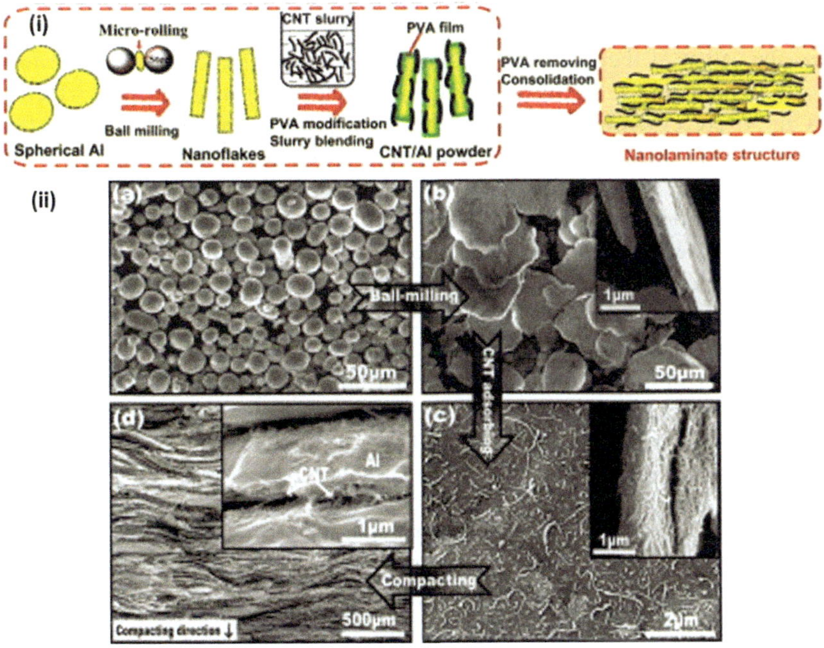

FIGURE 5.6 (i) Schematic of fabrication process of CNT-Al composite using flake powder metallurgy, (ii) FESEM images of (a) Al particle, (b) Al nanoflakes, (c) Al nanoflakes with embedded CNT, and (d) nanolaminates post compaction [80].

almost no damage to the structural integrity of the CNTs, as confirmed through Raman spectroscopy. As a result, a strong and ductile CNT-Al composite was fabricated. The tensile strength and plasticity was 375 MPa and 12%, respectively, with 1 vol.% CNT [80] and 435 MPa and 6%, respectively, with 2 vol.% CNT [74], which was better than that fabricated through conventional methods. Uniformly distributed CNTs throughout the nanolaminate could pin dislocations, thus reducing dynamic recovery post hot extrusion. This is confirmed from the dislocation density, which is much higher in case of flake PM nanocomposite, in comparison to conventional PM methods. Significant dislocation storage allows a high strain-hardening rate, which leads to larger uniform strains, while maintaining a high level of strength [86]. Thus, the increased ductility was achieved in the flake PM CNT-Al nanocomposites.

Liu et al. fabricated CNT-Cu composite by flake powder metallurgy, followed by hot rolling [81]. The composites showed a good balance in strength, ductility, and electrical conductivity. The UTS of 1 vol.% multi-walled CNT-Cu composite was 395 MPa, which is 87% higher than that of coarse-grained Cu, along with an electrical conductivity of 90% IACS and an elongation rate of >20% [81]. Homogeneous dispersion of multi-walled CNTs in Cu matrix, with minute structural damage of multi-walled CNTs, along with strong interfacial bonding between multi-walled CNTs and Cu contributed to this performance. X-ray tomography and in situ digital image correlation (DIC) revealed that the laminated structure is effective in reducing strain concentration, which is important to achieve strength-ductility balance [81].

Apart from balancing strength and ductility for nanolaminates, it is important to assess the effect of thermal cycling on mechanical properties. This is due to the increase in demand for multifunctional and higher performance MMCs which can work in severe environment. Severe thermal cycling generates huge interfacial stresses between reinforcement and matrix due to the huge difference in their coefficient of thermal expansion, which may induce micro damage and lead to catastrophic failure. Hence, the effect of thermal cycling on mechanical properties of composite is investigated in a temperature range of –196 °C to + 200 °C [87]. UTS, ductility, and Young's modulus of the CNT-Cu laminated composite decreased with increase in the number of thermal cycles. This is due to the gradual increase in dislocations density in the Cu matrix with increasing thermal cycles. Cracks start forming with further increase in the thermal cycles, which releases stress and thus reduces dislocation density. The major failure mode is delamination of CNT walls in the CNTs-Cu composites [87]. Flake powder metallurgy via shift-speed ball milling (SSBM) was also explored to fabricate laminated composite [82]. This technique combines two different milling speeds, namely low-speed ball milling (LSBM) and high-speed ball milling (HSBM), followed by sintering and hot extrusion, to fabricate CNT-Al composite. LSBM was done for 8 hr for slow flattening of spherical Al powders into flakes, giving enough time for CNTs to be uniformly dispersed on the surface of Al flakes, with little damage to the CNTs. Followed by this, 1 hr of HSBM was performed for embedding CNTs into Al flakes and cold welding of Al flakes. This methodology helps in achieving uniform CNT dispersion, while maintaining their structural integrity and interfacial interaction with the matrix. This effort resulted in strength comparable to composite fabricated though only LSBM and HSBM along with doubled and tripled ductility with 1.5 wt.% CNT/Al composites [82]. Thus, the fabrication of nanolaminates through flake powder metallurgy technique is most effective in achieving a strength-ductility balance.

5.7 OTHER METALS-CARBON NANOTUBE SYSTEMS

CNTs have also been used as reinforcement for metals, alloys, intermetallics, and bulk metallic glasses to improve the mechanical properties of the composite. Metal matrices like Ag, Co, W, Ni-Ti, Ti, Zn, Sn-alloys, stainless steel, and Si have also been explored. Si-CNT composites have been explored mainly as electrodes in lithium ion batteries. They have been prepared mainly by CVD deposition of Si on CNTs and are tabulated in Table 5.5. The electrical properties of Si-CNT composites are discussed in Chapter 8. Titanium and its alloys are important structural materials but have been explored less for synthesizing CNT reinforced composites. Intermetallics like Fe_3Al and metallic glasses based on Fe, Zr, and Ti have been studied to develop advanced composites with CNT reinforcements. CNTs have also been used to improve the properties of lead-free solders. The main method has been powder metallurgy due to its versatility. These are scattered efforts and hence will be discussed together in the following paragraphs. These have been summarized in Table 5.6.

TABLE 5.5
Table of the Summary of the Work Carried Out in an Al-CNT System

Composite Details	Composite Processing Technique	CNT Dispersion and Interface	Other Properties	Ref.
Si-CNT (anode material for Li-ion batteries)	Si powders ball milled CNTs were grown on Si powders by CVD (catalyst Ni-P	—	Better cyclic performance	[195]
Si-CNT (anode material for Li-ion batteries)	Ball milling	Strong interfacial contact is reported	For higher CNT content, reversible capacity decreases less with number of cycles	[196]
Si-CNT (composite electrode for Li secondary batteries)	CNTs were grown on Si powders by CVD, forms conductive buffer layer	—	Best performance, accommodates Si swelling and maintains conductivity network, minimizes charge transfer resistance at discharge	[197]
Si coating on CNT	CVD using silane (SiH_4) gas	Well covered and continuous film is found to form	Thermal stability of coated CNT improved, oxidation temperature of coated CNT is 949.3 K and of bare CNT is 844.2 K	[198]
Si particles coated with CNT	CNT coating on Si particle by decomposition of tetramethyl silane	–	Hydrogen storage capacity of composite improves with addition	[199]
Si-C-SWNT (28, 37 wt. %) composite, electrode for Li-ion batteries	Si-C powder prepared by ball milling, CNT dispersed ultrasonically, compacted and thermally treated (110°C, overnight, vacuum oven)	–	Maintains good electrical contact and excellent capacity retention with high reversible capacity	[200]

Note: — Data not available or not applicable.

TABLE 5.6
Table of the Summary of the Work Carried Out in Other Metal-CNT Systems

CNT Content	Composite Processing Technique	Tensile Test Sample Size	Mechanical Properties (UTS, YS, E, Strain to Failure, Hardness)		CNT Dispersion and Interface	Other Properties	Ref.
			With CNT	Without CNT			
$^{Fe}82^{P}18$ metallic glass – CNT (3 wt.%)	Raw material powders, melted and rapidly solidified to ribbons on a copper spun wheel	–	–	–	Good wettability, filling of CNTs with iron is reported along with no agglomeration of CNTs	–	[201]
Ti-CNT	Mechanical mixing, hot pressing in carbon die (935°C, 30 MPa, 2 h, vacuum)	–	H – 1216 VHN E* – 198 GPa (* measured using ultrasonic spectrum microscopy)	H – 221 VHN E* –120 GPa	–	–	[87]
Co-CNT	Co-plating formed by electroless deposition on CNT from $CoSO_4$ bath	–	–	–	Coating was found to be not uniform, although gaps or voids were decreased by annealing	–	[202]
W-CNT	W coating was formed on CNTs by PVD	–	–	–	Coated CNTs were not of uniform thickness	–	[203]
$^{Zr}52.5^{Cu}17.9$ $Ni_{14.6}Al_{10}$ Ti_5 – CNT (1, 2, 3, 4 vol.%)	Pre-alloyed ingots ground-mixed homogeneously with CNT, compressed into cylinders, melted and casted in mold (argon atmosphere)	–	$Zr_{52.5}Cu_{17.9}$ $Ni_{14.6}Al_{10}Ti_5$ – 3 vol.% CNT E* – 94.7 GPa H – 626 VHN (* by acoustic velocity)	E* –88.6 GPa H –579 VHN	Homogeneous dispersion is inferred from mechanical properties, ZrC formed at interface	–	[90]

Sb-$SnSb_{0.5}$ – CNT – anode for Li-ion batteries	Chemical reduction of metal-chloride salts in CNT suspensions	–	–	–	Semi-continuous Sb-layers on CNTs is observed	CNT-web works as effective stress absorber resulting in better reversible capacities	[204]
$Zr_{52.5}Cu_{17.9}$ $Ni_{14.6}Al_{10}$ Ti_5 – CNT (1, 3, 5, 7, 10 vol.%)	Pre-alloyed ingots ground-mixed homogeneously with CNT, compressed into cylinders (40 MPa, 60 min), melted and casted in mold (argon gas pressure 600 mbar)	–	$Zr_{52.5}Cu_{17.9}$ $Ni_{14.6}Al_{10}$ Ti_5 – 4 vol.% CNT E – 98.49 GPa H – 654 VHN (* by acoustic velocity)	E – 88.56 GPa H – 579 VHN	Some CNT clusters and some individually dispersed CNTs are observed, ZrC phase also is formed	–	[91]
Ag-CNT (10–25 vol.%)	Mixed by hand grinding (0.5 h), uniaxially pressed (320 MPa, 120 sec), isothermally sintered (700°C, 1 h), repressed (400 MPa)		Ag-9 vol.% CNT H – 84 HVN Ag – 8 vol.% CNT Bend strength – 465 MPa	H – 66 HVN Bend strength – 427 MPa	Some agglomerations at higher CNT addition are observed	Electrical resistivity increases slightly up to 10 vol.% CNT; after that sudden increase up to 25 vol.%	[205]
Au nanoparticle-CNT	CNT suspended in water-polyethyleneimine and $HAuCl_4$ added, heated, composite solution placed on glassy carbon electrode, dried to form coating	–	–	–	Gold particles are found tightly adhered to the surface and can resist washing and sonicating	Composite shows better electrocatalytic activity as dioxygen reducers than only CNT	[206]
95.8Sn-3.5 Al-0.7Cu – CNT (0.01, 0.04, 0.07 wt.%)	Mixed in blender (10 h, 50 rpm), uniaxially compacted (140 bar), sintered (175°C, 2 h, in argon), cold extruded (ratio – 20)	Diameter – 5 mm Gauge length – 25 mm	95.8Sn-3.5 Al-0.7Cu – 0.04 wt.% CNT $\sigma_{YS(T)}$ – 36 MPa σ_{TS} – 46 MPa H ~ 17 VHN	$\sigma_{YS(T)}$ – 31 MPa σ_{TS} – 35 MPa H ~ 16.3 VHN		Better wetting and smaller contact angle is reported with increase in CNT content Melting point does not change with CNT content	[207]

(Continued)

TABLE 5.6 (*Continued*)
Table of the Summary of the Work Carried Out in Other Metal-CNT Systems

CNT Content	Composite Processing Technique	Tensile Test Sample Size	Mechanical Properties (UTS,YS, E, Strain to Failure, Hardness)		CNT Dispersion and Interface	Other Properties	Ref.
			With CNT	Without CNT			
95.8Sn-3.5Al-0.7Cu – CNT (0.01, 0.04, 0.07 wt.%)	Mixed in blender, uniaxially compacted, sintered (175°C, 2 h, in argon atm.), cold extruded (ratio – 20)	Diameter – 5 mm Gauge length – 25 mm	95.8Sn-3.5Al-0.7Cu – 0.04 wt.% CNT $\sigma_{YS(T)}$ – 36 MPa σ_{TS} – 46 MPa H ~ 17 VHN	$\sigma_{YS(T)}$ – 31 MPa σ_{TS} – 35 MPa H ~ 16.3 VHN	Wetting gets better and contact angle smaller with increase in % CNT	Better wetting and smaller contact angle is reported with increase in CNT content Melting point does not change with CNT content CTE decreases with CNT content	[93]
Fe_3Al –CNT (~3 wt.%)	Fe_3Al powders prepared by ball milling and heat treatment, added to dispersed CNTs and the slurry is ball milled (24 h), dried, SPS (1150°C, 35 MPa)	Compressive test specimen – 4 × 4 × 7 mm^3	σ_{CS} – 1583 MPa Bending strength – 820 MPa H – 503 HVN	σ_{CS} – 1566 MPa Bending strength – 780 MPa H – 386 HVN	Some clusters of CNTs, straightly aligned CNTs, and some CNTs interlocked in Fe_3Al particles are observed	–	[96]
Fe_3Al –CNT (5 wt.%)	Fe_3Al powders prepared by ball milling and heat treatment, added to dispersed CNTs and the slurry is ball milled (12 h), dried, SPS (1000°C, 30 MPa)	Compressive test specimen – 4 × 4 × 7 mm^3	σ_{CS} – 1583 MPa H – 503 HVN	σ_{CS} – 1566 MPa H – 386 HVN	Woven cross-ply CNT structure in composite is observed	–	[98]
Ti-Ni coated SWNT (0.5, 1.5, 4.5 wt.%)	Powder metallurgy method	–	H – 900 VHN	H – 300 VHN	–	–	[88]

$Ti_{50}Cu_{28}Ni_{15}Sn_7$ – metallic glass – CNT (12 vol.%)	Elemental metallic powder and CNT ball milled, vacuum hot pressed (450°C, 1.2 GPa)	–	H – 9.34 GPa	H – 6.1 GPa	–	T_g and crystallization temperature increases, supercooled liquid range is the same	[89]
W-Cu (15 wt.%) alloy – Cu coated CNT (0–10 wt.%)	CNTs electroless coated with Cu, ball milled with W-Cu powder in isopropyl alcohol(5 h), dried (100°C, vacuum), hot press sintered (1400°C, 30 MPa, 2 h)	Transverse rupture test specimen 20 × 6.5 × 5.25 mm^3	Transverse rupture strength – 1440.8 MPa	Transverse rupture strength – 1220.8 MPa	Agglomeration of CNTs at higher content (>4 wt.%) is observed	Thermal conductivity increases with increase in CNT content	[208]
Zn-CNT	CNTs dispersed in electrolyte bath and electro-deposited	–	–	–	–	Better corrosion resistance and delayed white rust formation with CNT addition	[209]
Sn-3.8Ag-0.7Cu solder alloy–SWNT (0.01–1 wt.%)	Solder powder and CNTs mixed in blender (15 h, 50 rpm), cold compacted (120 bar, 30 sec), sintered (180°C, 3 h), cold extruded (ratio – 20)	Diameter – 5 mm Gauge length – 25 mm	Solder alloy – 1 wt.% CNT σ_{TS} – 57 MPa H – 187 MPa	Mechanical properties σ_{TS} – 38 MPa H – 159 MPa	CNTs distributed at Ag_3Sn grain boundaries, homogeneous distribution, and alignment of CNTs with good bonding to matrix is found at fracture surface	Melting temperature decreases with addition of CNT	[92]
Ni-Cu-P – CNT (3.68 wt.%)	Composite coating obtained by electroless plating, annealed	—	H – 1500 VHN	H – 600 VHN	Some agglomeration of CNTs are found	Corrosion resistance increases with CNT addition	[210]
Ti-based BMG– CNT (0–12 vol.%)	Mechanical alloying (8 h), vacuum hot pressing (450°C, 1.2 GPa)	Compressive test specimen 2 × 2 × 2 mm	Ti-based BMG – 12 vol.% CNT H – 9.34 GPa σ_{CS} – 1937 MPa	H – 6.85 GPa σ_{CS} – 1688GPa	—	Glass transition and crystallization temperature increases with CNT addition	[211]

(Continued)

TABLE 5.6 (*Continued*)
Table of the Summary of the Work Carried Out in Other Metal-CNT Systems

CNT Content	Composite Processing Technique	Tensile Test Sample Size	Mechanical Properties (UTS,YS, E, Strain to Failure, Hardness)		CNT Dispersion and Interface	Other Properties	Ref.
			With CNT	Without CNT			
Cr-CNT (1 mass.%)	Electrodeposited	—	H – 10.5 GPa	H – 8 GPa	Uniform dispersion of CNTs in the matrix is reported	Addition of CNT decreases wear loss and increases coefficient of Friction	[212]
Au-CNT-Pt	CNT dispersed in dimethylformamide by ultrasonication, dip coated on Au electrode, Pt is electrodeposited on the CNT coated electrode	–	–	–	–	Pt modified Au-CNT electrode shows better electrocatalytic activity and stability	[213]
Ti – CNT (0–0.35 wt.%)	Ti powder particle coated with CNT by dipping in CNT solution, dried (100°C) SPS in two steps (first, 600°C, 20 kN; second, 800°C, 41.6 kN), heated (1000°C), hot extruded (400°C, ratio – 37)	Diameter – 3 mm Gauge length – 15 mm	Ti-0.35 wt.% CNT σ_{TS} – 697 MPa $\sigma_{YS(T)}$ – 591 MPa H – 285 VHN	σ_{TS} – 472 MPa $\sigma_{YS(T)}$ – 591 MPa H – 261 VHN	Uniform distribution of CNT in Ti matrix and formation of TiC at interface is observed	–	[214]
Ti-CNT (3 wt.%)	CNTs electronegatively charged by surface treatment, mixed with Ti powder by coacervation technique, SPS (800/900/1000°C, 50 MPa, 5 min)	–	$\sigma_{YS(C)}$ – 145 MPa at sintering temperature of 800°C		Uniform distribution of CNT in Ti matrix and formation of TiC at interface is observed	–	[215]

Pd-Ni-CNT (0–5 wt.%)	Mixture of Pd, Ni, and CNT are dispersed in liquid and dropped on glassy carbon and dried, covered with Nafion solution and microwaved	–	–	–	–	With addition of 1 wt.% CNT and 1 wt.% Ni, the catalytic activity is maximum	[216]
Sn-Ag-Cu solder alloy – CNT (0.05 wt.%)	Blending of CNT and Sn-Ag-Cu solder alloy powder (10 h, 50 rpm), compaction and sintering (175°C, 2 h), cold extrusion (ratio – 20)	Diameter – 5 mm Gauge length – 25 mm	$\sigma_{YS(T)}$ ~ 49 MPa σ_{TS} ~ 55Pa	$\sigma_{YS(T)}$ ~ 43 MPa σ_{TS} – 47 MPa	Clusters of CNTs are found uniformly distributed in matrix	–	[217]
Sn-Ag-Cu solder alloy – CNT (0–0.07 wt.%)	Blending of CNT and Sn-Ag-Cu solder alloy powder (10 h), compaction and sintering (175°C, 2 h), cold extrusion (ratio – 20)	–	Solder alloy – 0.01 wt.% CNT Shear yield strength ~ 18 5 MPa Shear strength ~ 28 MPa	Shear strength Shear yield strength ~ 16.5 MPa Shear strength ~ 22 MPa	Presence of both dispersed CNTs and clusters found in matrix are reported	Intermetallics compound layer at solder alloy and Cu-pad joint grown slower upon aging with CNT addition	[218]

Note: E – elastic modulus. H – hardness; σ_{TS} – tensile strength; σ_{CS} – compressive strength; $\sigma_{YS(T)}$ – yield strength in tension; $\sigma_{YS(C)}$ – yield stress in compression; — Data not available or not applicable.

CNTs have been observed to be a good reinforcement for titanium matrix composites. The formation of TiC is a possibility with CNTs; however, it is not detrimental as compared to Al_4C_3 in Al-CNT composites. Ti-CNT composite, produced by vacuum hot pressing of mechanically mixed Ti-CNT powder, showed a 450% improvement in hardness and a 65% increase in elastic modulus although CNT content was not known [88]. It was observed that CNTs were better reinforcement than graphite and C60 allotropes. TiC formation was observed from XRD. A 200% increase in hardness has been observed in Ti-Ni shape memory alloy with 4.5 wt.% CNT addition [89]. Ti-based BMG-CNT composites, processed by powder metallurgy technique, have shown a 53% increase in hardness [90]. However, reinforcing Zr-based BMG with CNT has not been proven so successful. Such composites, prepared by melting and casting technique, show ~10% improvement in hardness and elastic modulus [91,92]. ZrC formation was observed in these composites. Due to their small length scales, CNTs are expected to influence phenomena like glass transition, crystallization, and phase separation in metallic glasses. Addition of 0.01 wt.% SWNT was shown to result in a 50% increase in the tensile strength of Sn-Ag-Cu solder [93], while addition of 0.04 wt.% CNT results in increase in tensile strength by 31% [94]. Reinforced Sn-Bi soldering alloy has also seen improvement in bending strength by 10% and toughness by 50%, which was attributed to the role played by pinning and bridging effect of CNT on fracture behavior of the structure [95]. Fatigue life of stainless steel (SS304) was also found improved with addition of CNT, due to dense arrangement of entangled CNT at the crack tip and a finer grain size [96].

CNTs have also been used as reinforcement for intermetallics. Addition of 3 wt.% CNTs to Fe_3Al by powder metallurgy technique resulted in an increase in the hardness and compressive strength by 30% and 11%, respectively [97,98].

It is observed in this chapter that CNTs have been successfully used as reinforcement material for several metallic or metal-based systems. The studies in systems like Al-, Ni-, and Cu-CNT are large in number, whereas other metal matrices such as Ti, Mg, and Si have been barely investigated. There is large scope for the improvement of properties and development of novel methods of fabrication for meeting the challenges. The processing methods used are still in the preliminary stage and need to be optimized. The improvement in the mechanical properties of the composites has been plotted vs. the CNT content of composites in Figure 5.7 [247]. It is observed that novel techniques that result in better dispersion are more successful in improving mechanical properties of the composites. Few cases of electroless deposited composites, which have shown very high increase in mechanical properties, also employ special treatment of CNTs prior to electroless plating. Hence, it is clear that processing routes are very important for the mechanical properties exhibited by the composite. Further research on modification of existing processing routes and innovation of new ones can assist in preparation of even stronger CNT-MMCs.

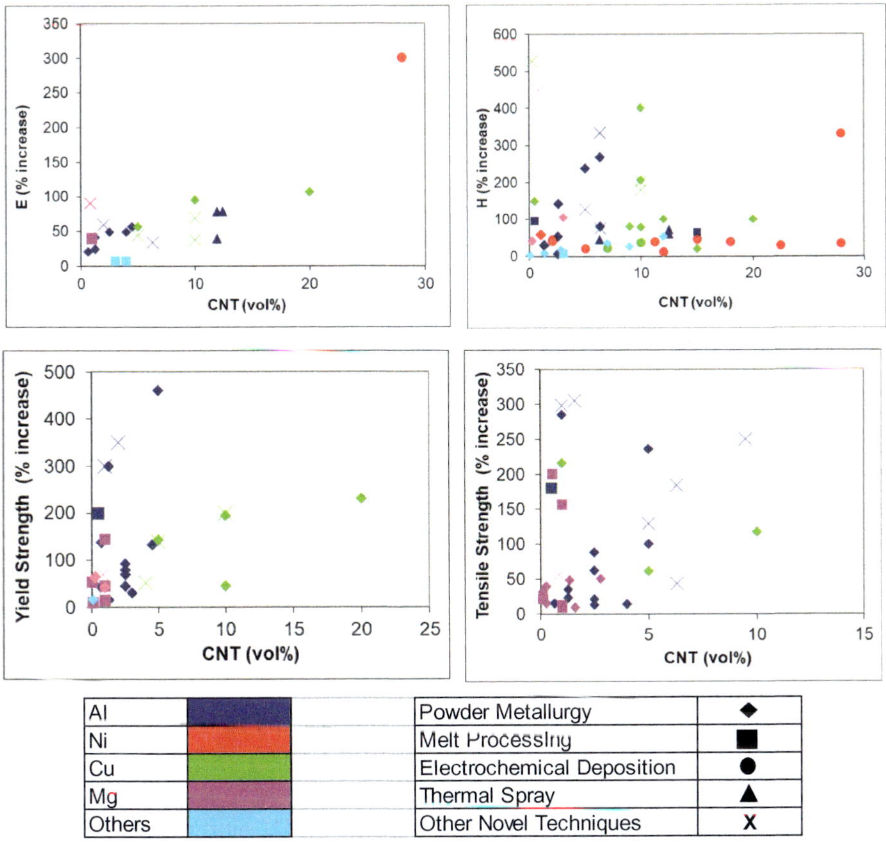

FIGURE 5.7 Improvement in the mechanical properties in CNT-MMCs as a function of CNT content and processing technique used [247]. (Reproduced with permission from Maney).

5.8 CHAPTER HIGHLIGHTS

This chapter provides a comprehensive summary of the work carried out on several CNT-MMCs. The tables provide a summary of CNT content, processing technique at powder preparation and consolidation stages, CNT dispersion, and the effect on the mechanical, thermal, and electrical properties as studied. The sample size for mechanical property measurement is also tabulated along with comments on the dispersion of CNTs. Most of the studies on Al-CNT and Cu-CNT composites have utilized powder metallurgy techniques. This is due to the different possibilities it provides in controlling the CNT dispersion. The use of SPS is becoming popular for CNT-MMC consolidation. This is due to the small sintering times and dense composite it produces. Most of the studies on Ni-CNT composites have utilized electrochemical deposition techniques to synthesize these composites in thin coatings or film form. Melt processing has been used for low melting metals like Mg and Al and some exotic materials like bulk metallic glasses.

REFERENCES

1. Kuzumaki, T., Miyazawa, K., Ichinose, H., and K. Ito. 1998. Processing of carbon nanotube reinforced aluminum composite. *J. Mater. Res.* 13: 2445–2449.
2. Laha, T., Agarwal, A., McKechnie, T., and S. Seal. 2004. Synthesis and characterization of plasma spray formed carbon nanotube reinforced aluminum composite. *Mater. Sci. Eng. A* 381: 249–258.
3. Rao, P. K. V., Raghukumar, B., Sudheer Kumar, B., Phanindra Swamy, G., Ganga Raju, Y., and P. Sajjad Ali Khan. 2018. Investigation of flexural and impact strength of carbon nanotube reinforced AA7075 metal matrix. *Int. J. Engg. Tech.* 7: 764–767.
4. Shin, S. E., and D. H. Bae. 2018. Fatigue behavior of Al2024 alloy-matrix nanocomposite reinforced with multi-walled carbon nanotubes. *Composites B.* 134: 61–68.
5. Bakshi, S. R., Singh, V., Seal, S., and A. Agarwal. 2009. Aluminum composite reinforced with multiwalled carbon nanotubes from plasma spraying of spray dried powders. *Surf. Coat. Tech.* 203: 1544–1554.
6. Laha, T., Agarwal, A., and T. McKechnie. 2005. Comparative evaluation of plasma and high velocity oxy-fuel spray formed carbon nanotube reinforced Al-based composite. In *Surface Engineering in Materials Science-III*. A. Agarwal, N. B. Dahotre, S. Seal, J. J. Moore, and C. Blue, Eds. Warrendale, PA: The Minerals, Metals and Materials Society.
7. Laha, T., Liu, Y., and A. Agarwal. 2007. Carbon nanotube reinforced aluminum nanocomposite via plasma and high velocity oxy-fuel spray forming. *J. Nanosci. Nanotech.* 7: 515–524.
8. Bakshi, S. R., Singh, V., Balani, K., McCartney, D. G., Seal, S., and A. Agarwal. 2008. Carbon nanotube reinforced aluminum composite coating via cold spraying. *Surf. Coat. Tech.* 202: 5162–5169.
9. Laha, T., Chen, Y., Lahiri, D., and A. Agarwal. 2009. Tensile properties of carbon nanotube reinforced aluminum nanocomposite fabricated by plasma spray forming. *Composites A* 40: 589–594.
10. S. R. Bakshi. Plasma and cold sprayed aluminum carbon nanotube composites: quantification of nanotube distribution and multi-scale mechanical properties. Ph.D. Dissertation. Florida International University, Miami, FL. http://digital-commons.fiu.edu/etd/97/.
11. Esawi, A. M. K., and M. A. E. Borady. 2008. Carbon nanotube-reinforced aluminum strips. *Composites Sci. Tech.* 68: 486–492.
12. Salas, W., Alba-Baena, N. G., and L. E. Murr. 2007. Explosive shock-wave consolidation of aluminum powder/carbon nanotube aggregate mixtures: optical and electron metallography. *Met. Mater. Trans. A* 38: 2928–2935.
13. He, C., Zhao, N., Shi, C., Du, X., Li, J., Li, H., and Q. Cui. 2007. An approach to obtaining homogeneously dispersed carbon nanotubes in Al powders for preparing reinforced Al-matrix composites. *Adv. Mater.* 19: 1128–1132.
14. Noguchi, T., Magario, A., Fukuzawa, S., Shimizu, S., Beppu, J., and M. Seki. 2004. Carbon nanotube/aluminum composites with uniform dispersion. *Mater. Trans.* 45: 602–604.
15. Kwon, H., Estili, M., Takagi, K., Miyazaki, T., and A. Kawasaki. 2009. Combination of hot extrusion and spark plasma sintering for producing carbon nanotube reinforced aluminum matrix composites. *Carbon* 47: 570–577.
16. Choi, H., Shin, J., Min, B., Park, J., and D. Bae. 2009. Reinforcing effects of carbon nanotubes in structural aluminum matrix nanocomposites. *J. Mater. Res.* 24: 2610–2616.
17. Laha, T., and A. Agarwal. 2008. Effect of sintering on thermally sprayed carbon nanotube reinforced aluminum nanocomposite. *Mater. Sci. Eng. A* 480: 323–332.
18. Kwon, H., Park, D. H., Silvain, J. F., and A. Kawasaki. 2010. Investigation of carbon nanotube reinforced aluminum matrix composite materials. *Compos. Sci. Technol.* 70: 546–550.

19. Pérez-Bustamante, R., Gómez-Esparza, C. D., Estrada-Guel, I., Miki-Yoshida, M., Licea-Jiménez, L., Pérez-García, S. A., and R. Martínez-Sánchez. 2009. Microstructural and mechanical characterization of Al–MWCNT composites produced by mechanical milling. *Mater. Sci. Eng. A*. 502: 159–163.
20. Bhat, A., Balla, V. K., Bysakh, S., Basu, D., Bose, S., and A. Bandopadhyay. 2011. Carbon nanotube reinforced Cu-10Sn alloy composites: mechanical and thermal properties. *Mater Sci. Eng. A*. 528: 6727–6732.
21. Dong, S. R., Tu, J. P., and X. B. Zhang. 2001. An investigation of the sliding wear behavior of Cu-matrix composite reinforced by carbon nanotubes. *Mater. Sci. Eng. A* 313: 83–87.
22. Tu, J. P., Yang, Y. Z., Wang, L. Y., Ma, X. C., and X. B. Zhang. 2001. Tribological properties of carbon-nanotube-reinforced copper composites. *Tribology Lett.* 19: 225–228.
23. Chen, W. X., Tu, J. P., Wang, L. Y., Gan, H. Y., Xu, Z. D., and X. B. Zhang. 2003. Tribological application of carbon nanotubes in a metal-based composite coating and composites. *Carbon*. 41: 215–222.
24. Chen, X. H., Li, W. H., Chen, C. S., Xu, L. S., Yang, Z., and J. Hu. 2005. Preparation and properties of Cu matrix composite reinforced by carbon nanotubes. *Trans. Nenferrous Met. Soc. China*. 15(2): 314–318.
25. Kim, K. T., Cha, S. I., and S. H. Hong. 2007. Hardness and wear resistance of carbon nanotube reinforced Cu matrix nanocomposites. *Mater. Sci. Eng. A* 449–451: 46–50.
26. Kim, K. T., Cha, S. I., Hong, S. H., and S. H. Hong. 2006. Microstructures and tensile behavior of carbon nanotube reinforced Cu matrix nanocomposites. *Mater. Sci. Eng. A* 430: 27–33.
27. Cha, S. I., Kim, K. T., Arshad, S. N., Mo, C. B., and S. H. Hong. 2008. Extraordinary strengthening effect of carbon nanotubes in metal-matrix nanocomposites processed by molecular-level mixing. *Adv. Mater.* 17: 1377–1381.
28. Singhal, S. K., Lal, M., Sharma, I., and R. B. Mathur. 2012. Fabrication of copper matrix composites reinforced with carbon nanotubes using a combination of molecular-level-mixing and high energy ball milling. *J. Composite Mater.* 47: 613–621.
29. Kim, H. N., Chang, S. N., and D. K. Kim. 2009. Enhanced microhardness of nanocrystalline carbon nanotube reinforced Cu composite using planar shock-wave compaction. *Scripta Mater.* 61: 871–874.
30. Li, H., Misra, A., Zhu, Y., Horita, Z., Koch, C. C., and T. G. Holesingerd. 2009. Processing and characterization of nanostructured Cu-carbon nanotube composites. *Mater. Sci. Eng. A* 523: 60–64.
31. Yang, Y. L., Wang, Y. D., Ren, Y., He, C. S., Deng, J. N., Nan, J., Chen, J. G., and L. Zuo. 2008. Single-walled carbon nanotube-reinforced copper composite coatings prepared by electrodeposition under ultrasonic field. *Mater. Lett.* 62: 47–50.
32. Li, Y-H., Housten, W., Zhao, Y., and Y. Q. Zhu. 2007. Cu/single-walled carbon nanotube laminate composites fabricated by cold rolling and annealing. *Nanotechnology* 18: 205607.
33. Daoush, W. M., Lim, B. K., Mo, C. B., Nam, D. H., and S. H. Hong. 2009. Electrical and mechanical properties of carbon nanotube reinforced copper nanocomposites fabricated by electroless deposition process. *Mater. Sci. Eng. A*. 513–514: 247–253.
34. Chu, K., Wu, Q., Jia, C., Liang, X., Nie, J., Tian, W., Gai, G., and H. Guo. 2010. Fabrication and effective thermal conductivity of multi-walled carbon nanotubes reinforced Cu matrix composites for heat sink applications. *Composites Sci. Tech.* 70: 298–304.
35. Singh, A. R. P., Hwang, J. Y., Scharf, T. W., Tiley, J., and R. Banerjee. 2010. Bulk nickel–carbon nanotube nanocomposites by laser deposition. *Mater. Sci. Tech.* 26: 1393–1400.

36. Deng, F. M., Chen, X. H., Chen, W. X., and W. Z. Li. 2004. Electroless plating Ni-P matrix composite coating reinforced by carbon nanotubes. *Trans. Nonferrous Met. Soc. China.* 14(4): 681–685.
37. Chen, W. X., Tu, J. P., Xu, Z. D., Chen, W. L., Zhang, X. B., and D. H. Cheng. 2003. Tribological properties of Ni–P-multi-walled carbon nanotubes electroless composite coating. *Mater. Lett.* 57: 1256–1260.
38. Tu, J. P., Zhu, L-P., Chen, W-X., Zhao, X-B., Liu, F., and X-B. Zhang. 2004. Preparation of Ni-CNT composite coatings on aluminum substrate and its friction and wear behavior. *Trans. Nonferrous Met. Soc. China.* 14: 880–884.
39. Wang, L. Y., Tu, J. P., Chen, W. X., Wang, Y. C., Liu, X. K., Olk, C., Cheng, D. H., and X. B. Zhang. 2003. Friction and wear behavior of electroless Ni-based CNT composite coatings. *Wear.* 254: 1289–1293.
40. Shen, G. R., Cheng, Y. T., and L. T. Tsai. 2005. Synthesis and characterization of Ni–P-CNT's nanocomposite film for MEMS applications. *IEEE Trans. Nanotech.* 4(5): 539–547.
41. Sun, Y., Sun, J., Liu, M., and Q. Chen. 2007. Mechanical strength of carbon nanotube–nickel nanocomposites. *Nanotechnology.* 18: 505704.
42. Dai, P. Q., Xu, W. C., and Q. Y. Huang. 2008. Mechanical properties and micro-structure of nanocrystalline nickel-carbon nanotube composites produced by electrodeposition. *Mater. Sci. Eng. A.* 483–484: 172–174.
43. Suarez, S., Ramos-Moore, E., and F. Muklich. 2013. A high temperature X-ray diffraction study of the influence of MWCNTs on the thermal expansion of MWCNT/Ni composites. *Carbon.* 51: 404–409.
44. Goh, C. S., Wei, J., Lee, L. C., and M. Gupta. 2006. Simultaneous enhancement in strength and ductility by reinforcing magnesium with carbon nanotubes. *Mater. Sci. Eng. A* 423: 153–156.
45. Li, Q., Viereckl, A., Rottmair, C. A., and R. F. Singer. 2009. Improved processing of carbon nanotube/magnesium alloy composites. *Composites Sci. Tech.* 69: 1193–1199.
46. Morisada, Y., Fujii, H., Nagaoka, T., and M. Fukusumi. 2006. MWCNTs/AZ31 surface composites fabricated by friction stir processing. *Mater. Sci. Eng. A* 419: 344–348.
47. Goh, C. S., Wei, J., Lee, L. C., and M. Gupta. 2006. Development of novel carbon nanotube reinforced magnesium nanocomposites using the powder metallurgy technique. *Nanotechnology* 17: 7–12.
48. Zeng, X., Liu, Y., Huang, Q. Y., Zeng, G., and G. H. Zhou. 2013. Effects of carbon nanotubes on the microstructure and mechanical properties of the wrought Mg–2.0Zn alloy. *Mater. Sci. Eng. A.* 571: 150–154.
49. Rashad, M., Pan, F., Tang, A., Asif, M., and M. Aamir. 2014. Synergetic effect of graphene nanoplatelets (GNPs) and multi-walled carbon nanotube (MW-CNTs) on mechanical properties of pure magnesium, *J. Alloys & Compounds.* 603: 111–118.
50. Hou, J., Du, W., Parande, G., Gupta, M., and S. Li. 2019. Significantly enhancing the strength & ductility combination of Mg-9Al alloy using multi-walled carbon nanotubes. *J Alloys and Compounds.* 790: 974–982.
51. Fukuda, H., Szpunar, J. A., Kondoh, K., and R. Chromik. 2010. The influence of carbon nanotubes on the corrosion behavior of AZ31B magnesium alloy. *Corros. Sci.* 52: 3917–3923.
52. Banhart, J. 2001. Manufacture, characterisation and application of cellular metals and metal foams. *Prog. Materi. Sci.* 46: 559–632.
53. Gibson, L. J., and M. F. Ashby. *Cellular Solids: Structure and Properties*, 2nd ed. Cambridge University Press (1997).
54. Duarte, I., and J. M. F. Ferreira, 2004. Composite and nanocomposite metal foams. *Materials* (2016).
55. Harris, P. J. F. 2013. Carbon nanotube composites. *Int. Mater. Rev.* 28;9(2): 79.

56. Tjong, S. C. 2013. Recent progress in the development and properties of novel metal matrix nanocomposites reinforced with carbon nanotubes and graphene nanosheets. *Mater. Sci. Eng. R: Rep.* 74: 281–350.
57. Duarte, I., Ventura, E., Olhero, S., and J. M. F. Ferreira, 2015. A novel approach to prepare aluminium-alloy foams reinforced by carbon-nanotubes. *Mater. Lett.* 160: 162–166.
58. Duarte, I., Ventura, E., Olhero, S., and J. M. F. Ferreira, 2015. An effective approach to reinforced closed-cell Al-alloy foams with multiwalled carbon nanotubes. *Carbon* 95: 589–600.
59. Zhang, Z., Ding, J. Xingchu, X., Sun, X., Song, K., Zhao, W., and B. Liao, 2015. Fabrication and characterization of closed-cell aluminum foams with different contents of multi-walled carbon nanotubes. *Mater. Des.* 88: 359–365.
60. Wang, J., Yang, X., Zhang, M., Li, J., Shi, C., Zhao, N., and T. Zou. 2015. A novel approach to obtain in-situ growth carbon nanotube reinforced aluminum foams with enhanced properties. *Mater. Lett.* 161: 763–766.
61. Yang, X., Yang, K., Wang, J., Shi, C., He, C., Li, J., and N. Zhao, 2017. Compressive response and energy absorption characteristics of in situ grown CNT-reinforced Al composite foams. *Adv. Eng. Mater.* 19: 1700431.
62. Yang, K., Yang, X., He, C., Liu, E., Shi, C., Ma, L., Li, Q., Li, J., and N. Zhao. 2017. Damping characteristics of Al matrix composite foams reinforced by in-situ grown carbon nanotubes. *Mater. Lett.* 209: 68–70.
63. Yang, K., Yang, X., Liu, E., Shi, C., Ma, L., He, C., Li, Q., Li, J., and N. Zhao, 2018. High strain rate dynamic compressive properties and deformation behavior of Al matrix composite foams reinforced by in-situ grown carbon nanotubes. *Mater. Sci. Eng. A* 729: 487–495.
64. Yang, X., Hu, Q., Li, W., Song, H., Zou, T., Zong, R., Sha, J., He, S., and N. Zhao, 2020. Compression-compression fatigue performance of aluminium matrix composite foams reinforced by carbon nanotubes. *Fatigue Fract. Eng. Mater. Struct.* 2020: 1–13.
65. Xu, J., Yang, X., He, C., Yang, K., Li, W., Sha, J., and N. Zhao, 2020. Crushing behavior and energy absorption property of carbon nanotube-reinforced aluminum composite foam-filled 6061 aluminum alloy tubes. *J. Mater. Sci.* 55: 7910–7926.
66. Ma, Y., Yang, X., He, C., Yang, K., Xu, J., Sha, J., Shi, C., Li, J., and N. Zhao, 2020. Fabrication of in-situ grown carbon nanotubes reinforced aluminum alloy matrix composite foams based on powder metallurgy method. *Mater. Lett.* 233: 351–354.
67. Muchhala, D., Yadav, B. N., Kumar, R., Mondal, D. P., and A. N. C. Venkat, 2019. Effect of SWCNTs content and relative density on the energy absorption capabilities of closed-cell Al-cenosphere-SWCNTs hybrid foam. *Compos. Part B Eng.* 176: 107304.
68. Pang, Q., Hu, Z. L., and J. S. Song, 2019. Preparation and mechanical properties of closed-cell CNTs-reinforced Al composite foams by friction stir welding. *Int. J. Adv. Manuf. Technol.* 103, 3125–3136.
69. Uzun, A. 2019. Production of aluminium foams reinforced with silicon carbide and carbon nanotubes prepared by powder metallurgy method. *Compos. Part B Eng.* 172: 206–217.
70. Akinwekomi, A. D., Law, W-C., Choy, M-T., Chen, L., Tang, C-Y., Tsui, G. C-P., and X S Yang,. 2018. Processing and characterisation of carbon nanotube-reinforced magnesium alloy composite foams by rapid microwave sintering. *Mater. Sci. Eng. A* 726: 82–92.
71. Huang, L. J., Geng, L., and H. X. Peng, 2015. Microstructurally inhomogeneous composites: is a homogeneous reinforcement distribution optimal? *Prog. Mater. Sci.* 71: 93–168.
72. Koyama, M., Zhang, Z., Wang, M., Ponge, D., Raabe, D., Tsuzaki, K., Noguchi, H., and C. C. Tasan, 2017. Bone like crack resistance in hierarchical metastable nanolaminate steels. *Science* 355: 1055–1057.

73. Jiang, L., Li, Z., Fan, G., Cao, L., and D. Zhang, 2012. The use of flake powder metallurgy to produce carbon nanotube (CNT)/aluminum composites with a homogenous CNT distribution. *Carbon* 50: 1993–1998.
74. Meng, L., Wang, X., Ning, J., Hu, X., Fan, G., and K. Wu, 2018. Beyond the dimensional limitation in bio-inspired composite: insertion of carbon nanotubes induced laminated Cu composite and the simultaneously enhanced strength and toughness. *Carbon.* 130: 222–232.
75. Chen, X., Tao, J., Yi, J., Li, C., Bao, R., Liu, Y., You, X., and S. Tan, 2018. Balancing the strength and ductility of carbon nanotubes reinforced copper matrix composites with microlaminated structure and interdiffusion interface. *Mater. Sci. Eng. A.* 712: 790–793.
76. Kang, T. J., Yoon, J. W., Kim, D. I., Kum, S. S., Huh, Y. H., Hahn, J. H., Moon, S. H., Lee, H-Y., and Y. H. Kim, 2007. Sandwich-type laminated nanocomposites developed by selective dip-coating of carbon nanotubes. *Adv Mater.* 19: 427–432.
77. Niu, Z., Ma, W., Li, J., Dong, H., Ren, Y., Zhao, D., Zhou, W., and S. Xie, 2012. High-strength laminated copper matrix nanocomposites developed from a single walled carbon nanotube film with continuous reticulate architecture. *Adv Funct Mater.* 22: 5209–5215.
78. Chen, X., Bao, R., Yi, J., Fang, D., Tao, J., and Y. Liu, 2019. Enhancing interfacial bonding and tensile strength in CNT-Cu composites by a synergetic method of spraying pyrolysis and flake powder metallurgy. *Materials* 12: 670.
79. Jiang, L., Li, Z., Fan, G., Cao, L., and D. Zhang, 2012. Strong and ductile carbon nanotube/aluminum bulk nanolaminated composites with two-dimensional alignment of carbon nanotubes. *Scr Mater.* 66: 331–334.
80. Liu, J., Xiong, D. B., Tan, Z., Fan, G., Guo, Q., Su, Y., Li, Z., and D. Zhang, 2018. Enhanced mechanical properties and high electrical conductivity in multiwalled carbon nanotubes reinforced copper matrix nanolaminated composites. *Mater Sci Eng A.* 729: 452–457.
81. Xu, R., Tan, Z., Xiong, D., Fan, G., Guo, Q., Zhang, J., Su, Y., Li, Z., and D. Zhang, 2017. Balanced strength and ductility in CNT/Al composites achieved by flake powder metallurgy via shift-speed ball milling. *Compos Part A Appl Sci Manuf.* 96: 57–66.
82. Yang, X., Liu, E., Shi, C., He, C., Li, J., Zhao, N., and K. Kondoh, 2013. Fabrication of carbon nanotube reinforced Al composites with well-balanced strength and ductility. *J Alloys Compd.* 563: 216–220.
83. Yang, X., Zou, T., Shi, C., Liu, E., He, C., and N. Zhao, 2016. Effect of carbon nanotube (CNT) content on the properties of in-situ synthesis CNT reinforced Al composites. *Mater Sci Eng A.* 660: 11–18.
84. Dresselhaus, M. S., Dresselhaus, G., and R. Saito, 1995. Physics of carbon nanotubes. *Carbon.* 33: 883–891.
85. Hu, C. M., Lai, C. M., Du, X. H., Ho, N. J., and J. C. Huang, 2008. Enhanced tensile plasticity in ultrafine-grained metallic composite fabricated by friction stir process. *Scr Mater.* 59: 1163–1166.
86. Liu, J., Xiong, D. B., Su, Y., Guo, Q., Li, Z., and D. Zhang, 2019. Effect of thermal cycling on the mechanical properties of carbon nanotubes reinforced copper matrix nanolaminated composites. *Mater Sci Eng A* 739: 132–139.
87. Kuzumaki, T., Ujiie, O., Ichinose, H., and K. Ito. 2000. Mechanical characteristics and preparation of carbon nanotube fiber-reinforced Ti composite. *Adv. Eng. Mater.* 2: 416–418.
88. Zeng, Q., Luna, J., Bayazitoglu, Y., Wilson, K., Imam, M. A., and E. V. Barrera. 2007. Metal coated functionalized single-walled carbon nanotubes for composites application. *Mater. Sci. Forum* 561–565: 655–658.
89. Hsu, C-F., and P-Y. Lee. 2007. Effect of carbon-nanotube addition on thermal stability of Ti-based metallic glass composites. *Mater. Sci. Forum* 534–536: 865–868.

90. Bian, Z., Pan, M. X., Zhang, Y., and W. H. Wang. 2002. Carbon-nanotube-reinforced $Zr_{52.5}Cu_{17.9}Ni_{14.6}Al_{10}Ti_5$ bulk metallic glass composites. *Appl. Phys. Lett.* 81: 4739–4741.
91. Bian, Z., Wang, R. J., Wang, W. H., Zhang, T., and A. Inoue. 2004. Carbon-nanotube-reinforced Zr-based bulk metallic glass composites and their properties. *Avd. Funct. Mater.* 14: 55–63.
92. Kumar, K. M., Kripesh, V., and A. A. O. Tay. 2008. Single-wall carbon nanotube (SWCNT) functionalized Sn–Ag–Cu lead-free composite solders. *J. Alloys Compd.* 450: 229–237.
93. Nai, S. M. L., Wei, J., and M. Gupta. 2006. Improving the performance of lead-free solder reinforced with multi-walled carbon nanotubes. *Mater. Sci. Eng. A* 423: 166–169.
94. Peng, H. E., Xiao-chun, L. U., Tie-song, L., Hai-xin, L. I., Jing, A. N., Xin, M. A., Ji-cai, F., Yan, Z., Qi, L., and Q. Yi-yu. 2012. Improvement of mechanical properties of Sn–58Bi alloy with multi-walled carbon nanotubes, *Trans Non-ferrous Soc China.* 22: S692–S696.
95. Syed, R., and W. Jiang. 2014. Fatigue crack behavior of stainless steel 304 by the addition of carbon nanotubes, *J. Nanomater. Article ID*: 863767, 6 pages. doi:10.1155/2014/863767.
96. Pang, L. X., Sun, K. N., Ren, S., Sun, C., and J. Q. Bi. 2007. Microstructure, hardness, and bending strength of carbon nanotube–iron aluminide composites. *J. Comp Mater.* 41: 2025–2031.
97. Xu, C. L., Wei, B. Q., Ma, R. Z., Liang, J., Ma, X. K., and D. H. Wu. 1999. Fabrication of aluminum–carbon nanotube composites and their electrical properties. *Carbon* 37: 855–858.
98. Pang, L-X., Sun, K-N., Ren, S., Sun, C., Fan, R-H., and Z-H. Lu. 2007. Fabrication and microstructure of Fe3Al matrix composite reinforced by carbon nanotube. *Mater. Sci. Eng. A* 447: 146–149.
99. Zhong, R., Cong, H., and P. Hou. 2003. Fabrication of nano-Al based composites reinforced by single-walled carbon nanotubes. *Carbon* 41: 848–851.
100. Tang, Y., Cong, H., Zhong, R., and H-M. Cheng. 2004. Thermal expansion of a composite of single-walled carbon nanotubes and nanocrystalline aluminum. *Carbon* 42: 3260–3262.
101. George, R., Kashyap, K. T., Rahul, R., and S. Yamdagni. 2005. Strengthening in carbon nanotube/aluminum (CNT/Al) composites. *Scripta Mater.* 53: 1159–1163.
102. Esawi, A., and K. Morsi. 2007. Dispersion of carbon nanotubes (CNTs) in aluminum powder. *Composites A* 38: 646–650.
103. Zhou, S-M., Zhang, X-B., Ding, Z-P., Min, C-Y., Xu, G-L., and W-M. Zhu. 2007. Fabrication and tribological properties of carbon nanotubes reinforced Al composites prepared by pressureless infiltration technique. *Composites A* 38: 301–306.
104. Deng, C. F., Zhang, X. X., Ma, Y. X., and D. Z. Wang. 2007. Fabrication of aluminum matrix composite reinforced with carbon nanotubes. *Rare Metals* 26: 450–455.
105. Deng, C. F., Wang, D. Z., Zhang, X. X., and A. B. Li. 2007. Processing and properties of carbon nanotubes reinforced aluminum composites. *Mater. Sci. Eng. A* 444: 138–145.
106. Morsi, K., and A. Esawi. 2007. Effect of mechanical alloying time and carbon nanotube (CNT) content on the evolution of aluminum (Al)–CNT composite powders. *J. Mater. Sci.* 42: 4954–4959.
107. Deng, C. F., Wang, D. Z., Zhang, X. X., and Y. X. Ma. 2007. Damping characteristics of carbon nanotube reinforced aluminum composite. *Mater. Lett.* 61: 3229–3231.
108. Deng, C. F., Zhang, X. X., Wang, D. Z., Lin, Q., and A. B. Li. 2007. Preparation and characterization of carbon nanotubes/aluminum matrix composites. *Mater. Lett.* 61: 1725–1728.
109. Deng, C. F., Zhang, X. X., and D. Z. Wang. 2007. Chemical stability of carbon nanotubes in the 2024Al matrix. *Mater. Lett.* 61: 904–907.

110. Deng, C. F., Ma, Y. X., Xhang, P., Zhang, X. X., and D. Z. Wang. 2008. Thermal expansion behaviors of aluminum composite reinforced with carbon nanotubes. *Mater. Lett.* 62: 2301–2303.
111. Kashyap, K. T., and R. G. Patil. 2008. On Young's modulus of multi-walled carbon nanotubes. *Bull. Mater. Sci.* 31: 185–187.
112. Choi, H. J., Kwon, G. B., Lee, G. Y., and D. H. Bae. 2008. Reinforcement with carbon nanotubes in aluminum matrix composites. *Scripta Mater.* 59: 360–363.
113. Tokunaga, T., Kaneko, K., and Z. Horita. 2008. Production of aluminum-matrix carbon nanotube composite using high pressure torsion. *Mater. Sci. Eng. A* 490: 300–304.
114. Uozumi, H., Kobayashi, K., Nakanishi, K., Matsunaga, T., Shinozaki, K., Sakamoto, H., Tsukada, T., Masuda, C., and M. Yoshida. 2008. Fabrication process of carbon nanotube/light metal matrix composites by squeeze casting. *Mater. Sci. Eng. A* 495: 282–287.
115. Bakshi, S. R., Singh, V., McCartney, D. G., Seal, S., and A. Agarwal. 2008. Deformation and damage mechanisms of multiwalled carbon nanotubes under high-velocity impact. *Scripta Mater.* 59: 499–502.
116. Lim, D. K., Shibayanagi, T., and A. P. Gerlich. 2009. Synthesis of multi-walled CNT reinforced aluminum alloy composite via friction stir processing. *Mater. Sci. Eng. A.* 507: 194–199.
117. Bakshi, S. R., Batista, R. G., and A. Agarwal. 2009. Quantification of carbon nanotube distribution and property correlation in nanocomposites. *Composites A* 40: 1311–1318.
118. Bakshi, S. R., Keshri, A. K., Singh, V., Seal, S., and A. Agarwal. 2009. Interface in carbon nanotube reinforced aluminum silicon composites: thermodynamic analysis and experimental verification. *J. Alloys Compd.* 481: 207–213.
119. Pérez-Bustamante, R., Estrada-Guel, I., Amézaga-Madrid, P., Miki-Yoshida, M., Herrera-Ramírez, J. M., and R. Martínez-Sánchez. 2009. Microstructural characterization of Al-MWCNT composites produced by mechanical milling and hot extrusion. *J Alloys Compd.* doi:10.1016/j.jallcom.2009.10.099.
120. Esawi, A. M. K., Morsi, K., Sayed, A. Gawad, A. A., and P. Borah. 2009. Fabrication and properties of dispersed carbon nanotube–aluminum composites. *Mater. Sci. Eng. A.* 508: 167–173.
121. Kim, I. Y., Lee, J. H., Lee, G. S., Baik, S. H., Kim, Y. J., and Y. Z. Lee. 2009. Friction and wear characteristics of the carbon nanotube–aluminum composites with different manufacturing conditions. *Wear.* 567: 593–598.
122. Lahiri, D., Bakshi, S. R., Keshri, A. K., Liu, Y., and A. Agarwal. 2009. Dual strengthening mechanisms induced by carbon nanotubes in roll bonded aluminum composites. *Mater. Sci. Eng. A* 523: 263–270.
123. Choi, H. J., Shin, J. H., Min, B. H., and D. H. Bae. 2010. Deformation behavior of Al–Si alloy based nanocomposites reinforced with carbon nanotubes. *Composites A.* 41: 327–329.
124. Wang, L., Heekyu Choi, H., Myoung, J. M., and W. G. Lee. 2009. Mechanical alloying of multi-walled carbon nanotubes and aluminum powders for the preparation of carbon/metal composites. *Carbon.* 47: 3427–3433.
125. He, C. N., Zhao, N. Q., Shi, C. S., and S. Z. Song. 2009. Mechanical properties and microstructures of carbon nanotube-reinforced Al matrix composite fabricated by in situ chemical vapor deposition. *J Alloys Compd.* 487: 258–262.
126. Sridhar, I., and K. R. Narayanan. 2009. Processing and characterization of MWCNT reinforced aluminum matrix composites. *J. Mater. Sci.* 44: 1750–1756.
127. Liao, J. Z., Tan, M. J., and I. Sridhar. 2009. Spark plasma sintered multi-wall carbon nanotube reinforced aluminum matrix composites. *Mater. Design.* doi:10.1016/j.matdes.2009.10.022.
128. Demchenko, I. I., and G. P. Stovpchenko. 2009. Effect of exogenous carbon nanoparticles on structure and reinforcement of aluminum matrix composite. *Mat-wiss. u. Werkstofftech.* 40(4): 326–330.

129. Bakshi, S. R., Lahiri, D., Patel, R. R., and A. Agarwal. 2010. Nanoscratch behavior of carbon nanotube reinforced aluminum coatings. *Thin Solid Films* 518: 1703–1711.
130. Morsi, K., Esawi, A. M. K., Lanka, S., Sayed, A., and M. Taher. 2010. Spark plasma extrusion (SPE) of ball-milled aluminum and carbon nanotube reinforced aluminum composite powders. *Composites A*. 41: 322–326.
131. Male, K. B., Hrapovic, S., Liu, Y., Wang, D., and J. H. T. Luong. 2004. Electrochemical detection of carbohydrates using copper nanoparticles and carbon nanotubes. *Analytica Chimica Acta* 516: 35–41.
132. Kim, K. T., Lee, K. H., Cha, S. I., Mo, C. B., and S. H. Hong. 2004. Characterization of carbon nanotubes/Cu nanocomposites processed by using nano-sized Cu powders. *Mater. Res. Soc. Symp. Proc.* 821: P3.25.1–P3.25.6.
133. Ngo, Q., Cruden, B. A., Cassell, A. M., Walker, M. D., Ye, Q., Koehne, J. E., Meyyappan, M., Li, J., and C. Y. Yang. 2004. Thermal conductivity of carbon nanotube composite films. *Mat. Res. Soc. Symp. Proc.* 812: 3.18.1–3.18.6.
134. Wang, J., Chen, G., Wang, M., and M. P. Chatrathi. 2004. Carbon-nanotube/copper composite electrodes for capillary electrophoresis microchip detection of carbohydrates. *Analyst* 129: 512–515.
135. Arribas, A. S., Bermejo, E., Chicharro, M., Zapardiel, A., Luque, G. L., Ferreyra, N. F., and G. A. Rivas. 2006. Analytical applications of a carbon nanotubes composite modified with copper microparticles as detector in flow systems. *Analytica Chimica Acta.* 577: 183–189.
136. Hrapovic, S., Majid, E., Liu, Y., Male, K., and J. H. T. Luong. 2006. Metallic nanoparticle-carbon nanotube composites for electrochemical determination of explosive nitroaromatic compounds. *Anal. Chem.* 78: 5504–5512.
137. Quang, P., Jeong, Y. G., Hong, S. H., and H. S. Kim. 2006. Equal channel angular pressing of carbon nanotube reinforced metal matrix nanocomposites. *Key Eng. Mater.* 326–328: 325–328.
138. Ferrer-Anglada, N., Gomis, V., El-Hachemi, Z., Weglikovska, U. D., Kaempgen, M., and S. Roth. 2006. Carbon nanotube based composites for electronic applications: cNT–conducting polymers, *CNT–Cu. Phys. Stat. Sol.* 203: 1082–1087.
139. Ping, C., Li, F., Jian, Z., and J. Wei. 2006. Preparation of Cu/CNT composite particles and catalytic performance on thermal decomposition of ammonium perchlorate. *Prop. Explos. Pyrotech.* 31: 452–455.
140. Dingsheng, Y., and L. Fingliang. 2006. Electroless deposition of Cu on multi-walled carbon nanotubes. *Rare Metals.* 25(3): 237–240.
141. Huang, W., Chen, H., and J-M. Zuo. 2006. One-dimensional self-assembly of metallic nanostructures on single-walled carbon-nanotube bundles. *Small* 2: 1418–1421.
142. Kang, X., Mai, Z., Zou, X., Cai, P., and J. Mo. 2007. A sensitive nanoenzymatic glucose sensor in alkaline madia with copper nanocluster/multimalled carbon nanotube-modified glassy carbon electrode. *Anal. Biochem.* 363: 143–150.
143. Chai, Y., Zhang, K., Zhang, M., Chan, P. C. H., and M. M. F. Yuen. 2007. Carbon nanotube/copper composites for via filling and thermal management. *Electron. Comp. Technol. Conf.*, 1224–1229.
144. Quang, P., Jeong, Y. G., Yoon, S. C., Hong, S. H., and H. S. Kim. 2007. Consolidation of 1 vol.% carbon nanotube reinforced metal matrix nanocomposites via equal channel angular pressing. *J. Mater. Proc. Tech.* 187–188: 318–320.
145. Quang, P., Jeong, Y. G., Yoon, S. C., Hong, S. I., and H. S. Kim. 2007. Carbon nanotube reinforced metal matrix nanocomposites via equal channel angular pressing. *Mater. Sci. Forum* 534–536: 245–248.
146. Xu, L., Chen, X., Pan, W., Li, W., Yang, Z., and Y. Pu. 2007. Electrostatic-assembly carbon nanotube-implanted copper composite spheres. *Nanotechnology* 18: 435607.

147. Valentini, F., Biagiotti, V., Lete, C., Palleschi, G., and J. Wang. 2007. The electro-chemical detection of ammonia in drinking water based on multi-walled carbon nanotube/copper nanoparticle composite paste electrodes. *Sensors and Actuators B* 128: 326–333.
148. Luque, G. L., Ferreyra, N. F., and G. A. Rivas. 2007. Electrochemical sensor for amino acids and albumin based on composites containing carbon nanotubes and copper microparticles. *Talanta* 71: 1282–1287.
149. Kim, K. T., Cha, S. I., Gemming, T., Eckert, J., and S. H. Hong. 2008. The role of interfacial oxygen atoms in the enhanced mechanical properties of carbon-nanotube-reinforced metal matrix nanocomposites. *Small* 4: 1936–1940.
150. Li, H., Misra, A., Horita, Z., Koch, C. C., Mara, N. A., Dickerson, P. O., and Y. Zhu, 2009. Strong and ductile nanostructured Cu-carbon nanotube composite. *Appl. Phys. Lett.* 95: 071907(1–3).
151. Kim, C., Lim, B., Kim, B., Shim, U. G., Oh, S., Sung, B., Choi, J. Ki, J., and S. Baik. 2009. Strengthening of copper matrix composites by nickel-coated single-walled carbon nanotube reinforcements. *Synthetic Metals.* 159: 424–429.
152. Sun, Y., and Q. Chen. 2009. Diameter dependent strength of carbon nanotube reinforced composite. *Appl. Phys. Lett.* 95: 021901(1–3).
153. Chen, X. H., Peng, J. C., Li, X. Q., Deng, F. M., Wang, J. X., and W. Z. Li. 2001. Tribological behavior of carbon nanotubes—reinforced nickel matrix composite coatings. *J. Mater. Sci. Lett.* 20: 2057–2060.
154. Chen, X. H., Cheng, F. Q., Li, S. L., Zhou, L. P., and D. Y. Li. 2002. Electrodeposited nickel composites containing carbon nanotubes. *Surf. Coat. Tech.* 155: 274–278.
155. Chen, X., Zhang, G., Chen, C., Zhou, L., Li, S., and X. Li. 2003. Carbon nanotube composite deposits with high hardness and high wear resistance. *Adv. Eng. Mater.* 5(7): 514–518.
156. Xu, Q., Zhang, L., and J. Zhu. 2003. Controlled growth of composite nanowires based on coating Ni on carbon nanotubes by electrochemical deposition method. *J. Phys. Chem. B.* 107(33): 8294–8296.
157. Arai, S., Endo, M., and N. Kaneko. 2004. Ni-deposited multi-walled carbon nanotubes by electrodeposition. *Carbon.* 42: 641–644.
158. Shi, Y. L., Yang, Z., Xu, H., Li, M. K., and H. L. Li, 2004. Preparation of electro-plated Ni-P-ultrafine diamond, Ni-P-carbon nanotubes composite coatings and their corrosion properties. *J. Mater. Sci.* 39: 5809–5815.
159. Yang, Z., Xu, H., Li, M. K., Shi, Y. L., Huang, Y., and H. L. Li. 2004. Preparation and properties of Ni/P/single-walled carbon nanotubes composite coatings by means of electroless plating. *Thin Solid Films.* 466: 86–91.
160. Wang, F., Arai, S., and M. Endo. 2005. The preparation of multi-walled carbon nanotubes with a Ni–P coating by an electroless deposition process. *Carbon*, 43: 1716–1721.
161. Zhao, G., and F. Deng. 2005. Electroless plating of Ni-P-CNTs composite coating. *Key Eng. Mater.* 280–283: 1445–1448.
162. Yang, Z., Xu, H., Shi, Y. L., Li, M. K., Huang, Y., and H. L. Li. 2005. The fabrication and corrosion behavior of electroless Ni–P-carbon nanotube composite coatings. *Mater. Res. Bull.* 40: 1001–1009.
163. Chen, X. H., Chen, C. S., Xiao, H. N., Cheng, F. Q., Zhang, G., and G. J. Yi. 2005. Corrosion behavior of carbon nanotubes–Ni composite coating. *Surf. Coat. Tech.* 191: 351–356.
164. An, B. G., Li, L. X., and H. X. Li. 2005. *Trans. Nenferrous Met. Soc. China.* 15(5): 1045–1048.
165. Majkic, G., and Y-C. Chen. 2006. *Processing of light-weight shape memory alloys using spark plasma sintering. Proc. 47th AIAA/ASME/ASCE/AHS/ASC Struct., Struct. Dynamics., Mater. Conf. 1767.*
166. Chen, C. S., Chen, X. H., Yang, Z., Li, W. H., Xu, L. S., and B. Yi. 2006. Effect of multi-walled carbon nanotubes as reinforced fibres on tribological behavior of Ni–P electroless coatings. *Diamond Relat. Mater.* 15: 151–156.

167. Arai, S., Endo, M., Sato, T., and A. Koide. 2006. Fabrication of nickel–multiwalled carbon nanotube composite films with excellent thermal conductivity by an electrodeposition technique. *Electrochem. Solid-State Lett.* 9(8): C131–C133.
168. Chao, T. Y., Shen, G. R., and Y. T. Cheng. 2006. Comparative study of Ni–P-diamond and Ni–P-CNT nanocomposite fi. *J. Electrochem. Soc.* 153(1): 98–104.
169. Shi, L., Sun, C. F., Gao, P., Zhou, F., and W. M. Liu. 2006. Electrodeposition and characterization of Ni–Co-carbon nanotubes composite coatings. *Surf. Coat. Tech.* 200: 4870–4875.
170. Chen, X. H., Chen, C. S., Xiao, H. N., Liu, H. B., Zhou, L. P., Li, S. L., and G. Zhang. 2006. Dry friction and wear characteristics of nickel/carbon nanotube electroless composite deposits. *Tribo. Inter.* 39: 22–28.
171. Li, Z. H., Wang, X. Q., Wang, M., Wang, F. F., and H. L. Ge. 2006. Preparation and tribological properties of the carbon nanotubes–Ni–P composite coating. *Tribology Int.* 39: 953–957.
172. Tan, J., Yu, T., Xu, B., and Q. Yao. 2006. Microstructure and wear resistance of nickel–carbon nanotube composite coating from brush plating technique. *Tribology Lett.* 21(2): 107–111.
173. Liu, Y. M., Sung, Y., Cheng Chen, Y. C., Lin, C. T., and Y. H. Chou and M. D. Ger. 2007. A method to fabricate field emitters using electroless codeposited composite of MWNTs and nickel. *Electrochem. Sol. State Lett.* 10(9): 101–104.
174. Xue, R. J., and Y. C. Wu. 2007. Mechanism and microstructure of electroless Ni-Fe-P plating on CNTs. *J. China Univ. Min. Tech.* 17(3): 424–427.
175. Arai, S., Saito, T., and M. Endo. 2007. Low-internal stress nickel multiwalled carbon nanotube composite electrodeposited from a sulfamate bath. *J. Electrochem. Soc.* 154(1): D530–D533.
176. Guo, C., Zuo, Y., Zhao, X., Zhao, J., and J. Xiong. 2007. The effects of pulse–reverse parameters on the properties of Ni–carbon nanotubes composite coatings. *Surf. Coat. Tech.* 201: 9491–9496.
177. Guo, C., Zuo, Y., Zhao, X., Zhao, J., and J. Xiong. 2008. The effects of electrodeposition current density on properties of Ni–CNTs composite coatings. *Surf. Coat. Tech.* 202: 3246–3250.
178. Arai, S., Fujimori, A., Murai, M., and M. Endo. 2008. Excellent solid lubrication of electrodeposited nickel-multiwalled carbon nanotube composite films. *Mater. Lett.* 62: 3545–3548.
179. Hwang, J. Y., Neira, A., Scharf, T. W., Tiley, J., and R. Banerjee. 2008. Laser-deposited carbon nanotube reinforced nickel matrix composites. *Scripta Mater.* 59: 487–490.
180. Deng, M., Ding, G., Wang, Y., Wu, H., Yao, Y., and L. Zhu. 2009. Fabrication of Ni-matrix carbon nanotube field emitters using composite electroplating and micromachining. *Carbon.* 47: 3466–3471.
181. Popp, A., Engstler, J., and J. J. Schneider. 2009. Porous carbon nanotube-reinforced metals and ceramics via a double templating approach. *Carbon.* 47: 3208–3214.
182. Huang, B. R., Lin, T. C., Yang, Y. K., and S. D. Tzeng. 2010. The stability of the CNT/Ni field emission cathode fabricated by the composite plating method. *Diamond Relat. Mater.* 19: 158–161.
183. Scharf, T. W., Neira, A., Hwang, J. Y., Tiley, J., and R. Banerjee. 2009. Self-lubricating carbon nanotube reinforced nickel matrix composites. *J. Appl. Phys.* 106: 013508(1–7).
184. Li, T., Qua, S., Li, Z., Tao, W., and M. Wang. 2009. Investigation of the mechanical properties of the Ni–P-CNTs coated copper composite materials: experiments and modeling. *Mater. Sci. Eng. A.* 500: 182–187.
185. Suzuki, Y., Arai, S., and M. Endo. 2010. Electrodeposition of Ni–P alloy-multiwalled carbon nanotube composite fi. *J. Electrochem Soc.* 157(1): D50–D53.
186. Chen, D., Chen, L., Liu, S., Ma, C. X., Chen, D. M., and L. B. Wang. 2004. Microstructure and hydrogen storage property of Mg/MWNTs composites. *J. Alloys Compd.* 372: 231–237.

187. Yang, J., and R. Schaller. 2004. Mechanical spectroscopy of Mg reinforced with Al2O3 short fibers and C nanotubes. *Mater. Sci. Eng. A* 370: 512–515.
188. Carreño-Morelli, E., Yang, J., Couteau, E., Hernadi, K., Seo, J. W., Bonjour, C., Forró, L., and R. Schaller. 2004. Carbon nanotube/magnesium composites. *Phys. Stat. Sol.* 201: R53–R55.
189. Li, S., Song, S., Yu, T., Chen, H., Zhang, Y., and H. Chen. 2005. Microstructure and fracture surfaces of carbon nanotubes/magnesium matrix composite. *Mater. Sci. Forum* 488–489: 893–896.
190. Huang, Z. G., Guo, Z. P., Calka, A., Wexler, D., and H. K. Liu. 2007. Effects of carbon black, graphite and carbon nanotube additives on hydrogen storage properties of magnesium. *J. Alloys Compd.* 427: 94–100.
191. Thakur, S. K., Gan, T. K., and M. Gupta. 2007. Development and characterization of magnesium composites containing nano-sized silicon carbide and carbon nanotubes as hybrid reinforcements. *J. Mater. Sci.* 42: 10040–10046.
192. Goh, C. S., Wei, J., Lee, L. C., and M. Gupta. 2008. Ductility improvement and fatigue studies in Mg-CNT nanocomposites. *Composites Sci. Tech.* 68: 1432–1439.
193. Paramsothy, M., Hassan, S. F., Srikanth, N., and M. Gupta. 2009. Adding carbon nanotubes and integrating with AA5052 aluminum alloy core to simultaneously enhance stiffness, strength and failure strain of AZ31 magnesium alloy. *Composites A.* 40: 1490–1500.
194. Schaller, R., Mari, D., Santos, S. M. D., Tkalcec, I., and E. Carreño-Morelli. 2009. Investigation of hydrogen storage in carbon nanotube–magnesium matrix composites. *Mater. Sci. Eng. A.* 521–522: 147–150.
195. Shu, J., Li, H., Yang, R., Shi, Y., and X. Huang. 2006. Cage-like carbon nanotubes/Si composite as anode material for lithium ion batteries. *Electrochem. Comm.* 8: 51–54.
196. Eom, J. Y., Park, J. W., Kwon, H. S., and S. Rajendran. 2006. Electrochemical insertion of lithium into multiwalled carbon nanotube/silicon composites produced by ball milling. *J. Electrochem. Soc.* 153: A1678–A1684.
197. Kim, T., Mo, Y. H., Nahm, K. S., and S. M. Oh. 2006. Carbon nanotubes (CNTs) as a buffer layer in silicon/CNTs composite electrodes for lithium secondary batteries. *J. Power Sources* 162: 1275–1281.
198. Wang, Y. H., Li, Y. N., Lu, J., Zang, J. B., and H. Huang. 2006. Microstructure and thermal characteristic of Si-coated multi-walled carbon nanotubes. *Nanotechnology* 17: 3817–3821.
199. Ishihara, T., Nakasu, M., Yasuda, I., and H. Matsumoto. 2006. Preparation of Si-carbon nanotube composite by decomposition of tetramethylsilane (TMS) and its hydrogen storage property. *Sci. Tech. Adv. Mater.* 7: 667–671.
200. Wang, W., and P. N. Kumta. 2007. Reversible high capacity nanocomposite anodes of Si/C/SWNTs for rechargeable Li-ion batteries. *J. Power Sources.* 172: 650–658.
201. Li, Y. B., Ya, Q., Wei, B. Q., Liang, J., and D. H. Wu. 1998. Processing of a carbon nanotube-Fe82P18 metallic glass composite. *J. Mater. Sci. Lett.* 17: 607–609.
202. Chen, X., Xia, J., Peng, J., Li, W., and S. Xie. 2000. Carbon-nanotube metal-matrix composites prepared by electroless plating. *Comp. Sci. Tech.* 60: 301–306.
203. Zhang, Y., Zhang, Q., Li, Y., Wang, N., and J. Zhu. 2000. Coating of carbon nanotubes with tungsten by physical vapor deposition. *Solid State Comm.* 115: 51–55.
204. Chen, W. X., Lee, J. Y., and Z. Liu. 2003. The nanocomposites of carbon nanotube with Sb and $SnSb_{0.5}$ as Li-ion battery anodes. *Carbon* 41: 959–966.
205. Feng, Y., Yuan, H. L., and M. Zhang. 2005. Fabrication and properties of silver-matrix composites reinforced by carbon nanotubes. *Mater. Char.* 55: 211–218.
206. Hu, X., Wang, T., Qu, X., and S. Dong. 2006. In situ synthesis and characterization of multiwalled carbon nanotube/Au nanoparticle composite materials. *J. Phys. Chem. B* 110: 853–857.

207. Nai, S. M. L., Wei, J., and M. Gupta. 2006. Lead-free solder reinforced with multiwalled carbon nanotubes. *J. Electronic Mater.* 35: 1518–1522.
208. Shi, X. L., Yang, H., Shao, G. Q., Duan, X. L., Yan, L., Xiong, Z., and P. Sun. 2007. Fabrication and properties of W–Cu alloy reinforced by multi-walled carbon nanotubes. *Mater. Sci. Eng. A* 457: 18–23.
209. Praveen, B. M., Venkatesha, T. V., Naik, Y. A., and K. Prashantha. 2007. Corrosion studies of carbon nanotubes–Zn composite coating. *Surf. Coat. Tech.* 201: 5836–5842.
210. Wu, Y., Ren, R., Wang, F., Yuan, Z., Wang, T., and X. Hu. 2008. Preparation and characterization of Ni–Cu–P/CNTs quaternary electroless composite coating. *Mater. Res. Bull.* 43(12): 3425–3432.
211. Hsu, C-F., Lin, H-M., and P-Y. Lee. 2008. Characterization of mechanically alloyed Ti-based bulk metallic glass composites containing carbon nanotubes. *Adv. Eng. Mater.* 10: 1053–1055.
212. Liu, B., Zeng, Z., and Y. Lin. 2009. Mechanical properties of hard Cr–MWNT composite coatings. *Surf. Coat. Technol.* 203: 3610–3613.
213. Huang, J., Xie, Q., Tan, Y., Fu, Y. Su, Z., Huang, Y., and S. Yao. 2009. Preparation of Pt/multiwalled carbon nanotubes modified Au electrodes via Pt–Cu co-electrodeposition/Cu stripping protocol for high-performance electrocatalytic oxidation of methanol. *Mater. Chem. Phys.* 118: 371–378.
214. Kondoh, K., Threrujirapapong, T., Imai, H., Umeda, J., and B. Fugetsu. 2009. Characteristics of powder metallurgy pure titanium matrix composite reinforced with multi-wall carbon nanotubes. *Compos. Sci. Technol.* 69: 1077–1081.
215. Xue, F., Jiehe, S., Yan, F., and C. Wei. 2010. Preparation and elevated temperature compressive properties of multi-walled carbon nanotube reinforced Ti composites. *Mater. Sci. Eng. A.* 527: 1586–1589.
216. Singh, R. N., Singh, A., and Anindota. 2009. Electrocatalytic activity of binary and ternary composite films of Pd, MWCNT, and Ni for ethanol electro-oxidation in alkaline solutions. *Carbon.* 47: 271–278.
217. Nai, S. M. L., Han, Y. D., Jing, H. Y., Zhang, K. L., Tan, C. M., and J. Wei. 2009. *Enhancing the properties of a lead-free solder with the addition of ni-coated carbon nanotubes. International Conference on Electronic Packaging Technology & High Density Packaging.* Xi'an, China, Aug. 16–19, 2010. IEEE. 540–543.
218. Nai, S. M. L., Wei, J., and M. Gupta. 2009. Interfacial intermetallic growth and shear strength of lead-free composite solder joints. *J Alloys Compd.* 473: 100–106.
219. Bakshi, S. R., Lahiri, D., and A. Agarwal, 2010. Carbon nanotube reinforced metal matrix composites—a review. *Int. Mater. Rev.* 55: 41–64.
220. Ci, L., Ryu, Z., Jin-Phillipp, N. Y., and M. Ruhle. 2006. Investigation of the interfacial reaction between multi-walled carbon nanotubes and aluminum. *Acta Mater.* 54: 5367–5375.
221. Deng, C. F., Zhang, X. X., Wang, D. Z., and Y. X. Ma. 2007. Calorimetric study of carbon nanotubes and aluminum. *Mater. Lett.* 61: 3221–3223.
222. Majid, M., Majzoobi, G. H., Noozad, G. A., Reihani, A., Mortazavi, S. Z., and M. S. Gorji. 2012. Fabrication and mechanical properties of MWCNTs-reinforced aluminum composites by hot extrusion, *Rate Metals.* 31: 372–378.
223. Kim, W. J., and Y. J. Yu. 2014. The effect of the addition of multiwalled carbon nanotubes of the uniform distribution of TiC nanoparticles in aluminum nanocomposites. *Scripta Mater.* 72–73: 25–28.
224. Kallip, K., Leparoux, M., Alogab, K. A., Clerc, S., Deguilhem, G., Arroyo, Y., and H. Kwon. 2015. Investigation of different carbon nanotube reinforcements for fabricating bulk AlMg5 matrix nanocomposites. *J. Alloys. Compnds.* 646: 710–718.
225. Merajin, M. T., Mashhadi, M. M., and A. A. Saheri. 2015. Improving mechanical properties of near-net-shape aluminum/MWCNT nanocomposites fabricated by plasma spray forming using electroless copper coating of MWCNT. *J. Comp. Mater.* 49: 131–139.

226. Jagannathan, M., Shankaran, S., and P. Haridoss. 2015. Microstructure and mechanical behavior of copper coated multiwalled carbon nanotubes reinforced aluminum composites. *Mater. Sci. Eng. A.* 638: 197–207.
227. Mansoor, M., Khan, S., Ali, A., and K. M. Ghauri. 2019. Fabrication of aluminum-carbon nanotube nano-composite using aluminum-coated carbon nanotube precursor. *J. Comp. Mater.* 53: 1–10.
228. Xue, Z. W., Wang, L. D., Zhao, P. T., Xu, S. C., Qi, J. L., and W. D. Fei. 2012. Microstructure and tensile behavior of carbon nanotubes reinforced Cu matrix composites with molecular-level dispersion. *Materials and Design.* 34: 298–301.
229. Tu, J. F., Rajule, N., Molian, P., and Y. Li. 2016. Laser synthesis of a copper–single-walled carbon nanotube nanocomposite via molecular-level mixing and non-equilibrium solidification. *J. Phys D, Appl. Phys.* 49: 495301 (11).
230. Pak, J. H., Kim, G. N., Hwang, S. G., Kim, B. S., Noh, J. P., and S. C. Huh. 2016. Mechanical properties of Cu matrix composite fabricated by extrusion process. *Trans. Nonferrous Met. Soc. China.* 26: 2679–2686.
231. Kondoha, K., Fukuda, H., Umeda, J., Imai, H., Fugetsu, B., and M. Endo. 2010. Microstructural and mechanical analysis of carbon nanotube reinforced magnesium alloy powder composites. *Mater. Sci. Eng. A.* 527: 4103–4108.
232. Srivatsan, T. S., Godbole, C., Paramsothy, M., and M. Gupta. 2012. On the influence of carbon nanotubes and processing on tensile response and fracture behavior of a magnesium alloy. *Adv. Mater. Res.* 410: 133–141.
233. Nai, M. H., Wei, J., and M. Gupta. 2014. Interface tailoring to enhance mechanical properties of carbon nanotube reinforced magnesium composites. *Mater & Design.* 60: 490–495.
234. Han, G., Wang, Z., Liu, K., Li, S., Du, X., and W. Du. 2015. Synthesis of CNT-reinforced AZ31 magnesium alloy composites with uniformly distributed CNTs. *Mater. Sci. Eng. A.* 628: 350–357.
235. Akinwekomi, A. D., Law, W. C., Tang, C. Y., Chen, L., and C. P. Tsui. 2016. Rapid microwave sintering of carbon nanotube-filled AZ61 magnesium alloy composites. *Composites B.* 93: 302–309.
236. Merino, C. A. I., Sillas, J. E. L., Meza, J. M., and J. M. H. Ramirez. 2017. Metal matrix composites reinforced with carbon nanotubes by an alternative technique. *J. Alloys Compd.* 707: 257–263.
237. Huang, Y., Li, J., Wan, L., Meng, X., and Y. Xie. 2018. Strengthening and toughening mechanisms of CNTs/Mg-6Zn composites via friction stir processing. *Mater. Sci. Eng. A.* 732: 205–211.
238. Massey, C., Umanzor, M., and G. Vasudevamurthy. 2016. Thermal and mechanical properties of mechanically alloyed 304LSS-CNT metal matrix composites. *J. Comp. Mater.* 51: 1019–1028.

6 Mechanics of CNT-MMCs

Fiber strengthening has been known for a very long time. Humankind has long utilized natural materials like jute and bamboo and they all can serve as fibrous reinforcement. The focus on understanding the mechanics of reinforcement by fibers has been underway for almost 50 years now [1–3]. Several models have been developed to estimate the reinforcement effect of hard particles and high strength fibers. The mechanical behavior of these composites can be described at multiple length scales by different models as summarized in Figure 6.1. At macroscale length (>1 mm), the average property of the material may be considered and material behavior is represented by constitutive laws like Hooke's law. At micro/meso-length scales (1 μm–1 mm), the deformation and stress around an inclusion, pore, and crack influence the mechanical behavior, which can be predicted by micromechanical models, dislocation plasticity models, and fracture mechanics. At nano-scale length, the mechanical behavior is influenced by the interaction between nano-reinforcment and the matrix at an atomic level, which can be studied by molecular dynamics. A single CNT of small size (a few microns in length and a diameter less than 100 nm) is expected to cause reinforcement in regions around it at the nano-scale. It is of basic interest to study whether the strengthening at the nanoscale level is translated into the micro- and macro-scale. The applicability of existing micromechanical models and development of new ones for strengthening in CNT composites started to receive attention almost a decade back [4–7]. Most of the research has been done on polymer-CNT composites, which, in principle, can be applied to metal matrices as well. Carbon nanotubes are different in the sense that strengthening caused by them reaches to very small length scales as compared to larger fibers. This would lead to new mechanisms for strengthening because they can interfere with the deformation mechanisms occurring at the nano-scale. In addition, there are several specific issues with CNT composites, namely clustering, degree of dispersion, alignment, curvature, and single/multi-wall, which influence the mechanical properties and need to be taken into account for bulk fiber reinforced composites. The grain size in the metal matrix composites may also be affected by the presence of CNTs, which will affect the strengthening behavior.

The addition of CNTs to metal matrix is expected to result in an increase in stiffness and strength of the composites. This is due to the extremely high strength and stiffness of CNTs as outlined in Table 1.2. Several models (Figure 6.1) have been developed for the enhancement of strength and stiffness in fiber-reinforced composites. This chapter briefly summarizes the micro-mechanical models available for fiber-reinforced composites. The applicability and assumptions of these models for CNT-reinforced composites is critically discussed and analyzed. This chapter also discusses the strengthening mechanism due to CNT addition. Some of these

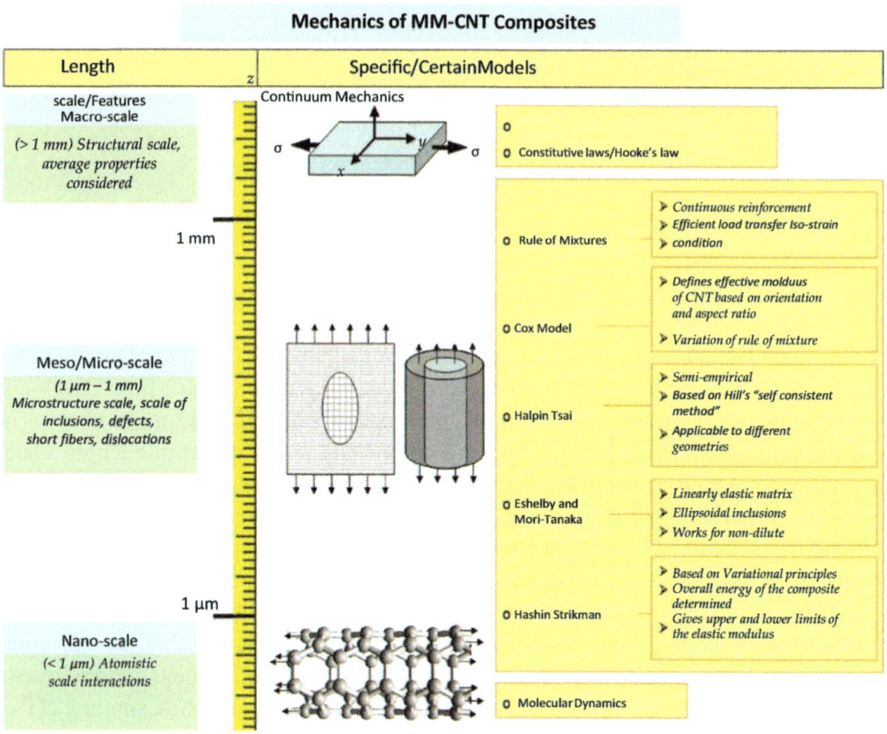

FIGURE 6.1 Schematic of the various models for describing elastic behavior of composites at different length scales.

mechanisms are unique only to CNT composites. The applicability of the strengthening models is also discussed.

6.1 ELASTIC MODULUS OF METAL MATRIX-CARBON NANOTUBE COMPOSITES

Metallic materials are known for their strength and toughness. However, their lower stiffness leads to large elastic deflections under loading. Large elastic deflections are unacceptable in structural applications where it could lead to deformation of the structure. A large elastic modulus translates into smaller deflections for a given load and enables structural members with smaller dimensions and is thus desirable due to economic reasons. A large resilience is also required in applications such as sporting equipment. Improvement in the stiffness of metallic materials has been carried out using fibers since the availability of carbon and ceramic fibers in the 1970s. Many micromechanical models have been developed to predict the enhancement of stiffness due to the presence of a fiber in a matrix. These models take into account the volume fraction, stiffness, distribution, and orientation of the fibers in the matrix. Whether these models can be applied to CNT composites where the stiffening occurs at the nanoscale is of fundamental interest. One way to find out is to compare the

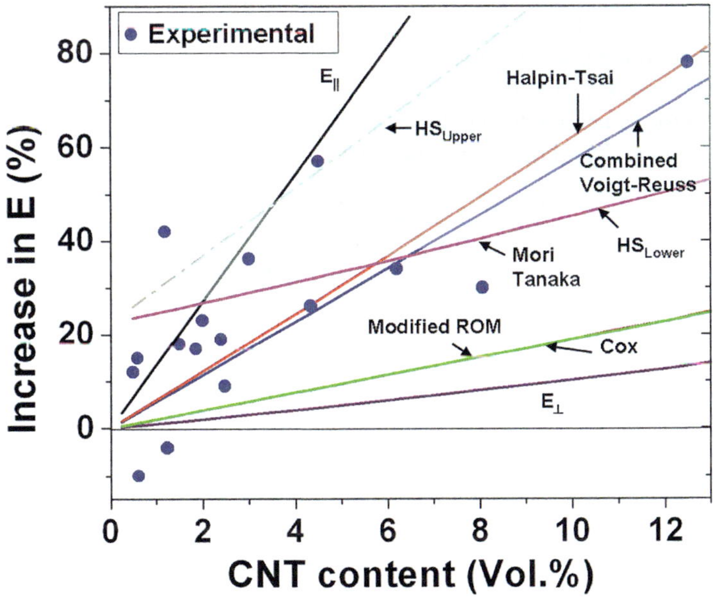

FIGURE 6.2 Variation of the percentage increase in elastic modulus of Al CNT composites as a function of their CNT content. The percentage increment in ***E*** as calculated from models has been super-imposed on the measured values.

increase in the stiffness predicted from the micromechanical models and the experimentally observed values. Several studies on CNT-MMCs have been reported over the last 20 years. Figure 6.2 shows the percentage increase in the experimentally measured elastic modulus values of Al-CNT composites as a function of the CNT content over the unreinforced aluminum matrix. The properties of the Al-CNT composite have been compared with unreinforced Al prepared by the same processing technique so that changes in the properties can be attributed only to CNT reinforcement. The values of the elastic modulus as predicted by various micromechanics models are also superimposed in Figure 6.2. For the model calculations, the values of $E_{Al} = 69$ GPa, $E_{CNT} = 1000$ GPa, $\nu_{Al} = 0.33$, $\nu_{CNT} = 0.3$, and aspect ratio l/d of CNT = 50 have been assumed. The micro-mechanics models as mentioned in Figure 6.2 are described in this section and their applicability in predicting the stiffening due to CNTs is discussed. In the equations that follow, E stands for elastic modulus, σ stands for yield strength, V stands for volume fraction, k stands for bulk modulus, μ stands for rigidity modulus, ν stands for Poisson's ratio, and the subscript m is used for properties corresponding to the matrix while subscript f is used for properties corresponding to the fiber (CNT).

6.1.1 Modified Rule of Mixtures

Rule of mixtures (ROM) is the simplest model. In this model, it is assumed that the fibers are parallel, continuous, and run throughout the length of the sample. When the load is applied parallel to the length of the fibers, it is known as a Voigt condition.

Assuming iso-strain condition, the load is distributed among the fibers and the matrix and the composite elastic modulus is represented by $E_{\parallel}$ (also known as ROM). When the load is applied normal to the length of the fibers, it is known as Reuss condition. The total strain is distributed among the fibers and the matrix in such a manner that the load is the same in both, and the composite elastic modulus is represented by $E_{\perp}$. The elastic modulus of the composites in both these cases is given by:

$$E_{\parallel} = V_f E_f + \left(1 - V_f\right) E_m \tag{6.1}$$

$$E_{\perp} = \frac{E_f E_m}{E_f \left(1 - V_f\right) + E_m V_f} \tag{6.2}$$

This model assumes efficient load transfer to the fibers since they run across the entire length (continuous) of the composite. In actual cases, the fibers are short and the load transfer is not efficient. Hence, a *modified* ROM has been proposed, which is as follows:

$$E_{\parallel} = \eta V_f E_f + \left(1 - V_f\right) E_m \tag{6.3}$$

where η is the load transfer efficiency factor and is generally equal to 1/5. The models described above have been applied in the case of some CNT-MMC systems. In the case of plasma sprayed Al-23 wt.% Si composite reinforced with 10 wt.% CNTs, the calculated value using ROM (~130 GPa) was found to be slightly higher than the measured value (120.4 GPa) because the model does not take porosity shape into account and assumes that the reinforcement is continuous [8]. However, a simple ROM has been found to fit experimentally obtained data, as shown in Figure 6.3, in the case of Al-CNT composites produced by hot rolling of ball milled powder mixtures for a range of CNT concentrations [9]. This indicates higher load transfer efficiency, which may be a result of good dispersion and high density of the samples.

However, in actual composites, the CNTs can be distributed randomly and in such a case, the resultant elastic modulus is a weighted average of both parallel and perpendicular types of loading geometries. The elastic modulus for the randomly oriented CNT composites can be given by the combined Voigt-Reuss Model, which is given as follows:

$$E = \frac{3}{8} E_{\parallel} + \frac{5}{8} E_{\perp} \tag{6.4}$$

The weights (coefficients) correspond to those used in Halpin-Tsai equations for strength prediction, which is described later. The combined Voigt-Reuss model, although not used in the literature, could be used to predict the stiffness in well-dispersed CNT systems.

As seen from Figure 6.2, a number of experimentally measured values, especially at low concentrations (<4 vol.%), agree well with the values predicted by the ROM ($E_{\parallel}$) values in the case of Al-CNT composites. It could mean that at low concentrations, processes like extrusion and rolling are able to disperse and possibly align the CNTs in the direction of deformation. However, at higher concentrations, the

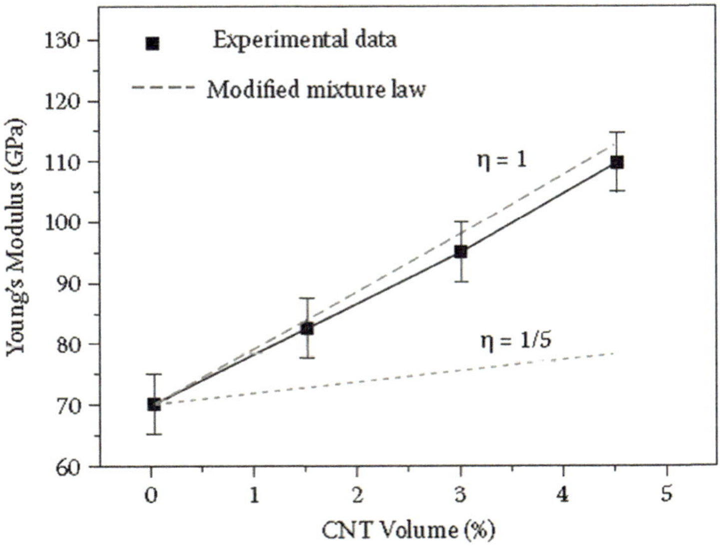

FIGURE 6.3 Graph showing the experimental value of elastic modulus of Al-CNT composite prepared by hot extrusion of ball-milled powders [9]. The ROM is observed to fit the data well. (Reproduced with permission from Materials Research Society).

combined Voigt-Reuss model gives better predictions, indicating a predominantly random orientation.

6.1.2 Cox Model

Cox has provided a model to take into account the orientation of fibers in strengthening and applied it to fibrous materials such as paper [1]. The fiber orientation with respect to the principal direction (θ) can be expressed using a distribution function $f(\theta)$ such that

$$\int_0^{\pi} f(\theta)\,d\theta = 1 \tag{6.5}$$

The orientation of the fibers then determines the effective elastic modulus of the fiber mat, which is incorporated within the composite. Elastic modulus of the composite according to this model is given by [1,2,4]

$$E = \eta_o \eta_L E_f V_f + E_m\left(1 - V_f\right) \tag{6.6}$$

Where,

$$\eta_o = \frac{1}{5}, \eta_L = 1 - \frac{\tan h(\beta s)}{\beta s}, s = \frac{2l}{r}, \beta = \frac{2\pi E_m}{E_f\left(1 + V_m\right)\ln\left(1/V_f\right)}$$

This model takes into account the orientation as well as the aspect ratio of the fibers. This model has been applied to Al-CNT composites prepared by hot rolling of ball-milled powders [9]. It was observed that the predicted and experimentally measured values were very close to each other for up to 2 vol.% CNT reinforcement. However, at higher loading, measured values were lower than those predicted by the Cox model, indicating this model is not valid for poorly dispersed systems. As seen from Figure 6.2, the Cox model results are very similar to those predicted by the modified ROM for $\eta = 5$.

6.1.3 Halpin-Tsai Model

Halpin and Tsai developed a semi-empirical formula based on Herman's solution of Hill's self-consistent model [2]. These equations showed that the reinforcement geometry (represented by aspect ratio, *l/d*) had a profound impact on stiffness properties of unidirectional fiber composites. These equations also suggest that variation of geometry from spherical particles to very long (aspect ratio approaching infinity) would lead to orders of magnitude increase in stiffening for the same volume fraction in both unidirectional and randomly oriented composites [10,11]. The Halpin-Tsai equations have been used to obtain the elastic modulus of randomly oriented composites as follows [12]:

$$\frac{E_{|C}}{E_M} = \frac{3}{8}\left[\frac{1+\left(2l/D\right)\eta_L V_f}{1-\eta_L V_f}\right] + \frac{5}{8}\left[\frac{1+2\eta_T V_f}{1-\eta_T V_f}\right] \tag{6.7}$$

Where

$$\eta_L = \frac{E_f/E_m - 1}{E_f/E_m + 2l/D}, \eta_T = \frac{E_f/E_m - 1}{E_f/E_m + 2},$$

and l and D represent the length and diameter of the CNT, respectively. The Halpin-Tsai model has been found to closely predict values in the case of small CNT concentrations in polymer-CNT composites [5,13]. In the case of cold sprayed Al-CNT composites, the values of elastic modulus obtained by nanoindentation were found to be closer to those calculated using Halpin-Tsai equations [14]. As seen from Figure 6.2, the Halpin-Tsai equations predict values that are slightly lower than the experimental values at low CNT concentrations. This is because these equations are developed for randomly oriented composites, while at low concentration there could be some alignment of the CNTs [9]. However, from Figure 6.2, it is clear that the Halpin-Tsai equations provide a conservative estimate for the stiffness of Al-CNT composites.

6.1.4 Hashin-Shtrikman Model

This model is based on variational principles [15,16] and provides the upper and lower bounds for the elastic modulus of a composite. The upper and lower bounds are basically maximum and minimum in the change in the strain energy and correspond to the values for the non-homogeneous and anisotropic condition (the fiber/CNT) vs. the isotropic and homogeneous condition (the matrix), respectively, under the same surface forces and displacements. It is independent of the shape of the particle. The upper and lower bound of the values of k and μ for the composite can be obtained from the following equations:

$$\frac{V_f}{1+\frac{(1-V_f)(k_f-k_m)}{k_m+k^-}} \le \frac{k-k_m}{k_f-k_m} \le \frac{V_f}{1+\frac{(1-V_f)(k_f-k_m)}{k_m+k^+}} \tag{6.8}$$

$$\frac{V_f}{1+\frac{(1-V_f)(\mu_f-\mu_m)}{\mu_m+\mu^-}} \le \frac{\mu-\mu_m}{\mu_f-\mu_m} \le \frac{V_f}{1+\frac{(1-V_f)(\mu_f-\mu_m)}{\mu_m+\mu^+}} \tag{6.9}$$

Where

$$k^- = \frac{4}{3}\mu_m, k^+ = \frac{4}{3}\mu_f, \mu^- = \frac{3}{2\left(\frac{1}{\mu_m}+\frac{10}{9k_m+8\mu_m}\right)}, \mu^+ = \frac{3}{2\left(\frac{1}{\mu_f}+\frac{10}{9k_f+8\mu_f}\right)}$$

The E and ν are related to k and μ by the classical relations

$$E = \frac{9k\mu}{3k+\mu} \tag{6.10}$$

$$\nu = \frac{3k-2\mu}{2(3k+\mu)} \tag{6.11}$$

This model is important in predicting the extent of strengthening that could be achieved. However, one should keep in mind that the properties of CNTs could be highly anisotropic and the bulk and shear modulus need to be calculated with caution. As seen from Figure 6.2, in most of the cases, the Hashin-Shtrikman upper bound (HS_{upper}) value provides high estimates and the experimental values are lower than this. So it can be used as an upper limit of the increase in elastic modulus. The Hashin-Shtrikman lower bound (HS_{lower}) values coincide with those obtained by the Mori Tanaka scheme for particulate composites [17].

6.1.5 Modified Eshelby Model

The Eshelby model has been used to derive the strain in the matrix and the inclusion due to the presence of an inclusion of different elastic modulus in an isotropic medium [18]. This analysis has been very popular and has been applied extensively to particulate reinforced composites. Chen et al. [19] have used the modified Eshelby model to relate the properties in CNT composites to the volume fraction of CNTs as well as porosity. The longitudinal elastic modulus value is given by the formula:

$$E_{11} = E_m \varepsilon_{11}^m \left(\varepsilon_{11}^m + V_f \varepsilon_{11}^{CNT} \right)^{-1} \tag{6.12}$$

The values predicted by the model were higher than were those observed experimentally in the case of Al-CNT composite coating, which is ascribed to the poor bonding between CNT and the matrix [14].

6.1.6 Dispersion-based Model

CNT cluster formation is inevitable in most processing techniques, especially at higher CNT concentrations. CNT clustering leads to incomplete utilization of their mechanical properties. However, completely infiltrated clusters still provide some strengthening in metal matrix composites. A model for calculating the elastic properties of clusters and hence the composite reinforced by these clusters has been recently developed by Villoria and Miravete [20]. This model is helpful in calculating the properties of metal infiltrated CNT clusters. The overall properties of the composite are obtained by considering it as a dilute suspension of the clusters (properties with subscript dsc) in the matrix:

$$k_{dsc} = k_m + \frac{\left(k_{Cluster} - k_m\right) C_c}{1 + \frac{\left(k_{Cluster} - k_m\right)}{k_m + 4\,\mu_m/3}} \tag{6.13}$$

$$\mu_{dsc} = \mu_m \left[1 - \frac{15\left(1 - v_m\right)\left(1 - \frac{\mu_{Cluster}}{\mu_m}\right) C_c}{7 - 5v_m + 2\left(4 - 5v_m\right)\frac{\mu_{Cluster}}{\mu_m}} \right] \tag{6.14}$$

where c_c refers to the volume fraction of clusters, which is related to the overall CNT fraction by the relation $V_f = c_f.c_c$. The elastic modulus and Poisson's ratio can then be computed using Eqs. 6.10 and 6.11.

$$E_{Cluster} = \frac{\left[E_{11} + 4\left(1 + v_{12}\right)^2 k_{23}\right]\left[E_{11} + \left(1 - 2v_{12}\right)^2 k_{23} + 6\left(\mu_{12} + \mu_{23}\right)\right]}{3\left[2E_{11} + \left(8v_{12}^2 + 12v_{12} + 7\right)k_{23} + 2\left(\mu_{12} + \mu_{23}\right)\right]} \tag{6.15}$$

$$v_{Cluster} = \frac{E_{11} + 2\left(2v_{12}^2 + 8v_{12} + 3\right)k_{23} - 4\left(\mu_{12} + \mu_{23}\right)}{2\left[2E_{11} + \left(8v_{12}^2 + 12v_{12} + 7\right)k_{23} + 2\left(\mu_{12} + \mu_{23}\right)\right]} \tag{6.16}$$

where E_{11}, k_{23}, μ_{12}, μ_{23}, and ν_{12} are the effective properties of a composite cylinder given by

$$E_{11} = c_f E_f + (1-c_f) E_m + 4c_f (1-c_f) \left[\frac{(V_f - V_m)^2}{\dfrac{(1-c_f)\mu_m}{k_f + \mu_f/3} + \dfrac{c_f \mu_m}{k_m + \mu_m/3} + 1} \right] \quad (6.17)$$

$$\nu_{12} = \nu_{13} = c_f \nu_f + (1-c_f)\nu_m + \left[\frac{c_f (1-c_f)(\nu_f - \nu_m)\left(\dfrac{\mu_m}{k_m + \mu_m/3} - \dfrac{c_f \mu_m}{k_f + \mu_f/3} \right)}{\dfrac{(1-c_f)\mu_m}{k_f + \mu_f/3} + \dfrac{c_f \mu_m}{k_m + \mu_m/3} + 1} \right] \quad (6.18)$$

$$k_{23} = k_m + \frac{\mu_m}{3} + \frac{c_f}{\dfrac{1}{k_f - k_m \dfrac{\mu_f - \mu_m}{3}} + \dfrac{1-c_f}{k_m + 4\dfrac{\mu_m}{3}}} \quad (6.19)$$

$$\mu_{12} = \frac{\mu_m \left[\mu_f (1+c_f) + \mu_m (1-c_f) \right]}{\mu_f (1-c_f) + \mu_m (1+c_f)} \quad (6.20)$$

$$\mu_{23} = \mu_m \left[1 + \frac{c_f}{\dfrac{\mu_m}{\mu_f - \mu_m} + \dfrac{\left(k_m + 7\mu_m/3\right)(1-c_f)}{2\left(k_m + 4\mu_m/3\right)}} \right] \quad (6.21)$$

Here, c_f refers to the volume fraction of the CNT in the clusters.

The model predicted the elastic modulus epoxy-SWNT composites very closely (2.924 GPa and 2.686 GPa as compared to experimental values of 2.909 GPa and 2.659 GPa, respectively, for 0.1 and 0.5 vol.% SWNT reinforcement) [20]. However, to predict nicely for CNT-MMCs, several requirements must be met. One of the factors is that the CNT cluster should be infiltrated with the molten metal and must

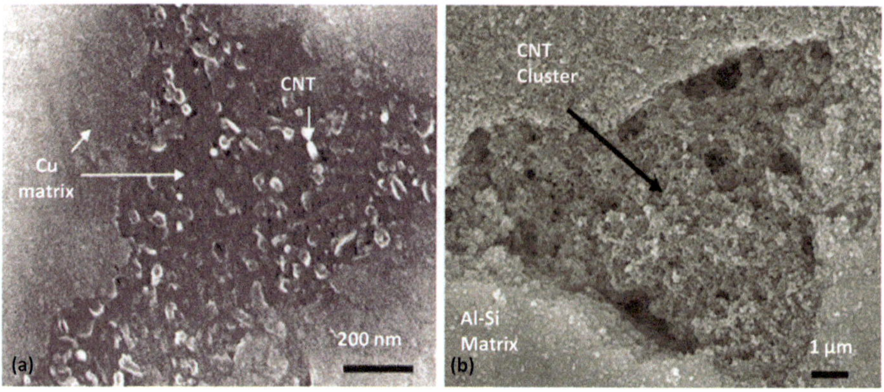

FIGURE 6.4 SEM images of polished cross-sections showing (a) infiltrated CNT cluster in Cu-CNT composite [21], and (b) a partially infiltrated cluster in Al-CNT composite. (Reproduced with permission from Elsevier).

behave like a homogeneous body. The interface between the CNT clusters and the matrix must be strong and without porosity. The elastic modulus of a cluster ($E_{Cluster}$) containing 50 vol.% CNTs and completely infiltrated by Al-12 wt.% Si alloy is found to be 229 GPa as calculated using the previous equations. This is high enough to bring about significant stiffness enhancement in the composite where clustering is observed. Figure 6.4 shows a CNT cluster in (a) Cu-CNT and (b) Al-CNT system. The cluster shown in Figure 6.4a is metal-infiltrated and likely to cause stiffness improvement, while the cluster in Figure 6.4b is unlikely. This is simply because the presence of porosity at interface and in the body of the cluster in Figure 6.4b would inhibit stress transfer to the CNTs.

Micromechanical models are useful in predicting the properties of the CNT composites without going through the process of experimentation. The micromechanical models also provide an understanding of the strengthening mechanisms along with serving as an extrapolation tool. The appropriate model must be chosen based on the microstructure of the CNT-MMC. A model such as ROM applied to a clustered composite would not predict the property accurately, as the model is developed for aligned fibers, not clusters or bundles of nano-fibers. Table 6.1 shows a summary of the models and the systems to which it has been applied and their accuracy in predicting the values. As seen from Figure 6.2, we can see that the ROM ($E_{||}$) matches with most of the values at low CNT concentrations, indicating that CNTs might be getting aligned during rolling and extrusion processes. More studies are needed to evaluate the efficacy of the model for CNT-MMCs processed by other techniques, especially those that do not lend themselves to inducing alignment. The Halpin-Tsai and the combined Voigt-Reuss models can be used to obtain a conservative value for the elastic modulus in most cases. The difference in E values at the same volume fraction as reported by several researchers is due to differences in processing, which leads to scatter in of densification, dispersion, and damage occurring to the CNTs.

TABLE 6.1
Comparison of experimental values of the elastic modulus of CNT-MMCs and those calculated using micromechanics models

Model Used with Assumptions	System/Ref.	CNT Vol.%	Elastic Modulus (GPa) Calculate	Elastic Modulus (GPa) Experimental	Remarks
Cox Model The effective elastic modulus of the fibrous reinforcement is determined and used in ROM	Al-CNT [22]	For CNT 0.5 0.5 with K_2ZrF_6 2 For SWNT 1 1 with K_2ZrF_6 2	 74.3 74.3 87.4 79.2 79.2 83.4	 78.1 75.2 84.85 70 93.7 79.3	Agreement good at low CNT concentration. Poor dispersion at higher loading for which the equations are not valid
1. Mori Tanaka Assumes linearly elastic matrix with ellipsoidal inclusions in non-dilute concentrations	Al-CNT [14]	0.6	75 85 HS Lower Bound – 85 HS Upper Bound – 88 72.5 71.5	Distribution in values obtained by Nanoindentation (mean 68.6)	Some large values were obtained by nanoindentation at places rich in CNT content. Halpin-Tsai and modified Eshelby models predict close to experimental values
2. Hashin-Strikman Based on variational principles, determines the lower and upper bound of reinforcement effect					
3. Halpin Tsai Semi-empirical equations developed by curve fitting to take into account aspect ratio of reinforcement as well as dispersion. Linear reinforcement is assumed.		1.2	81 86 Lower Bound – 86 Upper Bound – 91 75 73.5	Distribution in values obtained by nano-indentation (mean 68.1)	

(Continued)

TABLE 6.1 (*Continued*)
Comparison of experimental values of the elastic modulus of CNT-MMCs and those calculated using micromechanics models

Model Used with Assumptions	System/Ref.	CNT Vol.%	Elastic Modulus (GPa)		Remarks
			Calculate	Experimental	
4.Modified Eshelby Method Eshelby model modified for CNT geometry and assumes linear elastic behavior of CNTs and matrix.					
Rule of Mixture	Al-CNT	0.6	75	68.6	Same as above
Continuous reinforcement in loading	[14]	1.2	85	68.1	
direction with matrix and reinforcement in iso-strain	Al-Si-CNT [8]	5.67	130	120.4	Porosity and Si fraction taken into account from image analysis. Values a little low because analysis does not take pore shape into consideration
	Al-CNT	1.5	82.5	84	ROM predicts well for these
	[9]	3.0	95	98	composites due to good dispersion
		4.5	110	113	and high density

6.2 STRENGTHENING MECHANISMS IN METAL MATRIX-CARBON NANOTUBE COMPOSITES

The purpose of CNT addition to metal matrix is also to make use of its high tensile strength in order to increase the overall strength of the composite. This has been a major focus of CNT-MMC research and several researchers have investigated their tensile properties. CNTs being of higher elastic modulus can take a higher share of the total load for a given strain and hence the overall strength of the composite is increased. It is known that when the fibers are smaller and embedded in the composite, the load is transferred to the fibers through the interface and several new factors come into picture. Several models have been proposed to predict the strength of fiber-reinforced composites and have been applied to CNT-MMCs as well.

There is a lot of scatter in the strength of CNT-MMCs as shown in Figure 5.7. One of the reasons is that different processing techniques lead to variation in the microstructure and CNT distributions and hence, different degrees of strengthening. Another very important reason for the scatter in data is the differences associated with the experimental methods and sample sizes utilized for the mechanical testing. As, the data is based on miniaturized tensile specimens, nano-indentation and relatively large (≥15 mm) sample fabricated by different processes. Large samples are expected to include higher volume fraction of defects such as porosity and CNT clusters due to processing constraints and would result in poor mechanical properties as compared to miniaturized specimens. The stress state (e.g., plane stress or plane strain condition) may be different in non-standard thick or thin samples and may lead to brittle or ductile failure. Tensile tests conducted on relatively large (15 mm gauge length) Al-CNT composite samples, fabricated by different processes, showed different yield strength values. The yield strength of hot pressed and extruded Al reinforced with 5–10 vol.% CNT composite has been reported as 80 MPa [23], whereas yield strength for 1 wt.% CNT reinforcement fabricated by cold isostatic pressing and hot extrusion was 336 MPa [24,25]. It is evident from Figure 5.7 that the tensile properties of CNT-MMCs synthesized by different processes display a wide scatter in the mechanical properties attributed to the variance in the microstructural features, defects, CNT distribution, and porosity level caused by processing and lack of consistency in mechanical testing techniques and samples. Table 6.2 shows a summary of the bulk tensile test results on Al-CNT samples. It is observed that most of the focus has been on strengthening and hence the elastic modulus has not been reported. In Table 6.2, $\Delta\sigma_Y$, $\Delta\sigma_U$, ΔE, and $\Delta\varepsilon_f$ represent the percentage increase in yield strength, the fracture strength, Young's modulus, and the failure strain, respectively. Figure 6.5 shows the percentage increase in the tensile properties (yield strength and ultimate tensile strength) of Al-CNT composites as compared to unreinforced aluminum processed under identical conditions. The variation in the degree of strengthening for similar CNT content by different studies indicates that several factors affect strengthening in CNT composites. This section discusses the models for predicting the strength of CNT-MMCs and compares their applicability in the case of CNT composites.

TABLE 6.2
Mechanical properties of Al-CNT composites obtained from bulk tensile tests

Ref.	CNT Vol.%	Dispersion Method	Fabrication	Dispersion	Al_4C_3	Sample Size	ρ (% Th.)	$\Delta\sigma_Y$ (%)	$\Delta\sigma_U$ (%)	ΔE (%)	$\Delta\varepsilon_f$ (%)
[23]	5	Al powders (99.99% pure, 40 µm size) mixed with 5 and 10 vol.% CNT by stirring in ethanol	Hot compaction (at 873 K) → hot extrusion (25:1 ratio at 773 K)	Poor, CNT clusters	No	Cylindrical, 3 mm dia. 15-mm gauge length	94	—	−5	—	−32
	10			Poor, CNT clusters			96.2	—	−9	—	−59
[22]	0.5	Al powders (200 mesh) ball milled with CNT	Compacted and sintered (853 K) → hot extrusion (833 K)	Good	No	Bulk sized but not mentioned	98	8	NA	—	—
	2						98	24	NA	23	—
[26]	6.2	Al-Ni powders and 5 vol.% CNT ball milled	Pressed (600 MPa) →sintered (913 K for3 h)	Good	No	Dog bone shape, 20-mm gauge length, 5.5-mm wide, thickness not mentioned	95.4	—	52	34	−75
	6.2	Ni particles (1 wt%) on Al produced by precipitation calcination → CNT grown by CVD	repressing (2 GPa)	Very good			96.2	—	184	—	—
[24,25]	0.6	1 wt% CNT (refluxed with HNO_3) ball milled with 2024 Al powders (50 µm size)	Cold isostatic pressing (300 MPa) → hot extrusion (733 K to 12 mm dia. rods)	Good	No	Dog bone specimen of 15 mm gauge length, dia. not mentioned	98.8	—	29	15	−3
	1.2			Good			99.1	—	34	42	−5
	2.4			Poor			96.4	—	−10	19	−82

[27]	1.6	500 g (99.85% pure, 15 μm size), 10 g Mg (99.8% pure, 50 μm size) and 20 g CNT (dia. 15 nm, length 30 μm) mixed with natural rubber → heat in for 2 h	Nanoscale dispersed powder → hot extrusion at 673 K Extrusion ratio 10 and 20	CNTs at grain boundaries	No	Not mentioned, dia. = 4.7 and 3.4 mm for extrusion ratio =10 and 20 respectively	— —	— —	13 53	— —	−27 −45
[28]	0.6	Al powder (99.7% pure, 75 μm size – CNT (140 nm dia., 3–4 μm length) mixed in planetary mill (300 rpm)	Powder fill in copper can → cold rolling (50% reduction) → sintering in v(573 K for 3 h) → sintering in air (823 K for 45 min)	Good	No	Flat tensile specimen, 25 mm gauge length, 6-mm wide, 0.4-mm thick	99	43	4	—	−31
	1.2			Poor			99.2	−1	−20	—	
	1.2			Slightly improved			—	—	4	—	—
	2.4			Poor improved			97.5	−39	−52	—	−92
[29]	5	NSD Al-CNT powder	SPS (873 K max. temperature for 20 min at 50 MPa) → hot extrusion (at 673 K, extrusion ratio = 20)	Broken clusters	Yes	3-mm dia. samples	98	—	128	—	−39
[8]	12.5	Al-23 wt% Si powder (15–45 μm size) blended with 10 wt.% CNT	Plasma spraying	Poor	No	26-mm gauge length, 6-mm wide, 0.635-mm thick samples, slightly curved	—	—	4	78	−54

(Continued)

TABLE 6.2 (*Continued*)
Mechanical properties of Al-CNT composites obtained from bulk tensile tests

Ref.	CNT Vol.%	Dispersion Method	Fabrication	Dispersion	Al_4C_3	Sample Size	ρ (% Th.)	$\Delta\sigma_Y$ (%)	$\Delta\sigma_U$ (%)	ΔE (%)	$\Delta\varepsilon_f$ (%)
[30]	0.6	Al (99.9% pure, 325 mesh size)-CNT mixture→high energy milling (5 h)	Compaction and vacuum sintering (873 K for 3 h) →hot extrusion (into 10 mm dia. rods at 773 K, extrusion ratio = 16)	Good	Yes	Dog bone type, 30-mm gauge length	—	30	12.5	—	−39
	1.2						—	41	20	—	—
	1.8						—	77	49	—	−30
	2.4						—	70	57.5	—	−20
[31]	2.4	Al (99.7% pure, 200 mesh size) mixed with 2 wt.% CNT ball milled (200 rpm for 3 h and 6 h)	Cold compaction (475 MPa)→Hot extrusion (at 773 K, extrusion ratio = 4) →annealing (673 K and 773 K for 10 h)	Good	No	20-mm gauge length, 4-mm dia.	—	—	21	—	−34
	2.4						—	—	0	—	−6
[9]	1.5	Ball milling CNT and al mixture with 1 wt% stearic BPR 15:1 (500 RPM)	Ball-milled powders containerized and hot-rolled at 480 °C for 27 times with a 12% reduction	Good	No	Gage length 12.5 mm, 1.5 mm	100	47	35	18	−50
	3						100	84	69	36	−69
	4.5						100	133	114	57	−81

[32]	1.86 4.34 8.06	Ni particles (1 wt.%) on Al produced by precipitation calcination→CNT grown by CVD	Pressed at 600 MPa, sintered at 640 °C for 3 h in vacuum, repressed at 2 GPa and annealed at 850 °C for 2 h	Very good	No	Dog-bone shaped, gage length of 20 mm and width of 5.5 mm	99 97 94	— — —	54 141 130	17 26 30	−47 −60 −92
[33]	0.62 1.24 2.48	Mechanically mixed for 2 h at 200 rpm	CIP/Sintering/extrusion	Poor	No	Sintering of green compacts at 580 °C followed by cold extrusion (extrusion ratio = 2.25)	95 97 99	25 53 93	24.5 54 88	−10 −4 9	−12 24 10

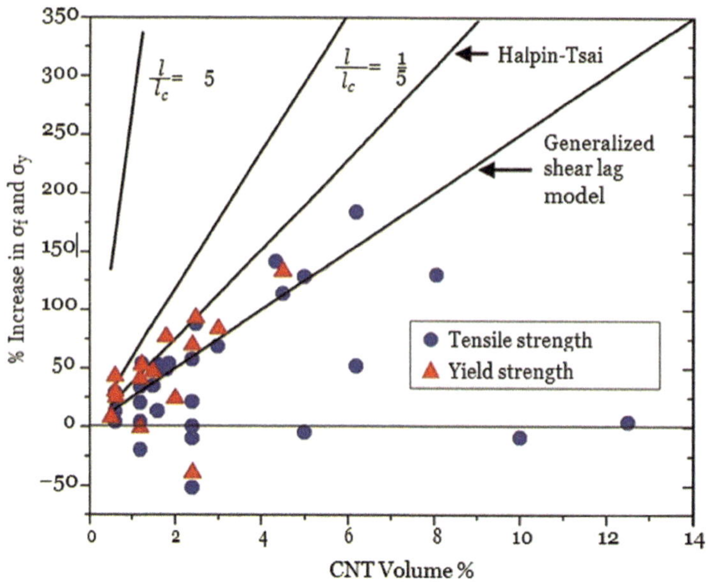

FIGURE 6.5 Variation of the experimental values of the increase in tensile (σ_f) and yield (σ_y) strength of Al-CNT composites and those calculated using micro-mechanics models.

6.2.1 Shear Lag Models

In shear lag models for short fiber composites, the load is transferred to the fibers through an interfacial shear stress. The tensile strength in the fiber varies from zero at the ends to the maximum at the center. The longer the fiber, the more strength that can be transmitted to it and hence the more efficient is the utilization of the fiber properties. There is a critical value of the fiber length (l_c) at which the maximum stress in the fiber at the center equals the fracture strength. When the length of the fiber is more than l_c, the fiber can be utilized to its maximum capability. The shear lag model proposed by Kelly-Tyson for short fiber composites with the CNT length lower than critical length ($l<l_c$) has been utilized for calculating the strength of Al-4 vol.% CNT composite [34] and is given by:

$$\sigma_c = \sigma_f V_f \left(\frac{l}{l_c}\right) + \sigma_m \left(1 - V_f\right) \tag{6.22}$$

where $l_c = \dfrac{\sigma_f d}{2\tau_m}$ the critical length of the fiber, V_f is the CNT volume fraction, l is the average CNT length, d is the CNT diameter, σ_m is matrix strength, τ_m is the matrix shear strength ($=\sigma_m/2$), and σ_f is the CNT strength. This model has been applied for calculating the compressive yield strength of Al-4 vol.% CNT composite having two grain sizes (72 and 200 nm) prepared by hot extrusion of ball-milled powders [34]. The values used were: σ_m = measured yield strength of 238 MPa for 200 nm grain size and 283 MPa for 72 nm grain size material, $V_f = 0.04$, l = 1 μm, $d = 35$ nm, $\sigma_f = 30$ GPa, and critical length l_c = calculated value of 3.7 and 4.4 μm for the 72 nm and 200 nm grain size composite. The calculated strength value for Al-4 vol.% CNT composite is found to be 434 MPa and 391 MPa for the 70 nm and 200 nm grain size material, respectively, while the experimentally measured values are ~403 and ~323 MPa [34]. The effect of different volume fractions of 1.5, 3, and 4.5% CNT reinforcement was further studied and a good agreement between the experimental and computed tensile yield strength values for up to 4.5 vol.% reinforcement was reported [9]. Figure 6.6 shows the experimentally measured and calculated yield strength for hot rolled Al-CNT composites [9]. The high degree of strengthening has been attributed to the well-dispersed CNTs obtained due to ball milling and the high density due to extrusion, which leads to extremely high load transfer efficiency to the CNTs. Figure 6.7a shows the TEM image of excellently dispersed CNTs in ball-milled Al powder after 6 h of milling. Figure 6.7b shows the corresponding stress strain curves for extruded samples prepared from the powders showing enhancement in the strength with CNT addition.

For the case where $l > l_c$, higher strengthening is expected and the Kelly-Tyson formula for strength of short fiber composites has been utilized for calculating the strength of Al-CNT composites by Kuzumaki et al. in the case of Al-CNT composites [23] and is given as:

$$\sigma_c = \sigma_f v_f \left(1 - \frac{l_c}{2l}\right) + \sigma_m^f \left(1 - V_f\right) \tag{6.23}$$

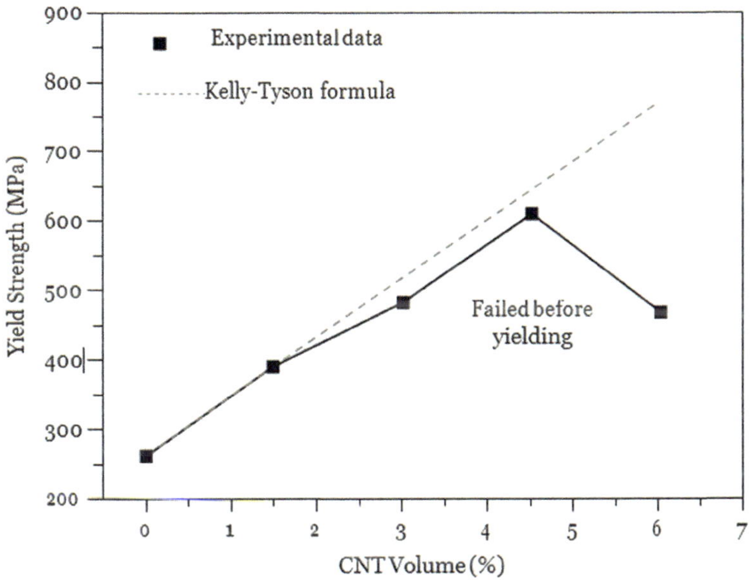

FIGURE 6.6 Variation of the yield strength of Al-CNT composites prepared by hot rolling of ball-milled powders as a function of CNT content [9]. (Reproduced with permission from Materials Research Society).

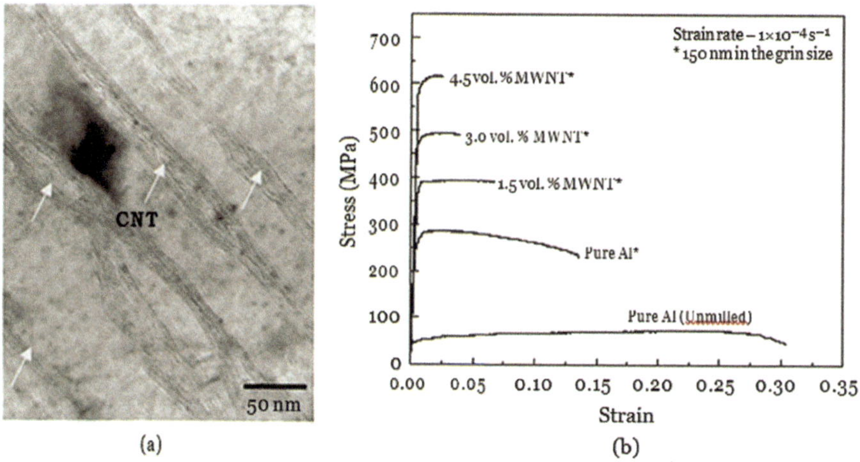

FIGURE 6.7 Showing (a) TEM image of a ball-milled (6 h) Al-CNT powder showing good dispersion of CNTs, and (b) stress–strain curves of samples obtained by hot extrusion of ball-milled powders [9]. (Reproduced with permission from Materials Research Society).

where σ_c is the composite fracture strength, σ_f is the fiber fracture strength (used for high modulus carbon fiber = 3 GPa), σ_m is the strength of the matrix at the failure strain of the composite (40 MPa), V_f is CNT volume fraction (0.05 and 0.1), l_c is the critical length of CNTs in the Al matrix (0.85 μm), and l is the average CNT length (2 μm). The estimated σ_c is about 270 MPa. The experimentally obtained strengths are much lower than this theoretical estimate due to CNT clustering [23].

Figure 6.5 shows the experimental values of the percentage increase in strength of Al-CNT composites. The values calculated using the two shear lag models represented by Eqs. 6.22 and 6.23 and assuming $\sigma_m = 100$ MPa (for Al) and $\sigma_f = 30{,}000$ MPa (for CNT) and $l/l_c = 5$ and 0.2 are also superimposed on the graph. From Figure 6.5, it can be seen that most of the experimental data are close to the values predicted for models where $l < l_c$. There is a high probability of CNT breakdown and damage during fabrication. CNT breakage and shortening of length was observed during cold spraying and ball milling of Al-CNT [9,14]. Hence, it needs to be kept in mind that starting CNTs may be longer than critical length, but the final length in the composite may be shorter due to damage while processing. The curved nature of CNTs is also expected to result in lower strengthening as compared to the computed strength predicted by the shear lag models, which assume straight fibers.

6.2.2 Strengthening by Interphase

In the case of a reaction product and new phase formation at the CNT/metal matrix interface, the strength of the composite is limited by the shear strength of the interfacial phase. Presence of a high strength interphase could help in the transfer of stress from the matrix to the CNTs. A model for the strength of the composite in the presence of an interfacial layer has been proposed by Coleman et al. according to which the strength of the composite is given as follows [35]:

$$\sigma_c = \left(1+\frac{2b}{D}\right)\left[\sigma_{Shear}\frac{l}{D}-\left(1+\frac{2b}{D}\right)\sigma_m\right]V_f+\sigma_m \qquad (6.24)$$

where σ_{Shear} is the shear strength of the interface, b is the width of the interphase, and D is the diameter of the CNT. This relation could be very useful for CNT-reinforced composites. However, it is difficult to obtain uniform dispersion with individual nanotubes separated from each other. In the case of CNT-MMCs, reaction between CNTs and the matrix might lead to interfacial carbide product (discussed at length in Chapter 7). When the applied stress exceeds the shear strength of the carbide, the facture occurs along the carbide layer leading to fiber pullout phenomena. Figure 6.8 shows a typical microstructure of CNT pullout [36]. Laha et al. have used Eq. 6.24 for determining theoretical strength of Al-CNT composites prepared by plasma spraying [8]. During plasma spraying of Al-23 wt.% Si alloy powders blended with 10 wt.% CNTs, a very fine layer of SiC (b = 5 nm) forms at the interface [37]. The calculated tensile strength value (226 MPa) using Eq. 6.24 was quite large compared to the experimentally measured value (83.1 MPa), which was attributed to porosity, inhomogeneous distribution, and clustering of CNTs in these composites [8].

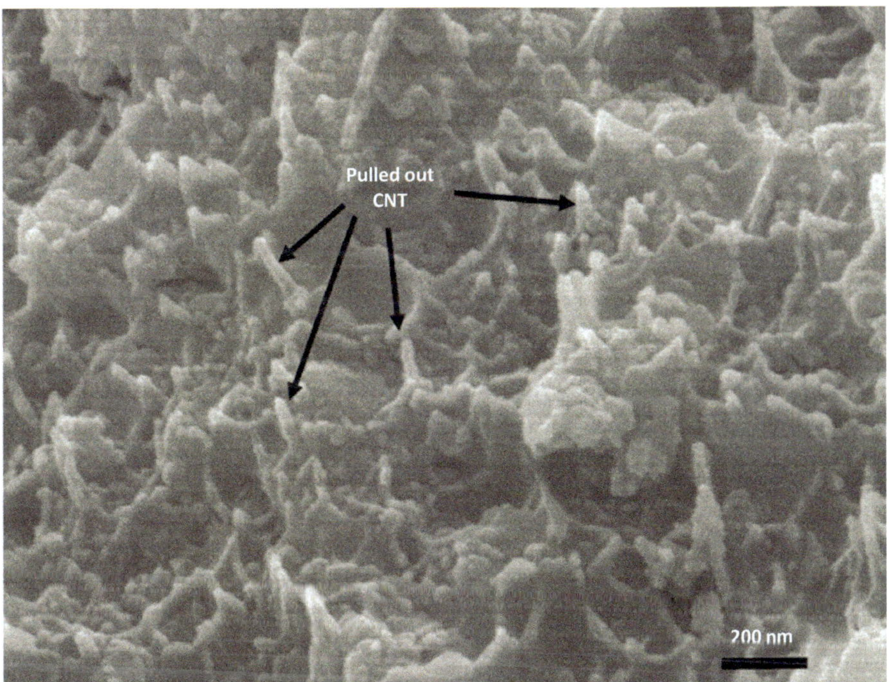

FIGURE 6.8 SEM image of the fracture surface of plasma-sprayed Al-12 wt.% Si alloy coating reinforced with 5 wt.% CNT showing pullouts [36]. (Reproduced with permission from Elsevier).

6.2.3 Strengthening by Carbon Nanotube Clusters

CNT clusters can also lead to strengthening if they are well infiltrated with metal and there is sufficient stress transfer to the cluster. In fact, a well-infiltrated CNT cluster is localized composite by itself and would have high strength. Elongated clusters of CNT were observed in the microstructure of Cu-CNT composites produced by SPS of ball-milled powders followed by cold rolling [21]. The composite showed a two-stage yielding process. Figure 6.9a and 6.9b show the microstructure and the corresponding stress strain curve, respectively, for Cu-CNT composites. The first yield strength ($\sigma_{y,1}$) corresponds to matrix yielding and the second ($\sigma_{y,2}$) yield point corresponds to CNT cluster yielding. Both yield points could be modeled by the following equations from a generalized shear lag model proposed by Ryu et al. [38].

$$\sigma_{y.1} = \frac{V_f \sigma_m}{2} S_{eff} + \sigma_m \tag{6.25}$$

Where

$$S_{eff} = S\cos^2\theta + \frac{(3\pi - 4)}{3\pi}\left(1 + \frac{1}{S}\right)\sin^2\theta$$

is the effective aspect ratio of an elongated CNT cluster oriented at an angle θ to the loading direction. The average S_{eff} is given as

$$S_{eff,Av} = \int_0^{\pi/2} S_{eff}(\theta)F(\theta)(2\pi\sin\theta)d\theta \tag{6.26}$$

where $F(\theta)$ is the probability distribution function of the misorientation of the CNT/Cu clusters, which was obtained by image analysis. The values of the first yield strength points for Cu-5 vol.% CNT and Cu-10 vol.% CNT composite were

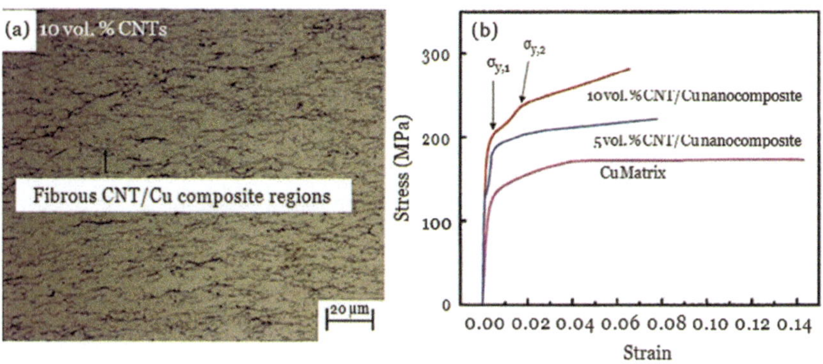

FIGURE 6.9 (a) Microstructure of CNT-Cu composites produced by SPS and cold rolling, and (b) stress–strain curves showing a two-stage yielding process [21]. (Reproduced with permission from Elsevier).

calculated to be 151 and 180 MPa, respectively, which were close to the experimentally measured values (149 and 197 MPa). During the first yielding, the CNT clusters were still elastic while the matrix started yielding.

When the stress exceeded the yield strength of the CNT cluster, secondary yielding was observed. The secondary yield point can be expressed by an ROM formula, which is given as:

$$\sigma_{y.2} = \sigma_{y.1}\left(1 - V_f\right) + \sigma_f V_f \tag{6.27}$$

The calculated values of the second yield point for Cu-5 vol.% CNT and Cu-10 vol.% CNT composite were found to be equal to 167 and 256 MPa, respectively, which were also close to the experimentally measured values (180 and 240 MPa). This study proves that the relations described previously can predict the mechanical behavior of composites with CNT clusters, which are well infiltrated with metal to cause effective load transfer to the CNT clusters.

Efficient strengthening has also been reported for Cu-CNT composites prepared by SPS of powders prepared by the molecular level mixing method [39]. The compressive yield strength of 5 and 10 vol.% CNT composites were found to be equal to 360 and 455 MPa, which was 2.3 and 3.1 times higher than Cu [39]. Eq. 6.25, which represents the generalized shear lag model, has been used for well-dispersed CNT composites like the study mentioned previously by replacing S_{eff} with S, the aspect ratio of the CNTs. A very good fit was obtained between calculated and predicted values for compressive yield strength of Cu-CNT composites [40]. For S = 40, a σ_c value of 300 and 450 MPa was obtained for Cu-5 vol.% CNT and Cu-10 vol.% CNT composite, respectively, using Eq. 6.25, which was very close to the experimental values of 360 and 455 MPa, respectively [40]. The values obtained from the generalized shear lag model (Eq. 6.25) for Al-CNT composite have also been represented in Figure 6.5. It is observed that many experimentally measured data points fall close to the generalized shear lag model at lower CNT concentrations, for which good dispersion can be easily obtained. However, at higher CNT loading, clustering reduces the strength. These studies prove the efficacy of the CNT cluster model suggested by Ryu et al. [38]. Clustering must be avoided as far as possible. Metal infiltrated CNT clusters may help in strengthening but the load transfer efficiency is not good and the strength of CNTs will be under-utilized. CNT clusters that are not infiltrated could be as bad as pores and are expected to reduce the strength drastically.

6.2.4 Halpin-Tsai Equations

The Halpin-Tsai expressions shown previously in Section 6.1.3 to discuss elastic modulus can also be used for calculating the strength of fiber-reinforced composites. A modified Halpin-Tsai equation was used to fit the properties of phenolic-based composites, which could also be used for CNT-MMCs [13] and is given as follows:

$$\sigma_c = \frac{1 + \xi\eta V_f}{1 - \eta V_f}\sigma_m \tag{6.28}$$

where η depends on (σ_f/σ_m) and is given by the expression

where

$$\eta = \frac{\alpha\left(\alpha_f / \alpha_f\right) - 1}{\alpha\left(\alpha_f / \alpha_f\right) + \xi'}$$

σ_m and σ_f being the strength of the matrix and the fiber, respectively. The coefficient ξ and α can be determined and are influenced by the degree of dispersion of the CNTs in the matrix. Where α is the orientation factor, for random orientation in thin sample (2-D) $\alpha = 1/3$, for a thick sample with small fibers (as in the case of CNTs), $\alpha = 1/6$ and

$$\xi = 2\left(\frac{l}{d}\right)$$

Halpin-Tsai equations are found effective in predicting the strength of polymer-CNT systems especially at low CNT concentrations [5]. At higher CNT loading, clustering and inefficient dispersion decreases the strengthening effect. The value of ξ has to be optimized in order to take care of the dispersion. The values of ξ in the case of phenolic-CNT systems is given by [13]:

$$\xi = 2\left(\frac{l}{d}\right)e^{\left(-68V_f - 1.1\right)} \text{ for networked CNTs} \tag{6.29}$$

$$\xi = 2\left(\frac{l}{d}\right)e^{\left(-40V_f - 1.0\right)} \text{ for dispersed CNTs} \tag{6.30}$$

These equations predict lower strengthening by addition of CNTs, which fit the experimental data in the case of phenolic-CNT systems. Such correlations must be developed for CNT-MMCs in order to understand the effect of dispersion at higher CNT content. From Figure 6.5, it is observed that for Al-CNT composites, Halpin-Tsai, and the generalized shear lag model models predict yield strength close to experimentally measured values. Halpin-Tsai equations provide a conservative estimate of the strength of Al-CNT composites.

6.2.5 Strengthening by Dislocations

Dislocations can be piled up against CNTs and dislocation forests can build up during deformation processing of CNT-MMCs. The formation of dislocation forests can be accentuated by the presence of CNTs and can cause strengthening. During sandwich processing of Al-CNT composites, it has been observed that the dislocation density increases due to CNT addition as shown in TEM images in Figure 6.10, and this leads to an increase in strengthening [41]. The strength can be expressed by the Taylor expression as:

$$\sigma = \sigma_0 + \alpha M^T G b \rho^{1/2} \tag{6.31}$$

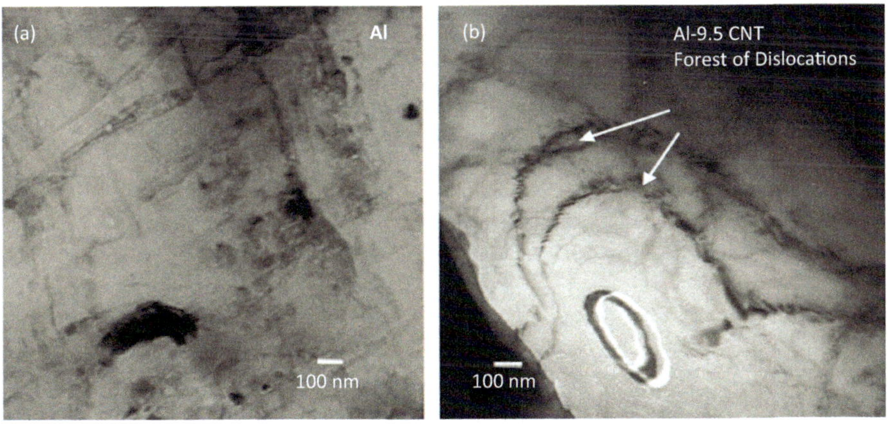

FIGURE 6.10 TEM images of sandwich processed (a) Al and (b) Al-CNT films showing the presence of dislocation networks in the CNT-containing films [41]. (Reproduced with permission from Elsevier).

where σ is the flow stress, σ_0 is the friction stress, α is a constant (1/3), M^T is Taylor factor (3 for untextured polycrystalline materials), G is the shear modulus, b is the Burgers vector of the dislocations, and ρ is the dislocation density. Dislocation density can also increase due to other factors such as thermal expansion mismatch. This dislocation density due to thermal mismatch (ρ_{th}) is given as [22]:

$$\rho_{th} = \frac{10V_f\varepsilon}{bt\left(1-V_f\right)} \tag{6.32}$$

where V_f is the volume fraction of CNT, ε is the thermal strain, b is the Burgers vector, and t is the diameter of the CNTs. CNTs can also act as precipitate and impede the motions of dislocations. Dislocations passing through the CNTs would then leave dislocation loops, which will create back-stress and repel dislocation motion, a mechanism known as Orowan looping. The increment in the shear strength of the composite due to Orowan looping mechanism is given as [22,42]:

$$\Delta\tau = \frac{A}{r\ln\left(2r/r_0\right)}Gbf^{1/2} \tag{6.33}$$

where A is a constant that is 0.093 for edge dislocation and 0.14 for screw dislocation, r_0 is the radius of the core of the dislocation, r is the volume equivalent radius for CNT, and G, b, and f have been previously defined. This mechanism is feasible; however, extensive TEM imaging is required to observe the dislocation loops around CNTs. Figure 6.11 shows the strengthening due to various reinforcement mechanisms including Orowan strengthening for AZ91 Mg alloy reinforced with CNTs [43]. It is seen that the experimental data is lower than the predicted value because not all the strengthening mechanisms may work or some may be less efficient than predicted by the model.

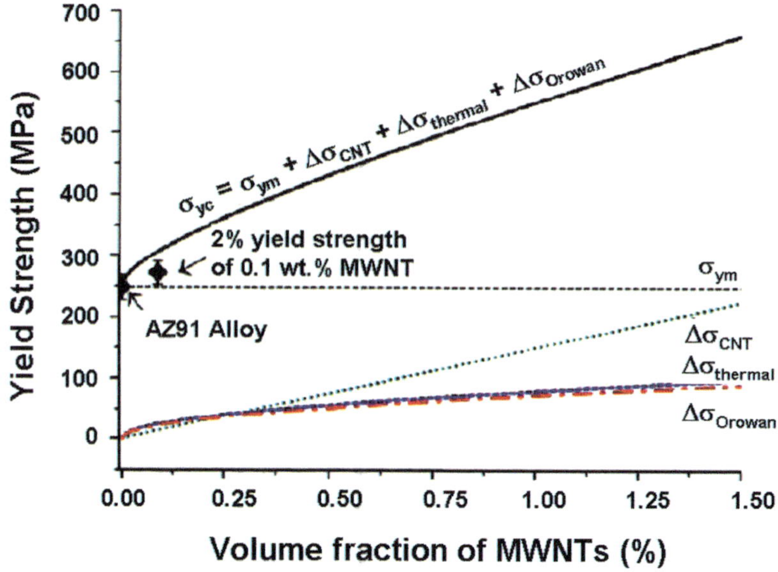

FIGURE 6.11 Graph showing the contributions of various mechanisms to the strengthening in AZ91 Mg alloy-CNT composite with experimental values [43]. (Reproduced with permission from Elsevier).

6.2.6 Strengthening by Grain Refinement

This is a very important and common mechanism for strengthening and was overlooked in some of the early studies on CNT-MMCs. The presence of CNTs in the matrix could lead to grain refinement in a number of ways. CNTs increasee work hardening, thermal conductivity, and act as secondary phases. All these lead to increased nucleation rates during recrystallization processes leading to fine grain structure. A few authors have estimated the effect of grain refinement in CNT composites [34,44–46]. Figure 6.12 shows that compressive yield strength of Al-4 vol.% CNT composites prepared by hot extrusion of ball-milled powder increases by ~100 MPa over aluminum although the grain sizes are similar [34]. Grain refinement and CNT reinforcement was found to result in a 520 MPa yield strength and 5% strain to failure in a Al-5 wt.% Si alloy reinforced with 3 vol.% CNTs produced by ball milling followed by hot rolling at 480 °C [9]. In the case of nanostructured Cu-CNT composites prepared by ball milling followed by high-pressure torsion [44,47], it was observed that the increase in microhardness was larger than that would be expected by grain size reduction alone (Figure 6.13) [44]. Similarly, the UTS of the Cu-CNT composite with 5 vol.% CNTs showed an improvement of ~85% in strength compared to that of unreinforced Cu. However, the UTS of the 10 vol.% CNT composite having smaller grain size was similar to that of the 5 vol.% CNT composite [47]. These results indicate that although there is grain size reduction in CNT composites, the strengthening achieved is higher than that caused by the grain refinement effect.

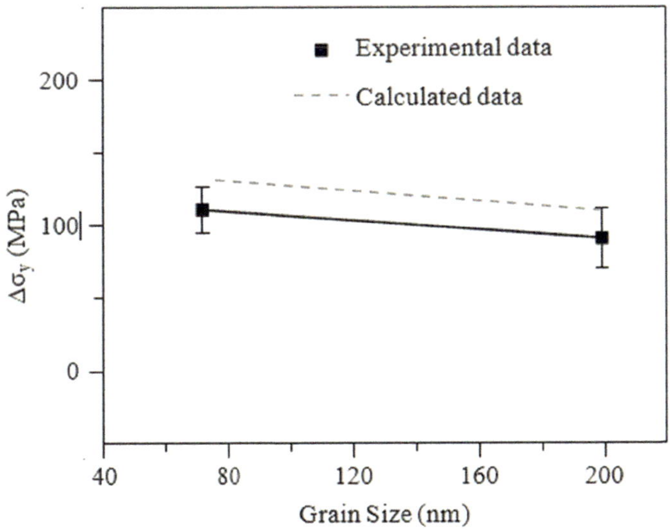

FIGURE 6.12 Increment of compressive yield strength of Al-4 vol.% CNT composite over Al for various grain sizes [34]. (Reproduced with permission from Elsevier).

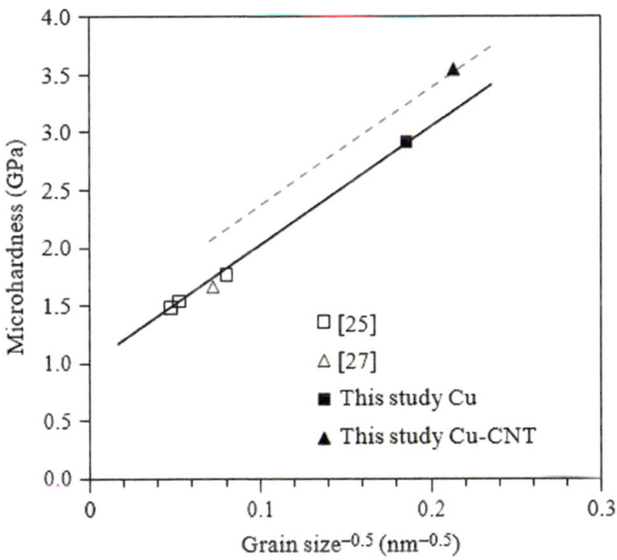

FIGURE 6.13 Variation of the microhardness of Cu-CNT composite containing 1 wt% CNT as a function of grain size indicating effect of CNTs [44]. (Reproduced with permission from Elsevier).

Table 6.3 summarizes the various models available for predicting the strength of CNT-MMCs along with examples where they have been applied. It is observed from Figure 6.5 that Halpin-Tsai and the generalized shear lag model predict experimental values closely. The Kelly-Tyson formula for short fiber composites ($l < l_c$) provides a value that has not been achieved experimentally, indicating that the load transfer efficiency to the CNTs might be very poor.

6.2.7 In Situ Observation of CNT Strengthening Mechanisms

The strengthening in a CNT–MMCs system is contributed by multiple mechanisms which include: load transfer from matrix to CNTs through interface [49,50], strain hardening (dislocation pile-up) [51–53], grain refinement [51], dispersion strengthening [54], solution strengthening [55], strengthening by carbide [55], grain boundary pinning or sub-grain formation [52], and thermal mismatch between CNTs and matrix [49]. The efficiency of these mechanisms is governed by dispersion of CNTs in the metal matrix and the interfacial bonding between the two. Various techniques, used for improving dispersion and its quantification, are discussed in detail in Chapter 2. However, as far as interfacial bonding is concerned, it plays very important role in strengthening the composite. The interfacial bonding between reinforcement and matrix is basically mechanical bonding that results from wettability between carbon and metals. Typically, such bonding is relatively weak due to the large surface energy difference between the matrix and CNTs. Interfacial chemical reaction has shown to be a promising approach for the CNT-metal bonding enhancement. However, it leads to carbide formation at the interface even when processed using rapid fabrication techniques, like SPS and plasma spraying, due the favorable thermodynamics and fast kinetics [56–58]. Formation of carbides can degrade mechanical properties, although suitable carbide reactions have shown to be a promising approach for the CNT-metal bonding enhancement [57,59,60].

The most common means to verify strengthening in a bulk composite is the indirect examination of CNT failure behavior on the fracture surface post-mortem. However, such evidence is not definitive as CNT pull-out may not only result from load transfer strengthening, as even a small load could result in CNT pull-out due to weak interface. Similarly, CNT fracture, which is a sign of efficient load transfer through strong interface, is difficult to observe through SEM and could have even happened during composite fabrication or could be due to excessive defects induced by harsh processing. Furthermore, analyzing the samples post-fracture makes it difficult to conclude which mechanisms dominate strengthening. Different empirical equations or models can be used to verify strengthening [61,62]. However, these models investigate the contribution of particular strengthening mechanism and not the synergistic effect of several strengthening mechanisms.

In situ testing, on the other hand, provides direct evidence of CNT-induced strengthening in MMCs. In situ SEM/TEM/AFM experiments have been applied to investigate CNT failure in polymer matrix composites (PMCs) [63,64] and ceramic matrix composites (CMCs) [65,66] for over two decades. However, the in situ studies reported on metal matrix composites have emerged over the past few years [68–71]. All the studies reported used tensile testing inside SEM to provide evidence for CNT

TABLE 6.3
Comparison of experimental values of the strength of MM-CNT composites and those calculated using micromechanics models

Model Used with Assumptions	System/Ref.	CNT Vol.%	Yield/Tensile Strength (MPa) Calculated	Experimental	Remarks
Kelly-Tyson formula ($l>l_c$) Applies to straight fiber composites, good load transfer efficiency	Al-CNT [43]	5	156	84	Disagreement due to clustering and poor dispersion.
		10	270	80	
Orowan looping	Al-CNT [39]	For CNT			Values match closely for CNT. CNTs might obstruct dislocation motion though this mechanism.
		0.5	91	86	
		0.5 with K_2ZrF_6	91	93	
		2	101	99	
		For SWNT			
		1	184	80	
		1 with K_2ZrF_6	184	99	
		2	227	91	
Thermal mismatch	Al-CNT [39]	For CNT			Predicted values too high. The mechanism is less likely to operate since coherent interfaces are needed to transmit thermal stresses.
		0.5	117	86	
		0.5 with K_2ZrF_6	117	93	
		2	197	99	
		For SWNT			
		1	471	80	
		1 with K_2ZrF_6	471	99	
		2	636	91	
Short fiber composites ($l<l_c$) Good dispersion with straight fibers assumed	Al-CNT [23]	4			Good dispersion due to ball milling and high density due to extrusion leads to better load transfer to CNTs.
		Grain size: 72 nm	~368	~358	
		Grain size: 200 nm	~413	~403	
	Al-CNT [9]	1.5	380	386	Good dispersion and high density lead the high load transfer efficiency.
		3.0	520	483	
		4.5	550	610	

(Continued)

TABLE 6.3 (*Continued*)
Comparison of experimental values of the strength of MM-CNT composites and those calculated using micromechanics models

Model Used with Assumptions	System/Ref.	CNT Vol.%	Yield/Tensile Strength (MPa) Calculated	Experimental	Remarks
Strengthening by interphase Uniform interfacial layer assumed over straight fibers that help in load transfer	A-Si-CNT [8]	5.67	226	83.1	Porosity, clustering, inhomogeneous distribution of CNTs not taken into account in the model.
Generalized shear lag model Assumes the linear elastic behavior of CNT clusters in plastic matrix	Cu-CNT [35]	First yield point			Calculated values of first yield strength are close to experimental at lower CNT content.
		5	151	149	
		10	180	197	
		Second yield (calculated through ROM)			
		5	167	180	
		10	256	240	
	Cu-CNT [21]	5	S = 40 S = 50		Strengthening is due to efficient load transfer to CNTs by interfacial oxygen atoms
		10	300 338	360	
			450 525	455	
Modified shear lag assumes load bearing by reinforcements in particulate reinforced composites	Cu-CNT [48]	2	148	139	Relative deviation between theoretical and experimental values at higher CNT content is due to the difficulty in dispersion
		4	179	186	
		8	213	188	

strengthening an Al matrix. Among all the active strengthening mechanisms, load transfer/sharing is considered the dominant contributor to the overall strengthening efficiency of CNT-MMCs, due to fiber-like structure and high aspect ratio of CNT. However, this has not been clearly verified in bulk MMCs, using post fracture imaging. In situ testing provides evidence supporting the theory behind load transfer strengthening mechanisms of CNTs in metal matrix composites. The in situ imaging of unreinforced SPS aluminum sample reveals presence of microcracking and the major crack post yielding. The propagation of major crack and the coalescence of microcracks with increase in load results in failure, which is also recorded in the corresponding stress–strain curve (Figure 6.14). Addition of 1 vol.% CNT to Al changes the stress strain response as seen in Figure 6.14. The yield strength and elastic modulus of Al increased by ~40% and ~65%, but at the expense of ductility [68]. The deformation and pullout of several CNTs from the matrix with a cheese-like flow pattern provides direct evidence of CNT-induced strengthening. Post fracture analysis shows the presence of telescopic fracture behavior at the ends of the CNT, which is due to transfer of load from the matrix to reinforcement and then gradual failure of outer layers of the multiwalled CNT. Further, the composite strength estimated using Kelly and Tyson shear lag model is comparable to in situ tensile test results, thus validating fiber strengthening [68].

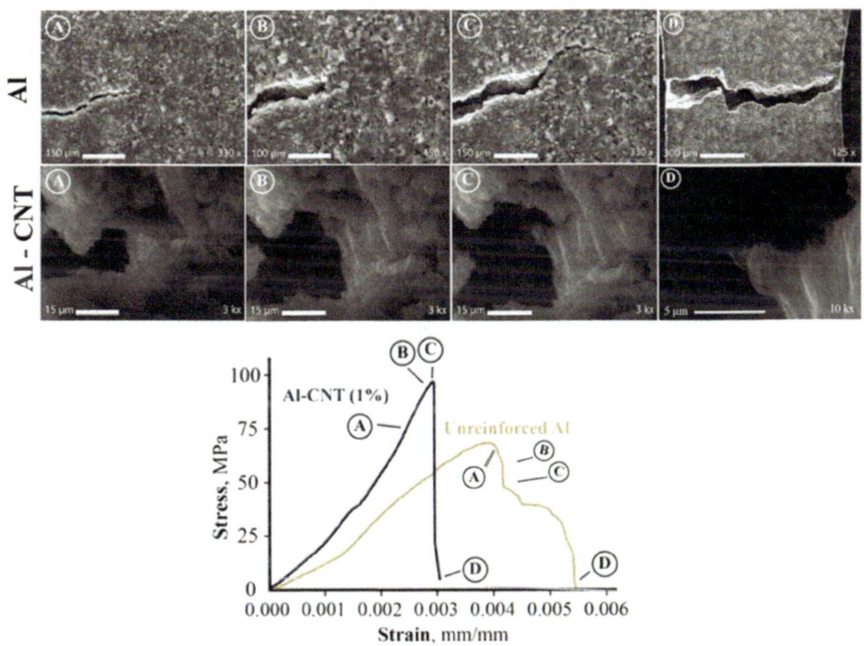

FIGURE 6.14 Results from in situ testing of the spark plasma sintered (SPS) Al and Al-CNT sample crack propagation in Al sample is shown from (A) to (D) and major crack propagation and evolution of damage during subsequent loading within the major crack in Al-CNT sample (A)–(D) [68].

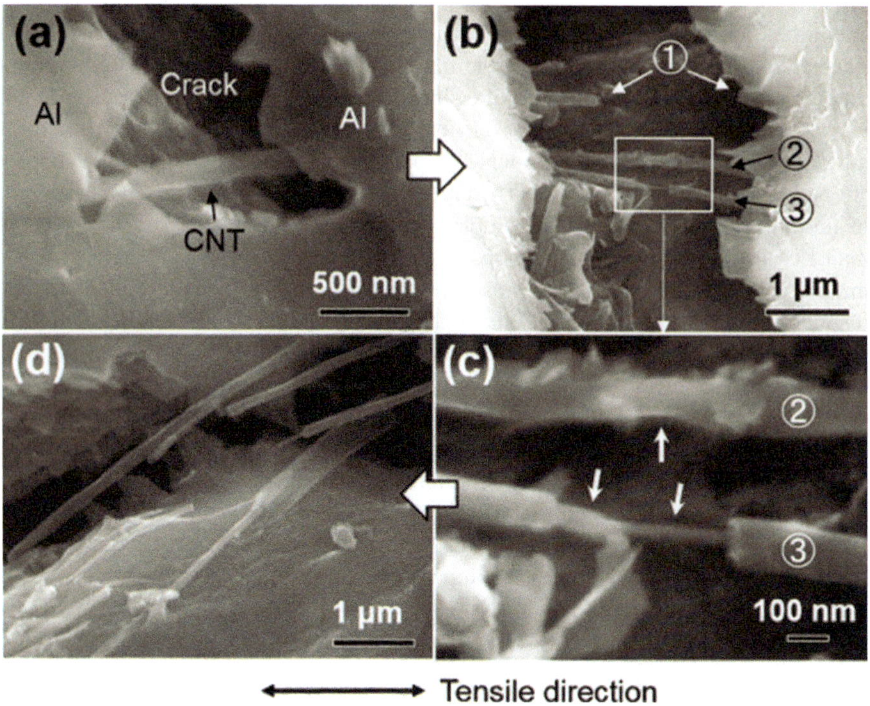

FIGURE 6.15 Different stages of MWCNT fracture as seen through in situ tensile test. (a) CNT prohibits crack growth in Al matrix by crack bridging, (b) and (c) shows CNTs (designated as CNT 1, 2, and 3) undergoing different wall fracture processes at different magnifications, and (d) fractured CNTs exposed on the composite surface post tensile test [69].

Similarly, another study reported in situ tensile testing within an SEM chamber of CNT-Al composite fabricated through sparking plasma sintering (SPS) and hot-extrusion [69]. It was observed from in situ tensile test that post-yielding, plastic deformation begin as a result microcracks were produced, which propagated in the composite, until it encountered a CNT. The CNTs were partially exposed and acted as bridges to restrain the growth of cracks. On further increase is displacement, the exposed CNTs fracture vertically into two, as seen in Figure 6.15. Through higher magnification images, it was observed that the outer walls of the CNTs are ruptured, while the inner part is undamaged and retains a columnar shape (Figure 6.15). The breaking of CNT walls suggests effective load transfer from Al matrix to CNTs during tensile failure [69].

Overall examination of various strengthening mechanisms through different mathematical models and equations revealed that load transfer contributed dominantly in high strengthening efficiency of CNT-Al composite, which was in agreement with the shear-lag model [69]. However, the nature of the CNT-Al interface leading to the effective load transfer was still unclear. Hence, the role played by interfacial carbide in improving the load transfer efficiency of CNTs in metal matrix (Al) is studied [70]. The reaction mechanism between CNTs and Al matrix was investigated using a range of sintering temperatures between 700 and 900 K [70]. Interfacial Al_4C_3 nanostructures were formed on partially damaged CVD produced CNTs in Al matrix (Figure 6.16). At sintering temperature below 800 K, CNTs were found to be stable in Al matrix, with no significant changes in YS and UTS. The quantity of Al_4C_3 increased between partially reacted CNT and Al matrix for temperature range 800–875 K (Figure 6.16(i) and 6.16(ii)) with noticeable increase of ~33% and ~31% in YS and UTS [70]. At 900 K, some mono-crystal rod-shaped Al_4C_3 phases were formed in Al matrix, resulting in decreased strengthening (Figure 6.16(iii)). All of this was confirmed through XRD, TEM, and ambient tensile testing [70].

In situ tensile testing was conducted for Al-CNT composite sintered at 875 K. As far as imaging is concerned, the role played by interfacial reaction products in strengthening the interface was not observed and the failure mechanism observed were similar to that in Ref. 64. However, interfacial strength and load transfer efficiency as estimated from empirical relationships (Eqs. 6.34 and 6.35) increased significantly compared to low temperature sintered composites [70]. It is therefore expected that partially reacted CNTs with interfacial carbides may be an ideal reinforcement structure for achieving high load transfer efficiency in metal matrix composites.

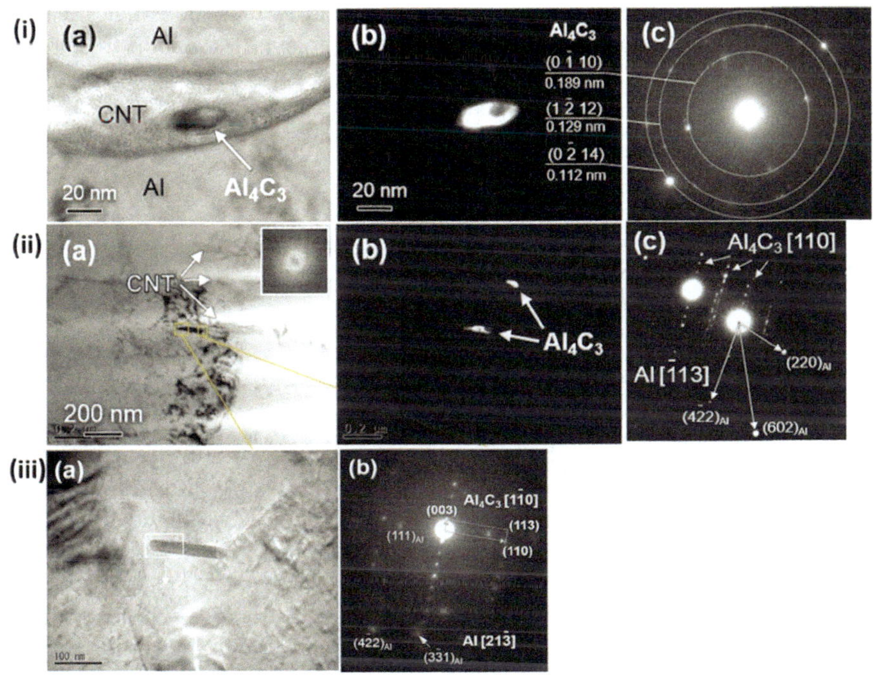

FIGURE 6.16 TEM observations a(i, ii and iii) bright field image, b(i, ii) dark field image and (c(i, ii) and b(iii)) SAD pattern [70] for CNT-Al composite processed at (i) 850 °C, (ii) 875 °C and (iii) 900 °C.

$$\tau = \frac{\Delta\sigma_{LT}}{SV} + \frac{\sigma_m}{S} \tag{6.34}$$

$$\delta_{LTE} = \frac{\Delta\sigma_{LT}}{\Delta\sigma_{LT}^{\max}} \tag{6.35}$$

Where τ interface strength, V is the volume fraction of CNTs (1.3 vol.%), and S is the aspect ratio of CNTs ($S = l/d$). σ_m is the UTS of matrix (~170 MPa). Load transfer efficiency (δ_{LTE}) of CNTs is defined as the ratio of $\Delta\sigma_{\mathrm{LT}}$, which is the improved load transfer strength to the maximum load transfer effect ($\Delta\sigma_{\mathrm{LT}}^{\max}$), when τ reached to its maximum value, the matrix shear yield strength (τ_m). τ_m is approximately equal to half of matrix tensile yield strength [70].

A rather new approach of enhancing the load sharing efficiency is introducing in situ nanoparticles at the interface [71]. This study reveals that the strong CNT-Al interface is resulted from the anchoring effect of interfacial Al_2O_3 nanoparticles [71]. The amorphous film, formed on Al powder surface by rocking ball milling process, transforms to stable and well-crystallized α- and γ-Al_2O_3 particles during the consolidation through SPS and hot extrusion. The role of interfacial nanoparticles in enhancing the strength of CNTs/Al interface and consequent load transfer effect is verified through in situ tensile tests. With Al-CNT composite having clean interface the phenomenon of CNT pulling out is often observed with composite failure. However, the CNT fracture mode in the composite with interfacial Al_2O_3 nanoparticles leads to CNT fracture, which produces comparatively short broken CNTs on the fracture surface [71]. For CNTs to fracture, its interfacial strength (τ) should be larger than a critical interfacial stress (τ_c), expressed as following:

$$\tau_c = \frac{\sigma_f}{2S} \tag{6.36}$$

Where σ_f is the fracture strength and S is the aspect ratio of CNT

$$\tau = \frac{V\sigma_f^{\,2}}{4S\left[V\sigma_f + (1-V)\sigma_m - \sigma_c\right]} \tag{6.37}$$

Where σ_c and σ_m are the strength of composite and matrix, respectively, and V is CNT volume content. These Al_2O_3 nanoparticles can be occasionally observed on the fractured surface of CNT as confirmed from their size through TEM (Figure 6.17). Also, with the presence of interfacial nanoparticles, the straight interface becomes tortuous. This is due to the increased resistance to CNT extension in Al matrix during the hot-extrusion process, generated by the pinning effect of hard Al_2O_3 nanoparticles [71]. Such tortuous morphology is favorable to produce high anchoring effect of nanoparticles that resists the sliding of CNTs under tensile loading thereby leading to effective load transfer from matrix to CNTs.

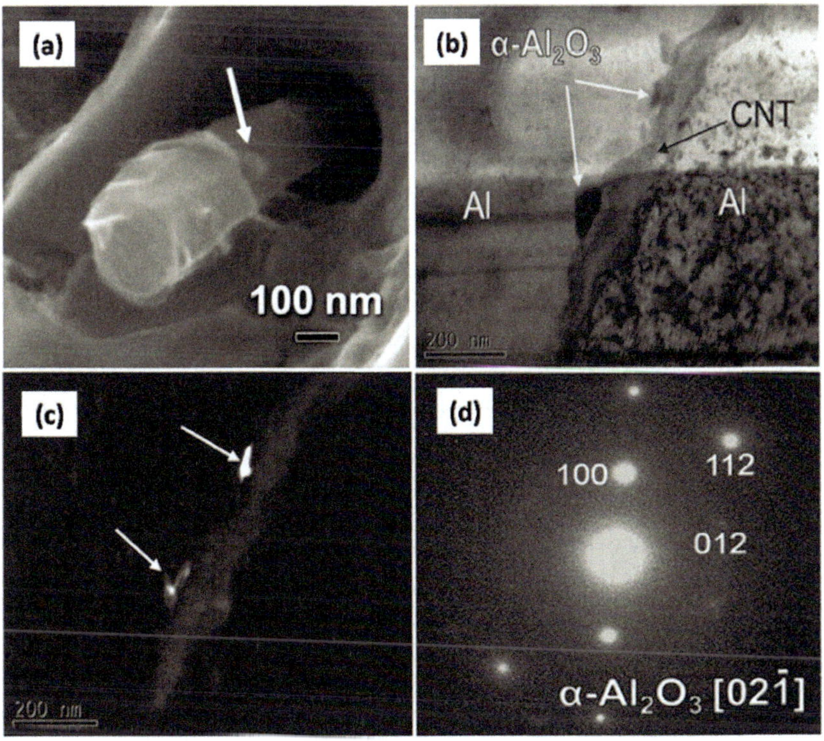

FIGURE 6.17 (a) SEM image of fractured surface of CNT Al composite during in situ tensile test showing interface with in situ Al_2O_3 nanoparticle, (b) bright field TEM image, (c) dark field TEM image, and (d) SAD pattern confirming Al_2O_3 nanoparticle at Al-CNT interface [71].

6.3 CHAPTER HIGHLIGHTS

In this chapter, the factors that bring about an increase in stiffness and strength in CNT-MMCs are discussed. The micromechanical models that help in understanding the mechanisms behind the strengthening process and predicting the strength without resorting to extensive experimentation are summarized. Also, the in situ tests that help confirm the CNT-induced strengthening and deformation mechanism in MMCs are detailed. Uniform dispersion of CNTs is the key to attaining properties predicted by the models. The large scatter in data on the improvement in properties for the same CNT content as observed in Figures 6.2 and 6.5 indicates that processing plays an important role. Samples produced by the same process under different processing conditions have different degrees of dispersion and lead to different amounts of strengthening. For example, in the case of dispersion by ball milling, the time of milling, speed, and ball-to-powder ratio would affect the dispersion attained. Final consolidation by hot extrusion would produce composites with different levels depending on the extrusion ratio and temperature. Clusters of CNTs are detrimental if not infiltrated with metal, in which case the load is not transferred to the clusters. They could be as bad as porosities. Processing methods must be developed or modified keeping these issues in mind. The properties of the CNT composites are also affected by nanotube waviness as suggested by some FEM simulations [67]. However, these effects are difficult to quantify and control in CNT composites. Most of the experimental data show the expected strengthening at low CNT content. Incorporating high volume fraction of CNTs in the metal matrix with good dispersion is still a challenge.

REFERENCES

1. Cox, H. L. 1952. The elasticity and strength of paper and other fibrous objects. *Brit. J. Appl. Phys.* 3: 72–79.
2. Hill, R. 1964. Theory of mechanical properties of fiber-strengthened materials: i. *Elastic behavior. J. Mech. Phys. Solids* 12: 199–212.
3. Kelly, A., and W. R. Tyson. 1965. Tensile properties of fibre-reinoforced metals: copper/tungsten and copper/molybdenum. *J. Mech. Phys. Solids* 13: 329–350.
4. Laborde-Lahoz, P., Maser, W., Matrinez, T., Benito, A., Seeger, T., Cano, P., de Villoria, R. G., and A. Miravete. 2005. Mechanical characterization of carbon nanotube composite materials. *Mech. Adv. Mater. Struct.* 12: 13–19.
5. Coleman, J. N., Khan, U., Blau, W. J., and Y. K. Gun'ko. 2006. Small but strong: a review of the mechanical properties of carbon nanotube–polymer composites. *Carbon* 44: 1624–1652.
6. Seidel, G. D., and D. C. Lagoudas. 2006. Micromechanical analysis of the effective elastic properties of carbon nanotube reinforced composites. *Mech. Mater.* 38: 884–907.
7. Zalamea, L., Kim, H., and R. B. Pipes. 2007. Stress transfer in multi-walled carbon nanotubes. *Composites Sci. Tech.* 67: 3425–3433.
8. Laha, T., Chen, Y., Lahiri, D., and A. Agarwal. 2009. Tensile properties of carbon nanotube reinforced aluminum nanocomposite fabricated by plasma spray forming. *Composites A* 40: 589–594.
9. Choi, H., Shin, J., Min, B., Park, J., and D. Bae. 2009. Reinforcing effects of carbon nanotubes in structural aluminum matrix nanocomposites. *J. Mater. Res.* 24: 2610–2616.
10. Halpin, J. C., and S. W. Tsai, 1969. Effects of environmental factors on composite materials. Air Force Materials Laboratory Technical Report AFRL-TR-67-423.
11. Halpin, J. C., and J. L. Kardos. 1976. The Halpin-Tsai equations: a review. *Polym. Eng. Sci.* 16: 344–352.
12. Qian, D. Dickey, E. C., Andrews, R., and T. Rantell. 2000. Load transfer and deformation mechanisms in carbon nanotube-polystyrene composites. *Appl. Phys. Lett.* 76: 2868–2870.
13. Yeh, M-K., Tai, N-H., and J-H. Liu. 2006. Mechanical behavior of phenolic-based composites reinforced with multi-walled carbon nanotubes. *Carbon* 44: 1–9.
14. Bakshi, S. R., Singh, V., Balani, K., McCartney, D. G., Seal, S., and A. Agarwal. 2008. Carbon nanotube reinforced aluminum composite coating via cold spraying. *Surf. Coat. Tech.* 202: 5162–5169.
15. Hashin, Z., and S. Strikman. 1962. On some variational principles in anisotropic and nonhomogeneous elasticity. *J. Mech. Phys. Solids* 10: 335–342.
16. Hashin, Z., and S. Strikman. 1992. Extremum principles for elastic heterogeneous media with imperfect interfaces and their application to bounding of effective moduli. *J. Mech. Phys. Solids* 40: 767–781.
17. Mori, T., and K. Tanaka. 1972. Average stress in matrix and average elastic energy of materials with misfitting inclusions. *Acta Metall.* 21: 571–574.

18. Eshelby, J. D. 1957. The determination of the elastic field of an ellipsoidal inclusion, and related problems. *Proc. Roy. Soc. London A* 241: 376–396.
19. Chen, Y., Balani, K., and A. Agarwal. 2007. Modified Eshelby tensor modeling for elastic property prediction of carbon nanotube reinforced ceramic nanocomposites. *Appl. Phys. Lett.* 91: 319031–319033.
20. de Villoria, R. G., and A. Miravete. 2007. Mechanical model to evaluate the effect of the dispersion in nanocomposites. *Acta Mater.* 55: 3025–3031.
21. Kim, K. T., Cha, S. I., Hong, S. H., and S. H. Hong. 2006. Microstructures and tensile behavior of carbon nanotube reinforced Cu matrix nanocomposites. *Mater. Sci. Eng. A* 430: 27–33.
22. George, R., Kashyap, K. T., Rahul, R., and S. Yamdagni. 2005. Strengthening in carbon nanotube/aluminium (CNT/Al) composites. *Scripta Mater.* 53: 1159–1163.
23. Kuzumaki, T., Miyazawa, K., Ichinose, H., and K. Ito. 1998. Processing of carbon nanotube reinforced aluminum composite. *J. Mater. Res.* 13: 2445–2449.
24. Deng, C. F., Zhang, X. X., Wang, D. Z., Lin, Q., and A. B. Li. 2007. Preparation and characterization of carbon nanotubes/aluminum matrix composites. *Mater. Lett.* 61: 1725–1728.
25. Deng, C. F., Wang, D. Z., Zhang, X. X., and A. B. Li. 2007. Processing and properties of carbon nanotubes reinforced aluminum composites. *Mater. Sci. Eng. A* 444: 138–145.
26. He, C., Zhao, N., Shi, C., Du, X., Li, J., Li, H., and Q. Cui. 2007. An approach to obtaining homogeneously dispersed carbon nanotubes in Al powders for preparing reinforced Al-matrix composites. *Adv. Mater.* 19: 1128–1132.
27. Yuuki, J., Kwon, H., Kawasaki, A., Magario, A., Noguchi, T., Beppu, J., and M. Seki. 2007. Fabrication of carbon nanotube reinforced aluminum composite by powder extrusion process. *Mater. Sci. Forum* 534–536: 889–892.
28. Esawi, A. M. K., and M. A. E. Borady. 2008. Carbon nanotube-reinforced aluminium strips. *Composites Sci. Tech.* 68: 486–492.
29. Kwon, H., Estili, M., Takagi, K., Miyazaki, T., and A. Kawasaki. 2009. Combination of hot extrusion and spark plasma sintering for producing carbon nanotube reinforced aluminum matrix composites. *Carbon* 47: 570–577.
30. Pérez-Bustamante, R., Gómez-Esparza, C. D., Estrada-Guel, I., Miki-Yoshida, M., Licea-Jiménez, L., Pérez-García, S. A., and R. Martínez-Sánchez. 2009. Microstructural and mechanical characterization of Al–MWCNT composites produced by mechanical milling. *Mater. Sci. Eng. A* 502: 159–163.
31. Esawi, A. M. K., Morsi, K., Sayed, A. Gawad, A. A., and P. Borah. 2009. Fabrication and properties of dispersed carbon nanotube–aluminum composites. *Mater. Sci. Eng. A.* 508: 167–173.
32. He, C., Zhao, N. Q., Shi, C. S., and S. Z. Song. 2009. Mechanical properties and microstructures of carbon nanotube-reinforced Al matrix composite fabricated by in situ chemical vapor deposition. *J Alloys Compd.* 487: 258–262.
33. Sridhar, I., and K. R. Narayanan. 2009. Processing and characterization of MWCNT reinforced aluminum matrix composites. *J. Mater. Sci.* 44: 1750–1756.
34. Choi, H. J., Kwon, G. B., Lee, G. Y., and D. H. Bae. 2008. Reinforcement with carbon nanotubes in aluminum matrix composites. *Scripta Mater.* 59: 360–363.
35. Coleman, J. N., Cadek, M., Blake, R., Nicolosi, V., Ryan, K. P., Belton, C., Fonseca, A., Nagy, J. B., Gunko, Y. K., and W. J. Blau. 2004. High-performance nanotube reinforced plastics: understanding the mechanism of strength increase. *Adv. Funct. Mater.* 14: 791–798.
36. Bakshi, S. R., Singh, V., Seal, S., and A. Agarwal. 2009. Aluminum composite reinforced with multiwalled carbon nanotubes from plasma spraying of spray dried powders. *Surf. Coat. Tech.* 203: 1544–1554.
37. Laha, T., Liu, Y., and A. Agarwal. 2007. Carbon nanotube reinforced aluminum nanocomposite via plasma and high velocity oxy-fuel spray forming. *J. Nanosci. Nanotech.* 7: 515–524.

38. Ryu, H. J., Cha, S. I., and S. H. Hong. 2003. Generalized shear-lag model for load transfer in SiC/Al metal-matrix composites. *J. Mater. Res.* 18: 2851–2858.
39. Cha, S. I., Kim, K. T., Arshad, S. N., Mo, C. B., and S. H. Hong. 2008. Extraordinary strengthening effect of carbon nanotubes in metal-matrix nanocomposites processed by molecular-level mixing. *Adv. Mater.* 17: 1377–1381.
40. Kim, K. T., Cha, S. I., Gemming, T., Eckert, J., and S. H. Hong. 2008. The role of interfacial oxygen atoms in the enhanced mechanical properties of carbon-nanotube-reinforced metal matrix nanocomposites. *Small* 4: 1936–1940.
41. Lahiri, D., Bakshi, S. R., Keshri, A. K., Liu, Y., and A. Agarwal. 2009. Dual strengthening mechanisms induced by carbon nanotubes in roll bonded aluminum composites. *Mater. Sci. Eng. A* 523: 263–270.
42. Orowan, E. 1934. Zur kristallplastizitat. III. Uber die mechanismus des gleitvorganges. *Z. Phys.* 89: 634–659.
43. Li, Q., Viereckl, A., Rottmair, C. A., and R. F. Singer. 2009. Improved processing of carbon nanotube/magnesium alloy composites. *Composites Sci. Tech.* 69: 1193–1199.
44. Li, H., Misra, A., Zhu, Y., Horita, Z., Koch, C. C., and T. G. Holesingerd. 2009. Processing and characterization of nanostructured Cu-carbon nanotube composites. *Mater. Sci. Eng. A* 523: 60–64.
45. Choi, H. J., Shin, J. H., Min, B. H., Bae, D. H. 2010. Deformation behavior of Al-Si alloy based nanocomposites reinforced with carbon nanotubes. *Compos. Part A Appl. Sci. Manuf.* 41: 327–329.
46. Choi, H. J., Shin, J. H., Bae, D. H. 2011. Grain size effect on the strengthening behavior of aluminum-based composites containing multi-walled carbon nanotubes. *Compos Sci Technol.* 71: 1699–1705.
47. Yoon, E. Y., Lee, D. J., Park, B., Akbarpour, M. R., Farvizi, M. 2013. Grain refinement and tensile strength of carbon nanotube-reinforced Cu matrix nanocomposites processed by high-pressure torsion. *Metals Mater. Int.* 19: 927–932.
48. Akbarpour, M. R. 2016. Analysis of load transfer mechanism in Cu reinforced with carbon nanotubes fabricated by powder metallurgy route. *J Mater Eng Perform.* 25: 1749–1756.
49. Stein, J., Lenczowski, B., Fréty, N., and Anglaret, E. 2012. Mechanical reinforcement of a high-performance aluminium alloy AA5083 with homogeneously dispersed multi-walled carbon nanotubes. *Carbon.* 50: 2264–2272.
50. Nam, D. H., Cha, S. I., Lim, B. K., Park, H. M., Han, D. S., and Hong, S. H. 2012. Synergistic strengthening by load transfer mechanism and grain refinement of CNT/Al-Cu composites. *Carbon.* 50: 2417–2423.
51. Kang, K., Bae, G., Won, J., and Lee, C. 2012. Mechanical property enhancement of kinetic sprayed Al coatings reinforced by multi-walled carbon nanotubes. *Acta Mater.* 60: 5031–5039.
52. Yoo, S. J., Han, S. H., and Kim, W. J. 2013. Strength and strain hardening of aluminum matrix composites with randomly dispersed nanometer-length fragmented carbon nanotubes. *Scr. Mater.* 68: 711–714.
53. Li, S., Sun, B., Imai, H., and Kondoh, K. 2013. Powder metallurgy Ti-TiC metal matrix composites prepared by in situ reactive processing of Ti-VGCFs system. *Carbon.* 61: 216–228.
54. Kang, K., Bae, G., Kim, B., and Lee, C. 2012. Thermally activated reactions of multi-walled carbon nanotubes reinforced aluminum matrix composite during the thermal spray consolidation. *Mater. Chem. Phys.* 133: 495–499.
55. Fukuda, H., Kondoh, K., Umeda, J., and Fugetsu, B. 2011. Interfacial analysis between Mg matrix and carbon nanotubes in Mg-6wt.% Al alloy matrix composites reinforced with carbon nanotubes. *Compos. Sci. Technol.* 71: 705–709.

56. Ci, L., Ryu, Z., Jin-Phillipp, N. Y., and Rühle, M. 2006. Investigation of the interfacial reaction between multi-walled carbon nanotubes and aluminum. *Acta Mater.* 54: 5367–5375.
57. Bakshi, S. R., and Agarwal, A. 2011. An analysis of the factors affecting strengthening in carbon nanotube reinforced aluminum composites. *Carbon* 49: 533–544.
58. Jiang, L., Li, Z., Fan, G., Cao, L., and Zhang, D. 2012. The use of flake powder metallurgy to produce carbon nanotube (CNT)/aluminum composites with a homogenous CNT distribution. *Carbon* 50: 1993–1998.
59. Wagner, H. D., Lourie, O., Feldman, Y., and Tenne, R. 1998. Stress-induced fragmentation of multiwall carbon nanotubes in a polymer matrix. *Appl. Phys. Lett.* 72: 188–190.
60. Deng, F., Ogasawara, T., and Takeda, N. 2007. Tensile properties at different temperature and observation of micro deformation of carbon nanotubes-poly(ether ether ketone) composites. *Compos. Sci. Technol.* 67: 2959–2964.
61. Yamamoto, G., Shirasu, K., Hashida, T., Takagi, T., Suk, J. W., An, J., Piner, R., and Ruoff, R. 2011. Nanotube fracture during the failure of carbon nanotube/alumina composites. *Carbon* 49: 3709–3716.
62. Xia, Z., Riester, L., Curtin, W. A., Li, H., Sheldon, B. W., Liang, J., Chang, B., and Xu, J. M. 2004. Direct observation of toughening mechanisms in carbon nanotube ceramic matrix composites. *Acta Mater.* 52: 931–944.
63. Boesl, B., Lahiri, D., Behdad, S., and Agarwal, A. 2014. Direct observation of carbon nanotube induced strengthening in aluminum composite via in situ tensile tests. *Carbon* 69: 79–85.
64. Chen, B., Li, S., Imai, H., Jia, L., Umeda, J., Takahashi, M., and Kondoh, K. 2015. Load transfer strengthening in carbon nanotubes reinforced metal matrix composites via in-situ tensile tests. *Compos Sci Technol.* 113: 1–8.
65. Chen, B., Shen, J., Ye, X., Imai, H., Umeda, J., Takahashi, M., and Kondoh, K. 2017. Solid-state interfacial reaction and load transfer efficiency in carbon nanotubes (CNTs)-reinforced aluminum matrix composites. *Carbon* 114: 198–208.
66. Chen, B., Kondoh, K., Umeda, J., Li, S., Jia, L., and Li, J. 2019. Interfacial in-situ Al_2O_3 nanoparticles enhance load transfer in carbon nanotube (CNT)-reinforced aluminum matrix composites. *J Alloys Compd.* 789: 25–29.
67. Anumandla, V., and R. F. Gibson. 2006. A comprehensive closed form micromechanics model for estimating the elastic modulus of nanotube-reinforced composites. *Composites A* 37: 2178–2185.

7 Interfacial Phenomena in CNT-MMCs

In composite materials, there is an interaction between the matrix and the reinforcement at the interface, and through these interactions, a degree of load transfer occurs and thereby strengthening is achieved. Hence, studying the matrix/reinforcement interface is extremely important in CNT-MMCs. Several aspects such as chemical stability of the CNTs in the metal matrix, the thermodynamic and kinetic aspects of carbide formation, and wettability of the CNTs with the matrix influence the processing and final properties of the composite in a significant manner. In this chapter, the focus is on the importance of the CNT/metal interface, the thermodynamics and mechanism of the metal-CNT reaction, and its impact on the properties of the composite. One of the most common react products at the interface is known to be aluminum carbides. The effect of CNT defects and processing temperature on carbide formation is elucidated. In addition to aluminum carbide, other carbides have been proposed as interphases to enhance load transfer and thereby endow the composites with more efficient reinforcing effects. Promising approaches for tailoring of interphases and carbide reactions are discussed in detail.

7.1 SIGNIFICANCE OF INTERFACIAL PHENOMENA

It is necessary to study the interfacial reaction phenomena and chemical stability of the CNTs in the metallic matrix for several reasons. The first and most important effect is on mechanical properties of the composite. The strength of the fiber matrix composites depends on the stress transfer at the interface [1] and the interfacial strength between the reinforcement and the matrix [2]. Much work has been carried out on carbon fiber reinforced aluminum alloys for applications as lightweight and high strength structural material [3–5]. It has been shown that the interfacial reactions and degree of wetting of the fibers affect the properties of the composite [6–8]. Aluminum carbide (Al_4C_3) forms at the matrix fiber interface in Al-7 wt.% Si [4] and Al-13 wt.% Si [9] composites reinforced with carbon fiber prepared by liquid metal infiltration technique. Reduction in the strength and premature failure has been observed in 75 vol.% carbon fiber reinforced A357 alloy due to formation of Al_4C_3 and the presence of brittle Si particles [10]. These studies concluded that Al_4C_3 formation needed to be avoided. On the contrary, there have been some reports of improvement of properties of Al-SiC$_p$ composites due to limited amounts of Al_4C_3 formation [11].

In the case of CNT-MMCs, it is important that the applied load must be transferred to the stronger CNTs. Figure 7.1 shows how the load is transferred in two possible cases when the CNTs are parallel and perpendicular to the load applied. Interfacial compounds can form at the surface and the ends of the CNTs as shown in

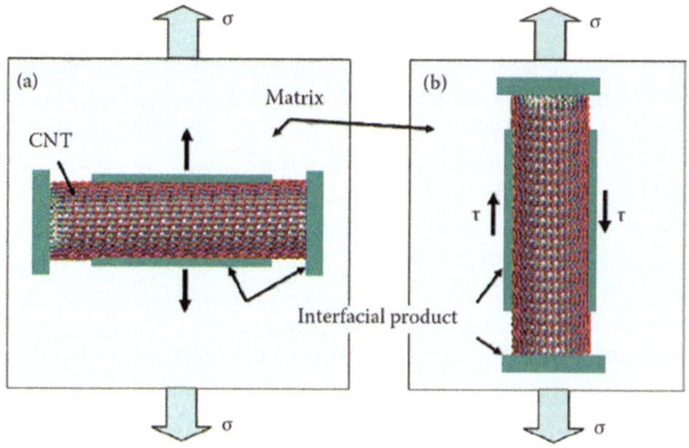

FIGURE 7.1 Schematic showing the different modes of stress transfer to the CNTs when the CNTs are aligned (a) normal, and (b) parallel to the loading direction.

Figure 7.1. It is also observed in Figure 7.1a that when the CNTs are normal to the applied stress, the load is transferred to the CNTs through the interfacial product, which serves as a link between the CNTs and the matrix. The strength of the composites in this case would then depend on the fracture strength of the interfacial compound, which could be a metal carbide formed due to reaction between metal matrix and the outer walls of the multiwall CNT reinforcement. SWNTs are not suitable for reinforcement in such cases as they may be fully destroyed upon reaction with the metal matrix. In the absence of an interfacial layer, the stress is transferred through weak attractive forces between CNT and the matrix. The wetting between CNTs and the matrix will then become very important, as poor wetting will lead to a bad interface with the porosity. In this situation, the property enhancement totally depends on the strength of the interface. The axial tensile strength of the CNTs is not used in strengthening. In the case where the CNTs are aligned parallel to the applied stress as shown in Figure 7.1b, the load is transferred to the CNTs through two mechanisms. First, it occurs through the interfacial reaction product at the ends of the CNTs, which helps in "pinning" the CNTs to the matrix. The second load transfer mechanism occurs through the reaction product on the CNT surface, which results in the interfacial shear stress generation as shown in the figure. The strength of the CNT-MMC would then be limited by the shear strength of the interfacial compound as described earlier in Section 5.2.2. Formation of the brittle interfacial carbide layer could lead to easy fracture, resulting in CNT pullout. Formation of a strong interface is essential so that the load can be transferred to strong CNTs to utilize their load bearing ability. CNT-MMCs have very often achieved increases in strength and stiffness at the expense of ductility. Brittle and stiff carbide interfaces were thought to be detrimental to ductility in the early days of CNT-MMC research, but more detailed investigations have elucidated more complexity on the role of carbide interfacial products. Chen et al. [12] were among the first to show simultaneous increase in strength and ductility via the careful tailoring of the carbide interfacial products. Increased sintering temperatures were shown to led to both the formation of carbides (*increasing strength*) as well as enhanced bond strength between aluminum grains, thereby promoting greater deformation prior to failure (*increasing ductility*). The details of carbide formation and the potential drawbacks and benefits of this type of interfacial product are discussed in detail in Section 7.3.

Wetting between the molten metal and CNTs is another important factor that affects the interface microstructure and determines the strength of the CNT/matrix interface. In the case where the CNT does not react with the matrix, the bonding is characterized by weak attractive van der Waals forces. For the nonwetting metals, using CNTs with pre-deposited interphases such as a Ni coating may help in improving the interfacial bonding. CNTs can be coated with Ni or other metals by electrodeposition or electroless deposition processes. Wetting also influences the dispersion behavior of CNTs in the matrix, especially when processing includes molten or semi-molten metal.

CNTs remain well dispersed and separate if they have an excellent wetting with the molten metal. In a non-wetting case, due to lower capillarity, the CNTs are brought together and can form clusters. This was observed in the case of plasma spraying of spray-dried Al-Si agglomerates containing well-dispersed CNTs [13]. Study of the single splat revealed that the CNT cluster formation started at the molten droplet formation stage as shown in the schematic in Figure 7.2a. CNTs are brought together by surface tension forces. The scanning electron micrographs in Figures 7.2b and 6.2c show the presence of CNT clusters in finger and disc parts of the plasma sprayed splat structure, respectively [13]. Wetting of the fibers/preform by the liquid metal plays a significant role in melt infiltration for the composite fabrication. In the case of pressure less infiltration where the infiltration is due to capillary action, the surface tension of the molten metal and interfacial energy between the molten metal

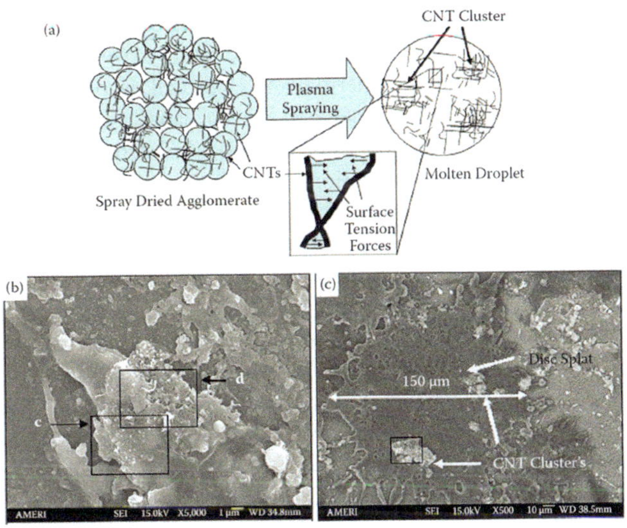

FIGURE 7.2 (a) Schematic showing CNT cluster formation when a spray-dried agglomerate containing well-dispersed CNTs undergoes melting. SEM micrographs of (b) splat finger showing CNT cluster infiltrated with metal marked by rectangles, and (c) disk portion of the splat showing a CNT-rich cluster [13]. (Microstructures reproduced with permission from Elsevier.)

and the CNT become important. A good wetting would lead to increased capillary action and infiltration. The effect is somewhat opposite in pressure-induced infiltration. It is difficult to push a liquid through a pore when the liquid wets the material around the pore. Therefore, the pressure required for infiltration would be higher. However, viscosity of molten metal affects the infiltration pressure in a more significant manner than wetting. Excellent wetting is desired for improved mechanical properties and cannot be compromised. Most molten metals do not wet graphite, which is one of the reasons for its use for making crucibles for handling molten metal. Interfacial reactions leading to formation of interfacial carbides can improve wetting if the liquid has a lower contact angle with the carbide forming at the interface. The extent and nature of fiber/matrix chemical reactions is dependent on the chemistry of the matrix [14] and can be controlled by using coatings on reinforcements [15,16] or by adding carbide-forming elements [17].

7.2 ENERGETICS OF CARBON NANOTUBE-METAL INTERACTION

Metals react with carbon to form carbide according to the following reaction:

$$\frac{x}{y}M + C = \frac{1}{y}M_xC_y \tag{7.1}$$

The near perfect structure of CNTs makes them exceptionally stable. The surface of the CNT is made up of the sp^2 bonded carbon atoms equivalent to the (0001) plane of graphite, which makes CNTs quite stable chemically. The bonding between the carbon atoms of the graphene sheet in-plane is a sigma bond, which gives rise to the high mechanical strength, while the pi bonds formed by overlapping of the out-of-plane orbitals gives rise to the high thermal and electrical properties through ballistic transport. The interaction between metal and SWNT can be in two manners as shown in Figure 7.3 [18]. Interaction with a multi-walled CNT can also be imagined in the same manner. The end contact interaction, through the sigma bond formation, could be very strong leading to formation of carbides at the interface. Weak interaction occurs on the side contact condition through weak bonds formed by out-of-plane orbitals. Carbide formation on the side contact is not thought to be favored due to the stable nature of the basal plane of a graphene sheet. However, defects in the CNT with sp^3 bonded carbon atoms and exposed (1010) planes may react with the metal forming side-contacted carbides. Figure 7.4 shows the Gibb's free energy of carbide formation per mole of carbon (graphite) for different metal carbides. The data shown is obtained from FactSage™ thermodynamic software and associated databases. This plot is similar to the Ellingham diagram for oxides. It is observed that the free energy for carbide formation is positive for certain metals like Mg and Ni, indicating that these carbides will not form or will not be thermodynamically stable at these temperatures. However, metals like Ti and Zr are strong carbide formers, which can be observed from the large negative free energy change for carbide formation. Al, Si, Cr, and W would also form carbides on reacting with CNT. This diagram can be used similar to the Ellingham diagrams in the sense that in the presence of two or more

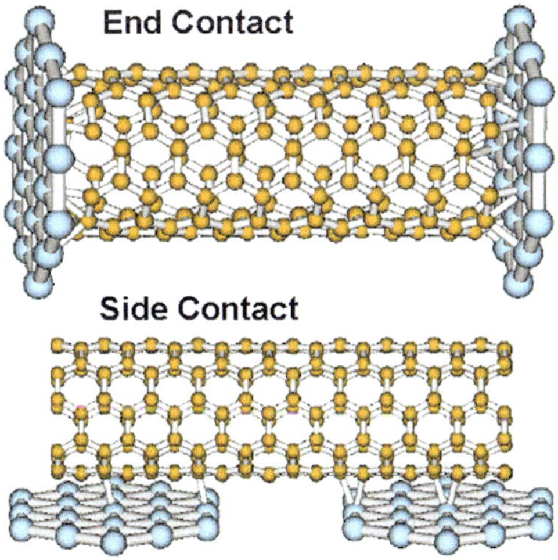

FIGURE 7.3 Schematic showing interaction between SWNT and metal in the end-contact and side-contact configurations [18]. (Reproduced with permission from American Physical Society).

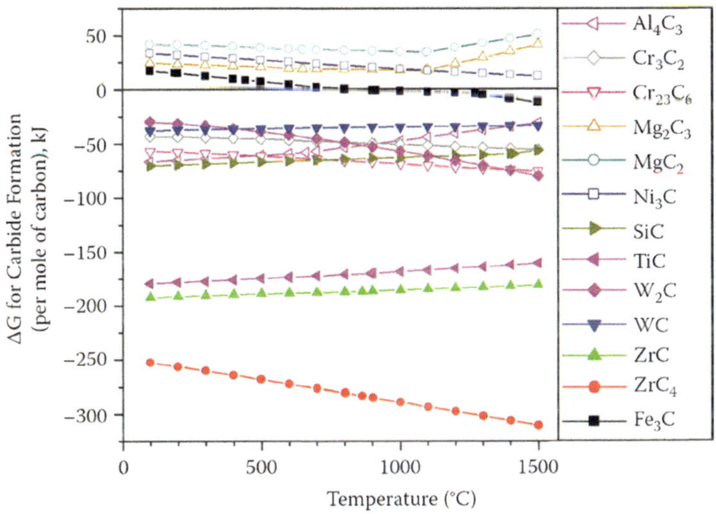

FIGURE 7.4 Variation of Gibbs free energy of formation of various carbides (per mole of graphite) as a function of temperature. (Computed by authors using thermodynamic database and software FactSage™).

metals, the metal with the lower curve will preferentially form carbide. This means that using a Ti-coated CNT in Al composites will lead to suppression of Al_4C_3 formation and TiC will preferentially form.

An important aspect of the composite fabrication by processes involving molten metals/alloys is wetting of the reinforcement. Wetting is related to the surface energies of the interacting species by the Young's equation and the Young–Dupre relation given as follows:

$$cos\theta = \frac{(\gamma_{SV} - \gamma_{SL})}{\gamma_{LV}} \tag{7.2}$$

$$W_A = \gamma_{LV}(1 + cos\theta) \tag{7.3}$$

Here θ is the contact angle, γ_{SV}, γ_{SL}, and γ_{LV} are the solid–vapor, solid–liquid, and liquid–vapor surface energies, and W_A is the work of adhesion between the liquid and the substrate. These surface forces have been shown in Figure 7.5a [19]. The smaller the contact angle, the better is the wetting. The surface tension of CNTs (γ_{SV}) has been reported to be 45.3 mJ m^{-2}, which is similar to carbon fiber [20]. It has been shown that liquids with surface tension between 100 and 200 mN m^{-1} results in good wetting with CNT [21,22]. This indicates that filling of the internal cavity of individual CNTs due to capillarity during composite fabrication is not likely. A large pressure would be required to fill CNTs with molten metals. The higher the work of adhesion, the higher the spreading, and the better the interfacial strength will be, even though the metal and CNT do not react. Thus, using Ni-coated CNTs as reinforcement in Cu-CNT composites may be helpful because copper and nickel are soluble in each other and the coating process leads to very good interaction between Ni and CNT. Possible favorable interphases are discussed further in Section 7.4. This may lead to better mechanical bonding and effective interfacial stress transfer between CNT and the matrix, thus improving the strength.

Carbide formation between the CNT and the metal matrix occurs by a nucleation and growth phenomena. As is true with all nucleation processes, there is a critical dimension associated with carbide formation [23]. The critical thickness for carbide nucleation can be written as:

$$t_{crit} = -V_M\left(\frac{\Delta\gamma}{\Delta G^f}\right) \tag{7.4}$$

where V_M is the molar volume of the carbide formed, ΔG^f is the free energy of formation per mole of carbide, and $\Delta\gamma = \gamma_{MC/CNT} + \gamma_{MC/Alloy} - \gamma_{Alloy/CNT}$ is the increase in the total surface energy as a result of formation of new interfaces, M_C in the subscript

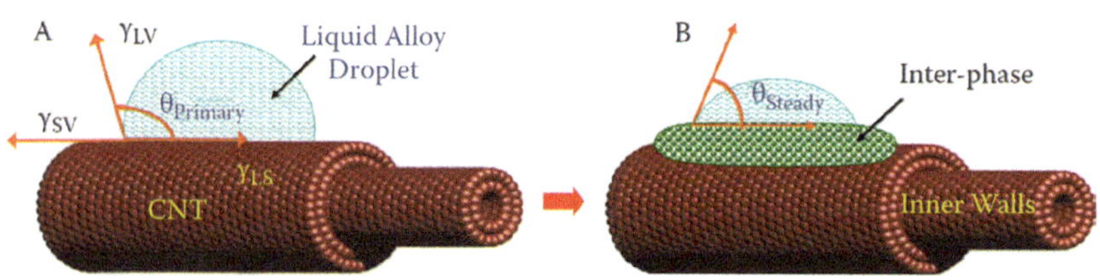

FIGURE 7.5 Schematic showing (a) initial wetting of a molten droplet of metal on CNT, and (b) reactive wetting with a decrease in contact angle because of formation of inter-phase [19]. (Adapted with permission from Elsevier).

being metal carbide. When carbide thickness reaches t_{Crit}, further growth is energetically favorable. Carbides not meeting the critical thickness would dissolve in the molten metal. The formation of carbide at the interface might result in the decrease in contact angle and improvement in wetting as shown in Figure 7.5b [19]. This phenomenon is called *reactive wetting* and this could be very helpful in improving wetting, aiding infiltration, and improving CNT-matrix bond strength and the mechanical properties of the composite. It is seen that the larger the free energy change of carbide formation and the smaller the increase in free energy (due to the formation of new interfaces), the smaller will be the value of t_{Crit}. Smaller t_{Crit} values indicate easy formation of carbide as well as better wetting.

The interface conditions discussed so far have assumed an equilibrium state. However, approaches utilizing a non-equilibrium amorphous interface have also been explored. Guo et al. [24] develop a tight interface between Al and CNTs by the interdiffusion of Al atoms and C atoms at the side walls of CNTs, that lead to a ~4 nm amorphous transition layer between the two phases. The interdiffusion is thought to be enabled by the higher surface energy at the prismatic planes, compared to the basal planes where no interfacial layer developed. The processing approach consisted of using fine Al powder, followed by spark plasma sintering at 630 °C, just below the melting point of Al, followed by hot extrusion. The same approach using coarser Al powder was found not to be as effective, indicating higher surface energy associated with finer Al powder also enabled interdiffusion to occur. The fast cooling rates associated with SPS are believed to contribute to the attainment of the non-equilibrium amorphous phase and the absence of Al_4C_3 phase. The composite achieved an impressive tensile strength of 300 MPa, among the highest reported in the literature.

7.3 CARBON NANOTUBE-METAL INTERACTION IN VARIOUS SYSTEMS

7.3.1 Formation and Effects of Carbides

The interface characteristics and behaviors between CNT and metal matrices have been poorly studied until recently. The last decade has seen a substantial increase in knowledge on the nature of interfaces, processing routes to control interface structures, and resulting effects on mechanical properties. Aluminum is one of the mostly widely studied systems for which the interfacial reaction with interfacial reactions with CNTs have been studied in detail. Molten aluminum silicon alloys have surface tension of ~800 mN m^{-1} [25]. Hence, it is expected that the wetting between Al-Si alloys and CNTs will be poor. From the sessile drop experiments, it has been observed experimentally that Al and Al-Si alloys do not wet graphite in the beginning and exhibit a large contact angle of ~160° [24,26,27]. However, the formation of Al_4C_3 and SiC reduces the contact angle to 45° and 38°, respectively, over a time of 104 s as shown in Figure 7.6. Hence, formation of interfacial carbides favors wetting, which will promote infiltration of liquid melt into CNT preforms. The reaction at the triple point between liquid alloy and CNT leads to formation of carbide and subsequent spreading of metal. Minimal reaction of CNT is desirable such that efficient

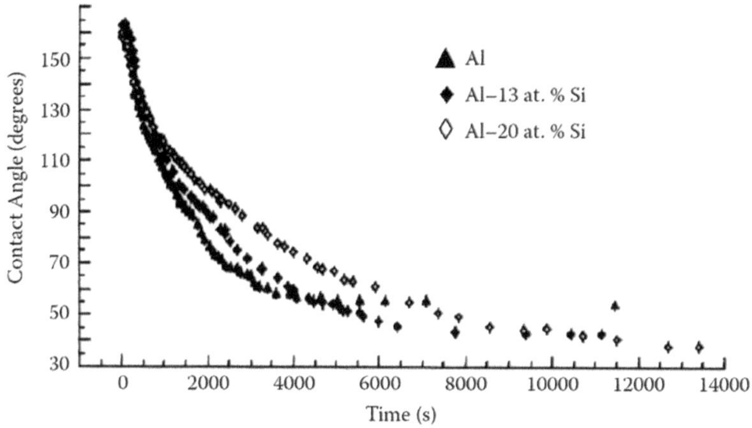

FIGURE 7.6 Variation of contact angle of aluminum and aluminum alloys with graphite as a function of time during the sessile drop experiment [23]. (Reproduced with permission from Elsevier).

stress transfer can occur without much damage to the CNT structure. Such strengthening cannot be achieved by SWNT as it will lose its tubular structure after carbide formation at the interface.

$$\frac{4}{3}\left[Al\right]+C=\frac{1}{3}Al_4C_3,\Delta G^f_{Al_2O_3}=\Delta G^0_{Al_4C_3}+\frac{1}{3}RT\ln\frac{a_{Al_4C_3}}{\left(a_{Al}\right)^4} \tag{7.5}$$

$$\left[Si\right]+C=SiC,\Delta G^f_{SiC}=\Delta G^0_{SiC}+RTln\frac{a_{SiC}}{a_{Si}} \tag{7.6}$$

Here the square brackets [ ] represent that Al and Si are in the molten Al-Si solution, ΔG_0 refers to the standard free energy of formation per mol of carbon, a denotes activity, R is the universal gas constant, and T is the absolute temperature at which the reaction takes place. Using FactSage™, the free energy of carbide formation (per mole of carbon) for Al_4C_3 and SiC was computed and plotted for different temperatures as shown in Figure 7.9a. The points of intersection represent the temperature and compositions for which both carbides are equally likely to form. Using these intersection points, a pseudo– phase diagram has been plotted as shown in Figure 7.9b that shows which carbide will form for a given temperature and composition of alloy [17].

Although several studies have observed the formation of reaction products at the Al-CNT interface, not many detailed investigations have been made. It has been observed from DSC (Figure 7.7) that Al reacts with CNTs above the melting temperature [28]. This is also supported by the fact that most of the studies involving solid state processing like extrusion have not observed formation of Al_4C_3 [29–31]. Figure 7.8 presents high-resolution TEM images showing reaction products at CNT interfaces in an aluminum matrix. In Al-CNT composites fabricated by hot pressing at 520 °C, phases with Al:C atom ratio of 1:1 and 1:2 were observed by EDS [34]. Aluminum carbide formation has been observed in CNTs on which aluminum was

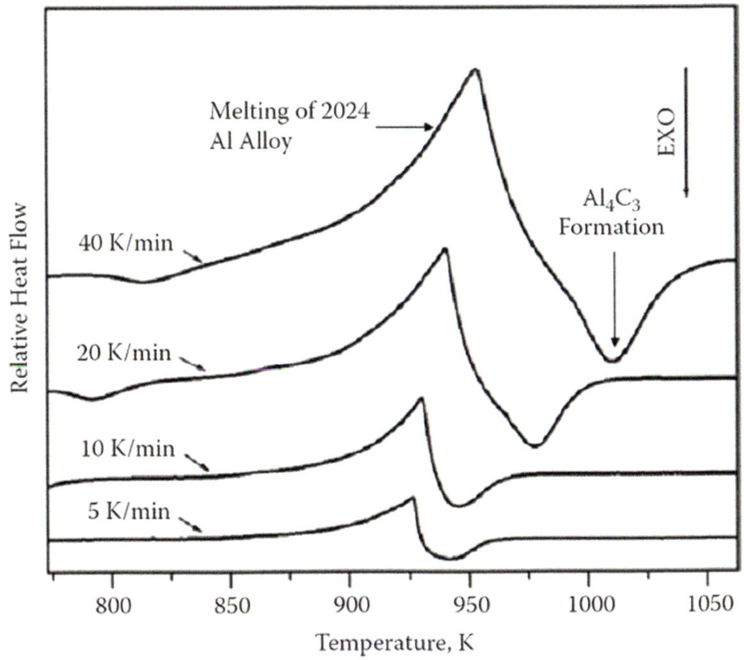

FIGURE 7.7 DSC plots for Al-2024 alloy powders mixed with 5 wt.% CNTs showing formation of Al_4C_3 after the melting of aluminum [28]. (Reproduced with permission from Elsevier).

deposited by the magnetic sputtering process, when they are annealed at temperatures above the melting point of aluminum [32] as shown in Figure 7.8a. Similar observations have been made in Al-CNT prepared by SPS followed by hot extrusion [33] as shown in Figure 7.8b. The reaction between Al-12 wt.% Si alloy and Al-23 wt.% Si alloy and CNTs led to the formation of Al_4C_3 [13,17] and SiC [19], respectively, during plasma spraying. Figure 7.8c and 7.8d are TEM images showing the formation of these carbides at the interfaces. This suggests that the matrix composition has a significant effect on the interfacial carbides forming in the composites. No orientation relationship between aluminum carbide and CNTs has been reported. An orientation relationship of (0002)C||(0003)Al_4C_3 has been observed in carbon fiber aluminum composites [5]. It is critical to understand the underlying thermodynamics and kinetics of this effect to control precisely the reactions at the interface between CNT and the alloy matrix. Different interfacial carbides may result in significantly different mechanical properties of the composites as the shear strength of the carbides determines the stress that could be transferred to the CNTs. An extensive thermodynamic analysis of the reaction between Al and Si in Al-Si alloy with CNT shows that the following reactions will occur.

Not only is thermodynamics important, but also kinetics. Figure 7.10 shows the t_{Crit} values for carbide formation as a function of alloy composition at 1700 K [17]. It also supports the experimentally observed results. The mechanism of growth of interfacial SiC layer has also been studied [19], which is shown by the schematic in

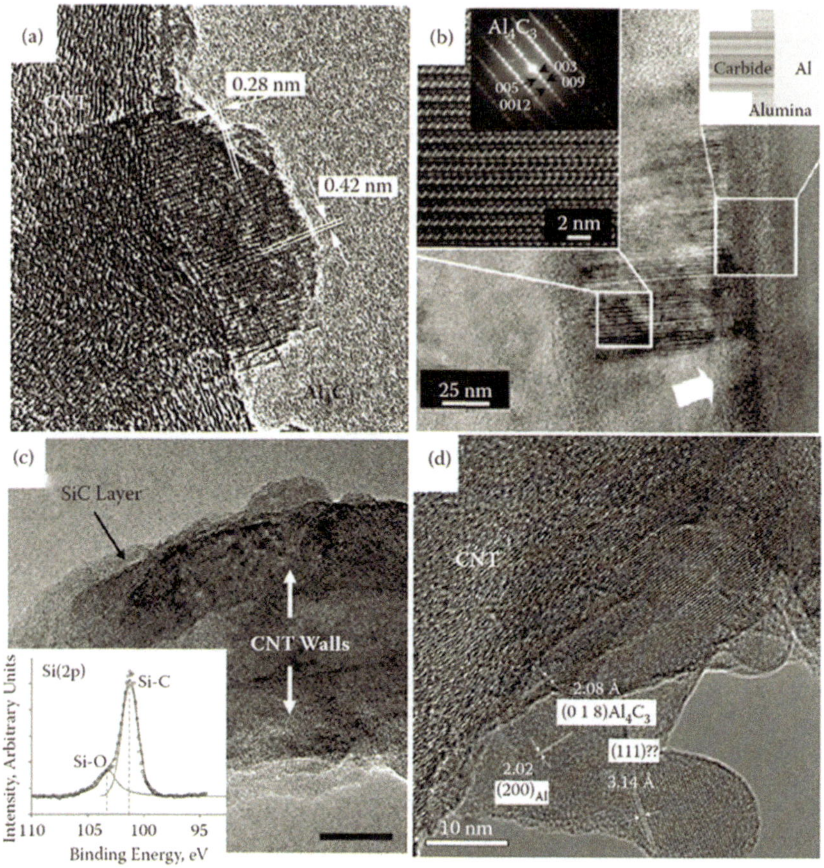

FIGURE 7.8 HRTEM showing formation of (a) Al_4C_3 in annealed Al films deposited on CNTs [32], (b) Al_4C_3 in SPS composites followed by extrusion [33], (c) SiC on CNTs in Al-23 wt.% Si alloy composite prepared by plasma spraying [19], and (d) Al_4C_3 on CNTs in Al-12 wt.% Si alloy composite prepared by plasma spraying [17]. (Reproduced with permission from Elsevier).

Figure 7.11. The growth of the silicon carbide will occur in two directions, that is (1) lateral growth on the CNT surface and (2) growth perpendicular to the carbide layer. The lateral growth will be caused by the reaction at the triple point, whereas the perpendicular growth is caused by diffusion of atoms through the SiC layer to the CNT interface. The diffusion of larger Si atoms in β-SiC crystal occurs by vacancy migration in through regular Si sites rather than through the tetrahedral and octahedral sites. However, smaller carbon atoms (0.77 Å) can easily diffuse through the interstitial sites in the β-SiC crystal. The activation energy required for the diffusion of carbon atoms is lower. The diffusivity of carbon atoms (~12.36 × 10–12 cm^2 s^{-1} at 2000 °C) is approximately two orders of magnitude higher than that of Si atoms (~8.66 × 10–14 cm^2 s^{-1} at 2000 °C) through β-SiC layer in the temperature range of 2283–2547 K [35]. Thus, the perpendicular growth of the SiC layer occurs by

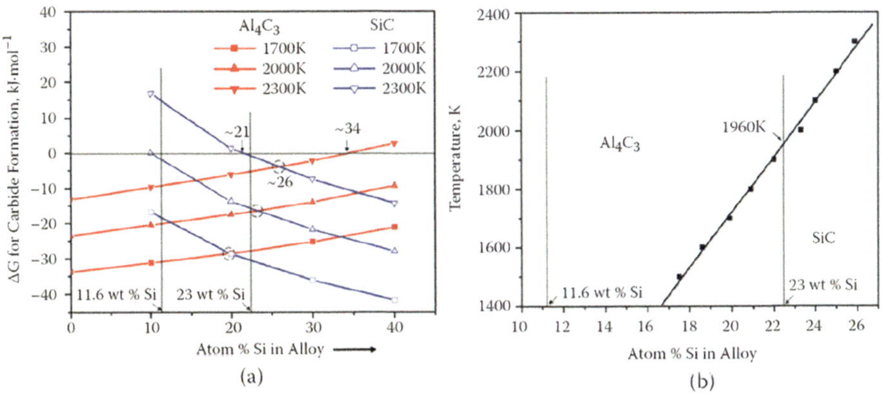

FIGURE 7.9 Graphs showing (a) variation of free energy of Al_4C_3 and SiC formation (per mole of graphite) as a function of Al-Si alloy composition at various temperatures, and (b) a pseudo-phase diagram that shows temperature and alloy composition that will form different carbides [17]. (Reproduced with permission from Elsevier).

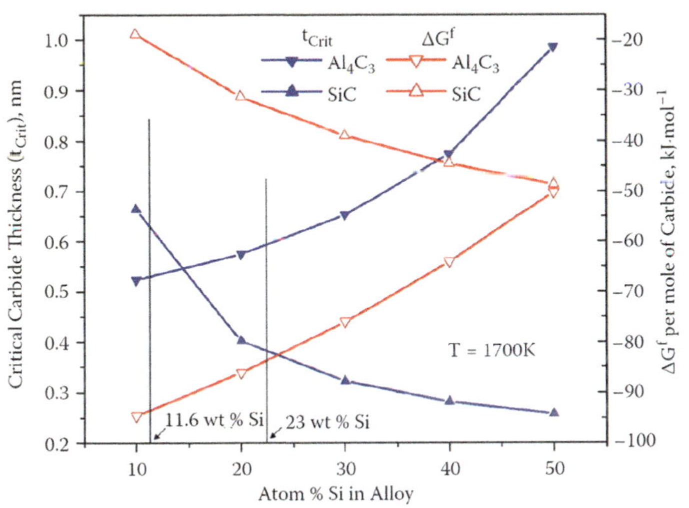

FIGURE 7.10 Graphs showing the variation of critical thickness for carbide formation for Al_4C_3 and SiC as a function of Al-Si alloy composition [17]. (Reproduced with permission from Elsevier).

diffusion of carbon atoms from the CNT surface to the interface of the SiC layer and the Al-Si matrix. However, the growth of SiC would be restricted by the rapid solidification involved in plasma spraying and the available carbon atoms from the very stable CNT structure. Growth of aluminum carbide is also governed by its crystal structure, which is rhombohedral (space group $R3m$) as shown in Figure 7.12 [17]. It is made up of alternating layers of Al_2C and Al_2C_2 with Al atoms having tetrahedral C coordination. C atoms have octahedral (C1 in Figure 7.12) and trigonal bipyramidal

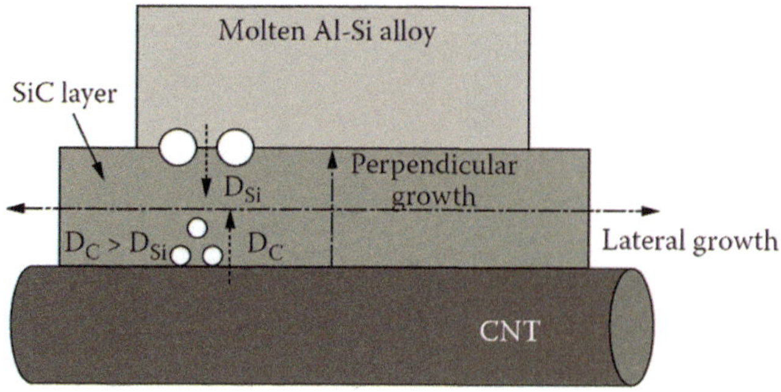

FIGURE 7.11 Mechanism of growth of a SiC layer on CNTs [19]. (Reproduced with permission from Elsevier).

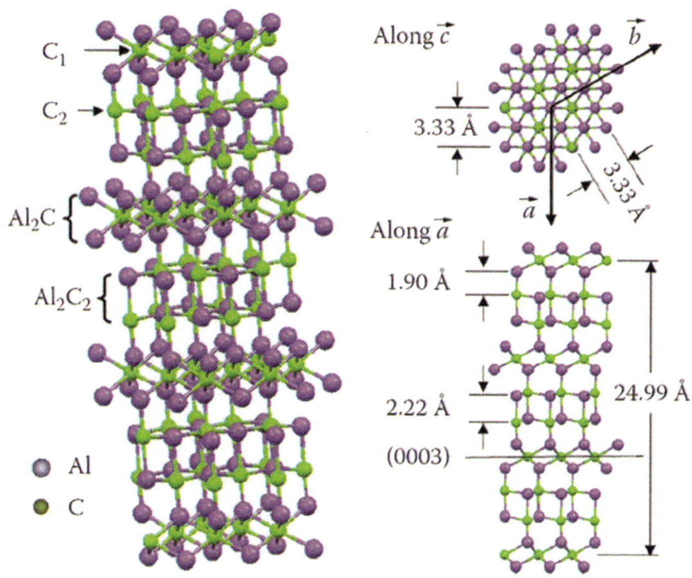

FIGURE 7.12 Schematic showing the crystal structure of Al_4C_3 and its projection along the a and c axes [17]. (Reproduced with permission from Elsevier).

(C2 in Figure 7.12) coordination with Al atoms [36]. The Al_2C layer is close packed with C in octahedral voids formed by close packed aluminum atoms. Therefore, it is expected that the lateral diffusion of carbon atoms by an interstitial mechanism would be favored through the Al_2C_2 layer. As seen from Figure 7.12, (0003) plane of Al_4C_3 has a hexagonal arrangement of carbon atoms similar to that in graphite. However, it is to be remembered that the C-C distance in graphite is 1.42 Å while it is 3.33 Å in Al_4C_3. Thus, Al_4C_3/CNT interface is expected to be strained without any orientation relationship.

Interfacial reactions between CNTs and Co, Ni, and Fe have also been studied by annealing at 1000 °C for 10 h [37]. The metals were deposited by dispersing CNTs

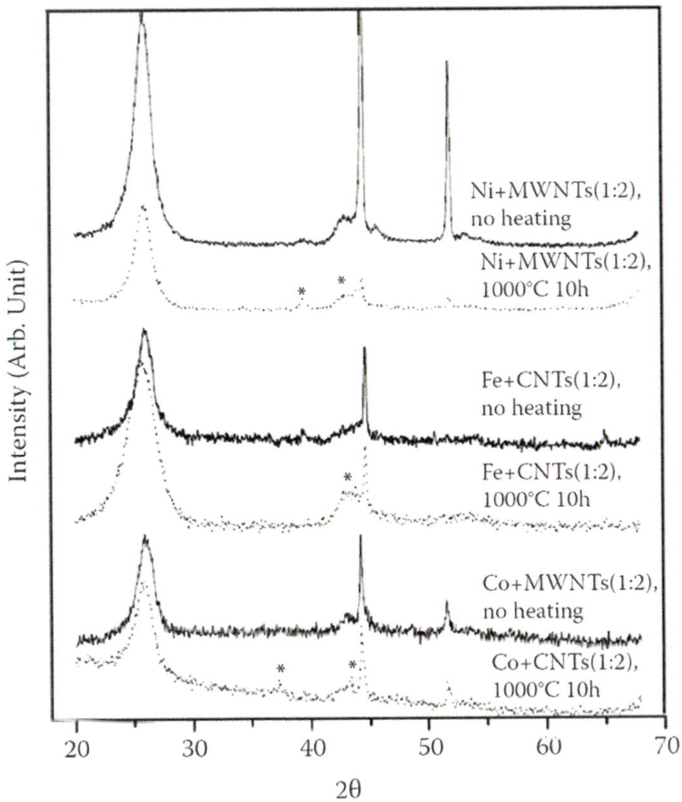

FIGURE 7.13 XRD plots taken from metal-CNT powders prepared by calcination of a salt mixture followed by H_2 reduction and subjected to an annealing treatment at 1000 °C for 10 h in argon atmosphere. The peaks marked with "*"are associated with the carbides of the respective metals formed as a reaction product during annealing [37]. (Reproduced with permission from Elsevier).

in a nitrite salt solution, followed by drying, calcination, and H_2 reduction. The XRD patterns (Figure 7.13) of the mixture reveal the peaks, marked by "*," denoting the presence of the reaction products after annealing. Formation of Co_2C and Ni_3C is observed. The formation of Ni_3C may not be thermodynamically stable as seen in Figure 7.4. In the case of Fe-CNTs, it was difficult to say whether the carbide was Fe_3C, Fe_7C_3, or $Fe_{2.2}C$. By observing the changes occurring in the intensity of (111) peak of the XRD pattern of cobalt after a 10-h annealing treatment with various forms of carbon at 1000 °C, the chemical activity of various forms of carbon can be gauged [37]. It was observed (Figure 7.14) that the interaction of layered graphite was the lowest followed by SWNT, multiwall CNT, and activated carbon in respective order. Layered graphite has a perfect structure of sp^2 hybridized carbon atoms arranged in ABABAB… stacking sequence, which would make it less reactive chemically. This shows that defects in activated carbon and in nanotubes provide sites for chemical reactions to occur. Ti and Zr are strong carbide formers. TiC formation has

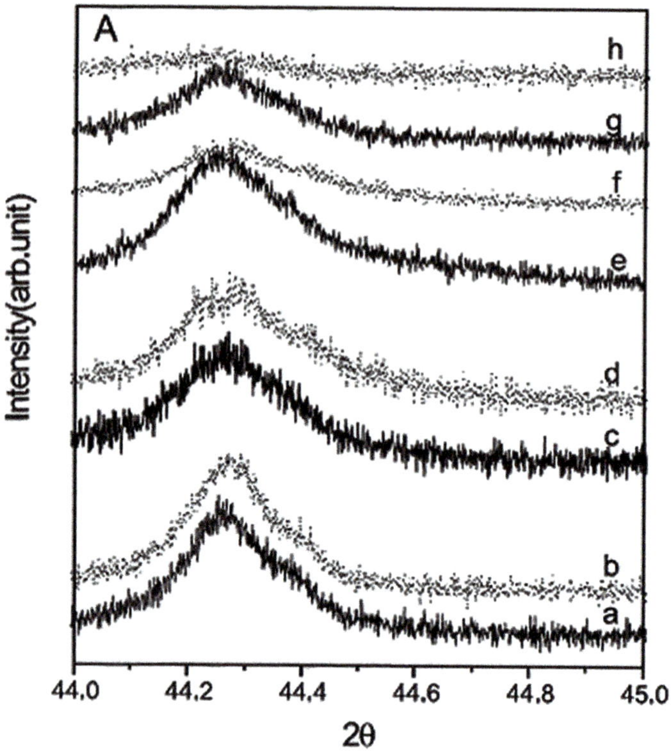

FIGURE 7.14 Slow scanning XRD plots showing the variation of the intensity of the Co (111) peak before and after annealing treatment at 1000 °C for various Co-carbon systems: Co-layered graphite (a and b), Co-SWCNT (c and d), Co-CNT (e and f), and Co-activated carbon (g and h) [37]. (Reproduced with permission from Elsevier).

been observed in hot pressed Ti-CNT composites [38]. ZrC formation has been observed in composites of $Zr_{52.5}Cu_{17.9}Ni_{14.6}Al_{10}Ti_5$ bulk metallic glass reinforced with CNTs formed by melting and copper mold casting. Reactions between volatile oxides/halides and CNTs have been suggested as a means for synthesis of carbide nanorods, namely TiC, NbC, Fe_3C, SiC, and BCx [39]. Tungsten carbide-CNT composites that have been synthesized by reduction and carbonization of WO_3 precursors followed by calcination also indicate the formation of reaction product at interface in a W-CNT system [40].

Extraordinary strengthening was observed under compression in the case of Cu-CNT composites prepared by SPS of molecular level mixed powders [41]. TEM and EDS indicated presence of O at the interface of the Cu and CNTs as shown in Figure 7.15 [42]. The oxygen comes from residual Cu_2O that might have been left from incomplete reduction of the calcined powders. The strengthening was attributed to the presence of oxygen atoms, which aids in stress transfer. However, the strengthening needs to be ascertained in tension as well. Composites of Mg [43–45] and Ag

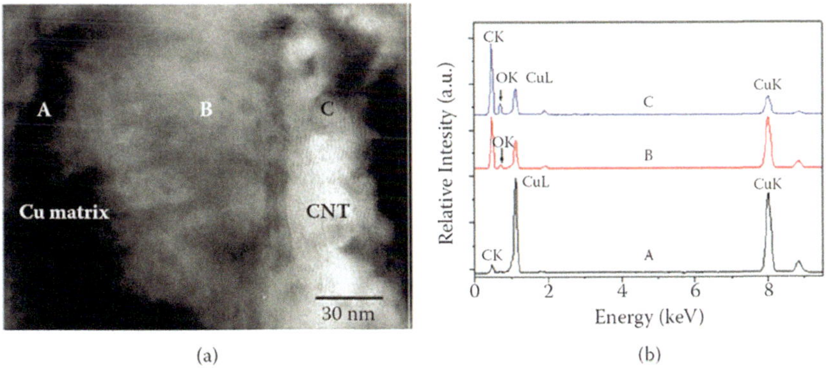

FIGURE 7.15 (a) TEM image of the Cu-CNT composite showing embedded CNTs in a Cu matrix in composite prepared by SPS of molecular level mixed powders, and (b) the corresponding EDS spectra taken from points A, B, and C showing the presence of oxygen at the interface [42]. (Reproduced with permission from Wiley).

[46] have not shown any formation of carbides with CNTs as predicted from their positive free energy change for carbide formation. Liquid Ni has high solubility for C but solid solubility of C in Ni is low. In addition, as seen from Figure 7.4, Ni does not form stable carbide. Consequently, carbide formation has not been observed in Ni-CNT composites prepared mostly by electrochemical methods and laser deposition [47].

It is observed that studying the interfacial phenomena in CNT-MMCs is very crucial. It plays an important role in strengthening of the composite. Failure as well as successful fabrication of CNT-MMC would require mastering the interface and engineering the dimension as well as composition to control its properties. Processing parameters will have to be optimized to control the extent of interfacial reactions. Application of coatings on CNTs might improve the wettability and aid dispersion of CNTs in metal matrix. One can measure the CNT-matrix interfacial strength by in situ pullout experiments in an electron microscope. In this way, the effect of various parameters on bonding can be studied. Therefore, there is a large scope for further research in interface engineering in CNT-MMCs.

7.3.2 Effect of CNT Defects on Al_4C_3 Nucleation

It was noted earlier that prismatic planes of CNTs have more surface energy than the basal planes and are thus more reactive. Aside from these intrinsic structural features, defects on CNTs (of which there are several types) can also lead to higher energy sites that can affect reactivity and the resulting nucleation and growth of Al_4C_3 interfacial products. Zhou et al. [48] conducted a systematic study where defects were intentionally introduced into the MWCNT structure and the resulting effects on Al_4C_3 nucleation were characterized. The study utilized an acid treatment to induce defects onto pristine CNTs, it consisted of a 4 h exposure at 50 °C to a 3:1 mixture of H_2SO_4/HNO_3. It should be noted that acid treatments are often utilizing during

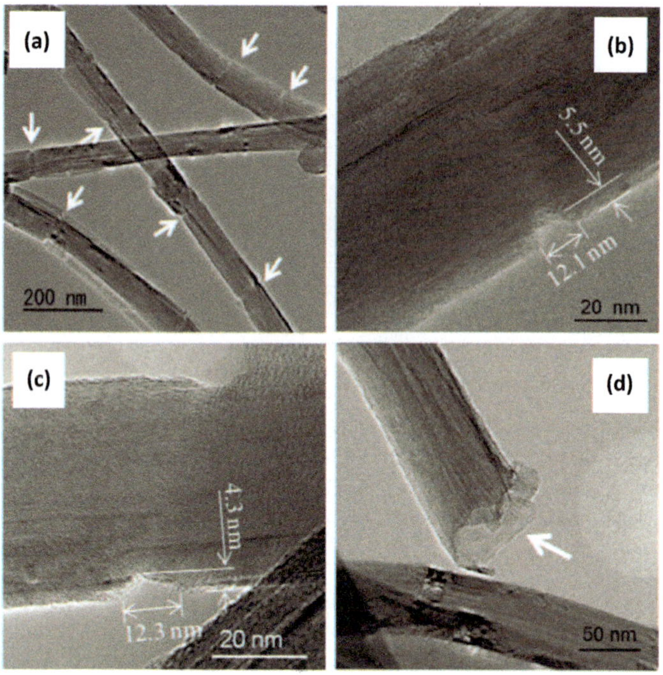

FIGURE 7.16 TEM characterization of acid-treated MWCNTs showing (a) nanodefects on walls of MWCNT indicated by arrows, (b) and (c) higher magnification TEM images of nanodefects that traverse several walls, and (d) open tip defects on the MWCNT, these are typically present even without an acid treatment [48]. (Reproduced with permission from Elsevier).

CNT-MMC processing in order to purify and/or remove amorphous carbon from MWCNTs. The resulting defected CNTs are shown in Figure 7.16. It can be seen that the acid-treated MWCNTs exhibit circumferential defects as well as tip defects. The circumferential defects have widths of 12 nm and depth of 5 nm, indicated they breach several walls of the MWCNT. The open tip defects are seen to expose individual sheets of graphene. These defects expose dangling bonds that become high energy sites prone to reaction. SPS was used to fabricate Al-MWCNT composites using the acid-treated MWCNTs, and heat treatments were conducted at temperatures ranging from 600–640°C for up to 2 h were used to initiate formation of carbide interfacial products.

The resulting carbide formation on the defected MWCNTs is shown in Figure 7.17. The carbide formation appears to be initiated at the open tips of the CNTs; this is consistent with the higher surface area available at the open ends, that enabled Al_4C_3 to nucleate at these defect sites. Furthermore, Al_4C_3 appears to continue to grow along the axial direction (<100> directions) of the MWCNTs. The Al/Al_4C_3 interface exhibits an orientation relationship of Al(111)//Al_4C_3(001). The activation energy of Al_4C_3 formation is 195.8 kJ mol^{-1} and the growth mechanism is driven by Al diffusion through Al_4C_3. Stress contrast can be seen to develop at the Al/Al_4C_3 interface indicating that there is tight interfacial adhesion that is expected to promote effective

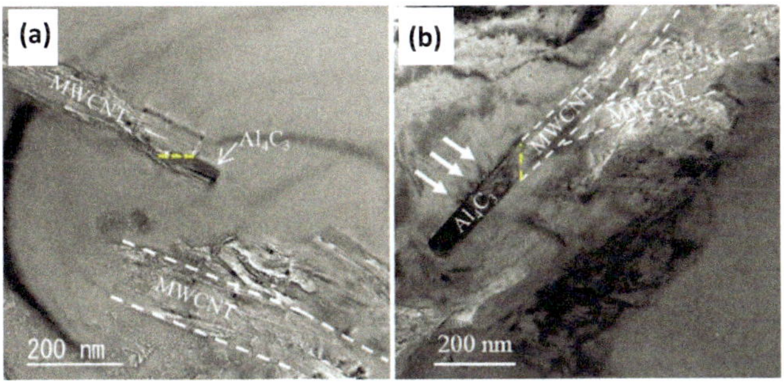

FIGURE 7.17 TEM images of Al_4C_3 formed at nanodefects on MWCNT edges after 1 h heat treatment at (a) 600 ° C, (b) 610 °C [48]. (Reproduced with permission from Elsevier).

load transfer, thus suggesting that the Al_4C_3 interfacial product will be beneficial for enhancing load transfer and strengthening. Carbides also formed at circumferential nanodefects on the walls of MWCNTs and grew simultaneously along the <110> and <100> directions, effectively forming out carbide shells on the MWCNTs. The observed orientation relationship of carbides formed at the side walls was MWCNT(002)//Al_4C_3(001). The CNTs used by Zhou et al. [48] had largely defect free inner walls due to the CVD process used to synthesize them. As a result, the defect free inner walls restricted the epitaxial growth of Al_4C_3 in the <001> direction perpendicular to the MWCNT tubular axis.

The strong interface between carbides and MWCNTs yields several benefits for load transfer and strengthening [48]. It is noted that carbides that nucleate both at edge/wall nanodefects and open tips were single crystals and often exhibited twinning as the result of compressive stresses during growth. This is in part due to the volume expansion of Al_4C_3, relative to MWCNT, the result volume expansion and compressive stresses after the transformation into carbide serve to anchor the MWCNTs into the Al matrix. The nano carbides also increase the interfacial shear resistance between MWCNT and Al, hindering CNT pullout. And lastly, strong covalent bonding between MWCNT and carbides at open tips may bridge the outer and inner walls of MWCNT, thereby hindering peeling of outer walls, often referred to as telescoping or "sword-in-sheath" deformation mechanism. This may increase the effective area of walls in the MWCNT that contribute to load transfer.

7.3.3 Effect of Temperature on Al_4C_3 Formation

The importance of the aluminum carbide phase formation and the myriad processes techniques used to develop Al-CNT composites has led a few researchers to systematically study the effect of temperature on Al_4C_3 formation. Chen et al. [49] investigated the effect of sintering temperatures in the range of 427–627 °C (75–95% of Al melting point) on the interfacial product evolution. Al-CNT MMCs are consolidated via SPS at variable temperatures, followed by hot extrusion at 627 °C. Up to a sintering temperature of 527 °C, there appears to be no carbide interfacial production formation as shown in the TEM micrograph in Figure 7.18a. At 577 °C most CNTs still

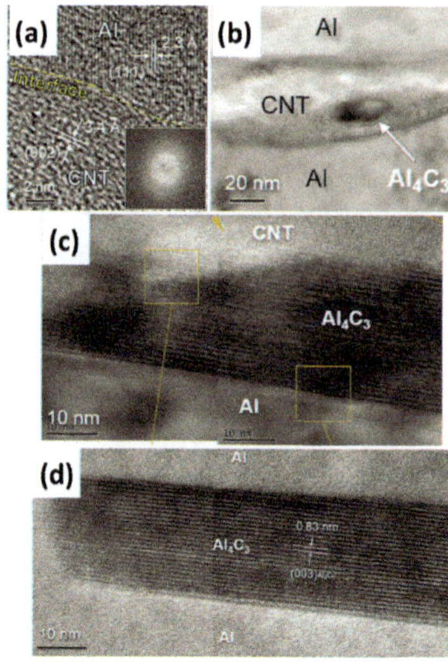

FIGURE 7.18 TEM images of interfaces formed at different SPS temperatures, (a) 527 °C shows no reaction product between Al and CNT, (b) 577 °C shows initial nucleation of Al_4C_3 phase, (c) 602 °C shows increased size and amount of Al_4C_3 reaction, however CNTs are still only partially reacted, (d) 627 °C shows stand alone Al_4C_3 phases embedded tightly in Al matrix, indicated complete reaction/consumption of CNTs. The remaining CNTs at this temperature show no signs of reaction. [49]. (Reproduced with permission from Elsevier).

appear to be stable and unreactive in the Al matrix, however, a few ~20 nm Al_4C_3 nanoparticles are found to nucleate in the vicinity of the Al/CNT interface (Figure 7.18b). The extent of partial reaction steadily increases as temperature is increased to 602 °C, with both the amount and size of Al_4C_3 particles increasing. The carbide formation occurs at both edges and tips but is preferred at the tips. The carbides formed at 602 °C had much higher aspect ratio (~5) (Figure 7.18c), as compared to those nucleated at 577 °C (~1). The orientation relationship at the interface was found to be Al_4C_3[110]//Al[−113]. As discussed earlier, initiation sites are thought to be those higher energy sites such as areas covered with amorphous carbon, or at CNT defect sites. At the highest temperature used, 627 °C, formation of standalone carbides was observed, along with unreacted CNTs. The carbide phases had high aspect ratios of ~5–15 and formed tight interfaces with Al as seen in Figure 7.18d. The standalone carbides reveal that once initiated, carbide formation can consume entirety of CNTs, and that growth is preferred along the [110] direction (axial growth), relative to the [001] (radial) direction. The orientation relationship with the Al matrix is Al_4C_3[1–10]//Al[213]. While other studies have detected carbide formation at lower temperatures, Chen et al. [49] used relatively low defect concentration CNTs synthesized by CVD and attributed the initial low defect state in CNTs along with the solid state processing route with the ability to tailor the interfacial product formation rate and extent by varying sintering temperature.

The role of interfacial product reaction extent on tensile strength is demonstrated in Figure 7.19a. The initiation of carbide formation at 577 °C coincides with an

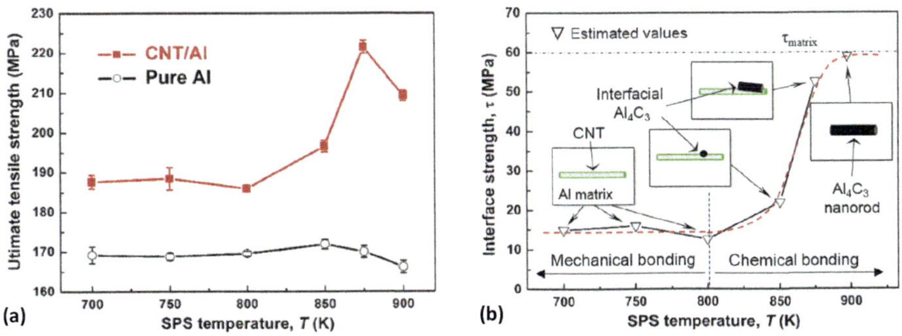

FIGURE 7.19 (a) UTS as a function of SPS temperature for Al-CNT composites, (b) calculated interfacial shear strength as a function of SPS temperature for Al-CNT composites [49]. (Reproduced with permission from Elsevier).

increase in strength relative to Al-CNT composites sintered at lower temperature (at all temperatures UTS of Al-CNT was higher than Al). The highest strengthening occurred at sintering temperature of 602 °C, where Al_4C_3 formation was robust, however, no standalone (complete CNT reaction) Al_4C_3 particles were found. The formation of standalone carbides result in slightly lower strengthening, but still higher than for the case where carbide formation was just beginning (577 °C). The carbide formation is thought to enhance CNT-metal bonding by changing the nature of the bonding from purely mechanical interlocking to a chemical bond, resulting in higher interfacial strength. Higher sintering temperature was calculated to lead to higher interfacial strength as shown in Figure 7.19b. The interfacial shear strength was calculated using the following equation:

$$\tau = \frac{\Delta\sigma_{LT}}{SV} + \frac{\sigma_m}{S} \tag{7.7}$$

Which is based on shear lag theory, where τ is the interfacial shear strength, $\Delta\sigma_{LT}$ is the improved strength (composite strength minus matrix strength) due to load transfer, S is the CNT aspect ratio, V is volume, and σ_m is the strength of the aluminum matrix. The load transfer efficiency of the strongest sample (602 °C) was estimated at 86%, which was an increase of 310% over the efficiency obtained in the samples sintered at 427 °C.

Temperature can may also play a role during powder mixing processes such as high energy ball milling (HEBM). Fan et al. [50] investigated the effect of an additional HEBM step to the flake powder metallurgy (FPM) method, compared to either standalone FPM or HEBM. The motivation behind the HEBM step was to disrupt or remove the Al_2O_3 scale inherent on Al powders, so that CNTs could have direct contact with Al phase and thereby promote a better interface. The HEBM step would also promote embedment of the CNTs into the bulk of the Al flakes. The additional HEBM step consisted of 2 h after FPM. The HEBM-only processed material was ball milled for 6 h. Figure 7.20 shows TEM micrographs of the results attained with the three approaches after consolidation via hot pressing and hot extrusion. Composites

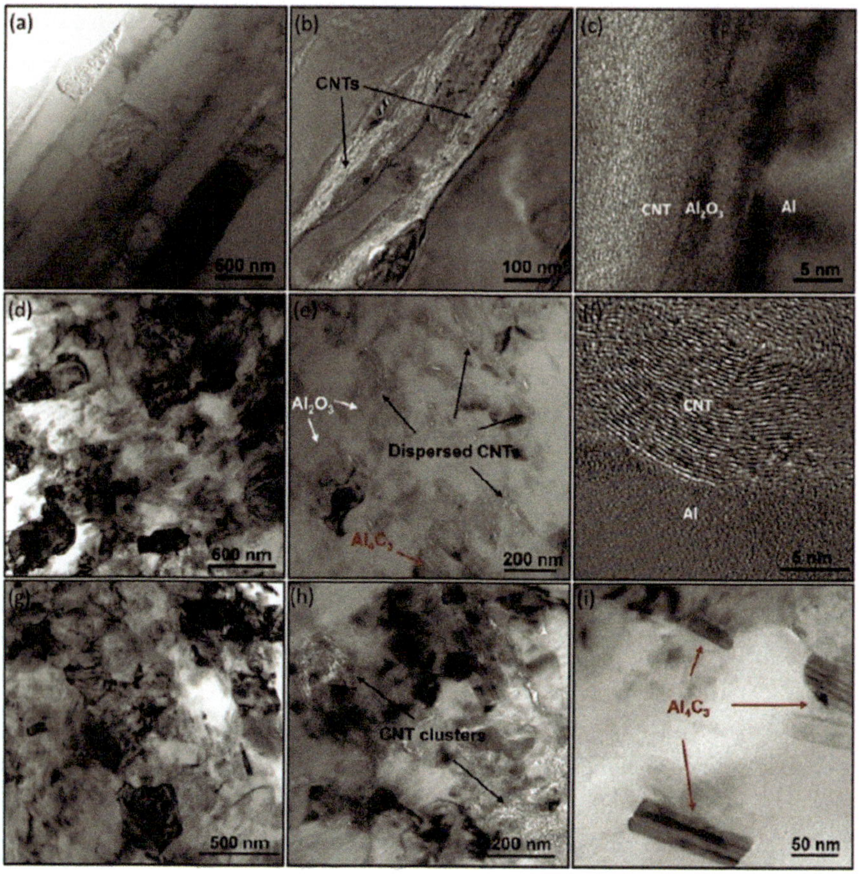

FIGURE 7.20 TEM images of Al-CNT composites fabricated by different processing routes: (a–c) FPM only, (d–f) FPM + 2h HEBM, (g–i) HEBM only for 6 h. [50]. (Reproduced with permission from Elsevier).

processed via FPM only showed Al_2O_3 oxide layers between Al and CNTs, as well as aggregated CNTs. The FPM + HEBM samples exhibited tight interfaces with between CNT and Al, with no oxide interfacial product as well as minimal carbide formation. The 2 h HEBM step is thought to successful break up the native alumina skin on Al without significantly increasing the number of defects present on CNT. In contrast, the HEBM-only samples exhibited significant carbide formation as well as CNT clusters, indicating the CNTs became damaged due to the prolonged mechanical agitation and higher temperature experienced, and not sufficiently dispersed via HEBM.

7.3.4 Degree of Interfacial Reaction at CNT/Metal Interface

The importance and varying characteristics of interfaces and interfacial products observed in Al/CNT system demonstrates the need to quantify the degree of interaction at the interface during processing. Yu et al. [51] quantified the interfacial reaction degree (IRD) and investigated its effects on Young's modulus. The IRD consists of the ratio between the reacted carbon in the composite to the total carbon (CNT) content. Reacted carbon was measured using electrochemical dissolution–gas chromatography. The Al-CNT composites studied had 1.5 wt.% CNT and were processed using flake powder metallurgy, uniaxial pressing, vacuum sintering (510 °C), and hot extrusion (450 °C). The IRD of the as-processed Al-CNT composites was 14%, likely due to reaction of amorphous carbon impurities or severely defected CNTs. The IRD was further controlled by exposing the samples to heat treatments at 500 °C for varying annealing times. The effect of annealing time and IRD on Young's Modulus on Al-CNT and control Al samples is shown in Figure 7.21. A 2 h heat treatment led to the coexistence of CNT and Al_4C_3 phases, while a 4.5 h heat treatment appeared to consume all the CNTs and only Al_4C_3 phase was observed. Despite the drastic change in reinforcement phases present, the modulus appears to gradually increase with increasing heat treatment time. The CNTs used had significant defects and the formation of the carbide phase is able to compensate for the loss of CNTs. The effective modulus of the individual reinforcements in this study is calculated to be 310 GPa for CNTs and 231 GPa for Al_4C_3. Calculations are based on comparisons between phases observed and modulus measurements, where it is understood that at 100% IRD, 1.9 vol.% of CNTs has been converted to 6.7 vol.% Al_4C_3. The modulus values of CNT are lower than theoretical values due to high quantity of defects present in industrial grade CNTs used, and carbide values may be lower due to random alignment of carbide rods. It is calculated that when using CNTs with a modulus of 600 GPa or greater, the formation of carbide phase will not be able to compensate for the loss of CNT-induced stiffening enhancement.

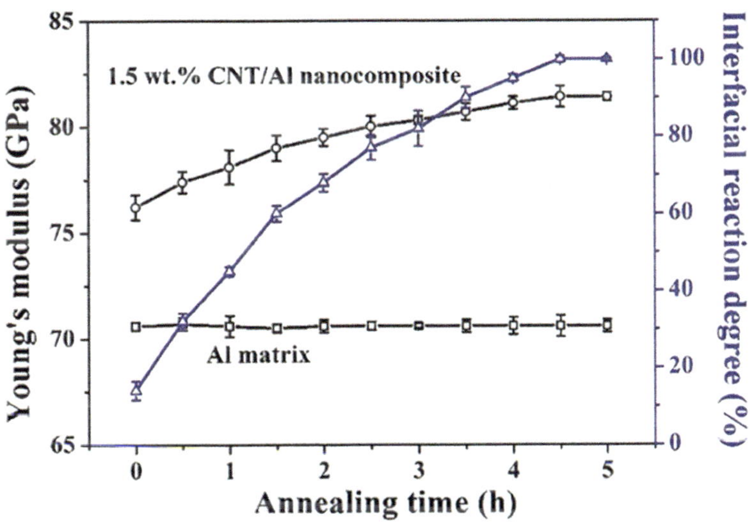

FIGURE 7.21 Variation in Young's modulus as a function of interfacial reaction degree and heat treatment time for Al and Al-CNT composite [51]. (Reproduced with permission from Elsevier).

7.4 INCORPORATION OF INTERPHASES

The importance of interfacial products has led to the deliberate incorporation and development of interphases in CNT-MMCs other than Al/CNT system. Cho et al. [52] incorporate a chromium carbide interphase into a Cu-CNT composite by adding small amounts of Cr. The use of interphase is especially important in Cu system, which has poor wetting with CNT. The interfacial reaction was aided by modifying CNTs with sulfuric and nitric acid to induce defects on the sidewalls of CNTs. After SPS processing, the formation of epitaxial grown chromium carbides is observed on CNT sidewalls. Defects such as radially unzipped walls provided initiation sites for chromium carbides, with the most common carbide observed being Cr_7C_3 (Figure 7.22). The hexagonal structure of Cr_7C_3 is thought to favor it over the thermodynamically favored $Cr_{23}C_6$ (at 800 °C). Cr_7C_3 is also more thermodynamically stable than Cr_3C_2. While Cr_2O_3 is thermodynamically more stable than the carbides, the reducing environment during SPS is thought to prevent the facile oxidation of Cr that is in solution with Cu matrix. Chu et al. [53] also incorporate CuCr powder but do no pretreat the CNTs with acid, and the resulting interphase consisted of Cr_3C_2. The Cr is thought to interact with pre-existing defects at CNT tips as well as amorphous carbon, hence the growth mechanism is not epitaxial and the hexagonal structure of Cr_7C_3 is not favored, however it is unclear why the thermodynamically favored phase ($Cr_{23}C_6$) was not encountered. Nonetheless, the carbide formation significantly decreases contact angle (Figure 7.23) and hence favors improved wetting and adhesion between CNTs and matrix. The addition of Cr leads to improved hardness and yield strength over Cu-CNT composite without Cr due to improved wetting and

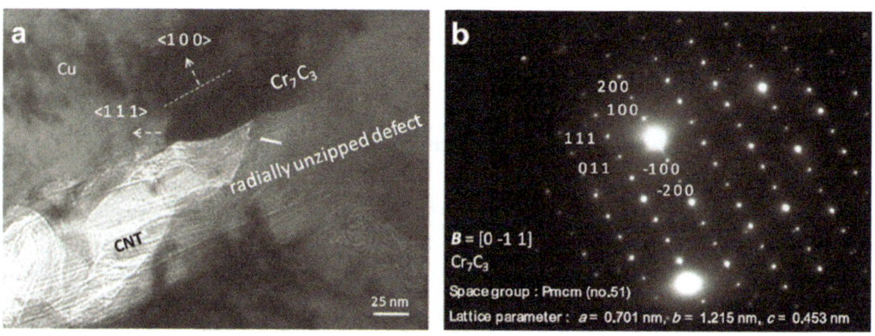

FIGURE 7.22 (a) Cr_7C_3 formed at radially unzipped CNT wall, (b) selected area diffraction pattern confirming Cr_7C_3 phase [52].(Reproduced with permission from Elsevier)

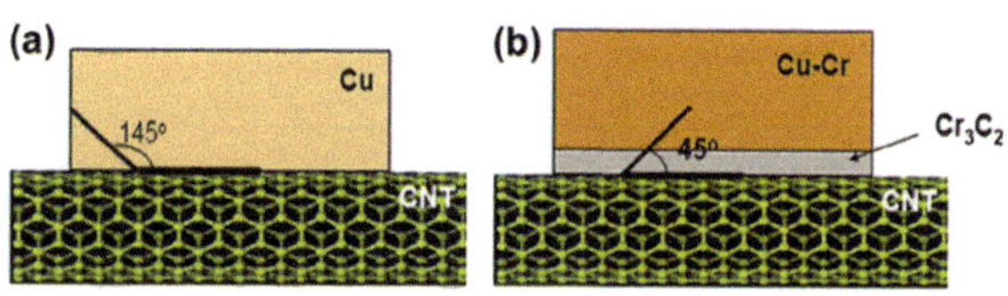

FIGURE 7.23 Schematic of wetting in (a) Cu-CNT, and (b) Cu-Cr-CNT systems [53]. (Reproduced with permission from Elsevier).

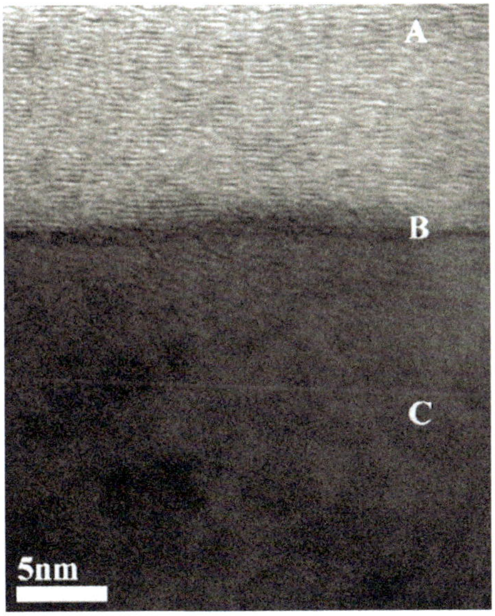

FIGURE 7.24 TEM image showing tight interface between Cu and Cu-coated CNTs [55]. (Reproduced with permission from Elsevier).

locking of CNTs into the Cu matrix by carbide formation, which enhance interfacial stress transfer. To improve load transfer efficiency in Cu/CNT system, Cheng et al. [54] utilize another carbide former, Ti, to form a TiC interphase in Cu-CNT composites fabricated by ball milling and spark plasma sintering. The formed interphase exhibits an orientation relationship of Cu(002)//TiC(002). Utilizing Cu-coated CNTs has also shown promising results for enhancing Cu-CNT properties when combined with SPS processing and hot-extrusion that led to CNTs being embedded within Cu grains as well as at grain boundaries [55]. The use of electro-deposited Cu served as the interphase and led to a tight atomic-level interface as seen in Figure 7.24; however, the property enhancement was not as great as when carbides are formed.

7.5 CHAPTER HIGHLIGHTS

It is clear from this chapter that the interface between CNT and the metal matrix has significant repercussions on the mechanical properties and microstructure evolution of the composite. CNTs are chemically stable but defects on the surface provide favorable reaction sites. Interfacial phenomena in SWNT amounts to destruction of the structure, while in the case of multiwall CNT, it leads to formation of an interfacial product layer. Often the interfacial layer formed is expected to improve the wetting of the CNT with the matrix as exemplified studied on Al and Al-Si alloys. Improvement in wetting is necessary to inhibit segregation of CNTs and cluster formation. Improved wetting and flow will lead to better coating of the CNTs and improvement in the work of adhesion and hence bonding. This improves the load transfer efficiency to the CNTs as shown by studies in Al-CNT and Cu-CNT composites. Reaction products formed at the tip of open CNTs might have a pinning effect, which will further facilitate load transfer to CNTs thereby contributing to strengthening. The type of carbide forming in the case of alloys can be modeled thermodynamically. The kinetics of carbide formation at metal/CNT interface has been investigated, focusing on the effect that CNT defect tailoring can have on the formation and location of carbide phase formation. Temperature can also be tailored to control the interfacial reaction degree, leading to completely reacted CNTs, partially reacted CNTs, or unreacted CNT/metal interfaces. It appears that there is now consensus that the formation of a limited amount of carbide is effective in enhancing load transfer and pinning CNTs onto the matrix, provided the carbide phase remains nanometric where its inherent brittleness does not dominant the mechanical behavior. The investigation and tailoring of other interphases for CNT-MMCs is a promising and ongoing field of study.

REFERENCES

1. Dilandro, L., Dibenedetto, A. T., and J. Groeger. 1988. The effect of fiber-matrix stress transfer on the strength of fiber reinforced composite materials. *Polym. Composites* 9: 209–221.
2. Jonathan, N., Coleman, J. N., Khan, U., Blau, W. J., and Y. K. Gun'ko. 2006. Small but strong: a review of the mechanical properties of carbon nanotube–polymer composites. *Carbon* 44: 1624–1652.
3. Ureña, A., Rams, J., Escalera, M. D., and M. Sánchez. 2005. Characterization of interfacial mechanical properties in carbon fiber/aluminium matrix composites by the nanoindentation technique. *Compos. Sci. Technol.* 65: 2025–2038.
4. Yang, M., and V. D. Scott. 1991. Carbide formation in a carbon fiber reinforced aluminum composite. *Carbon* 29: 877–879.
5. Diwanji, A. P., and I. W. Hall. 1992. Fibre and fibre-surface treatment effect in carbon/aluminum metal matrix composites. *J. Mater. Sci.* 27: 2093–2100.
6. Piggot, M. R. 1989. The interface in carbon fiber composites. *Carbon* 27: 657–662.
7. Chen, H., and A. T. Alpas. 1996. Wear of aluminium matrix composites reinforced with nickel-coated carbon fibres. *Wear* 192: 186–198.
8. Poteet, C. C., and I. W. Hall. 1997. High strain rate properties of a unidirectionally reinforced C/Al metal matrix composite. *Mater. Sci. Eng. A* 222: 35–44.
9. Sanctis, M. D., Pelletier, S., Bienvenu, Y., and M. Guigon. 1994. On the formation of interfacial carbides in a carbon fiber reinforced aluminum composite. *Carbon* 32: 925–930.
10. Vidal-Setif, M. H., Lancin, M., Marhic, C., Valle, R., Raviart, J. L., Daux, J. C., and Rabinovitch. 1999. On the role of brittle interfacial phases on the mechanical properties of carbon fiber reinforced Al-based matrix composites. *Mater. Sci. Eng. A* 272: 321–333.
11. Tham, L. K., Gupta, M., and L. Cheng. 2001. Effect of limited matrix-reinforcement interfacial reaction on enhancing the mechanical properties of aluminum silicon composites. *Acta Mater.* 49: 3243–3253.
12. Chen, B., Kondoh, K., Imai, H., Umeda, J., Takahashi, M. 2016. Simultaneously enhancing strength and ductility of carbon nanotube/aluminum composites by improving bonding conditions. *Scripta Materialia* 113: 158–162.
13. Bakshi, S. R., Singh, V., Seal, S., and A. Agarwal. 2009. Aluminum composite reinforced with multiwalled carbon nanotubes from plasma spraying of spray dried powders. *Surf. Coat. Technol.* 203: 1544–1554.
14. R. Y. Lin. 1995. Interface evolution in aluminum matrix composites during fabrication. *Key. Eng. Mater.* 104–107: 507–522.

15. Aggour, L., Fitzer, E., Heym, M., and E. Ignatowitz. 1977. Thin coatings on carbon fibers as diffusion barriers and wetting agents in Al composites. *Thin Solid Films* 40: 97–105.
16. Himbeault, D. D., Varin, R. A., and K. Piekarski. 1989. Tensile properties of titanium carbide coated carbon fiber-aluminum alloy composites. *Composites* 20: 471–477.
17. Srinivasa, R. B., Keshri, A. K., Singh, V., Seal, S., and A. Agarwal. 2009. Interface in carbon nanotube reinforced aluminum silicon composites: thermodynamic analysis and experimental verification. *J. Alloys. Compd.* 481: 207–213.
18. Palacios, J. J., Perez-Jimenez, A. J., Louis, E., SanFabian, E., and J. A. Verges. 2003. First-principles phase-coherent transport in metallic nanotubes with realistic contacts. *Phys. Rev. Lett.* 90: 106801(1–4).
19. Laha, T., Kuchibhatla, S., Seal, S., Li, W., and A. Agarwal. 2007. Interfacial phenomena in thermally sprayed multiwalled carbon nanotube reinforced aluminum nanocomposite. *Acta Mater.* 55: 1059–1066.
20. Nuriel, S., Liu, L., Barber, A. H., and H. D. Wagner. 2005. Direct measurement of multiwall nanotube surface tension. *Chem. Phys. Lett.* 404: 263–266.
21. Dujardin, E., Ebbesen, T. W., Hiura, H., and H. K. Tanigaki. 1994. Capillarity and wetting of carbon nanotubes. *Science* 265: 1850.
22. Ebbesen, T. W. 1996. Wetting, filling and decorating carbon nanotubes. *J. Phys. Chem. Solids* 57: 951–955.
23. Landry, K., Rado, C., Voitovich, R., and N. Eustathopoulos. 1997. Mechanisms of reactive wetting: the question of triple line configuration. *Acta Mater.* 45: 3079–3085.
24. Guo, B., Song, M., Yi, J., Ni, S., Shen, T., Du, Y. 2017. Improving the mechanical properties of carbon nanotubes reinforced pure aluminum matrix composites by achieving non-equilibrium interface. *Materials and Design* 120: 56–65.
25. Anson, J. P., Drew, R. A. L., and J. E. Gruzleski. 1999. The surface tension of molten aluminum and Al-Si-Mg alloy under vacuum and hydrogen atmospheres. *Met. Trans. B* 30: 1027–1032.
26. Landry, K., and N. Eustathopoulos. 1996. Dynamics of wetting in reactive metal/ceramic systems: Linear spreading. *Acta Mater*. 44: 2923–2932.
27. Landry, K., Kalogeropoulou, S., and N. Eustathopoulos. 1998. Wettability of carbon by aluminum and aluminum alloys. *Mater. Sci. Eng A* 254: 99–111.
28. Deng, C., Wang, D. Z., Zhang, X. X., and A. B. Li. 2007. Processing and properties of carbon nanotubes/aluminum matrix composites. *Mater. Sci. Eng. A* 444: 138–145.
29. Choi, H. J., Kwon, G. B., Lee, G. Y., and D. H. Bae. 2008. Reinforcement with carbon nanotubes in aluminum matrix composites. *Scripta Mater.* 59: 360–363.
30. George, R., Kashyap, K. T., Rahul, R., and S. Yamadagni. 2005. Strengthening in carbon nanotube/aluminium (CNT/Al) composites. *Scripta Mater.* 53: 1159–1163.
31. Eswai, A. M. K., and M. A. El Borady. 2008. Carbon nanotube reinforced aluminum strips. *Compos. Sci. Technol.* 68: 486–492.
32. Lijie Ci, L., Ryu, Z., Jin-Phillipp, N. Y., and M. Rühle. 2006. Investigation of the interfacial reaction between multi-walled carbon nanotubes and aluminum. *Acta Mater.* 54: 5367–5375.
33. Kwon, H., Estili, M., Takagi, K., Miyazaki, T., and A. Kawasaki. 2009. Combination of hot extrusion and spark plasma sintering for producing carbon nanotube reinforced aluminum matrix composites. *Carbon* 47: 570–577.
34. Xu, C. L., Wei, B. Q., Ma, R. Z., Liang, J., Ma, X. K., and D. H. Wu. 1999. Fabrication of aluminum–carbon nanotube composites and their electrical properties. *Carbon* 137: 855–858.
35. Hon, M. H., and R. F. Davis. 1980. Self-diffusion of 30Si in polycrystaline β-SiC. *J. Mater. Sci.* 15. 2073–2080.
36. Solozhenko, V. L., and O. O. Kurakevych. 2005. Equation of state of aluminum carbide Al_4C_3. *Solid State Comm.* 133: 385–388.

37. Zhong, Z., Liu, B., Sun, L., Ding, J., Lin, J., and K. L. Tan. 2002. Dispersing and coating of transition metals Co, Fe and Ni on carbon materials. *Chem. Phys. Lett.* 362: 135–143.
38. Kuzumaki, T., Ujiie, O., Ichinose, H., and K. Ito. 2000. Mechanical characteristics and preparation of carbon nanotube fiber-reinforced Ti composite. *Adv. Eng. Mater.* 2(7): 416–418.
39. Dai, H., Wong, E. W., Lu, Y. Z., Fan, S., and C. M. Lieber. 1995. Synthesis and characterization of nanorods. *Nature* 375: 769–772.
40. Shi, X., Yang, H., Sun, P., Shao, G., Duan, X., and X. Zhen. 2007. Synthesis of multi-walled carbon nanotube–tungsten carbide composites by the reduction and carbonization process. *Carbon* 45: 1735–1742.
41. Cha, S. I., Kim, K. T., Arshad, S. N., Mo, C. B., and S. H. Hong. 2005. Extraordinary strengthening effect of carbon nanotubes in metal-matrix nanocomposites processed by molecular-level mixing. *Adv. Mater.* 17: 1377–1381.
42. Kim, K. T., Cha, S. I., Gemming, T., Eckert, J., and S. H. Hong. 2008. The role of interfacial oxygen atoms in the enhanced mechanical properties of carbon nanotube-reinforced metal matrix nanocomposites. *Small* 4(11): 1936–1940.
43. Carreño-Morelli, E., Yang, J., Couteau, E., Hernadi, K., Seo, J. W., Bonjour, C., Forró, L., and R. Schaller. 2004. Carbon nanotube/magnesium composites. *Phys. Stat. Sol.* 201(8): R53–R55.
44. Goh, C. S., Wei, J., Lee, L. C., and M. Gupta. 2006. Development of novel carbon nanotube reinforced magnesium nanocomposites using the powder metallurgy technique. *Nanotechnology* 17: 7–12.
45. Li, Q., Viereckl, A., Rottmair, C. A., and R. F. Singer. 2009. Improved processing of carbon nanotube/magnesium alloy composites. *Compos. Sci. Technol.* 69: 1193–1199.
46. Feng, Y., Yuan, H. L., and M. Zhang. 2005. Fabrication and properties of silver matrix reinforced by carbon nanotubes. *Mater. Character.* 55: 211–218.
47. Hwang, J. Y., Neira, A., Scharf, T. W., Tiley, J., and R. Banerjee. 2008. Laser-deposited nanotube reinforced nickel matrix composite. *Scripta Mater.* 5: 487–490.
48. Zhou, W., Bang, S., Kurita, H., Miyazaki, T., Fan, Y. 2016. Interface and interfacial reactions in multi-walled carbon nanotube-reinforced aluminum matrix composites. *Carbon* 96: 919–928.
49. Chen, B., Shen, J., Ye, X., Imai, H., Umeda, J., Takahashi, M., Kondoh, K. 2017. Solid-state interfacial reaction and load transfer efficiency in carbon nanotubes (CNTs)-reinforced aluminum matrix composites. *Carbon* 114: 198–208.
50. Fan, G., Jiang, Y., Tan, Z., Guo, Q., Xiong, D-B., Su, Y., Lin, R., Hu, L., Li, Z., Zhang, S. 2018. Enhanced interfacial bonding and mechanical properties in CNT/Al composites fabricated by flake powder metallurgy. *Carbon* 130: 333–339.
51. Yu, Z., Tan, Z., Fan, G., Xiong, D-B., Guo, Q., Lin, R., Hu, L., Li, Z., Zhang, D. 2018. Effect of interfacial reaction on Young's modulus in CNT/Al nanocomposite: a quantitative analysis. *Materials Characterization* 137: 84–90.
52. Cho, S., Kikuchi, K., Miyazaki, T., Kawasaki, A., Arami, Y., Silvain, J. F. 2013. Epitaxial growth of chromium carbide nanostructures on multiwalled carbon nanotubes (MWCNTs) in MWCNT-copper composites. *Acta Materialia* 61: 708–716.
53. Chu, K., Jia, C-C., Jiang, L-K., Li, W-S. 2013. Improvement of interface and mechanical properties in carbon nanotube reinforced Cu-Cr matrix composites. *Materials and Design* 45: 407–411.
54. Cheng, B., Bao, R., Yi, J., Li, C., Tao, J., Liu, Y., Tan, S., You, X. 2017. Interface optimization of CNT/Cu composite by forming TiC nanoprecipitation and low interface energy structure via spark plasma sintering. *Journal of Alloys and Compounds* 722: 852–858.
55. Wang, H., Zhang, Z-H., Hu, Z-Y., Song, Q., Yin, S-P., Kang, Z., Li, S-L. 2018. Improvement of interfacial interaction and mechanical properties in copper matrix composites reinforced with copper coated carbon nanotubes. *Materials Science & Engineering A* 715: 163–173.

8 Tribological, Transport, and Functional Properties & Behavior

This chapter explores tribological, transport, and functional properties of CNT-MMCs. These properties are often related to the reinforcing effects of CNTs, for example, providing greater hardness and strength generally enhances wear resistance. Likewise, the formation of continuous reinforcement networks often promotes thermal and electrical transport. Applications focusing on thermal and electrical applications often benefit from reduced concerns over using higher volume contents of reinforcements. The studies presented in this chapter are critical to many applications where concerns over a specific functionality (electrical/thermal conductivity, corrosion resistance, etc.) may dominate over the documented mechanical reinforcement effects. In addition, this chapter discusses CNT reinforcement effects under combined thermal and mechanical loadings – thermomechanical behavior of CNT-MMCs has only received attention in the last couple of years.

8.1 TRIBOLOGICAL PROPERTIES

CNTs have the capability to significantly influence the tribological behavior of metal matrix composites. The tribological property of a material is often governed by its mechanical properties. It has been already discussed in previous chapters that reinforcement of CNTs, with excellent elastic modulus and tensile strength, increases the elastic modulus and tensile/compressive strength in metal matrix composites. Increase in the elastic modulus and tensile/compressive strength helps in decreasing the volume loss in CNT-MMC, causing the increase in wear resistance. In addition, the graphene layers of CNT offer lubrication during wear, which causes the lowering of the coefficient of friction (CoF). With decrease in CoF, the effective lateral force for wear also decreases, resulting in further reduction of volume loss. These are the reasons that make CNTs a very attractive reinforcement for CNT-MMCs in terms of tribological properties.

There are several studies [1–5] on the tribological properties of Ni-CNT composite coatings, mostly processed through electroless deposition or electrodeposition techniques. All of these studies reported an increase in the wear resistance (i.e., a decrease in wear loss volume) and a decrease in CoF with CNT addition to the Ni matrix [1–6]. Deng and his research group show a high of 83% increase in the wear resistance and 60% decrease in CoF with only 0.52 wt.% CNT addition to Ni coating, fabricated through electrodeposition technique [4]. Figure 8.1a and 8.1b show

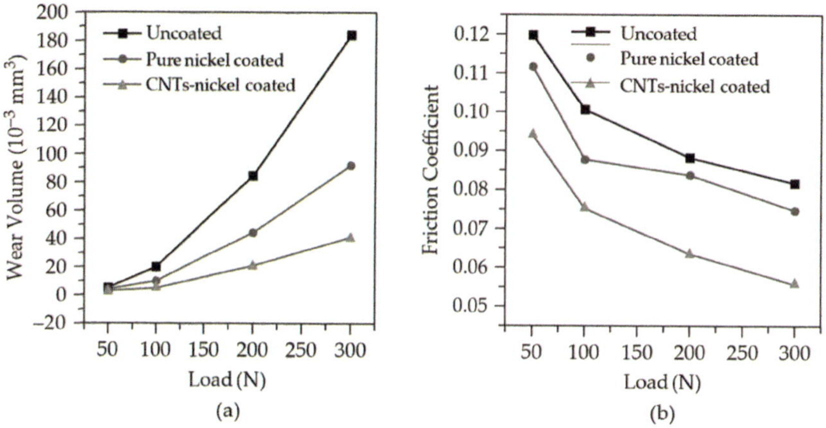

FIGURE 8.1 Variations of (a) wear volume and (b) coefficient of friction with load for uncoated, pure nickel-coated, and Ni-CNTs-coated carbon steel under lubrication [1]. (Reproduced with permission from Springer Nature).

the effect of CNT addition on wear volume and CoF with the increasing load for Ni-based composite coating [1].

The increase in the wear resistance in the presence of CNT has been attributed to the stiffening of the metal matrix. CNTs reinforce the metal matrix by forming a network that helps in load bearing and improving the matrix toughness [2]. In addition, CNTs present in the matrix hinder the dislocation movement during plastic flow and cause strain hardening [1,4,5]. Due to restricted plastic deformation, wear volume loss decreases in Ni-CNT composite coatings. Furthermore, CNTs gradually released during wear act as spacers between the probe and the surface, and reduce the contact and wear volume. Graphene layers, which are often released from the CNT surface, lubricate the wear track and decrease CoF.

CNT-reinforced Cu composites also show increase in the wear resistance and decrease in CoF with CNT addition [6–13]. Figure 8.2 shows increase in wear resistance as high as 300% with 10 vol.% CNT addition for an SPS-processed Cu-CNT composite [10]. CNTs perform better than carbon fibers in terms of increasing wear resistance because of their high elastic modulus and tensile strength. However, the effect for CNT and carbon fiber is similar on CoF for the Cu-based composite, as both can offer lubrication through the graphitic structure [7]. The reason for an improvement in tribological properties of Cu-CNT composite is similar to the Ni-CNT system. The wear in the Cu-CNT composite is mainly dominated by the plastic deformation at lower load, whereas cracking and spalling takes over at higher load [8]. The presence of CNT clusters results in increased wear rate at higher load as the clusters can be worn off easily due to poor bonding with the metal matrix [8]. The decrease in CoF with increasing CNT content in the Cu-CNT matrix matches well with the (estimated) values calculated using ROM [11]. Guiderdoni et al. [11]* investigated Cu-CNT composites with CNTs with variable number of walls, from 3 to 20 walls, the vol. fraction was kept constant at 5 vol.%. All composites exhibited decreased friction and wear, with no strong dependence on CoF on number of CNT

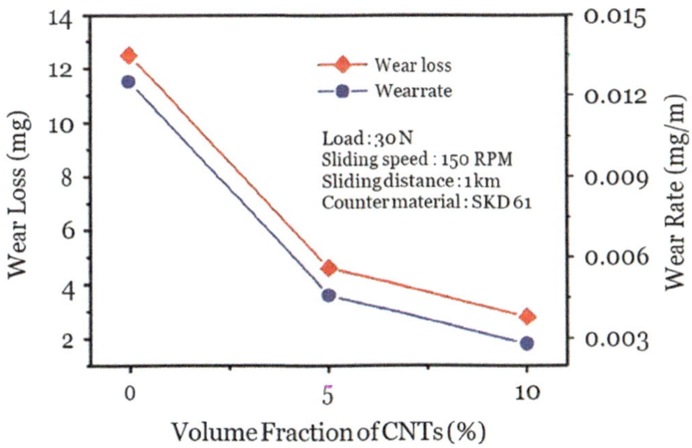

FIGURE 8.2 The variation in wear loss and wear rate of Cu-CNT nanocomposite with increasing volume fraction of carbon nanotubes [10]. (Reproduced with permission from Elsevier).

walls. Cu-CNT with 20 walls had higher wear than those with 2, 3, or 8 walls [12]. Rajkumar et al. [13] investigate the tribological behavior of Cu-CNTs with up to 20 vol.% CNT processed by microwave sintering and observe a carbonaceous tribofilms to be responsible for lower friction and wear rates.

Tribological studies on Al-CNT composites also show an increase in the wear resistance and decrease in CoF due to the reasons similar to other CNT-MMC [14–18]. There is only one study on electrodeposited Cr-CNT composite coating. Cr-CNT composite coating also shows an increase in wear resistance like other CNT-MMC structures. However, unlike other CNT-MMC, CoF increases slightly with CNT content in this Cr-CNT coating. Discontinuity in the oxide layer on Cr due to the presence of CNT is held responsible for such behavior of Cr-CNT composite coating [19]. However, CNTs have been generally found to provide lubrication and decrease CoF in other CNT-MMCs. Both Al- and Cr-based composite systems undergo abrasive wear [10,19].

Most of the tribological studies discussed until now investigate wear at the macro-scale using conventional techniques like ball-on disc, pin-on-disc, ring-on plate, etc. Bakshi et al. have compared the macro- and nano-scale tribological properties of plasma-sprayed Al-CNT composite coating using conventional pin-on-disk techniques and the nano-scratch technique, respectively [16,17]. SPM images of scratches at different loads clearly show the decrease in depth and mass/volume removal with increase in CNT content of the composite coating (Figure 8.3). An increase in the wear resistance by 300% is observed with 10 wt.% CNT addition, which has been attributed to the stiffening effect of CNT and an increase in the elastic recovery for the Al matrix. CNT contributes significantly toward the elastic recovery of the Al matrix. CNT-rich regions are found to produce more elastic recovery than CNT-deficient regions (Figure 8.4). No significant change in CoF is found in the case of nano-wear. The reason for such a discrepancy from the case of macro-wear is the absence of a graphitic layer on the scratch path due to a negligible amount of damage

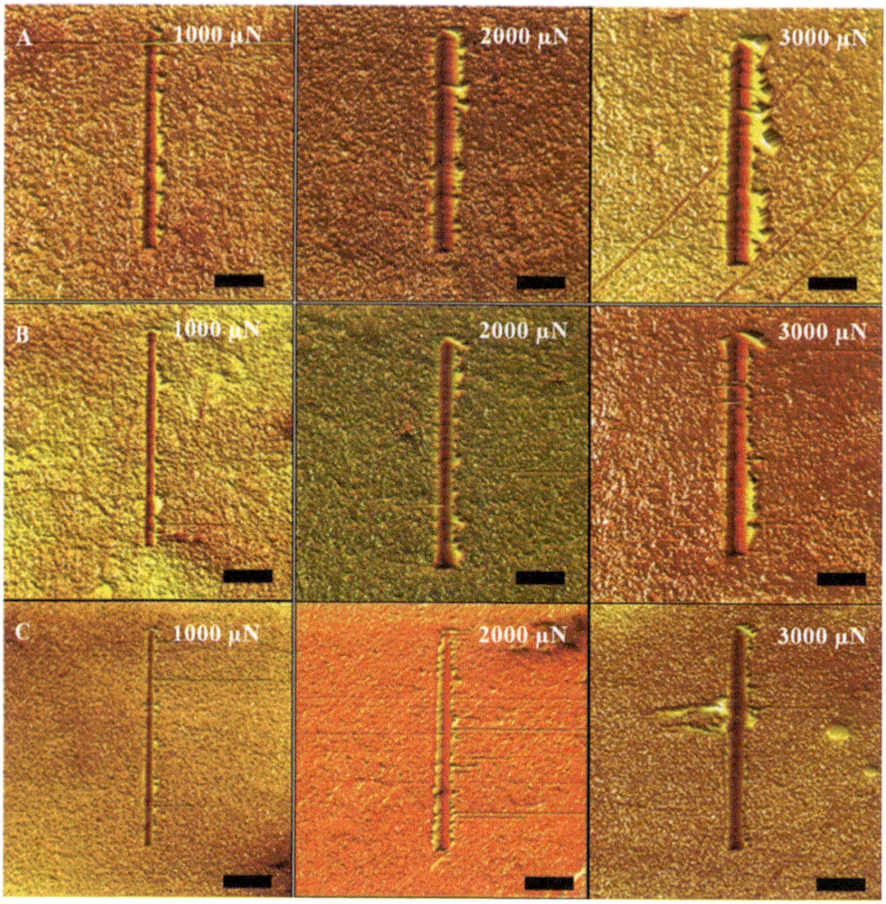

FIGURE 8.3 SPM of the scratches obtained for (a) Al-Si, (b) Al-5 CNT, and (c) Al-10 CNT coatings at different loads showing decrease in scratch depth with increasing CNT content. The scale bar shown corresponds to 5 μm [16]. (Reproduced with permission from Elsevier).

caused to the CNT during nano-wear. Nano-scratch experiments are performed at very small loads (1000–3000 μN), which are insufficient to cause separation of the graphene layer from the CNT. This is unlike macro-wear, where loads of several Newtons are applied, which results in damage to the CNT and release of the graphene layer to promote lubrication.

Dispersion of CNTs and their interfacial bonding with the metal matrix is extremely important for improving mechanical properties and wear resistance of MMC-CNT [2,3,6,8–19]. Processing is the critical stage to ensure excellent CNT distribution and reinforcement as highlighted in Chapters 2 and 3. The MLM process followed by SPS is proven very effective in excellent distribution of CNT and excellent interfacial bonding with the Cu matrix, resulting in an increase in wear resistance [10]. Other studies have reported a decrease in the wear resistance for CNT-MMCs at higher CNT content due to clustering and poor distribution of CNTs

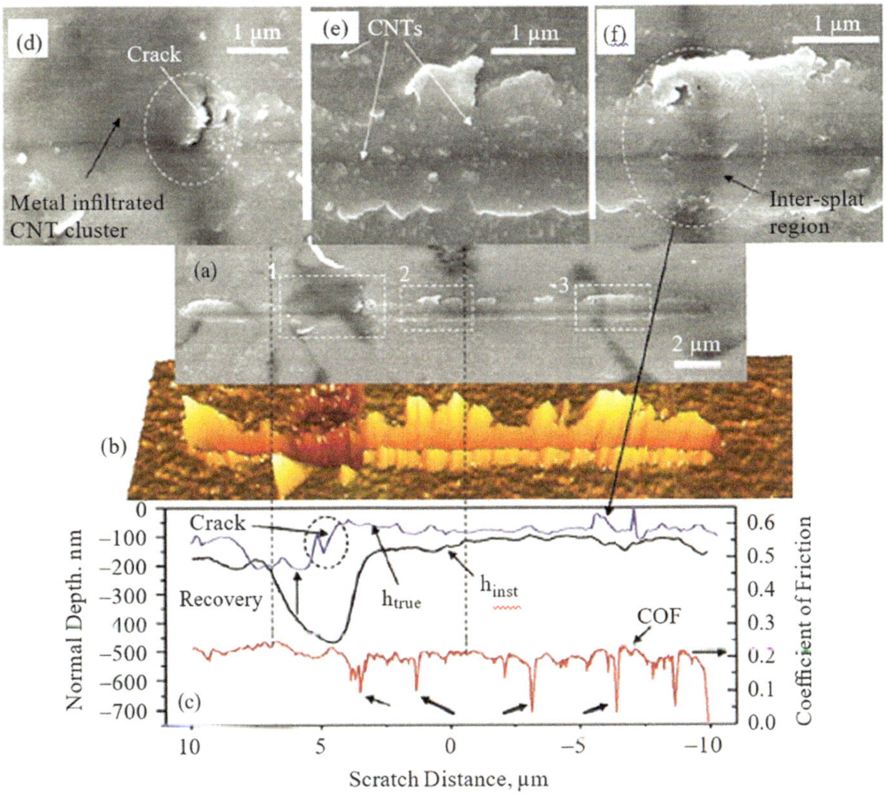

FIGURE 8.4 (a) SEM and (b) SPM image of the scratch on Al-5CNT coating at 2000 μN load, (c) the variation of the instantaneous and true depth for the same scratch showing more elastic recovery at a CNT-rich region, and (d–f) high-magnification SEM images from the wear track are shown [16]. (Reproduced with permission from Elsevier).

in the matrix [3,6,9,16–19]. CoF is not significantly affected by the quality of dispersion of CNTs, as the presence of the graphite layer in the wear track is dominated by CNT content and not dispersion. Hence, CoF has been found to decrease linearly with CNT content in most of the CNT-MMC systems. Tribological behavior of CNT-MMCs is directly related to the mechanical property. This probably is the reason for several tribological studies on MM-CNT systems, as compared to their electrochemical, corrosion, and sensor properties. The majority of the tribological studies have been done on Ni-CNT composites as they are largely synthesized as coatings and thin films for non-structural applications. A very few studies are available on tribological behavior of Cu- and Al-based CNT composite systems. Hence, more research is required to investigate optimization of CNT content and its distribution to obtain the best wear resistance and CoF for CNT-MMCs. At the same time, wear properties at the service condition (e.g., high temperature or corrosive atmosphere) should also be studied to propose real-life application of such CNT-MMCs with improved tribological properties.

8.2 THERMAL PROPERTIES

One of the potential applications of CNT-MMCs is as high thermal conductivity materials for excellent thermal management. Carbon nanotubes possess excellent thermal conductivity and very low CTE. The thermal conductivity of SWNT at room temperature could vary between 1800 and 6000 W m^{-1} K^{-1} [20,21]. The reason for unusually high thermal conductivity in carbon nanotubes is due to the presence of very strong sp^2 bonds. CNTs are seamless and atomically perfect graphitic cylinders of a few nanometers diameter. The rigid structure combined with the virtual absence of atomic defects or coupling to soft phonon modes of the embedding medium makes CNTs very efficient thermal conductors. The higher range of thermal conductivity values has been associated with the large phonon mean free paths in CNT [21], which leads to ballistic conduction. The low thermal coefficient of expansion (CTE $\approx$ 0) in CNT arises from the lattice shrinkage of CNT with an increasing temperature up to 673 K. There are two probable explanations for such behavior of CNT. The distance between in-plane carbon atoms of graphite slightly contracts up to 673 K, causing shrinkage in CNTs. Further, desorption of molecules from the surface of CNTs also causes shrinkage to the structure. As a result, CNTs show negative CTE up to 673 K [22]. Both of these properties and especially low CTE have made CNTs very attractive reinforcement for metal matrix composites intended for application where thermal stability is very important. Electronic packaging materials and interconnects need higher rates of heat dissipation and lower volume expansion for efficient performance. Such application demands materials with high thermal conductivity and low CTE. The metals used for such application (Cu, Al, etc.) are good thermal conductors, but at the same time, they possess higher CTE. CNTs with very low CTE and high thermal conductivity are suitable second-phase reinforcement for Al and Cu in such applications [23–25]. Figure 8.5 shows a 65% decrease in CTE of Al-matrix with 15 vol.% CNT addition [24].

Lead-free solder, used for interconnect joints in integrated circuits, is also very sensitive to thermal expansion as it can result in loosening and breaking of the delicate joints. CNT reinforcement effectively reduces the CTE of Sn-Ag-Cu solder alloy matrix [26]. CNTs have also been found very effective reinforcement for increasing the dimensional stability at elevated temperatures for Mg-based composites with potential application in aerospace, automotive, and sports equipment industries [27,28].

The high thermal conductivity of CNT has also been taken advantage of in W-Cu alloy used in electrical contacts, welding electrodes, and thermal management devices. W-Cu alloy is suitable for such applications due to high arc erosion resistance and low CTE of W coupled with high electrical and thermal conductivity of Cu. Still, the thermal conductivity of W-Cu alloy is much lower than that of CNT. Figure 8.6 shows a 27.8% increase in the thermal conductivity of a W-Cu matrix with 10 wt.% CNT addition [29]. However, no significant increase in thermal conductivity of the Cu matrix has been observed with the addition of CNT. Rather, the thermal conductivity of the composite decreases with CNT content increasing beyond 10 vol.%. Thermal resistance at the Cu-CNT interface along with the presence of porosities and CNT clusters is responsible for low thermal conductivity [30]. A similar decrease in thermal conductivity by a factor of 3 was observed in the case of Al-12 wt.% Si alloy coating containing 10 wt.% CNTs (Al-10 CNT) prepared by plasma

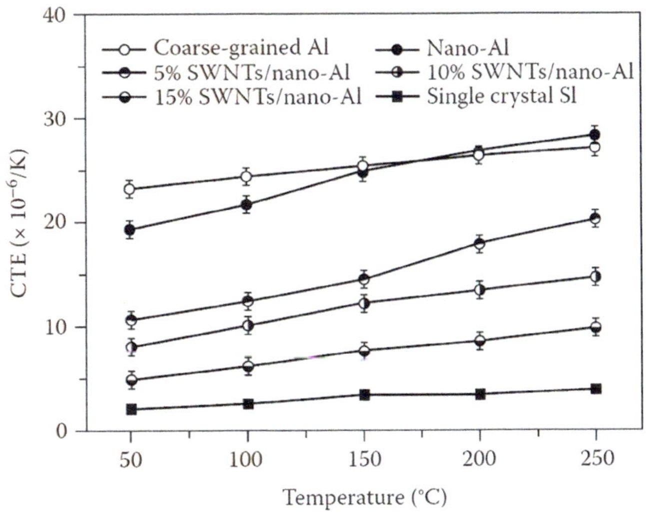

FIGURE 8.5 Variation in CTE plotted against temperature in the range of 50–250 °C shows decreases in CTE with increasing CNT content in an Al-CNT composite [24]. (Reproduced with permission from Elsevier).

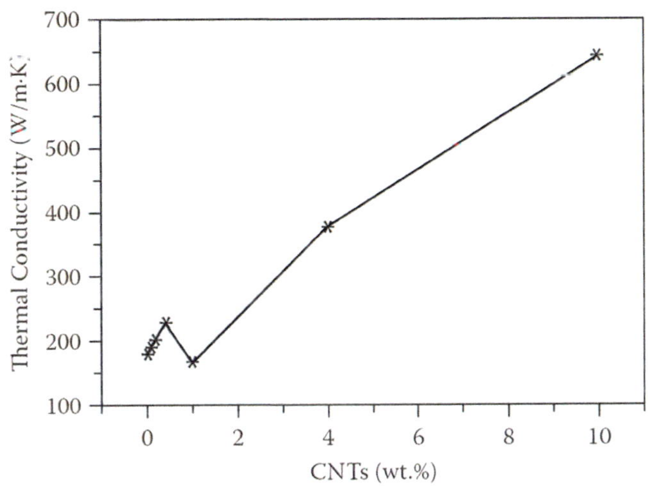

FIGURE 8.6 Variation of thermal conductivity of W-Cu-coated CNT composites with varying CNT weight fraction [29]. (Reproduced with permission from Elsevier).

spraying. Figure 8.7 shows the optical micrograph of the polished cross-section of the Al-10 CNT coating showing the presence of CNT clusters. The thermal conductivity of the Al-Si coating was found to be around 74 W m^{-1} K^{-1}, which is half the value reported for dense Al-Si produced by casting. This difference is due to the splat structure of the coating, which consists of fine inter-lamellar porosity that reduces the actual area of contact for thermal conduction.

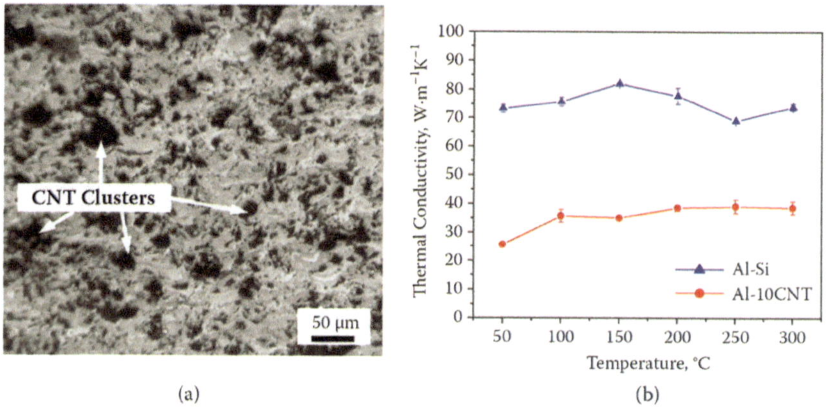

FIGURE 8.7 Optical micrograph of plasma-sprayed Al-12 wt.% Si coating containing 10 wt.% CNT (Al-10 CNT) coating showing the presence of CNT clusters, and (b) the variation of thermal conductivity of the Al-Si and Al-10 CNT coating with temperature.

Several theoretical models have been proposed based on the effective medium approach (EMA) of Maxwell-Garnett and take the linear nature of the CNTs into account. These models are ROM [31], Nan et al. [32,33], and Xue et al. [34], which can be represented by the following equations.

ROM [31]:

$$\frac{k_{composite}}{k_m} = 1 + \frac{f_{CNT} k_{CNT}}{3k_m} \tag{8.1}$$

Nan et al. [32,33]:

$$\frac{k_{composite}}{k_m} = \frac{3 + f_{CNT}\left(\beta_x + \beta_z\right)}{3 - 2\beta_x} \tag{8.2}$$

Where

$$\beta_x = \frac{2\left(k_{11}^{CNT} - k_m\right)}{k_{11}^c + k_m}, \beta_z = \frac{k_{33}^{CNT}}{k_m} - 1, k_{11}^c = \frac{k_c}{1 + \left(\frac{2a_k}{d}\right)\left(\frac{k_{CNT}}{k_m}\right)} .and\, k_{33}^{CNT}$$

$$= \frac{k_{CNT}}{1 + \left(\frac{2a_k}{L}\right)\left(\frac{k_{CNT}}{k_m}\right)}$$

Xue et al. with good CNT dispersion [34]:

$$1-f_{CNT}+\left(\frac{4f_{CNT}}{\pi}\right)\sqrt{\frac{K_{cnt}}{k_m}}\times\frac{k_{composite}}{k_m}=\tan^{-1}\left(\pi/4\sqrt{\frac{k_{CNT}}{k_m}}\right)/ \left(1-f_{CNT}+\left(\frac{4f_{CNT}}{\pi}\right)\sqrt{\frac{k_m}{k_{CNT}}}\right)\times\tan^{-1}\left(\frac{\pi}{4\sqrt{\frac{k_{CNT}}{k_m}}}\right) \tag{8.3}$$

In the previous equations, k_m is the thermal conductivity of the matrix, k_{CNT} is the thermal conductivity of CNT, f_{CNT} is the volume fraction of CNT in the microstructure, L is the average length of the CNTs in the microstructure, d is the average diameter, and a_k is the Kapitza radius, which is given as $a_k = R_k \times k_m$ where R_k is the Kapitza resistance or the thermal boundary resistance.

These equations predict the thermal conductivity of CNT composites with very good distribution of CNTs. In the case of curled CNTs, Song et al. have proposed that the heat transfer efficiency may be reduced because the effective length of the CNTs in the heat conduction direction decreases [35]. In the presence of CNT clusters, it was observed that the thermal conductivity reduces [31] as shown in Figure 8.7. For such cases, the microstructure can be assumed to be made of the matrix phase and the CNT cluster phase. Models developed for the thermal conductivity of two-phase materials can be utilized in such cases, which are represented by the following equations.

Garnet's relation [36]:

$$k_{Composite}=\frac{k_m\left(Lk_{cluster}+(1-L)k_m+f(1-L)\left(k_{Cluster}-k_m\right)\right)}{\left(Lk_{Cluster}+(1-L)k_m-fL\left(k_{cluster}-k_m\right)\right)} \tag{8.4}$$

where L is a depolarization factor that is 1/3 for spheroids, between 0 and 1/3 for prolate ellipsoids, and between 1/3 and 1 for oblate ellipsoids.

Bruggeman's formula based on Maxwell's relation for non-dilute systems [37]:

$$\frac{k_{Composite}}{k_m}-\frac{k_{Cluster}}{k_m}=(1-f)\left[\left(\frac{k_{Composite}}{k_m}\right)^{\frac{1}{3}}\left(1-\left(\frac{k_{Cluster}}{k_m}\right)\right)\right] \tag{8.5}$$

Landauer [38]:

$$k_{Composite}=\frac{1}{4}\left[\begin{array}{l}k_{Cluster}(3f-1)+k_m(2-3f)\\+\left\{\left[k_{Cluster}(3f-1)+k_m(2-3f)\right]^2+8k_mk_{Cluster}\right\}^{\frac{1}{2}}\end{array}\right] \tag{8.6}$$

Here, $k_{Composite}$, km, and $k_{Cluster}$ are thermal conductivity of the composite, the matrix, and the CNT cluster, respectively, while f is the CNT cluster volume fraction. The overall thermal conductivity of the composite is dependent on the thermal conductivity of the CNT clusters. The decrease in thermal conductivity of the plasma-sprayed Al-10 CNT coating by a factor of 3 as compared to Al-Si coating indicates that the CNT clusters must have a thermal conductivity up to 3 orders of magnitude lower than individual CNTs (Figure 8.7). The reason for such low conductivity of the CNT clusters is attributed to the lack of proper metal infiltration of the CNT clusters, which results in large porosity of the clusters as was shown in Figure 5.4.

Metals have also been used as a filler material in a CNT preform, in order to increase the thermal conductivity of the composite system. Vertically aligned CNT-based interconnects are perfect examples of such applications. CNTs show excellent thermal conductivity along an axial direction. However, localized heating, caused by high power consumption, requires faster dissipation of heat in lateral directions as well. Void space in a vertically aligned CNT-forest could be partially filled by Cu deposition to prepare a Cu-CNT composite with effective lateral heat conduction [39,40]. Decrease in thermal resistance as high as 70% has been recorded with 40 vol.% Cu deposition in a CNT preform [40].

CNTs are effective in increasing thermal stability of metal matrix composites. CNTs can efficiently increase the thermal conductivity of the metallic structure in cases where the matrix possesses low thermal conductivity. However, in cases of metals with high thermal conductivity, interfacial resistance becomes crucial for the improvement in thermal conductivity of the composite. In light of the current findings and achievements, it is clear that homogeneous dispersion of CNT and good bonding with the matrix is very important for excellent thermal properties of the composite. Good bonding at the interface ensures better thermal conductivity at the interface. Homogeneous distribution of CNT in the matrix aids in uniform decrease of CTE for the entire composite structure. Some of the studies in this field have taken care of CNT dispersion at the processing stage by ball milling the starting constituents [28]. The bonding between matrix and CNT has also been improved in the case of a W-Cu-CNT composite, by coating the CNT with Cu prior to mixing in the matrix [29]. Adverse effects of CNT clustering and poor bonding on thermal properties of the composite have been observed in certain cases [24,30]. Dispersion of CNT and its good bonding with the matrix is governed mainly by the processing stage of the composite. Hence, modifications of the processing route and optimization of CNT content should be the two most important factors at this stage to improve further the effect of CNT addition on thermal properties of CNT-MMCs.

8.3 ELEVATED TEMPERATURE MECHANICAL BEHAVIOR

Several studies since the early 2000s have investigated the effect of CNTs on the coefficient of thermal expansion (CTE) of Al-CNT composites [24,25,41–46]. It is generally accepted that this is due to the low or even negative CTE of CNTs; however, a detailed mechanistic explanation of load transfer and the role of interface and microstructural architectures has only been discussed in a few salient papers.

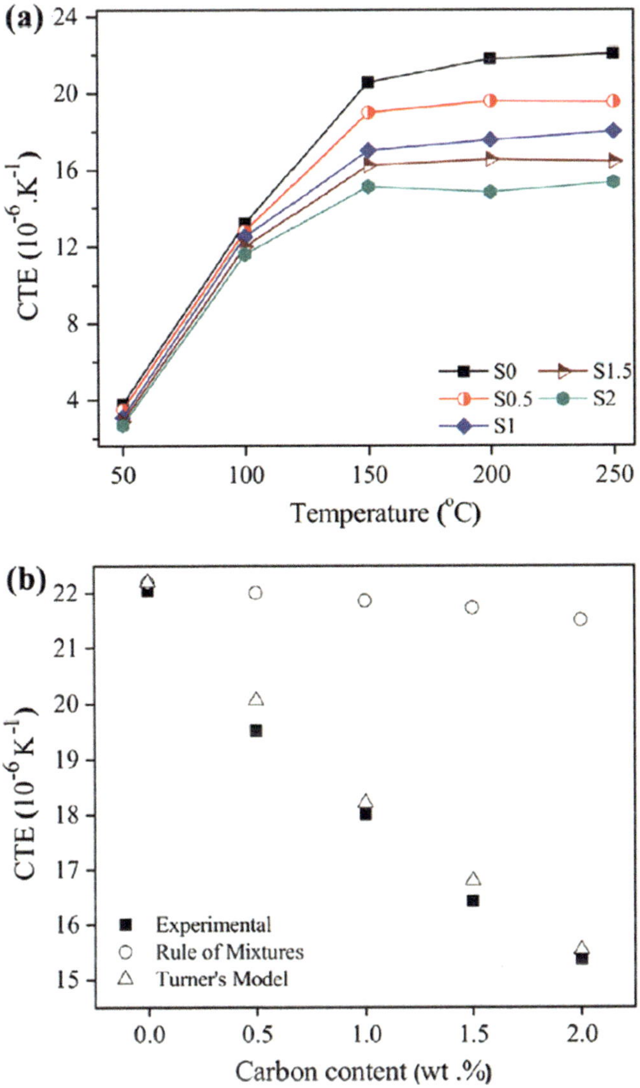

FIGURE 8.8 (a) CTE of CNT/Al composites versus temperature, (b) comparison between calculated and experimental CTE versus carbon content in the temperature range of 200–250 °C [46]. (Reproduced with permission from Elsevier).

Van Trinh et al. [46] observe that increasing CNT content led to a reduction in CTE, up to 30% reduction with 2.9 vol.% CNT in Al-CNT composites prepared by flake powder metallurgy. CTEs values measured from parallelepiped specimens from 50 to 250 °C are presented in Figure 8.8. The CTE value measured at the highest range decreased steadily with increasing CNT content, the values measured in this study are the lowest CTE values for Al-CNT MMCs that do not utilize nanometric Al as the matrix.

CTE data appears to follow the Turner model for continuous reinforcement (Eq. 8.7) below, much better than the rule of mixtures model (Eq. 8.8).

$$\alpha_C = \frac{\left(E_f \alpha_f V_f + E_M \alpha_M \left(1 - V_f\right)\right)}{\left(E_f V_f + E_M \left(1 - V_f\right)\right)} \tag{8.7}$$

$$\alpha_C = \alpha_f V_f + \alpha_M \left(1 - V_f\right) \tag{8.8}$$

where E_f is the Young's modulus of MWCNTs, α_f is the CTE of the MWCNTs along the axial direction, V_f is the volume fraction of MWCNTs, E_M is the Young's modulus of the matrix, and α_M is the CTE of the matrix. It should be noted that the thermal expansion behavior of CNTs themselves has been reported to be highly dependent on their chirality, diameter, and inter-atomic potentials [47]. The exceptionally low CTE values are attributed to the presence of CNT-rich regions between Al-CNT flakes, which may be providing a pinning effect for expansion.

While the formation of the reaction product aluminum carbide (Al_4C_3) has generally, although not exclusively, been shown to be detrimental to mechanical properties of CNT-MMCs, studies since 2017 on elevated temperature mechanical behavior have shown mixed results on the role of Al_4C_3. Zhou et al. [48] investigate the role of a deliberately formed aluminum carbide interfacial product on the thermal expansion behavior of spark plasma sintered Al-1.5 vol.% CNT composites. CNT-MMCs were extruded and machined into cylindrical specimens along the direction of CNT alignment. One set of specimens was heat treated at 600 °C for up to 0.5 h in order to induce formation of Al_4C_3 interfacial compounds. The CTE was measured from 50 to 400 °C for as-processed and heated treated Al-CNT composites, and compared with analytical models as shown in Figure 8.8b. The CTE of the composites in the longitudinal direction (α_c) can be calculated using Eq. 8.7. The CTE of both composites decreases with increasing CNT content (up to 1.5 vol.%), and the CTE values for the heat treated CNT-MMC are lower than those of extruded samples. The dashed red line in Figure 8.9 represents the extreme case where MWCNTs act as pores due to the absence of load transfer at the CNT/Al interface. The upper and lower bounds presented in Figure 8.9 are calculated using Eq. 8.7 using CNT Young's modulus values of 270 GPa and 950 GPa, based on experimental work by Yu et al. [49]. The as-processed samples show CTE values close to those assuming a loose interface, while the heat treated samples fall in the range expected from the calculated CTE values. Zhou et al. [50] calculated load transfer efficiency to be as high as 60% and attribute the enhanced load transfer efficiency and resulting lower CTE values to the presence of the Al_4C_3 interfacial layer. Thermal expansion curves are presented in Figure 8.10 and show that the heat treated sample with Al_4C_3 interfacial products exhibits a linear and reversible response, indicating the CNT/Al interface has remained intact, in contrast to the hysteresis observed in the extruded material. TEM observations in Figure 8.11 verify that nanovoids occur during thermal cycling due to slippage at the CNT/Al interface in the as-processed material. In contrast, no such voids are observed in the heat treated sample, and instead stress contrast at the interface is observed, indicative of frictional stresses due to a strong interface.

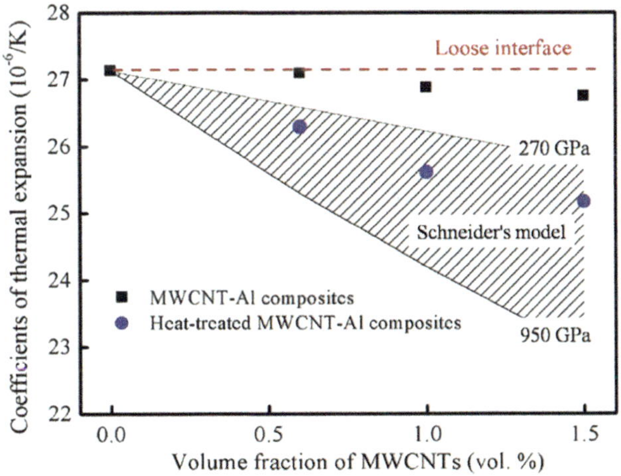

FIGURE 8.9 Comparison in the CTE values of the composites with various MWCNT concentrations as prepared (black squares) and (blue circles) HT at 873 K for 0.1 h, with the estimated CTE values using Schneider's model; the dotted red line shows the assumption of loose MWCNT-Al interfaces, the upper and lower black lines are estimated using the Young's modulus of 270 Gpa and 950 GPa for MWCNTs, respectively [48]. (Reproduced with permission from Elsevier).

Elevated temperature (up to 650 °C) tensile behavior of Cu-CNT composites fabricated by hot pressing and hot rolling are investigated by Liu et al. [51]. Mechanical properties of the composites decrease as temperature increases, with the failure mode transitioning from interfacial debonding at the CNT/Cu interface to failure within the coarsened Cu matrix at elevated temperature, as illustrated in Figure 8.12. TEM analysis also showed that CNT/Cu interface to remain intact at elevated temperatures. The extent of grain coarsening within the Cu-CNT composite can be seen in Figure 8.13. Determination of the stress exponent (n = 5) indicated the primary deformation mechanism at 500 °C was dislocation climb controlled by lattice diffusion, which is consistent with the larger grain size and failure within the matrix, while pipe-diffusion (n = 7) was the mechanism at 300–410 °C. The activation energy in each temperature range was higher for the Cu-CNT composite as compared to Cu.

Liu et al. [52] investigate the thermal cycling behavior of Cu-CNT composites developed by the flake powder metallurgy technique in the range of −196 to 200 °C, a temperature range relevant to satellite applications in which components can be either exposed to direct sunlight or in the shadow of the earth. After thermal cycling, tensile testing is conducted, the data is shown in Figure 8.14. For a given amount of thermal cycles, the CNT reinforced composites exhibit higher strength, but lower ductility. It also appears the magnitude of decreased strength and ductility in the Cu-CNT composites is less severe than in the pure Cu specimens. Figure 8.14b shows that Cu-CNT composites exhibit sharp drops during the early stages of loading, and the onset occurs earlier with increasing amount of thermal cycles. Initial stress drops are attributed to cracks and microvoids initiated at the CNT/Cu interface during the thermal cycles. Analysis of fractured samples showed interfacial sliding

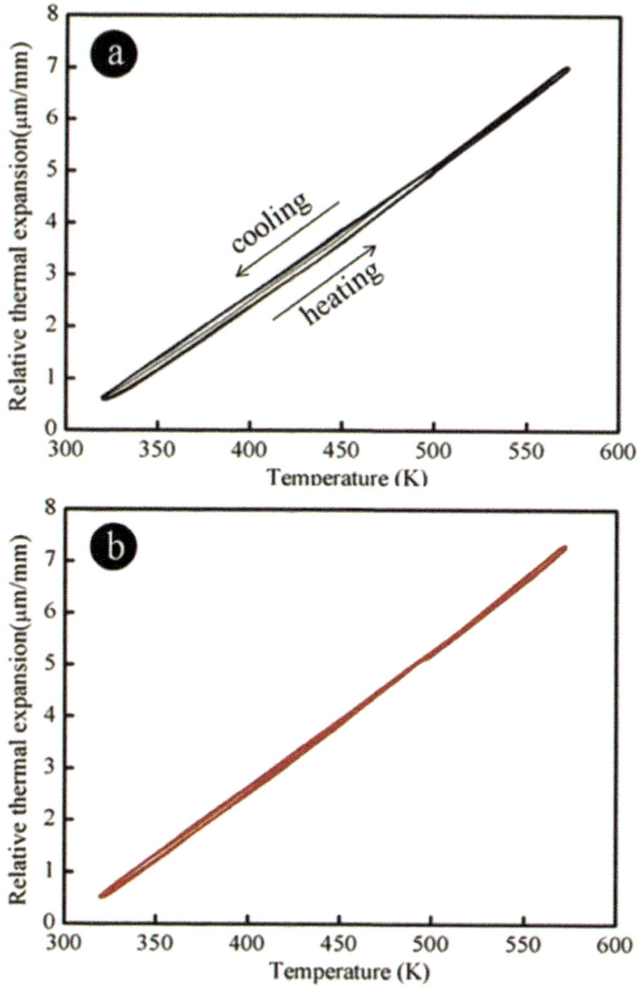

FIGURE 8.10 Cyclic thermal expansion behavior of 1.5 vol.% MWCNT/Al composite (a) as prepared and (b) heat treated at 873 K for 0.1 h [48]. (Reproduced with permission from Elsevier).

and debonding within the CNTs themselves, spacing between carbon layers was seen to increase with increased thermal cycling, and failure often occurred within the MWCNT walls. This study is promising for developing adequate CNT/metal interfaces but indicates the quality of CNTs (pre-existing defects, number of walls) may need to be optimized for thermal cycling applications.

Huang et al. [53] investigated the superplastic deformation behavior of Al-CNT composites fabricated by flake powder metallurgy. A shift speed ball milling approach was used to disperse CNTs, where first a low energy ball milling cycle is used to disperse the CNTs onto the surface of Al, followed by a high energy ball milling

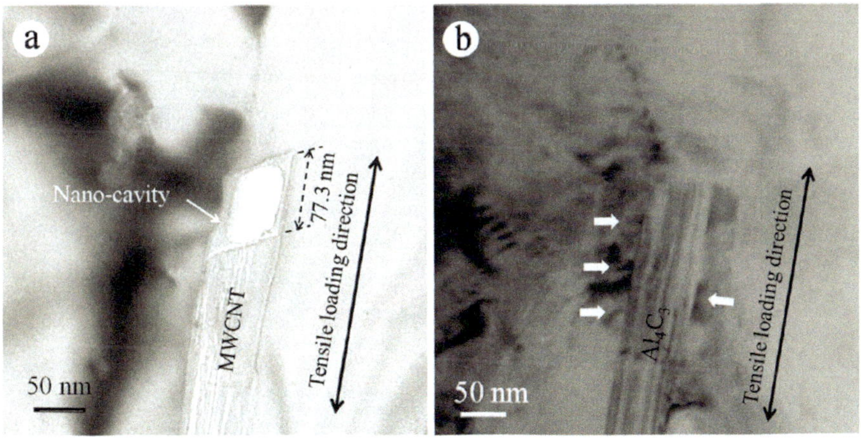

FIGURE 8.11 HRTEM showing a MWCNT in Al matrix after tensile testing, (a) as prepared, (b) HT at 873 K for 0.1 h [48].(Reproduced with permission from Elsevier)

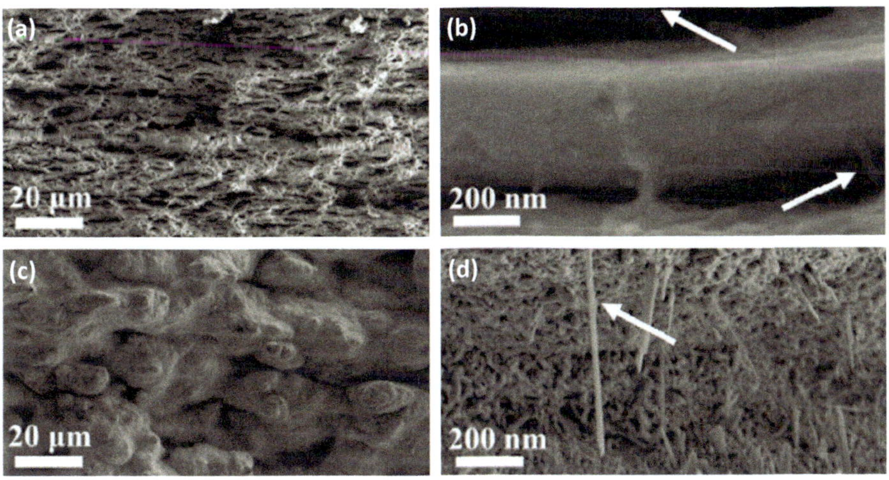

FIGURE 8.12 SEM of fractured surfaces of CNT/Cu composite at different temperatures: (a) 293 K, (b) higher magnification of 298 K, (c) 923 K, and (d) higher magnification of 923 K. [51]. (Reproduced with permission from Elsevier).

cycles to embed the CNTs within the Al matrix. Samples were sintered and extruded, and the resulting grain size was about 400 nm. Grain size was stable after being heating at 400 °C for 30 min, as well as after tensile tests. Figure 8.15a shows true stress–strain curves of the Al-CNT composites tested at 400 °C at different strain rates. The Al-CNT composites exhibit increased maximum total elongation as temperature was increased from 400 °C (89% elongation at failure) to 580 °C (106% elongation at failure), as shown in Figure 8.16, both maximum elongation rates were

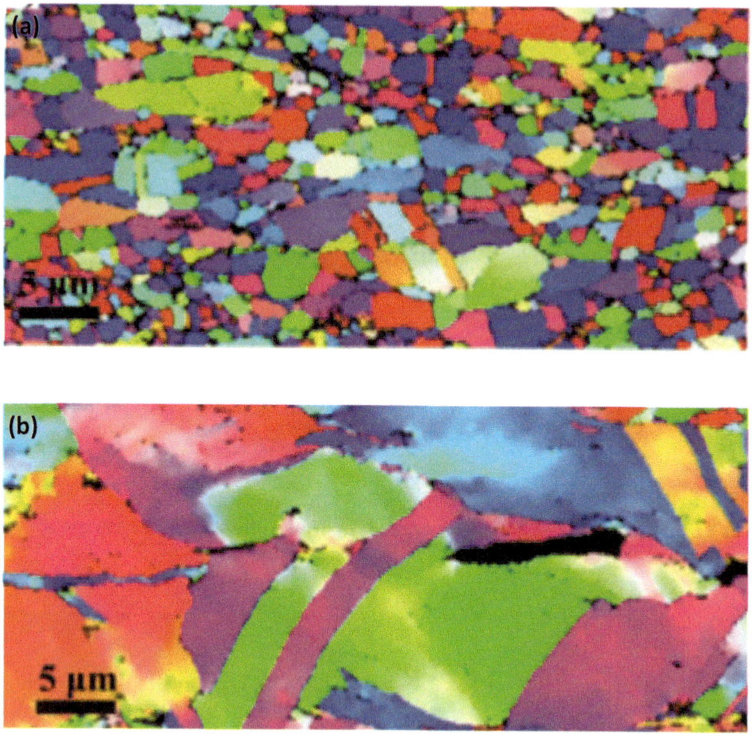

FIGURE 8.13 (a) EBSD map of CNT/Cu composite at (a) 293 K, and (b) 923 K [51]. (Reproduced with permission from Elsevier).

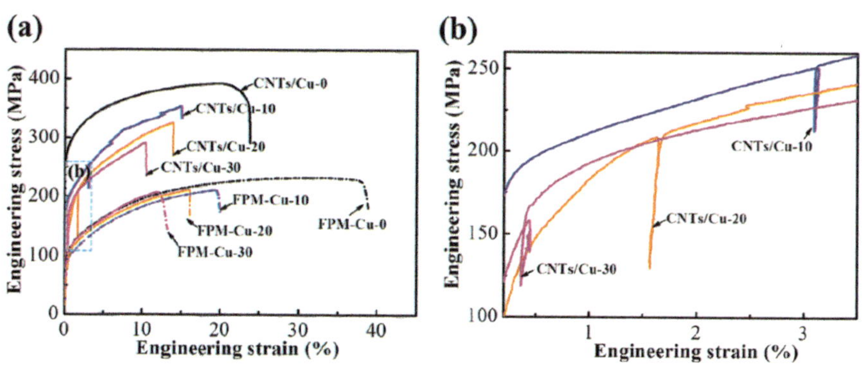

FIGURE 8.14 (a) Engineering stress–strain curves of CNT/Cu composites and flake powder metallurgy Cu subjected to different thermal cycles, (b) selected area (dotted square in (a)) showing sharp stress dots [52]. (Reproduced with permission from Elsevier).

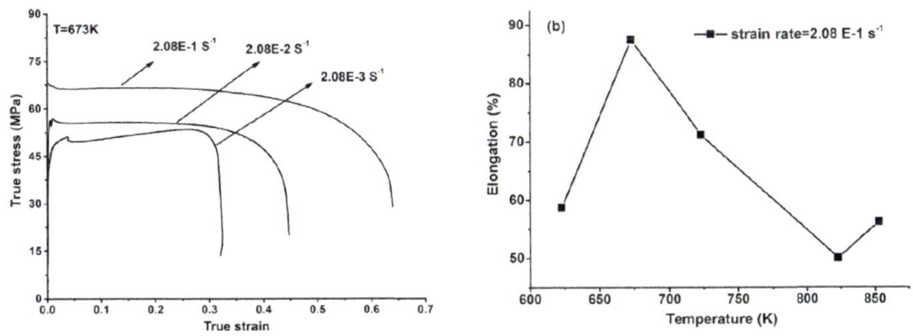

FIGURE 8.15 (a) True stress-strain curve at 673 K at different initial strain rates, and (b) elongation to failure against temperature [53]. (Reproduced with permission from Elsevier).

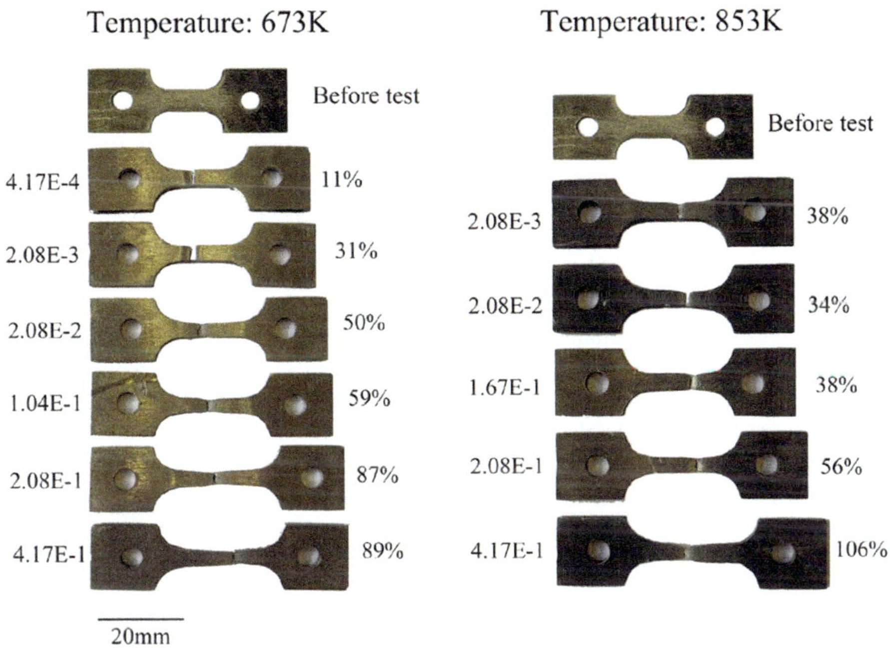

FIGURE 8.16 Tensile specimens after deformation to failure at 673 K and 853 K [53]. (Reproduced with permission from Elsevier).

attained at strain rates of 0.417 s^{-1}. Figure 8.16 shows that at a given intermediate strain rate of 0.208 s^{-1}, the relationship between strain to failure and temperature is not linear, with a maximum elongation occurring at 400 °C. The superplastic deformation is attributed primary to grain boundary sliding aided by a semi-solid matrix at the elevated temperatures used.

8.4 ELECTRICAL PROPERTIES

The electrical properties of carbon nanotubes make them very attractive. At low temperature, CNTs show virtually zero resistance to the electron motion due to the absence of scattering centers in its structure [54]. Such kind of ideal transport process, known as ballistic transport, has been made possible by the defect-free hexagonal structure of carbon, avoiding the presence of scattering centers like grain boundary and impurities. CNTs also show highly interesting variation in electronic properties from semiconductor to metallic. Diameter and chirality of CNTs mostly govern the electronic nature of the CNTs, although reports have shown that the presence of curvature or strain in the CNTs can lead to a change in their electronic behavior [55]. The excellent electron emission properties and electrical conductivity of CNTs in synergy with their mechanical properties (high elastic modulus and tensile strength) have created interest in studying their potential as field emitters [56,57], electronic packaging materials [23], interconnects [40], Li-ion batteries [58–61], micro-electromechanical systems (MEMS) [62,63], and electrical contact materials [64].

Silicon-CNT composite is very attractive for the application as anode material in Li-ion batteries [58–61]. Silicon has the highest specific capacity for intercalation of Li-ions among all the possible anode materials for Li-ion batteries. However, Si being an insulator in nature and its significant volume change between forward and reverse cycling cause its detachment from the conductive matrix/connector of the anode and thus, cause a huge reduction in reversible capacity of the battery. A composite of Si-CNT has been found very effective in playing the role of a conductive buffering layer. Due to the high electrical conductivity of CNT, it successfully maintains a conductive network between matrix/connector and Si even after the volume change and thus maintains the performance of the battery with increasing cycle numbers [58–61]. Figure 8.17 shows better performance of Si-CNT composite over Si anode in a Li-ion battery with an increasing number of cycles [60]. The composite electrode was prepared by electroless deposition of Ni on Si and then CNTs grown by CVD on it. A mixed electrode is just a physical mixture of Si and CNT.

Cu-CNT composites have been proposed for applications in interconnects and electronic packaging. Cu is a very good candidate due to its excellent electrical and thermal conductivity that helps in faster heat dissipation. However, such applications also demand the material with low CTE and better mechanical properties that can sustain electrical and thermal cycling stress. Cu-CNT composite is a promising candidate for these applications, as the presence of CNT effectively decreases the CTE and improves the stiffness of Cu-matrix without reducing the electrical conductivity [23,40,65–68]. The effective resistivity of a Cu-SWNT composite was found similar to pure Cu, which has become possible due to similar or better electrical conductivity of SWNT (Figure 8.18). However, there is always a chance of increased scattering during electron transfer at the Cu-CNT interface, which has a negative effect on the conductivity [65].

CNT-reinforced Ni composites have been found to enhance the performance of MEMs actuators and electron field emitters. Better performance of Ni in MEMs application requires modification of certain material properties, like reduced CTE and enhanced hardness and stiffness, and electrical conductivity. All these properties

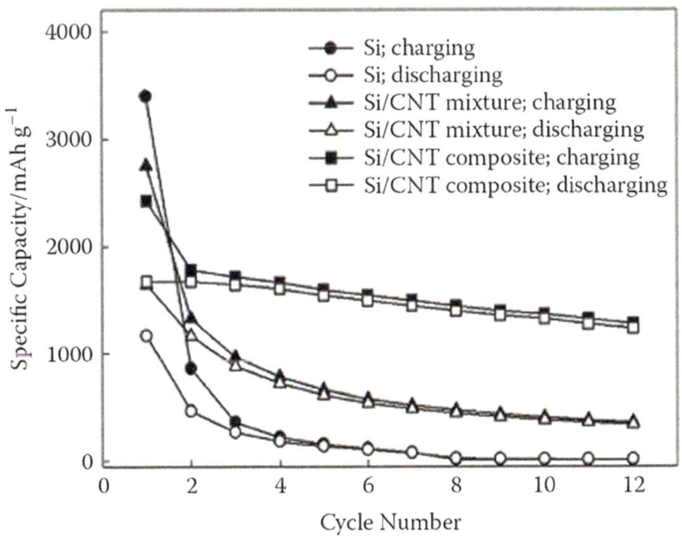

FIGURE 8.17 Galvanostatic charge/discharge voltage plots showing better cyclic performance of Si/CNT composite electrodes compared with those for bare Si and Si/CNT mixed electrodes [60]. (Reproduced with permission from Elsevier).

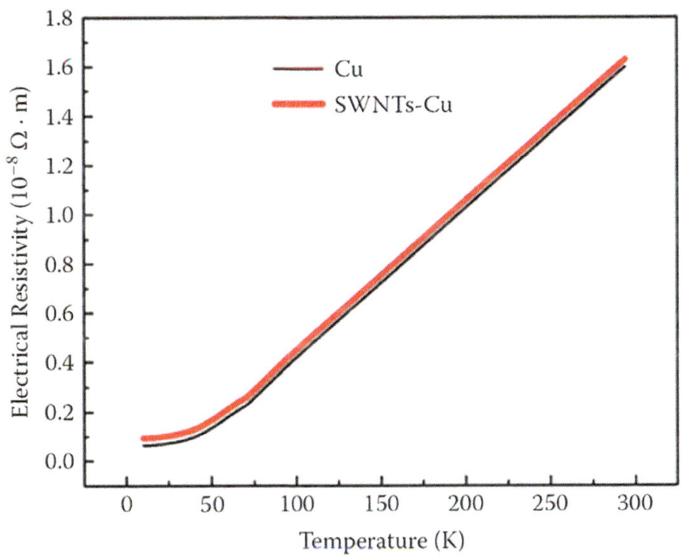

FIGURE 8.18 Plot of electrical resistivity in Cu and Cu-SWCNT composites [65]. (Reproduced with permission from Elsevier).

can be achieved by using CNT as a second-phase reinforcement to Ni. Studies have shown improvement in the device strength and power efficiency of the electrothermal microactuator made from Ni-P-CNT composite [62,63]. Ni-CNT composite shows significantly higher displacement with similar power input [62]. Ni-P-CNT actuators can show approximately four times higher efficiency than only Ni-based actuators in terms of displacement [63].

Ni-CNT composite has also been fabricated by a co-deposition process (electroplating and electroless-plating) for field-emitter application. CNTs are good candidates for field-emission applications due to their high aspect ratio, low threshold voltage, and good emission stability. Dispersion of CNT in the Ni matrix plays a major role in performance of the emitters. Researchers have attributed the excellent field-emission performance of Ni-CNT field emitter to the uniform distribution and strong adhesion of CNT to the Ni matrix [56,57]. Uniform dispersion of CNT in a Ni matrix could be achieved by purification of the CNTs through annealing and acid immersion prior to co-deposition [56,57]. Figure 8.19 shows better field-emission properties (e.g., lower turn on voltage and higher emission current) for the composite of Ni with CNTs dispersed in matrix than the one having clusters of CNTs. Uniform dispersion of CNT in an Ni matrix in this study has been achieved by surface treatment of CNTs prior to electrochemical co-deposition [56].

Apart from Ni, Cu, and Si matrices, Al- and Ag-based CNT composites also have been studied for their electrical properties [64,65,69]. The combination of high strength, wear resistance, chemical stability, and thermal conductivity along with good electrical conductivity offered by CNTs have been found attractive for Ag-CNT composites as electrical contact material. However, both Al and Ag-CNT composites show a drop in the electrical conductivity at and above room temperature because of CNT addition. Decrease in the electrical conductivity of the composite structures, in spite of excellent electrical conductivity of CNTs, is due to the high surface-to-volume ratio of nanotubes, creating a larger reinforcement-matrix

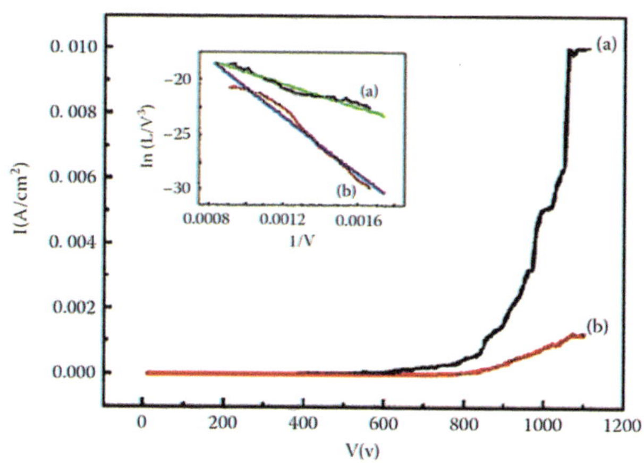

FIGURE 8.19 Field-emission properties of CNT-Ni co-deposited film with (a) uniformly distributed surface-treated CNTs and (b) CNT clusters formed due to the absence of surface treatment [56]. (Reproduced with permission from the Electrochemical Society).

interfacial region. Interfaces cause scattering during electron transfer, resulting in increased resistivity. The presence of carbon nanotubes also causes lattice strain in the metal matrix that decreases the conductivity of matrix. Xu et al. have observed an abrupt drop in electrical resistivity of the Al-CNT composite at lower temperature (80 K), although a suitable explanation for such behavior remains unclear [69].

The excellent electrical properties of CNTs along with their mechanical and thermal properties have shown them to be a promising reinforcement for metal matrix composites utilized in electrical and electronic applications. Thorough investigation is required in terms of optimization of CNT content and effect of CNT distribution. In some cases, the conductivity of the composite is hampered due to scattering of electron from the CNT/matrix interface. The conductivity of the CNT-MMC can be increased by aligning the CNTs in the matrix because the electrical conductivity of CNTs is the maximum in the axial (longitudinal) direction.

8.5 CORROSION PROPERTIES

The effect of CNT addition on corrosion properties has been studied primarily in Ni, Zn, Al, and Co-W systems, most studies have investigated coatings. The initial idea of adding CNT to metallic coatings was not for increasing the corrosion resistance. Rather, the CNT reinforcement was added to take advantage of its high elastic modulus and lubrication from self-lubricating graphene layers to increase the wear resistance of such coatings. The chemical inertness of CNT justifies its application for enhancing the corrosion resistance of CNT-MMCs. CNTs have unique chemical structures that prevent the formation of defects in composites. The sp^2 hybrid C C covalent bond in CNT is one of the most stable chemical bonds. By virtue of the chemical stability, CNT can be used to prepare composites with promising corrosion resistance [70]. A higher oxidation start (>873 K) temperature of CNTs also made them effective in increasing the corrosion resistance of the CNT-MMC.

The corrosion behavior of CNT-MMC has been predicted mainly based on potentiodynamic polarization tests and electrochemical impedance measurement, as discussed in Section 4.10. Several studies have found that CNTs increase the corrosion resistance of Ni-based composite [70–74]. Figure 8.20a shows a 300% decrease in the corrosion rate with CNT addition [74]. Corrosion potential increases by 75% toward positive value for Ni-CNT composites, indicating better corrosion resistance. Figure 8.20b shows a polarization curve for the pitting corrosion studied [74]. The higher pitting potential and lower anodic current density indicates the role of CNT in increasing the pitting corrosion resistance [74]. Figure 8.20c shows polarization curves for electrodeposited Co-W coatings with 0.72 wt.% CNT content and without any CNT. For nearly the entire range of potentials examined, the Co-W-CNT composite exhibits lower current, i.e., corrosion rate. Corrosion potential values are found to be intermediate between those for amorphous and crystalline Co-W films, suggesting a mixed microstructure in the composite coating [75].

Three major factors are responsible for improvement in the corrosion resistance of Ni-CNT composite coatings. First, due to the chemically inert nature, CNTs act as a physically passive barrier. Second, the nano-size of CNTs makes them suitable candidates for filling up crevices, micro-pores, cracks, defects, and gaps of the

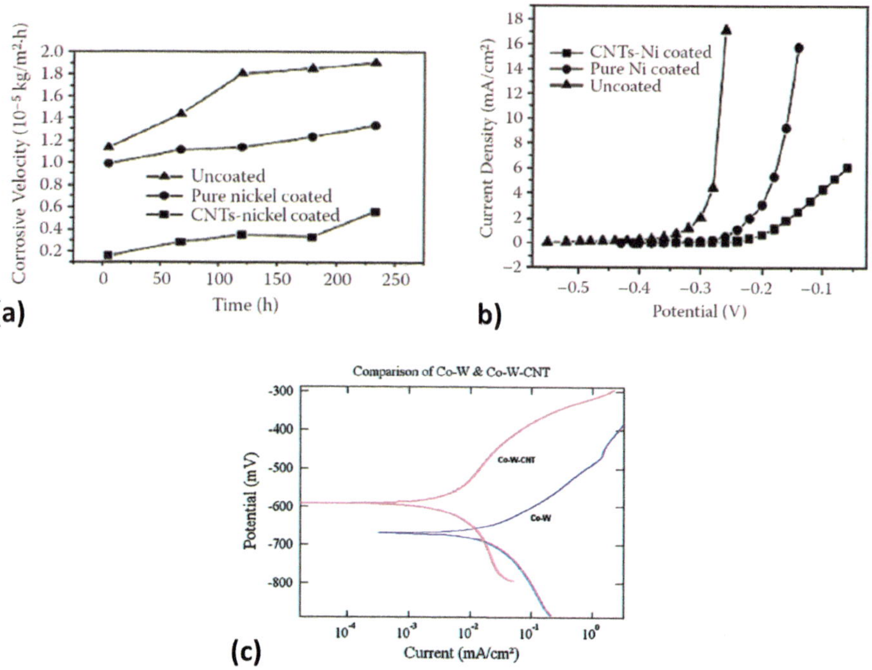

FIGURE 8.20 (a) Variation of the corrosion rate with immersion time [72], (b) anode polarization curves for uncoated, pure Ni-coated, and CNT-Ni coated carbon steel substrates in 3.5 wt.% NaCl solution [72] (Reproduced with permission from Elsevier), and (c) polarization curves for Co-W and Co-W-CNT coatings in 3.5 wt.% NaCl solution. (Reproduced with permission from ASM International).

electrodeposited Ni-surface, leaving fewer sites for the onset of pitting corrosion [74,76]. Third, an exposed CNT network provides a protective barrier after selective corrosion of the metal surface. An exposed CNT network hinders the infiltration of the electrolyte solution to the metal matrix beneath. The electrolyte solution can only infiltrate into the interspaces between CNTs. Thus, the corrosion rate of the CNT-MMC coating is decreased. Processing parameters have also been found to affect the corrosion resistance of the CNT-MMCcomposite coating. An increase in the corrosion resistance of the electrodeposited Ni-CNT composite coating was observed by altering the frequency and reverse ratio of the pulse current [73]. This occurs due to an increase in the CNT content of the coating caused by reverse ratio of pulsed current [73]. CNTs have been found effective in increasing the corrosion resistance of Zn-based composite coating on the steel substrate. The formation of white rust, which denotes the start of corrosion, takes double the time in a Zn-CNT composite than in only a Zn coating. This observation indicates enhancement of the service life of Zn-CNT composite coatings. The reason for the increase in corrosion resistance of a Zn-CNT composite coating is similar to that of Ni [76]. One study on AA5083 with up to 1.75 wt.% also showed improved corrosion resistance of the composite when immersed in dilute HCl solution for 24 h, the primary mechanism is also attributed to CNT filling of pores and voids [77].

Distribution and content of CNT in the metal matrix are critical factors for corrosion resistance in the MMC-CNT. Homogeneous distribution of CNT in the matrix resists localized corrosion and promotes uniform removal of mass during corrosion. On the contrary, CNT clusters increase defects and porosities in the matrix, which act as sites for the start of pitting corrosion. Hence, all of the studies have focused on the processing technique to ensure better distribution of CNT in the metal matrix for improving the corrosion resistance. CNT content also needs to be optimized to obtain the best corrosion resistance. The CNT-metal interface is a probable corrosion site due to the potential difference between metal and CNT. Thus, very high content of CNT decreases the corrosion resistance of the composite [71,73]. With the limited publications on corrosion behavior of MMC-CNT, the mechanism of improvement in corrosion resistance and the role of CNT are not very clear. Given the demonstrated capability of CNT-MMCs for structural applications, the corrosion behavior of CNT-MMC requires more attention in order to ensure the longevity and reliability of these new materials in operation.

8.6 HYDROGEN STORAGE

Mg-CNT composites have been investigated for hydrogen storage properties. Magnesium is a very attractive and promising material for hydrogen storage application due to its high theoretical hydrogen storage capacity of 7.6 wt.% [78]. However, its application is still limited due to its high temperature of operation (673 K) and slow kinetics of hydrogen absorption and desorption. Different forms of carbon, that is, graphite, carbon nano-fibers, and carbon nanotubes are found attractive for hydrogen storage application because of the presence of uniform pores, high surface area, and surface potential suitable for absorbing hydrogen [78,79]. Some research groups have studied the effect of CNT reinforcement on the hydrogen storage property of Mg [78,79]. Mg-5 wt.% CNT composite has shown the best hydrogen storage capacity at lower temperature among all the Mg-based composites. Mg-CNT composite shows improved cyclic performance, better hydrogen absorption (80% of capacity) and desorption, and increased kinetics [78]. Several reasons have been attributed to the improvement of the hydrogen storage property of Mg-CNT composite. Absorption of C on an Mg surface hinders the formation of a passive oxide/hydroxide layer and keeps the Mg surface readily available for hydrogen storage. Mg and CNT also form nano-size composites, which provide storage sites for hydrogen. Further, absorption of a monolayer of hydrogen on a CNT surface increases the hydrogen storage capacity of the Mg-CNT composite. Schaller et al. have claimed that CNT-reinforced Mg-Ni alloy is better for hydrogen storage material than Mg-CNT composite, with a storage capacity as high as 6.1 wt.%. This has become possible due to the presence of Mg_2Ni, which is an active phase in the hydrogen absorption. CNTs work as diffusion short circuits and assist in the transportation of hydrogen atoms [80]. However, CNT content should be optimized to obtain the best improvement in hydrogen storage property. The presence of more carbon in the storage material increases the chance of formation of hydrocarbons. The hydrocarbons have higher bonding energy

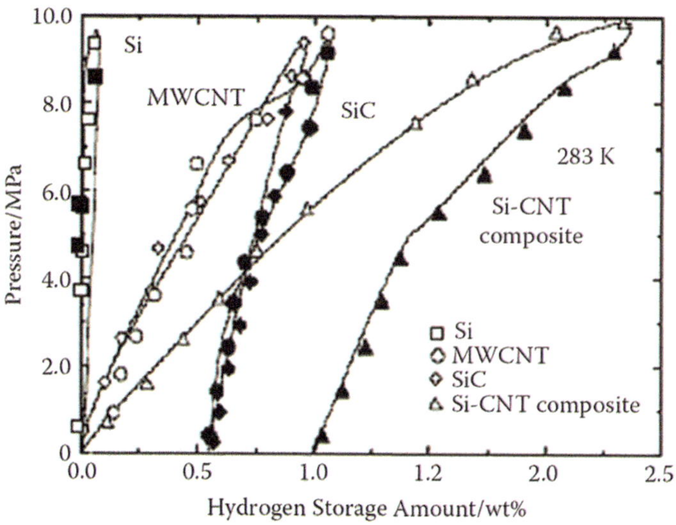

FIGURE 8.21 Adsorption isotherms for H_2 in Si, CNT, SiC, and Si-CNT composite. The open symbol is for adsorption and the closed symbol is for desorption [81]. (Reproduced with permission from IOP Science).

than metallic hydrides and hence their decomposition during a desorption cycle becomes difficult, reducing the cyclic efficiency of hydrogen storage. Decrease in the efficiency of hydrogen storage with increasing CNT content (>5 wt.%) has been observed for Mg-based composite [78]. Figure 8.21 shows better hydrogen storage capacity for Si-CNT as compared to only Si or CNT [81]. The mechanism of improvement in hydrogen storage capacity for Si-CNT composite is yet to be explained.

The last decade has attracted minimal research activity on the hydrogen storage capability of CNT-MMCs, available in open literature. As such, the role of CNTs in improving the hydrogen storage ability of CNT-MMCs is poorly understood. The proposed mechanisms rely on the physical properties of CNTs, without direct observation of their role within the composite. Once the role of CNT is established, further studies will be needed to optimize the CNT content for better performance. The effect of CNT distribution, if any, on the hydrogen storage property of CNT-MMCs is yet to be established.

8.7 SENSORS AND CATALYTIC PROPERTIES

CNT-MMCs have been proposed as electrodes in sensors and catalytic reactions. CNTs are suitable for such applications because of their high surface-to-volume ratio, linear structure, chemical stability, enhanced electro-active surface area, excellent absorptive properties, and high electrical conductivity. In addition, functionalization of CNTs helps in molecular level detection at very low concentrations, which increases sensitivity of the composite [82–84]. MMC-CNT used for sensor and catalysis have largely been in the form of a paste of metallic nanoparticles and CNT in a mineral oil. The paste is coated and dried into a film on the substrate to be used as an

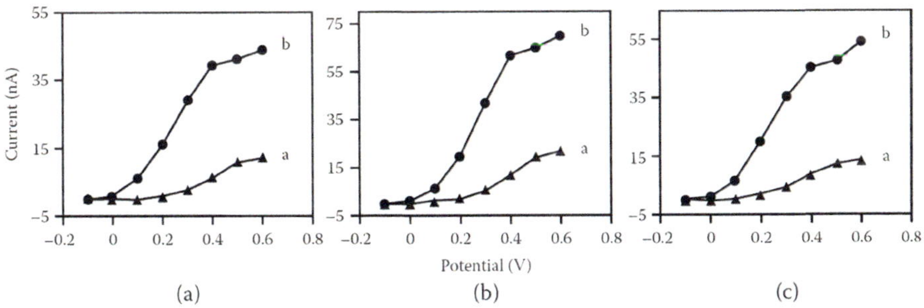

FIGURE 8.22 Hydrodynamic voltammograms for (a) 500 mM sucrose, (b) galactose, and (c) fructose at the (A) copper and (B) Cu-CNT composite paste detectors. Electrode composition ratios for mineral oil:CNT:Cu are (A) 1:0:1 and (B) 1:1:2 [85]. (Reproduced with permission from RSC).

electrode [82–88]. In a few cases, metallic nanoparticles have been electrodeposited on CNTs for sensing application [84–89]. Cu-CNT composites have been studied for the detection of carbohydrates and glucose [84,85], albumins and amino acids [87,88], and ammonia [83]. The presence of CNTs in Cu increases the linear range of detection, sensitivity, and detection limit and improves the detection speed for carbohydrate and glucose. The hydrodynamic voltammograms (Figure 8.22) for the oxidation of different carbohydrates show higher current at the same potential indicating improved sensitivity for amperometric detection [85]. The reproducibility and long-term stability of Cu-CNT composite electrodes have made them even more attractive for carbohydrate detection.

Cu-CNT composite has also shown efficiency in amino acid and albumin sensing due to detection capacity in the micro-molar level. These composites also displayed reproducibility and stability for capillary zone electrophoresis. Increased catalytic oxidation of amino acids in the presence of Cu and CNT has made the composite electrodes more effective sensors [88,89]. During detection of ammonia, the Cu-CNT composite shows better efficiency in terms of higher sensitivity, lower detection limit, good stability, and reproducibility.

CNT-MMCs, with other nano-sized metallic particles, for example, Pt and Au, have shown enhanced performance as sensors/detectors for trinitrotoluene (TNT) and nitroaromatic compounds and catalyst for methanol electro-oxidation [82,86] in fuel cells [89]. Pt-CNT composite detector shows higher sensitivity, lower detection limit, longer range of detection, and reproducibility for TNT. The role of CNT in better performance as sensors has been attributed to several factors. These include excellent absorptive property and higher electrical conductivity, which enhances the electroactive surface area, making the detection very sensitive toward the lower range. The synergistic electro-catalytic activity of Cu and CNT for glucose detection has been documented [84]. Strong complex forming activity of amino acid toward Cu helps in better performance of a Cu-CNT electrode in sensor [88].

Most of the CNT-MMCs have been made as the paste for sensor applications. Hence, not much attention has been paid toward the dispersion behavior of CNT and its correlation with sensing properties. Huang et al. have mentioned the positive effect of better CNT dispersion on the performance of Pt-CNT-modified Au electrode for methanol oxidation [94]. A majority of the studies on application of MM-CNT sensors still dwell on the enhancement in detection, without much focus on elucidating the mechanism by which CNT aids sensing. It is expected that more studies would be performed in this area of research with an emphasis on understanding the mechanism and correlating with CNT dispersion and processing.

8.8 CHAPTER HIGHLIGHTS

The effect of CNT reinforcements on tribological, thermal, thermo-mechanical, electrical, corrosion, hydrogen storage, and catalytic properties of CNT-MMCs is described in this chapter. The presence of CNT plays a dual role in the tribological performance of the CNT-MMCs by improving the wear resistance and reducing the coefficient of friction. Peeled-off graphene layers from the CNT surface provide lubrication and decrease the CoF for the CNT-MMC during wear. Good thermal conductivity and low CTE of CNT are accountable for increases in thermal conductivity and decreases in CTE for CNT-MMC. However, the effective thermal conductivity in CNT-MMC is hampered to some extent by scattering at CNT-metal interfaces and the presence of porosities and CNT clusters. Thermal expansion behavior largely control CNT-MMC thermal cycling behavior, as CTE mismatch disrupts initial interfacial state that can result in adverse effects on thermo-mechanical properties. High electrical conductivity CNTs along with their mechanical strength are effective in improving the performance of Si-based material in Li-ion batteries, performance of interconnects and electronic packaging material for Cu-CNT, and micro-actuation and field emission properties of Ni-CNT composites. Such behavior of CNT-MMCs is useful for interconnect and electronic packaging applications, as well as in the automotive and aerospace industries. Chemical inertness of CNTs improves the corrosion resistance for most CNT-MMCs. CNTs protect the metallic surface from corrosion and pitting by forming a passive film or filling up the micropores. However, with higher CNT content, corrosion may get aggravated due to localized micro-cell formation between metal and the CNT. CNT-MMCs also show improved hydrogen storage capability, although the exact mechanism is still under investigation. CNT-MMCs show superior performance as sensors and catalysts for various organic and inorganic chemical detection/reactions. However, the composites used for such purposes are mostly in the form of mixtures/pastes/aggregates of CNTs with metallic powders. One common factor for all CNT-MMC systems in different applications is the quality of dispersion of CNT in the metal matrix. Effective improvements in the targeted properties in CNT-MMCs are associated with the homogeneous dispersion of individual CNTs in the matrix, while the reverse is true for the presence of CNT clusters. Overall, the properties for CNT-MMCs, other than mechanical properties, have not been investigated thoroughly. Especially properties like corrosion behavior and hydrogen storage need a lot of attention for successful application of CNT-MMCs for such purposes.

REFERENCES

1. Chen, X. H., Peng, J. C., Li, X. Q., Deng, F. M., Wang, J. X., and W. Z. Li. 2001. Tribological behavior of carbon nanotubes—reinforced nickel matrix composite coatings. *J. Mater. Sci. Lett.* 20: 2057–2060.
2. Chen, X. Zhang, G., Chen, C., Zhou, L., Li, S., and X. Li. 2003. Carbon nanotube composite deposits with high hardness and high wear resistance. *Adv. Eng. Mater.* 5(7): 514–518.
3. Chen, W. X., Tu, J. P., Wang, L. Y., Gan, H. Y., Xu, Z. D., and X. B. Zhang. 2003. Tribological application of carbon nanotubes in a metal-based composite coating and composites. *Carbon 41*: 215–222.
4. Deng, F. M., Chen, X. H., Chen, W. X., and W. Z. Li. 2004. Electroless plating Ni–P matrix composite coating reinforced by carbon nanotubes. *Trans. Nonferrous Met. Soc. China* 14(4): 681–685.
5. Zhao, G., and F. Deng. 2005. Electroless plating of Ni-P-CNTs composite coating. *Key Eng. Mater.* 280–283: 1445–1448.
6. Tu, J. P., Zhu, L. P., Chen, W. X., Zhao, X. B., Liu, F., and X. B. Zhang. 2004. Preparation of Ni-CNT composite coatings on aluminum substrate and its friction and wear behavior. *Trans. Nonferrous Met. Soc. China* 14(5): 880–884.
7. Dong, S. R., Tu, J. P., and X. B. Zhang. 2001. An investigation of the sliding wear behavior of Cu-matrix composite reinforced by carbon nanotubes. *Mater. Sci. Eng. A* 313: 83–87.
8. Tu, J. P., Yang, Y Z., Wang, L. Y., Ma, X. C., and X. B. Zhang. 2001. Tribological properties of carbon-nanotube-reinforced copper composites. *Tribol. Lett.* 10(4): 225 228.
9. Chen, X. H., Li, W. H., Chen, C. S., Xu, L. S., Zhi, Y., and J. Hu. 2005. Preparation and properties of Cu matrix composite reinforced by carbon nanotubes. *Trans. Nonferrous Met. Soc. China.* 15(2): 314–318.
10. Kim, K. T., Cha, S. I., and S. H. Hong. 2007. Hardness and wear resistance of carbon nanotube reinforced Cu matrix nanocomposites. *Mater. Sci. Eng. A.* 449–451: 46–50.
11. Trinh, P. V., Trung, T. B., Thang, N. B., Thang, B. H., Tinh, T. X., Quang, L. D., Phuong, D. D., and P. N. Minh. 2010. Calculation of the friction coefficient of Cu matrix composite reinforced by carbon nanotubes. *Comput. Mater. Sci.* doi:10.1016/j.commatsci.2010.01.035.
12. Guiderdoni, C., Pavlenko, E., Turq, V., Weibel, A., Puech, P., Estournes, C., Peigney, A., Bacsa, W., Laurent, C. 2013. The preparation of carbon nanotube (CNT)/copper composites and the effect of the number of CNT walls on their hardness, friction and wear properties. *Carbon* 58: 185–197.
13. Rajkumar, K., Aravindan, S. 2011. Tribological studies on microwave sintered copper-carbon nanotube composites. *Wear* 270: 613–621.
14. Zhou, S. M., Zhang, X. B., Ding, Z. P., Min, C. Y., Xu, Z. L., and W. M. Zhu. 2007. Fabrication and tribological properties of carbon nanotubes reinforced name is correct. Al composites prepared by pressureless infiltration technique. *Composites A* 38: 301–306.
15. Kim, I. Y., Lee, J. H., Lee, G. S., Baik, S. H., Kim, Y. J., and Y. Z. Lee. 2009. Friction and wear characteristics of the carbon nanotube–aluminum composites with different manufacturing conditions. *Wear* 267: 593–598.
16. Bakshi, S. R., Lahiri, D., Patel, R. R., and A. Agarwal. 2010. Nanoscratch behavior of carbon nanotube reinforced aluminum coatings. *Thin Solid Films* 518: 1703–1711.
17. Bastwros, M. M. H., Esawi, A. M. K., Wifi, A. 2013. Friction and wear behavior of Al-CNT composites. *Wear* 307: 164–173.

18. Bakshi, S. R., Keshri, A. K., Agarwal, A. 2011. A comparison of mechanical and wear properties of plasma sprayed carbon nanotube reinforced aluminum composites at nano and microscale. *Materials Science and Engineering A* 528: 3375–3384.
19. Liu, B., Zeng, Z., and Y. Lin. 2009. Mechanical properties of hard Cr–MWNT composite coatings. *Surf. Coat. Technol.* 203: 3610–3613.
20. Che, J., Cagin, T., and W. A. Goddard. 2000. Thermal conductivity of carbon nanotubes. *Nanotechnology* 11: 65–69.
21. Berber, S., Kwon, Y. K., and D. Tomanek. 2000. Unusually high thermal conductivity of carbon nanotubes. *Phys. Rev. Lett.* 84(20): 4613–4616.
22. Y. Yoshida. 2000. High-temperature shrinkage of single-walled carbon nanotube bundles up to 1600 K. *J. Appl. Phys.* 87(7): 3338–3341.
23. Ferrer-Anglada, N., Gomis, V., El-Hachemi, Z., Weglikovska, U. D., Kaempgen, M., and S. Roth. 2006. Carbon nanotube based composites for electronic applications: cNT–conducting polymers, CNT–Cu. *Phys. Stat. Sol.* 203(6): 1082–1087.
24. Tang, Y., Cong, H., Zhong, R., and H. M. Cheng. 2004. Thermal expansion of a composite of single-walled carbon nanotubes and nanocrystalline aluminum. *Carbon* 42: 3260–3262.
25. Deng, C. F., Ma, Y. X., Zhang, P., Zhang, X. X., and D. Z. Wang. 2008. Thermal expansion behaviors of aluminum composite reinforced with carbon nanotubes. *Mater. Lett.* 62: 2301–2303.
26. Nai, S. M. L., Wei, J., and M. Gupta. 2006. Improving the performance of lead-free solder reinforced with multi-walled carbon nanotubes. *Mater. Sci. Eng. A* 423: 166–169.
27. Goh, C. S., Wei, J., Lee, L. C., and M. Gupta. 2006. Development of novel carbon nanotube reinforced magnesium nanocomposites using the powder metallurgy technique. *Nanotechnology.* 17: 7–12.
28. Thakur, S. K., Kwee, G. T., and M. Gupta. 2007. Development and characterization of magnesium composites containing nano-sized silicon carbide and carbon nanotubes as hybrid reinforcements. *J. Mater. Sci.* 42: 10040–10046.
29. Shi, X. L., Yang, H., Shao, G. Q., Duan, X. L., Yan, L., Xiong, Z., and P. Sun. 2007. Fabrication and properties of W–Cu alloy reinforced by multi-walled carbon nanotubes. *Mater. Sci. Eng. A* 457: 18–23.
30. Chu, K., Wu, Q., Jia, C., Liang, X., Nie, J., Tian, W., Gai, G., and H. Guo. 2010. Fabrication and effective thermal conductivity of multi-walled carbon nanotubes reinforced Cu matrix composites for heat sink applications. *Compos. Sci. Technol.* 70: 298–304.
31. Sivakumar, R., Guo, S., Nishimura, T., and Y. Kagawa. 2007. Thermal conductivity in multi-wall carbon nanotube/silica-based nanocomposites. *Scripta Mater.* 56: 265–268.
32. Nan, C-W., Shi, Z., and Y. Lin. 2003. A simple model of thermal conductivity of nanotube-based composites. *Chem. Phys. Lett.* 375: 666–669.
33. Nan, C-W., Liu, G., Lin, Y., and M. Li. 2004. Interface effect on thermal conductivity of carbon nanotube composites. *Appl. Phys. Lett.* 85: 3549–3551.
34. Xue, Q. Z. 2005. Model for thermal conductivity of carbon nanotube based composites. *Physica B* 368: 302–307.
35. Song, P. C., Liu, C. H., and S. S. Fan. 2006. Improving the thermal conductivity of nanocomposites by increasing the length efficiency of loading carbon nanotube. *Appl. Phys. Lett.* 88: 153111.
36. Bjorneklett, A., Haukeland, L., Wigren, J., and H. Kristiansen. 1994. Effective medium theory and the thermal conductivity of plasma-sprayed ceramic coatings. *J. Mater. Sci.* 29: 4043–4050.
37. Cernuschi, F., Ahmaniemi, S., Vuoristo, P., and T. Mantyla. 2004. Modelling of thermal conductivity of porous materials: application to thick thermal barrier coatings. *J. Eur. Cer. Soc.* 24: 2657–2667.

38. Landauer, R. 1952. The electrical resistance of binary metallic mixtures. *J. Appl. Phys.* 23: 779–784.
39. Ngo, Q., Cruden, B. A., Cassell, A. M., Walker, M. D., Ye, Q., Koehne, J. E., Meyyappan, M., Li, J., and C. Y. Yang. 2004. Thermal conductivity of carbon nanotube composite films. *Mat. Res. Soc. Symp. Proc.* 812: 3.18.1–3.18.6.
40. Chai, Y., Zhang, K., Zhang, M., Chan, P. C. H., and M. M. F. Yuen. 2007. *Carbon nanotube/copper composites for via filling and thermal management. 57th Electronic Components and Technology Conference Proceedings.* 1224–1229.
41. Uddi, S. M., Mahmud, T., Wolf, C., Glanz, C., Kolaric, I., Hulman, M., et al. 2009. Thermal expansion co-efficient of nanotube-metal composites. *Phys Status Solidi B* 246: 2836–2839.
42. Liu, Z. Y., Xiao, B. L., Wang, W. G., Ma, Z. Y. 2012. Elevated temperature tensile properties and thermal expansion of CNT/2009Al composites. *Compos Sci Technol* 72: 1826–1833.
43. Shin, S. E., Ko, Y. J., Bae, D. H. 2016. Mechanical and thermal properties of nanocarbon-reinforced aluminum matrix composites at elevated temperatures. *Compos Part B* 106: 66–73.
44. Sharma, M., Sharma, V. 2016. Chemical, mechanical, and thermal expansion properties of a carbon-nanotube-reinforced aluminum nanocomposite. *Int J Miner Metall Mater* 23: 222–233.
45. Yang, X., Zou, T., Shi, C., Liu, E., He, C., Zhao, N. 2016. Effect of carbon nanotube (CNT) content on the properties of in-situ synthesis CNT reinforced Al composites. *Mater Sci Eng A* 660: 1–8.
46. Trinh, P. V., Luan, N. V., Phuong, D. D., Minh, P. N., Weibel, A., Mesguich, D., Laurent, C. 2018. Microstructure, microhardness and thermal expansion of CNT/Al composites prepared by flake powder metallurgy. *Composites: Part A* 105: 126–137.
47. Jiang, H., Liu, B., Huang, Y., Hwang, K. C. 2004. Thermal expansion of single wall carbon nanotubes. *J. Eng. Mater-T.* 126: 265–270.
48. Zhou, W., Yamaguchi, T., Kikuchi, K., Nomura, N., Kawasaki, A. 2017. Effectively enhanced load transfer by interfacial reactions in multi-walled carbon nanotube reinforced Al matrix composites. *Acta Materialia* 125: 369–376.
49. Yu, M. F., Lourie, O., Dyer, M. J., Moloni, K., Kelly, T. F., Ruoff, R. S. 2000. Strength and breaking mechanism of multiwalled carbon nanotubes under tensile load. *Science* 287: 637–640.
50. Zhou, W., Bang, S., Kurita, H., Miyazaki, T., Fan, Y., Kawasaki, A. 2016. Interface and interfacial reactions in multi-walled carbon nanotube-reinforced aluminum matrix composites. *Carbon* 96: 919–928.
51. Liu, J., Fan, G., Tan, Z., Guo, Q., Su, Y., Li, Z., Xiong, D-B. 2019. Mechanical properties and failure mechanisms at high temperature in carbon nanotube reinforced copper matrix nanolaminated composite. *Composites Part A* 116: 54–64.
52. Liu, J., Xiong, D-B., Su, Y., Guo, Q., Li, Z., Zhang, D. 2019. Effect of thermal cycling on the mechanical properties of carbon nanotubes reinforced copper matrix nanolaminated composites. *Materials Science & Engineering A* 739: 132–139.
53. Huang, H., Fan, G., Tan, Z., Xiong, D-B., Guo, Q., Guo, C., Li, Z., Zhang, D. 2017. Superplastic behavior of carbon nanotube reinforced aluminum composites fabricated by flake powder metallurgy. *Materials Science & Engineering A* 699: 55–61.
54. Saito, R., Dresselhaus, G., and M. S. Dresselhaus. 2007. *Physical properties of carbon nanotubes*. London: Imperial College Press, pp. 137–143.
55. Yao, Y., Dai, X., Feng, C., Zhang, J., Liang, X., Ding, L., Choi, W., Choi, J., Kim, J. M., and Z. Liu. 2009. Crinkling ultralong carbon nanotubes into serpentines by a controlled landing process. *Adv. Mater.* 21: 4158–4162.

56. Liu, Y. M., Sung, Y., Chen, Y. C., Lin, C. T., Chou, Y. H., and M. D. Ger. 2007. A method to fabricate field emitters using electroless codeposited composite of MWNTs and nickel. *Electrochem. Sol. Stat. Lett.* 10(9): 101–104.
57. Deng, M., Ding, G., Wang, Y., Wu, H., Yao, Y., and L. Zhu. 2009. Fabrication of Ni-matrix carbon nanotube field emitters using composite electroplating and micromachining. *Carbon*, 47: 3466–3471.
58. Shu, J., Hong, L., Yang, R., Shi, Y., and X. Huang. 2005. Cage-like carbon nanotubes/Si composite as anode material for lithium ion batteries. *Electrochem. Comm.* 8: 51–54.
59. Eom, J. Y., Park, J. W., Kwon, H. S., and S. Rajendran. 2006. Electrochemical insertion of lithium into multiwalled carbon nanotube/silicon composites produced by ballmilling. *J. Electrochem. Soc.* 153(9): 1678–1684.
60. Kim, T., Mo, Y. H., Nahm, K. S., and S. M. Oh. 2006. Carbon nanotubes (CNTs) as a buffer layer in silicon/CNTs composite electrodes for lithium secondary batteries. *J. Power Sources* 162: 1275–1281.
61. Wang, W., and P. N. Kamta. 2007. Reversible high capacity nanocomposite anodes of Si/C/SWNTs for rechargeable Li-ion batteries. *J. Power Sources* 172: 650–658.
62. Shen, G. R. Cheng, Y. T. C., and L. N. Tsai. 2005. Synthesis and characterization of Ni–P-CNT's nanocomposite film for MEMS applications. *IEEE Trans. Nanotechnol.* 4(5): 539–547.
63. Chao, T. Y., Shen, G. R., and Y. T. Cheng. 2006. Comparative study of Ni–P-diamond and Ni–P-CNT nanocomposite films. *J. Electrochem. Soc.* 153(1): 98–104.
64. Feng, Y., Yuan, H. L., and M. Zhang. 2005. Fabrication and properties of silver-matrix composites reinforced by carbon nanotubes. *Mater. Character.* 55: 211–218.
65. Yang, Y. L., Wang, Y. D., Ren, Y., He, C. S., Deng, J. N., Nan, J., Chen, J. G., and L. Zuo. 2008. Single-walled carbon nanotube-reinforced copper composite coatings prepared by electrodeposition under ultrasonic field. *Mater. Lett.* 62: 47–50.
66. Liu, L., Bao, R., Yi, J., Li, C., Tao, J., Liu, Y., Tan, S., and You, X. 2017. Well-dispersion of CNTs and enhanced mechanical properties in CNTs/Cu-Ti composites fabricated by molecular level mixing. *Journal of Alloys and Compounds* 726: 81–87.
67. Singh, A., Prabhu, T. R., Sanjay, A. R., and Koti, V. 2017. An overview of processing and properties of Cu/CNT Composites. *Materials Today: Proceedings* 4: 3872–3881.
68. Wang, H., Zhang, Z-H., Hu, Z-Y., Song, Q., Yin, S-P., Kang, Z., and Li, S-L. 2018. Improvement of interfacial interaction and mechanical properties in copper matrix composites reinforced with copper coated carbon nanotubes. *Materials Science and Engineering A* 715: 163–173.
69. Xu, C. L., Wei, B. Q., Ma, R. Z., Liang, J., Ma, X. K., and D. H. Wu. 1999. Fabrication of aluminum-carbon nanotube composite and their electrical properties. *Carbon.* 37: 855–858.
70. Yang, Z., Xu, H., Shi, Y. L., Li, M. K., Huang, Y., and H. L. Li. 2005. The fabrication and corrosion behavior of electroless Ni–P-carbon nanotube composite coatings. *Mater. Res. Bull.* 40: 1001–1009.
71. Shi, Y. L., Yang, Z., Li, M. K., Xu, H., and H. L. Li. 2004. Electroplated synthesis of Ni–P–UFD, Ni–P-CNTs, and Ni–P–UFD–CNTs composite coatings as hydrogen evolution electrodes. *Mater. Chem. Phys.* 87: 154–161.
72. Chen, X. H., Chen, C. S., Xiao, H. N., Cheng, F. Q., Zhang, G., and G. J. Yi. 2005. Corrosion behavior of carbon nanotubes–Ni composite coating. *Surf. Coat. Technol.* 191: 351–356.
73. Guo, C., Zuo, Y., Zhao, X., Zhao, J., and J. Xiong. 2007. The effects of pulse–reverse parameters on the properties of Ni–carbon nanotubes composite coatings. *Surf. Coat. Technol.* 201: 9491–9496.

74. Wu, Y., Ren, R., Wang, F., Yuan, Z., Wang, T., and X. Hu. 2008. Preparation and characterization of Ni–Cu–P/CNTs quaternary electroless composite coating. *Mater. Res. Bull.* 43(12): 3425–3432.
75. Anand, E. E., and S. Natarajan. 2015. Effect of carbon nanotubes on corrosion and tribological properties of pulse-electrodeposited Co-W Composite coatings. *Journal of Materials Engineering and Performance* 24: 128–135.
76. Praveen, B. M., Venkatesha, T. V., Naik, Y. A., and K. Prashantha. 2007. Corrosion studies of carbon nanotubes–Zn composite coating. *Surf. Coat. Technol.* 201: 5836–5842.
77. Kumar, P. S. S. R., Smart, D. S. R., and S. J. Alexis. 2017. Corrosion behavior of aluminum metal matrix reinforced with multi-wall carbon nanotube. *Journal of Asian Ceramic Societies* 5: 71–75.
78. Chen, D., Chen, L., Liu, S., Ma, C. X., Chen, D. M., and L. B. Wang. 2004. Microstructure and hydrogen storage property of Mg/MWNTs composites. *J. Alloys Compd.* 372: 231–237.
79. Huang, Z. G., Gou, Z. P., Calka, A., Wexler, D., and H. K. Liu. 2007. Effects of carbon black, graphite and carbon nanotube additives on hydrogen storage properties of magnesium. *J. Alloys Compd.* 427: 94–100.
80. Schaller, R., Mari, D., Marques Dos Santos, S., Tkalcec, I., and E. Carreno-Morelli. 2009. Investigation of hydrogen storage in carbon nanotube-magnesium matrix composites. *Mater. Sci. Eng. A* 521–522: 147–150.
81. Ishihara, T., Nakasua, M., Yasuda, I., and H. Matsumoto. 2006. Preparation of Si-carbon nanotube composite by decomposition of tetramethylsilane (TMS) and its hydrogen storage property. *Sci. Tech. Adv. Mater.* 7: 667–671.
82. Hrapovic, S., Majid, E., Liu, Y., Male, K., and J. H. T. Luong. 2006. Metallic nanoparticle-carbon nanotube composites for electrochemical determination of explosive nitroaromatic compounds. *Anal. Chem.* 78(15): 5504–5512.
83. Valentini, F. Biagiotti, V., Lete, C., Palleschi, G., and J. Wang. 2007. The electro-chemical detection of ammonia in drinking water based on multi-walled carbon nanotube/copper nanoparticle composite paste electrodes. *Sensors and Actuators B* 128: 326–333.
84. Kang, X., Mai, Z., Zou, X., Cai, P., and J. Mo. 2007. A sensitive nanoenzymatic glucose sensor in alkaline madia with copper nanocluster/multimalled carbon nanotube-modifi glassy carbon electrode. *Anal. Biochem.* 363: 143–150.
85. Wang, J., Chen, G., Wang, M., and M. P. Chatrathi. 2004. Carbon-nanotube/copper composite electrodes for capillary electrophoresis microchip detection of carbohydrates. *The Analyst* 129: 512–515.
86. Hu, X., Wang, T., Qu, X., and S. Dong. 2005. In situ synthesis and characterization of multiwalled carbon nanotube/Au nanoparticle composite materials. *J. Phys. Chem. B* 110: 853–857.
87. Arribas, A. S., Bermejo, E., Chicharro, M., Zapardiel, A., Luque, G. L., Ferreyra, F., and G. A. Rivas. 2006. Analytical applications of a carbon nanotube composite modified with copper microparticles as detector in flow systems. *Anal. Chimica Acta.* 577: 183–189.
88. Luque, G. L., Ferreyra, N. F., and G. A. Rivas. 2007. Electrochemical sensor for amino acids and albumin based on composites containing carbon nanotubes and copper microparticles. *Talanta* 71: 1282–1287.
89. Huang, J., Xie, Q., Tan, Y., Fu, Y., Su, Z., Huang, Y., and S. Yao. 2009. Preparation of Pt/multiwalled carbon nanotubes modified Au electrodes via Pt–Cu co-electrodeposition/Cu stripping protocol for high-performance electrocatalytic oxidation of methanol. *Mater. Chem. Phys.* 118(2–3): 371–378.

9 Synergistic Interactions with Secondary Reinforcements

This chapter presents the use of secondary reinforcement in CNT-MMCs. The use of dual reinforcements has been explored in order to harness the properties of other reinforcement particles showing promise, as well as to compensate for weaknesses in the CNT-reinforcement approach. Secondary reinforcements in CNT-MMCs have mostly consisted of ceramic particles, particularly silicon carbide (SiC) and boron carbide (B_4C). A few others have explored the use of dual nanocarbon reinforcements, incorporating either nanodiamonds (ND) or graphene alongside CNT reinforcements. This chapter looks at the interactions of CNTs with these secondary reinforcements and how the dual-reinforcement approach affects the overall microstructural evolution, mechanical properties, and functional properties of the metallic composites.

9.1 CERAMIC PARTICLE-CNT DUAL REINFORCEMENT

Silicon carbide (SiC) has been the most widely use dual reinforcement for CNT-MMCs, in part due to its success and prevalence in the conventional MMC field, over other ceramic reinforcements. Most studies have focused on the use of nano SiC particles (nSiC), including the first study on dual reinforcement in CNT-MMCs by Kwon et al. in 2012 [1]. Kwon et al. [1] incorporated 10 vol.% nSiC (~20–30 nm) and 10 vol.% CNT (~20 nm diameter, ~10 µm length) into a commercially pure Al matrix using planetary ball milling under an Argon atmosphere. The major finding in that study was that nSiC promotes more homogenous distribution of CNTs. The Al-nSiC-CNT composite showed no signs of CNT agglomeration, while a Al-10 vol.% CNT composite contained agglomerated CNT clusters. The Al-nSiC-CNT composites exhibited greater hardness and finer crystallite sizes than both the pure Al and the Al-CNT composite. It should be noted that in this along with many of the early studies, a systematic effort to *isolate* the dual reinforcement effect and efficiency was not conducted. For example, the work by Kwon et al. [1] does not definitively address whether the addition of 10 vol.% nSiC is indeed more beneficial than adding an additional 10 vol.% of CNT. However, it was shown that the beneficial effect of adding nSiC saturated at about 5 vol.%, as shown in the hardness values presented in Table 9.1. It should also be noted that while CNT dispersion appears to improve, the elastic modulus values measured still fell well short of the values predicted by the Rule of Mixtures [5] for both Al-nSiC-CNT composites with 5 vol.% and 10 vol.% nSiC. In subsequent studies, Kwon et al. [6] showed that the effectiveness of nSiC may saturate at perhaps as low as 1 vol.%. A Al-nSiC-CNT composite

TABLE 9.1
Dually Reinforced Composites with CNT and Ceramic Reinforcements Listing Processing Method, Amount of Primary and Secondary Reinforcements and Enhancement Relative to Base Unreinforced or Single Reinforcement CNT-MMC

Hybrid Composite Composition	CNT Content	Ceramic Content	Processing Technique	Property Evaluated	Baseline Composite (Property) for Comparison with Hybrid Composite	% Change Relative to Baseline (Hybrid Composite Property)	Comments	References
Al-CNT-nSiC	10 vol.%	5 vol.% nSiC	Planetary Ball Milling, Sintering at 550 °C, Uniaxial Pressing	Crystallite Size	Al-10 vol.% CNT–1 vol.% SiC (60.9 nm)	– 26% (45.3 nm)	Comparison with an equivalent amount of CNT not conducted	[1]
Al-CNT-nSiC	10 vol.%	5 vol.% nSiC	Planetary Ball Milling, Sintering at 550 °C, Uniaxial Pressing	Vickers Hardness	Al-10 vol.% CNT–1 vol.% SiC (257 HV)	+ 12% (289 HV)	Effect of SiC addition appears to saturate at ~ 5 vol.%	[1]
Al-CNT-nSiC	10 vol.%	10 vol.% nSiC	Planetary Ball Milling, Sintering at 550 °C, Uniaxial Pressing	Crystallite Size	Al-10 vol.% CNT (51.4 nm)	– 17% (42.9 nm)	Pure Al had crystallite size of 183.4 nm	[1]
Al-CNT-nSiC	10 vol.%	10 vol.% nSiC	Planetary Ball Milling, Sintering at 550 °C, Uniaxial Pressing	Vickers Hardness	Al-10 vol.% CNT (221 HV)	+53% (338 HV)	Pure Al had hardness of 42 HV	[1]
Al-CNT-nSiC	10 vol.%	10 vol.% nSiC	Planetary Ball Milling, Sintering at 550 °C, Uniaxial Pressing	Elastic Modulus	Al-10 vol.% CNT–5 vol.% SiC (100.65 GPa)	22% (122.99 GPa)	Pure Al had Elastic Modulus of 2.70 GPa	[1]
Al-CNT-nSiC	10 vol.%	30 vol.% nSiC	Planetary Ball Milling, Spark Plasma Sintering	Vickers Hardness	Al-10 vol.% CNT (102 HV)	+ 51% (154 HV)	Comparison with an equivalent amount of CNT not conducted	[2]

Cu-CNT-nSiC	4 vol.%	2 vol.% nSiC	Planetary Ball Milling, Hot Pressing	Vickers Hardness	Cu-2 vol.% CNT–2 vol.% SiC (92.3 HV)	+ 15%(105.9 HV)	Pure Cu had hardness of 63 HV	[3]
Cu-CNT-nSiC	4 vol.%	2 vol.% nSiC	Planetary Ball Milling, Hot Pressing	Elastic Modulus	Cu-2 vol.% CNT–2 vol.% SiC (125.5 GPa)	+ 11% (139.3 GPa)	Pure Cu had elastic modulus of 105.7 GPa	[3]
Cu-CNT-nSiC	4 vol.%	2 vol.% nSiC	Planetary Ball Milling, Hot Pressing	Yield Strength (Compression)	Cu-2 vol.% CNT–2 vol.% SiC (257.7 MPa)	+ 11% (285.2 MPa)	Pure Cu had compressive yield strength of 129.6 MPa	[3]
Cu-CNT-nSiC	6 vol.%	2 vol.% nSiC	Planetary Ball Milling, Hot Pressing	Vickers Hardness	Cu-2 vol.% CNT–2 vol.% SiC (92.3 HV)	+ 18% (108.5 HV)	Comparison with an equivalent amount of CNT not conducted	[3]
Cu-CNT-nSiC	6 vol.%	2 vol.% nSiC	Planetary Ball Milling, Hot Pressing	Elastic Modulus	Cu-2 vol.% CNT–2 vol.% SiC (125.5 GPa)	+ 12% (140.2 GPa)	Effect of initial 2 vol.% CNT cannot be discerned	[3]
Cu-CNT-nSiC	6 vol.%	2 vol.% nSiC	Planetary Ball Milling, Hot Pressing	Yield Strength (Compression)	Cu-2 vol.% CNT–2 vol.% SiC (257.7 MPa)	+ 8% (279.6 MPa)	Addition of CNT beyond 4 vol.% appears to be detrimental	[3]
Al-CNT-nSiC	0.25 vol.%	0.25 vol.% nSiC	Planetary Ball Milling, Spark Plasma Sintering, Hot Extrusion	Grain Size	Al-0.5 vol.% CNT (3.95 μm)	– 15% (3.34 μm)	Equivalent amounts of total reinforcement used	[4]
Al-CNT-nSiC	0.25 vol.%	0.25 vol.% nSiC	Planetary Ball Milling, Spark Plasma Sintering, Hot Extrusion	Tensile Strength	Al-0.5 vol.% CNT (225 MPa)	– 28% (162 MPa)	Equivalent amounts of total reinforcement used	[4]

(*Continued*)

TABLE 9.1 (*Continued*)

Dually Reinforced Composites with CNT and Ceramic Reinforcements Listing Processing Method, Amount of Primary and Secondary Reinforcements and Enhancement Relative to Base Unreinforced or Single Reinforcement CNT-MMC

Hybrid Composite Composition	CNT Content	Ceramic Content	Processing Technique	Property Evaluated	Baseline Composite (Property) for Comparison with Hybrid Composite	% Change Relative to Baseline (Hybrid Composite Property)	Comments	References
Al-CNT-nSiC	0.25 vol.%	0.25 vol.% nSiC	Planetary Ball Milling, Spark Plasma Sintering, Hot Extrusion	Elongation	Al-0.5 vol.% CNT (11.5%)	+ 137% (27.3%)	Equivalent amounts of total reinforcement used	[4]
Al-CNT-nSiC	0.50 vol.%	0.50 vol.% nSiC	Planetary Ball Milling, Spark Plasma Sintering, Hot Extrusion	Grain Size	Al-1 vol.% CNT (1.86 μm)	– 24% (1.42 μm)	This hybrid composition yielded greatest reduction in grain size relative to CNT-only counterpart	[4]
Al-CNT-nSiC	0.50 vol.%	0.50 vol.% nSiC	Planetary Ball Milling, Spark Plasma Sintering, Hot Extrusion	Tensile Strength	Al-1 vol.% CNT (217 MPa)	+ 14% (247 MPa)	This was only hybrid composition that yielded high tensile strength than CNT-only counterpart	[4]
Al-CNT-nSiC	0.50 vol.%	0.50 vol.% nSiC	Planetary Ball Milling, Spark Plasma Sintering, Hot Extrusion	Elongation	Al-1 vol.% CNT (7.2%)	+ 188% (20.7%)	This hybrid composition yielded greatest increase in elongation relative to CNT-only counterpart	[4]

Al-CNT-nSiC	0.75 vol.%	0.75 vol.% nSiC	Planetary Ball Milling, Spark Plasma Sintering, Hot Extrusion	Grain Size	Al-1.5 vol.% CNT (1.54 μm)	– 10% (1.38 μm)	Pure Al had grain size of 6.29 μm	[4]
Al-CNT-nSiC	0.75 vol.%	0.75 vol.% nSiC	Planetary Ball Milling, Spark Plasma Sintering, Hot Extrusion	Tensile Strength	Al-1.5 vol.% CNT (155 MPa)	+ 3% (160 MPa)	Pure Al had tensile strength of 127 MPa	[4]
Al-CNT-nSiC	0.75 vol.%	0.75 vol.% nSiC	Planetary Ball Milling, Spark Plasma Sintering, Hot Extrusion	Elongation	Al-1.5 vol.% CNT (24.0%)	– 11% (21.3%)	Pure Al had elongation of 29.8%	[4]

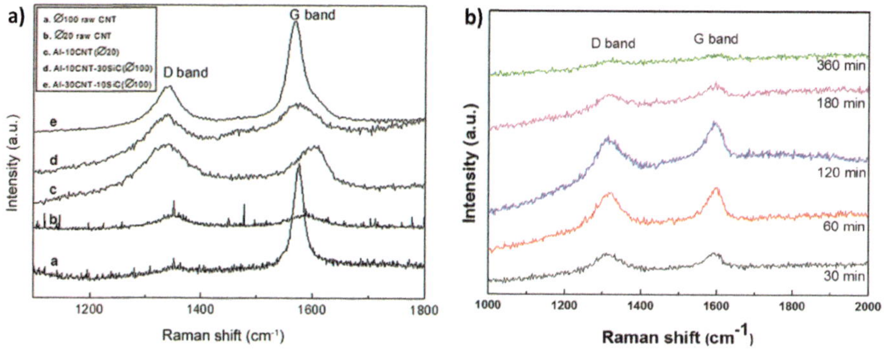

FIGURE 9.1 Raman spectra of Al-CNT-nSiC composites, (a) different fractions of nSiC to CNT [2], (b) effect of milling times with 1 vol.% nSiC and 6 wt.% CNT [7].

with 1 vol.% nSiC and 5 vol.% CNT exhibited higher hardness than composite with 10 vol.% of nSiC and 10 vol.% CNT.

Kwon et al. [2,7] performed Raman spectroscopy measurements (Figure 9.1) that showed that using too high a ratio of nSiC to CNT appeared to be detrimental to maintaining the CNT structure. The characteristic CNT peaks in the Raman spectrum are hardly visible in the Al-10 CNT–30 nSiC material, in comparison to the Al-30 CNT–10 nSiC material. The former sample did exhibit a higher hardness value than the sample with greater CNT content. However, hardness is not an indicator of overall performance and most of the composites produced by Kwon et al. [1,2,6,7] exhibited formation of Al_4C_3, which may be contributing to the higher hardness values. The effect of extended milling times was also investigated [2]. It was observed that milling for over 30 min was likely too aggressive a processing route, even with only 1 vol.% nSiC, as CNTs could not be found when milling times exceeded 30 min. The hardness and bending strength of the consolidated composites increased with increasing powder milling time; however, this may be due to the higher density in the samples milled for longer duration and again to some extent due to increased carbide formation.

Cu-nSiC-CNT composites were investigated by Akbarpour et al. [3] using 2 vol.% SiC (~40 nm) and 2–6 vol.% CNT (~60 nm diameter and ~3–4 μm length) in commercially pure Cu matrix. The use of nSiC led to better dispersion of 2 vol.% CNT (as compared to Al-2 vol.% CNT); however, some CNT clustering is initiated at 4–6 vol.% CNT. Compression tests were performed and showed that dual reinforced composites had improved elastic modulus and yield strength as compared to pure Cu material, as shown in Table 9.1. Furthermore, elastic modulus values matched up well with Rule of Mixtures calculations for Cu-2SiC–2CNT and Cu-2SiC–4CNT materials.

Overall, the works by Akbarour et al. [3] and Kwon et al. [1,2,6,7] showed the nSiC had a synergistic effect with CNT in that it acted as both an additional reinforcement as well as a mixing agent. The small size of SiC enables them to serve as nano milling media that can penetrate the small space between agglomerated CNT bundles and provide shearing forces to break apart the bundles/clusters. As with other milling media, excessive nSiC content appears to be detrimental due to excessive damage to the CNTs size and structure.

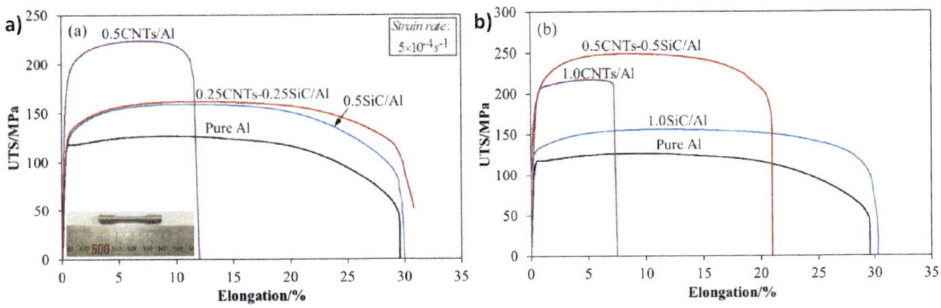

FIGURE 9.2 Tensile test curves for Al-CNT-SiC composites with (a) 0.5 vol.% total reinforcements, and (b) 1.0 vol.% total reinforcement [4]. (Reproduced with permission from Elsevier).

Zhang et al. [4] conduct a systematic study comparing CNT, SiC, and dual reinforcements in Aluminum by keeping the total volume fraction of reinforcement constant, at either 0.5, 1.0, or 1.5 vol.%. This allows for systematic comparisons, for example, for 1.0 vol.%, samples were made with 1 vol.% CNT, 1 vol.% SiC, and another with dual reinforcements set at 0.5 vol.% CNT + 0.5 vol.% SiC. This allows for a systematic comparison between the individual effects of the reinforcements, and whether combining reinforcements yields any improvement under the same reinforcement loadings. Tensile test data for specimens reinforced with 0.5 and 1.0 vol.% (total reinforcement) is shown in Figure 9.2. The Al-0.5 CNT + 0.5 SiC sample exhibited a 94% increase in tensile strength over pure aluminum, which was a 14% improvement over Al-1 CNT and 56% improvement compared to Al-1 SiC. These improvements over samples with equivalent reinforcement loading demonstrate a tangible synergy effect, which prior studies had failed to show in a systematic way. Furthermore, Zhang et al. [4] show that the synergistic effect is related to two effects associated with the modified interface of the Al-CNT/SiC composites. The first effect is an increased pull-out force needed to debond the SiC decorated CNTs due to increased roughness and friction associated with the presence of SiC particles. The second effect is associated with modified thermodynamics of Al_4C_3 formation, considering the following reactions:

(1) $4Al_{(s)} + 3C_{(s)} = Al_4C_{3(s)}$;
$\Delta G^{\theta}_{r} = -28.93\ln T + 1.121 \times 10^{2}T^{2} + 1.155 \times 10^{6}T^{-1} + 5.153 \times 10^{-6}T^{3} + 131.33T - 222{,}379.8$

(2) $4Al_{(s)} + 3SiC_{(s)} = Al_4C_{3(s)} + 3Si_{(s)}$
$\Delta G^{\theta}_{r} = -2650 + 25.7767T$

Calculating the Gibbs free energy change in the range of 25–660°C shows the Al_4C_3 formation is thermodynamically favorable for the reaction between aluminum and carbon, but not for the reaction between aluminum and silicon carbide. Hence, the SiC particles serve to inhibit the formation of Al_4C_3 by effectively blocking contact of CNT with Al and not forming any aluminum carbide phases itself. Formation of Al_4C_3 has been a common challenge in CNT-MMC field due to the brittle nature of this interfacial product. This effect was verified using XRD, where Al_4C_3 peaks were larger in Al-0.5 CNT than in Al-0.5 CNT + 0.5 SiC, and non-existent in Al-0.5 SiC. The reduction in Al_4C_3 formation may also impact the electrical conductivity. Electrical conductivity was highest in the Al-SiC samples at all reinforcement volume loadings, followed by Al-CNT-SiC, and lastly Al-CNT sample. This needs further investigation, as electrical conductivity was highest in the pure Al sample, and past studies have shown CNT to effectively enhance electrical conductivity.

Li et al. [8,9] take a different approach to SiC-CNT dual reinforcement, in which CNTs are directly grown via CVD onto micro SiC particles ranging in size from 2.5 to 13 μm. Micron sized particles are typically much easier to disperse into a metal matrix than nanoparticles, and as such the SiC particles are used as a carrier or vehicle to uniformly disperse the CNTs. The composite powders are processed by first mixing the SiC particles in $Ni(NO_3)_2 \cdot 6H_2O$ at a Ni:SiC ratio of 5:95. NaOH is then added to obtain a binary colloid of $Ni(OH)_2$/SiC. This colloid is then filtered, dried, and calcined at 350 °C for 4 h to form NiO covered SiC particles. The particles are then placed in a furnace where NiO is reduced and subsequent growth of CNTs occurs at 700 °C by introducing a flow of $CH_4/H_2/Ar$. This technique provides a uniform distribution of Ni catalyst particles on the SiC surface (Figure 9.3a) that results

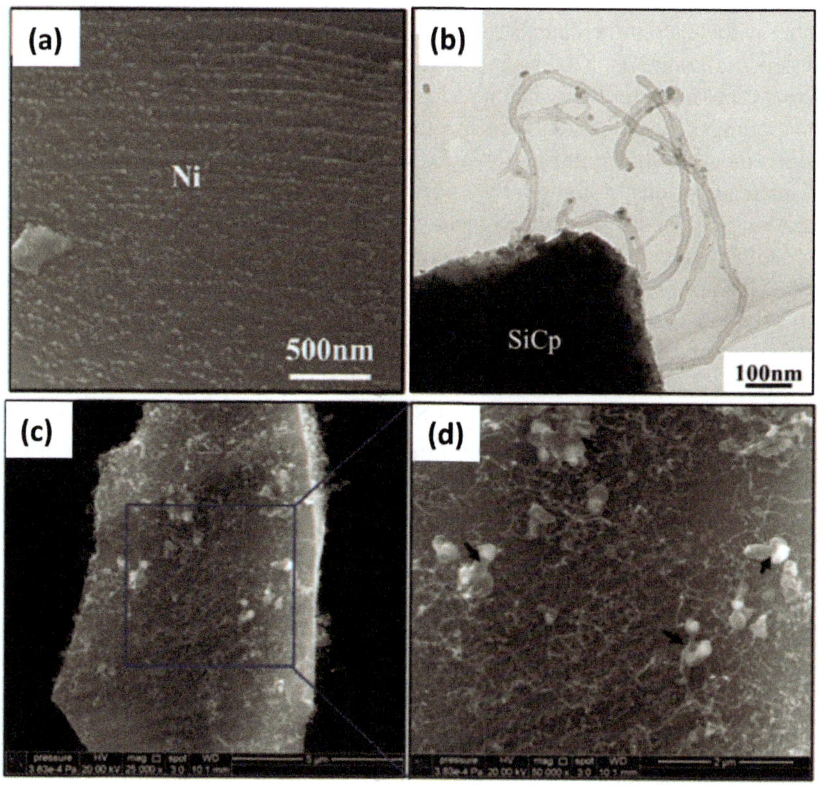

FIGURE 9.3 (a) Ni catalyst uniformly distributed on SiC particle [8], (b) TEM of CNTs grown on SiC [9], (c) SEM showing many CNTs grown on SiC surface [9], and (d) higher magnification SEM of CNT covered SiC particle surface [9]. (Reproduced with permission from Elsevier).

in CNT uniformly covered SiC particles (Figure 9.3b–d). The hybrid SiC/CNT particles are then mixed with either commercially pure Al [8] or 6061Al [9] by a ball milling route. The Al-SiC/CNT hybrid composite shows improved elastic modulus and tensile strength over both unreinforced Al as well as a conventionally prepared (ball milling alone) Al-SiC-CNT composite. The composites with the in situ grown CNTs exhibits better ductility as compared to the composites with standalone SiC and CNT reinforcements, as shown in Figure 9.4. Figure 9.5 shows fracture surfaces

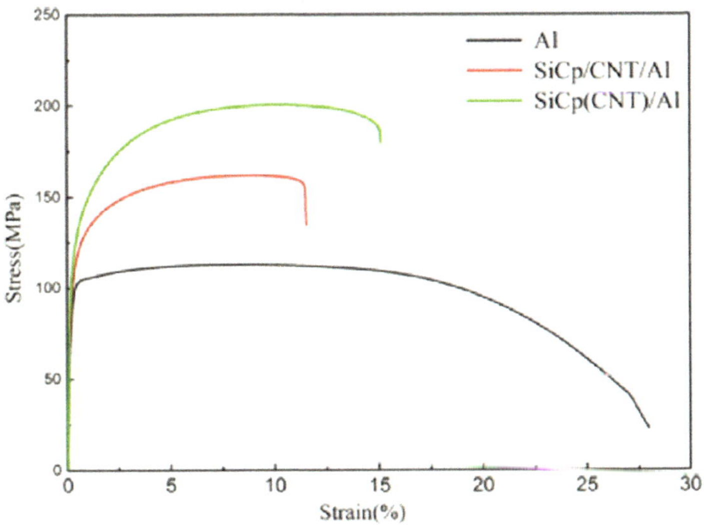

FIGURE 9.4 Tensile test curves for Al, Al-SiC-CNT conventional composite, and Al-CNT-SiC using in situ grown CNT on SiC particles [8].(Reproduced with permission from Elsevier)

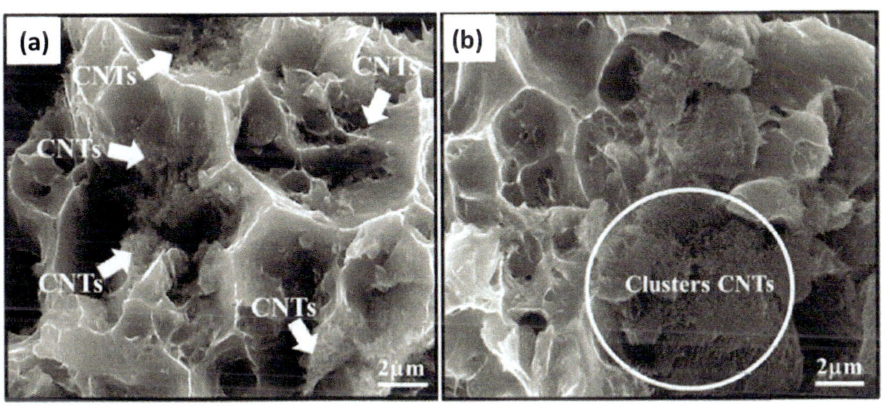

FIGURE 9.5 Fracture surfaces of (a) Al-SiC-CNT composite with in situ grown CNT on SiC, showing CNTs embedded in dimple-like features, and (b) conventionally mixed Al-SiC-CNT showing CNT clusters, that can serve as initiation sites for crack growth and propagation [8]. (Reproduced with permission from Elsevier).

of the composites, CNTs can be seen uniformly embedded in dimple-like features in the hybrid composite, while the conventionally mixed composite shows CNT clusters. These clusters are well known to serve as sites for crack initiation and propagation. Shiheng et al. [8,9] also showed that smallest SiC particles (2.5 μm) were not as effective carriers of CNT, as CNT clusters were observed in the powder, this led to clusters in the consolidated samples, as well as poorer densification and mechanical properties. The smaller particles were observed to cluster together prior to CNT growth, making it difficult for CNT to grow homogenously. Once grown, the CNT length is also close to that of 2.5 μm SiC particles, and hence entanglement of CNTs on neighboring SiC particles also occurs.

The tribological behavior of dual reinforced CNT-SiC MMCs has been investigated by Padmavathi et al. [10] and Hekner et al. [11]. Padmavathi et al. [10] investigate the effect of adding 0.5 and 1.0 vol.% CNT to Al-15 vol.% SiC composites fabricated by stir casting. Both loadings of CNT decrease the coefficient of friction during pin-on-disc sliding wear tests. Notably, CNTs lower the CoF most under lower applied normal loadings. Hekner et al. [11] studied the effect of 1 wt.% CNT on Al-15 wt.% nanoSiC composites fabricated by mechanical alloying and hot pressing. Addition of CNT led to higher CoF at both room temperature (RT) and high temperature (450 °C.), while wear rate was unaffected at RT and decreased at 450 °C.

A few studies have also investigated B_4C-CNT dual reinforced MMCs [12–15]. However, these studies have not systematically investigated the benefits of adding one reinforcement particle over the other, or any possible synergistic behavior. These studies generally show that adding more B_4C content is favorable for mechanical performance. However, CNTs are not dispersed well in these studies, especially compared to the more easily dispersed micron-size B_4C used in these studies.

9.2 NANOCARBON-CNT DUAL REINFORCEMENT

The exploration of other forms of nanocarbons as dual reinforcements is in its nascent stages. One study has explored the use of nanodiamonds [16] as dual reinforcements, while most studies utilize graphene as the nanocarbon of choice [17–19]. Some of these studies, including the lone work utilizing nanodiamond (ND), have the pitfall of using different amounts of reinforcement in the dually reinforced material, thereby making comparisons with the single reinforcement materials difficult or impossible. For example, Kwon et al. [16] investigate aluminum composites with 1 vol.% CNT, 1 vol.% ND, another with 1 vol.% CNT + 1 vol.% ND. It is found that the dually reinforced composites exhibit the highest values of hardness and flexural strength, up to 20% and 45% higher, respectively, than the samples with single nanocarbon reinforcement. However, any information on possible synergistic effects or the value of adding different reinforcements cannot be accessed since the composites with dual reinforcements have doubled the volume fraction of nanocarbon reinforcement (2 vol.% total), than the composites with only 1 vol.%. The study does not compare how the composites would behave with either 2 vol.% ND only, or 2 vol.% CNT only, which would be necessary to access the value of adding two different types of reinforcement.

The first study to utilize GNP and CNT as dual reinforcements in MMCs was conducted by Rashad et al. [18] on Magnesium MMCs processed by compaction,

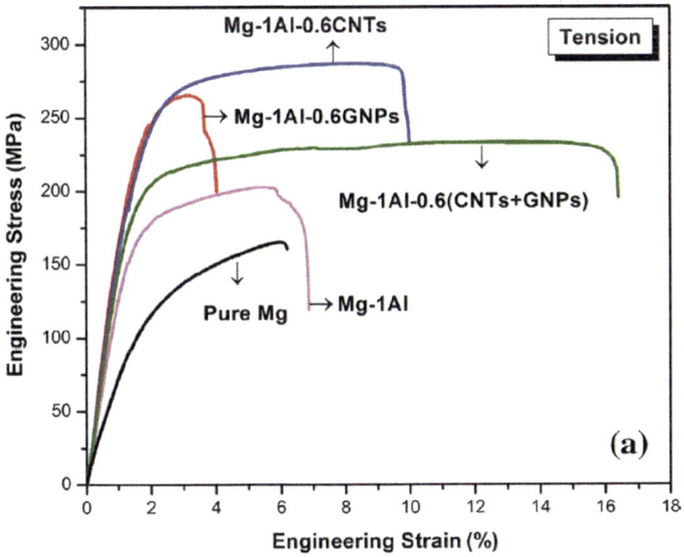

FIGURE 9.6 Tensile test curves for Pure Mg, Mg-1Al, and composites utilizing CNT, GNP, or both. Hybrid composite with dual reinforcement of CNT and GNP exhibits the greatest failure strain, and greater strength than unreinforced materials [18]. (Reproduced with permission from Elsevier).

sintering, and extrusion. The composite powder processing was done through a wet chemistry route, where individual solutions of GNP and CNT in ethanol were ultrasonicated and then mixed together, followed by introduction of Mg matrix. Rashad et al. [18] looked at composite with a total carbon reinforcement of 0.60 wt.%, and the dual reinforcement composite has a CNT to GNP ratio of 1:5. The lower amount of CNT was justified by the fact that it was desired to intercalate just a small amount of CNTs between the graphene sheets. All composites exhibited improved properties in both tension and compression, over pure magnesium or Mg-1Al. 1 wt.%, as seen in Figure 9.6 and Table 9.2. Al was added to all samples (and one control sample) to promote a better interface between Mg and carbon reinforcements. The composite with dual carbon reinforcement exhibited the highest failure strains. The composites with only one reinforcement (CNT or GNP) exhibited higher hardness, elastic modulus, yield strength, and ultimate tensile strength. Authors propose CNT intercalation alleviates GNP agglomeration that results in improved failure strains. The reasons for the hybrid composite attaining poorer strength than singly reinforced composites are not clear in this study.

Li et al. [19] investigated the role of total carbon content and CNT:graphene ratios in aluminum MMCs with dual reinforcement. Composite powder processing is based on earlier works that show graphene-CNT mixtures form spontaneous networks due to strong π–π interactions [20]. In this study, water was used as the medium for forming individual reduced graphene oxide (rGO) and CNT solutions that are then mixed together. The CNT-rGO suspensions are seen to remain stable (no agglomeration) for at least one month. The hybrids suspensions are then introduced to a solution with

TABLE 9.2
Dually reinforced composites with CNT and graphene listing processing method, amount of primary and secondary reinforcements and enhancement relative to base unreinforced or single reinforcement CNT-MMC

Hybrid Composite Composition	CNT Content	Graphene Content	Processing Technique	Property Evaluated	Baseline Composite (Property) for Comparison with Hybrid Composite	% Change Relative to Baseline (Hybrid Composite Property)	Comments	References
Mg-1 Al-CNT-GNP	0.1 wt.%	0.5 wt.% GNP	Wet Chemistry, Sintering, Extrusion	Elastic Modulus	Mg-1 wt.% Al-0.6 wt.% CNT (15.7 GPa)	– 4% (15 GPa)	Unreinforced Mg-1 Al has elastic modulus of 12.8 GPa	[18]
Mg-1 Al-CNT-GNP	0.1 wt.%	0.5 wt.% GNP	Wet Chemistry, Sintering, Extrusion	Tensile Yield Strength	Mg-1 wt.% Al-0.6 wt.% CNT (210 MPa)	– 12% (185 MPa)	Unreinforced Mg-1 Al has tensile yield strength of 155 MPa	[18]
Mg-1 Al-CNT-GNP	0.1 wt.%	0.5 wt.% GNP	Wet Chemistry, Sintering, Extrusion	Ultimate Tensile Strength	Mg-1 wt.% Al-0.6 wt.% CNT (287 MPa)	– 18% (234 MPa)	Unreinforced Mg-1 Al has ultimate tensile strength of 202 MPa	[18]
Mg-1 Al-CNT-GNP	0.1 wt.%	0.5 wt.% GNP	Wet Chemistry, Sintering, Extrusion	Strain to Failure	Mg-1 wt.% Al-0.6 wt.% CNT (10%)	+ 64% (16.4%)	Unreinforced Mg-1 Al has strain to failure of 6.9%	[18]
Mg-1 Al-CNT-GNP	0.1 wt.%	0.5 wt.% GNP	Wet Chemistry, Sintering, Extrusion	Vickers Hardness	Mg-1 wt.% Al-0.6 wt.% CNT (61 HV)	– 8% (56 HV)	Unreinforced Mg-1 Al has hardness of 50 HV	[18]
Al-CNT-rGO	0.43 vol.%	0.07 vol.% rGO	Flake Powder Metallurgy, Hot Pressing, Hot Extrusion	Elastic Modulus	Al-0.5 vol.% CNT (75.1 GPa)	+ 2% (76.5 GPa)	Equivalent amounts of total reinforcement used	[19]
Al-CNT-rGO	0.43 vol.%	0.07 vol.% rGO	Flake Powder Metallurgy, Hot Pressing, Hot Extrusion	Yield Strength	Al-0.5 vol.% CNT (220 GPa)	+ 23% (271 MPa)	Equivalent amounts of total reinforcement used	[19]
Al-CNT-rGO	0.43 vol.%	0.07 vol.% rGO	Flake Powder Metallurgy, Hot Pressing, Hot Extrusion	Tensile Strength	Al-0.5 vol.% CNT (275 GPa)	+ 16% (320 MPa)	Equivalent amounts of total reinforcement used	[19]

Al-CNT-rGO	0.43 vol.%	0.07 vol.% rGO	Flake Powder Metallurgy, Hot Pressing, Hot Extrusion	Ultimate Strain	Al-0.5 vol.% CNT (14.8%)	− 12% (13.0%)	Equivalent amounts of total reinforcement used	[19]
Al-CNT-rGO	1.29 vol.%	0.21 vol.% rGO	Flake Powder Metallurgy, Hot Pressing, Hot Extrusion	Elastic Modulus	Al-1.5 vol.% CNT (80.4 GPa)	+ 3% (83.0 GPa)	This hybrid composition has greatest increase in elastic modulus relative to CNT-only counterpart	[19]
Al-CNT-rGO	1.29 vol.%	0.21 vol.% rGO	Flake Powder Metallurgy, Hot Pressing, Hot Extrusion	Yield Strength	Al-1.5 vol.% CNT (260 MPa)	+ 38% (358 MPa)	This hybrid composition has greatest increase in yield strength relative to CNT-only counterpart	[19]
Al-CNT-rGO	1.29 vol.%	0.21 vol.% rGO	Flake Powder Metallurgy, Hot Pressing, Hot Extrusion	Tensile Strength	Al-1 5 vol.% CNT (326 MPa)	+ 27% (415 MPa)	This hybrid composition has greatest increase in tensile strength relative to CNT-only counterpart	[19]
Al-CNT-rGO	1.29 vol.%	0.21 vol.% rGO	Flake Powder Metallurgy, Hot Pressing, Hot Extrusion	Ultimate Strain	Al-1.5 vol.% CNT (12.6%)	− 35% (8.2%)	This hybrid composition has greatest decrease in failure strain relative to CNT-only counterpart	[19]
Al-CNT-rGO	2.57 vol.%	0.43 vol.% rGO	Flake Powder Metallurgy, Hot Pressing, Hot Extrusion	Elastic Modulus	Al-3.0 vol.% CNT (87.5 GPa)	− 4% (84.1 GPa)	Pure Al has elastic modulus of 70.5 GPa	[19]
Al-CNT-rGO	2.57 vol.%	0.43 vol.% rGO	Flake Powder Metallurgy, Hot Pressing, Hot Extrusion	Yield Strength	Al-3.0 vol.% CNT (326 MPa)	+24% (405 MPa)	Pure Al has yield strength of 205 MPa	[19]
Al-CNT-rGO	2.57 vol.%	0.43 vol.% rGO	Flake Powder Metallurgy, Hot Pressing, Hot Extrusion	Tensile Strength	Al-3.0 vol.% CNT (370 MPa)	+ 16% (430 MPa)	Pure Al has tensile strength of 250 MPa	[19]
Al-CNT-rGO	2.57 vol.%	0.43 vol.% rGO	Flake Powder Metallurgy, Hot Pressing, Hot Extrusion	Ultimate Strain	Al-3.0 vol.% CNT (4.6%)	− 30% (3.2%)	Pure Al has ultimate strain of 17.3%	[19]

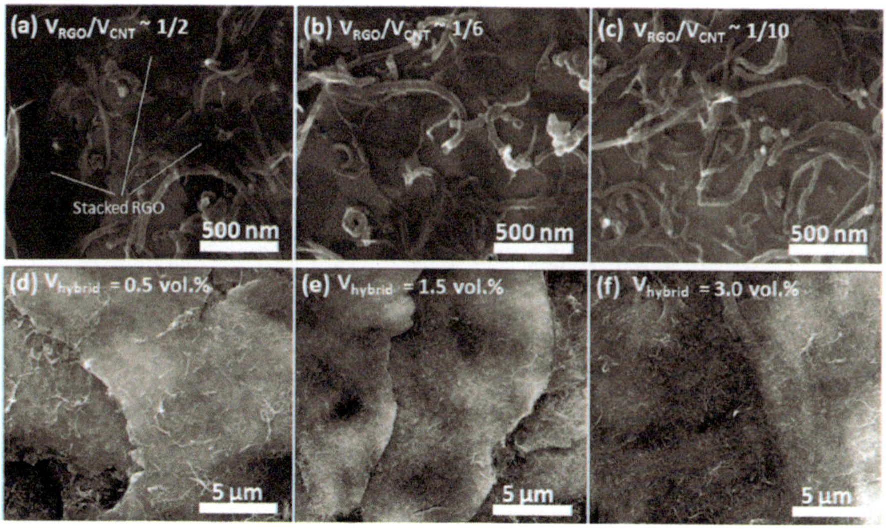

FIGURE 9.7 (a)–(c) rGO/CNT-Al composites with 1.5 vol.% total carbon, and variable ratios of rGO volume (V_{RGO}) to CNT volume (V_{CNT}), (d)–(f) rGO/CNT-Al composites with V_{CNT}/V_{rGO} ratio of 6, and variable amounts of total carbon [19]. (Reproduced with permission from Elsevier).

aluminum flakes. Composite powders are consolidated by compaction, hot pressing, and hot extrusion. Micrographs of composite hybrid powders with different CNT:graphene ratios, as well as total carbon contents ranging from 0.5 to 3.0 vol.%, are shown in Figure 9.7. The intercalation of CNTs prevents the stacking of rGO sheets seen in samples with only rGO. The hybrid composites also appear to form a more continuous network, as compared to that seen with only CNT reinforcement, where many CNTs remain isolated. This continuous CNT-rGO network promotes adhesion with the high aspect ratio Al flakes. It should be noted that in this study, the CNT:graphene ratios (V_{CNT}/V_{Gr} ~ 2–10) are kept much higher than in the study by Rashad et al. [18]. Li et al. [19] take into account the differences in specific surface area of each nanoreinforcement, ~2000 $m^2\ g^{-1}$ for rGO and ~<100 $m^2\ g^{-1}$ for CNT. They surmise that in order for there to be sufficient CNT coverage in between rGO layers, the total volume fraction of CNT must be higher.

The dual reinforcement approach taken by Li et al. [19] yields higher strength and modulus than the singly reinforced samples but also suffers lower fracture strains, as shown in Figure 9.8. This is opposite to the trend observed by Rashad et al. [18]. It may be that higher GNP content (as in Rashad et al. [18]) promotes greater ductility due to the greater area available for interlayer sliding, while higher CNT content (as in Li et al. [19]) promotes the formation of continuous reinforcement networks. Li et al. [19] propose that π–π interactions in these networked structures promote load transfer over a greater area, thereby expanding the deformation zones and the load transfer efficacy. It should be noted that the overall reinforcement efficiency in Li et

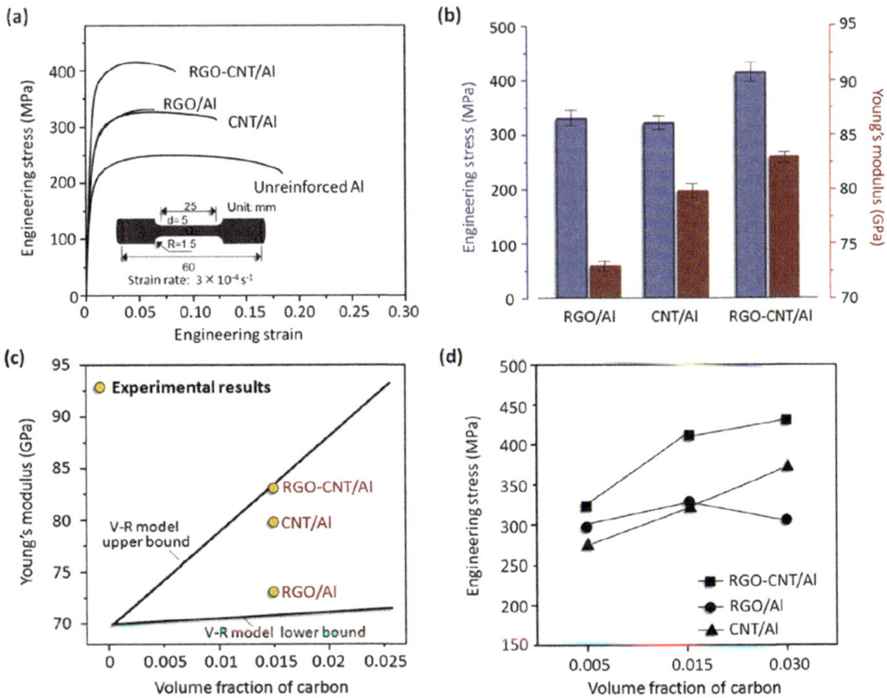

FIGURE 9.8 (a) Tensile test curves of singly reinforced and dually reinforced rGO/CNT-Al composites, (b) engineering stress and Young's modulus values for singly reinforced and dually reinforced rGO/CNT-Al composites, (c) comparison of Young's modulus values with those predicated from the Voigt-Reuss model, and (d) comparison of composite strength as the total carbon reinforcement content is increased [19].

al. [19] is impressive, Young's modulus values are seen to approach the upper bound predicted by the Voigt-Reuss model (see Section 6.1.1) (Figure 9.8). Also, unlike rGO-Al samples, Young's modulus increases at volume fractions as high as 3 vol.%. The composite strengthening performance is also seen to adhere well to that predicted by the Kelly and Tyson (see Section 6.2.1) modified shear lag model for short fiber reinforced composites. The interconnected CNT-rGO structures are believed to increase the effective length '*l*' in the model that is responsible for the strengthening enhancement.

While not actual MMCs, a few other studies have studied hybrid CNT-graphene structures for functional performance in metal-based systems. Dong et al. [21] grow CNT-graphene films on Cu substrates decorated with SiC particles using CVD and attain improvements in electrical conductivity over a pure graphene film. The SiC particle enables CNTs to grow and sprout out, while a graphene film grows steadily simultaneously. Kim et al. [22] synthesized CNTs on Cu substrates with graphene films using CVD to form a hybrid CNT-graphene nanostructures for supercapacitor applications.

9.3 CHAPTER HIGHLIGHTS

This chapter has presented the exciting efforts to enhance the performance of CNT-MMCs through the incorporation of secondary particles intended to provide a synergistic reinforcement effect. While still in the nascent stages of research, the use of ceramic nanoparticles as well as other nanocarbon dual reinforcements has shown promising results for enhancement of CNT-MMC properties. The most salient findings include:

- The use of ceramic nanoparticles alleviates agglomeration issues in CNT-MMCs due to the small size of the particles that enables them to break-up CNT nanoclusters and bundles.
- Decorating CNTs with SiC nanoparticles promotes favorable interfaces, as SiCs are less prone to forming brittle Al_3C_4. Additionally, CNTs decorated with SiC have rougher interfaces that provide greater resistance for CNT pull-out.
- CNTs have been grown directly on micro SiC particles, thus taking advantage of the relative ease of dispersing microparticles, micro SiC particles serve as effective carriers for harder to disperse CNTs.
- CNT-graphene dual reinforcements have shown promise for enhancing fracture strain (with high graphene content) and enhancing strength and elastic modulus (with high CNT content). Curiously, no study has investigated incorporating equal volume fractions of CNT and graphene, clearly there is much remaining scope for tailoring of properties.

REFERENCES

1. Kwon, H., Cho, S., Leparoux, M., Kawasaki, A. 2012. Dual-nanoparticulate-reinforced aluminum matrix composite materials. *Nanotechnology* 23: 225704.
2. Kwon, H., Leparoux, M., Kawasaki, A. 2014. Functionally graded dual-nanoparticulate-reinforced aluminium matrix bulk materials fabricated by spark plasma sintering. *Journal of Materials Science & Technology* 30: 736–742.
3. Akbarpour, M. R., Salahi, E., Hesari, F. A., Simchi, A., Kim, H. S. 2013. Fabrication, characterization and mechanical properties of hybrid composites of copper using the nanoparticulates of SiC and carbon nanotubes. *Materials Science & Engineering A* 572: 83–90.
4. Zhang, X., Li, S., Pan, D., Pan, B., Kondoh, K. 2018. Microstructure and synergistic-strengthening efficiency of CNTs-SiC_p dual-nano reinforcements in aluminum matrix composites. *Composites: Part A* 105: 87–96.
5. Rangaraj, S. S., Bhaduri, S. B. 1994. A modified rule-of-mixtures for prediction of tensile strengths of unidirectional fibre-reinforced composite materials. *Journal of Materials Science* 29: 2795–2800.
6. Kwon, H., Lee, G-G., Leparoux, M., Kawasaki, A. 2013. Functionally graded dual-nanparticulate-reinforced aluminum matrix composite materials. *Journal of Physics: Conference Series* 419: 012004.
7. Kwon, H., Saarna, M., Yoon, S., Weidenkaff, A., Leparoux, M. 2014. Effect of milling time on dual-particulate-reinforced aluminum alloy matrix composite materials. *Materials Science & Engineering A* 590: 338–345.

8. Li, S., Su, Y., Ouyang, Q., Zhang, D. 2016. In-situ carbon nanotube-covered silicon carbide particle reinforced aluminum matrix composites fabricated by powder metallurgy. *Materials Letters* 167: 118–121.
9. Li, S., Su, Y., Zhu, X., Jin, H., Ouyang, Q., Zhang, D. 2016. Enhanced mechanical behavior and fabrication of silicon carbide particles covered by in-situ carbon nanotube reinforced 6061 aluminum matrix composites. *Materials and Design* 107: 130–138.
10. Padmavathi, K. R., Ramakrishnan, R. 2014. Tribological behavior of aluminium hybrid metal matrix composite. *Procedia Engineering* 97: 660–667.
11. Hekner, B., Myalski, J., Valle, N., Botor-Probierz, A., Sopicka-Lizer, M., Wieczorek, J. 2017. Friction and wear behavior of Al-SiC(n) hybrid composites with carbon addition. *Composites Part B* 108: 291–300.
12. Alizadeh, A., Abdollahi, A., Biukani, H. 2015. Creep behavior and wear resistance of Al 5083 based hybrid composites reinforced with carbon nanotubes (CNTs) and boron carbide (B4C). *Journal of Alloys and Compounds* 650: 783–793.
13. Dhandapani, S., Rajmohan, T., Palanikumar, K., Charan, M. 2016. Synthesis and characterization of dual particle (MWCT+B_4C) reinforced sintered hybrid aluminum matrix composites. *Particulate Science and Technology* 34: 255–262.
14. Gowda, A. C., Koppad, P. G., Sethuram, D., Keshavamurthy, R. 2018. Morphology studies on mechanically milled aluminium reinforced with B_4C and CNTs. *Silicon.* doi:10.1007/s12633-018-9905-1.
15. Gnanavelbabu, A., Saravanan, P., Rajkumar, K., Karthikeyan, S., Baskaran, R. 2018. Effect of abrasive waterjet machining parameters on hybrid AA6061-B_4C-CNT composites. *Materials Today: Proceedings* 5: 13438–13450.
16. Kwon, H., Lee, G-G., Kim, S-G., Lee, B-W., Seo, W-C. 2015. Mechanical properties of nanodiamond and multi-walled carbon nanotubes dual-reinforced aluminum matrix composite materials. *Materials Science & Engineering A* 632: 72–77.
17. Ghasali, E., Sangpour, P., Jam, A., Rajaei, H., Shirvanimoghaddam, K., Ebadzadeh, T. 2018. Microwave and spark plasma sintering of carbon nanotube and graphene reinforced aluminum matrix composite. *Archives of Civil and Mechanical Engineering* 18: 1042–1054.
18. Rashad, M., Pan, F., Tang, A., Asif, M., Aamir, M. 2014. Synergetic effect of graphene nanoplatelets (GNPs) and multi-walled carbon nanotube (MW-CNTs) on mechanical properties of pure magnesium. *Journal of Alloys and Compounds* 603: 111–118.
19. Li, Z., Fan, G., Guo, Q., Li, Z., Su, Y., Zhang, D. 2015. Synergistic strengthening effect of graphene-carbon nanotube hybrid structure in aluminum matrix composites. *Carbon* 95: 419–427.
20. Zhang, C., Ren, L., Wang, X., Liu, T. 2010. Graphene oxide-assisted dispersion of pristine multiwalled carbon nanotubes in aqueous media. *The Journal of Physical Chemistry C* 114: 11435–11440.
21. Dong, X., Li, B., Wei, A., Cao, X., Chan-Park, M. B., Zhang, H., Li, L-J., Huang, W., Chen, P. 2011. One-step growth of graphene-carbon nanotube hybrid materials by chemical vapor deposition. *Carbon* 49: 2944–2949.
22. Kim, Y-S., Kumar, K., Fisher, F. T., Yang, E-H. 2012. Out-of-plane growth of CNTs on graphene for supercapacitor applications. *Nanotechnology* 23: 015301.

10 Computational Studies in CNT-MMCs

Computational studies on CNT-MMCs have been focused on studying the mechanical, electrical, and thermal properties of isolated carbon nanotubes using molecular dynamics (MD) simulations and first principles [1–8]. Although some of these studies are very useful in understanding the behavior of carbon nanotubes in bending and buckling modes [1,4,9], the key findings cannot be easily extrapolated to predict the properties of CNT-MMCs. The first objective of this chapter is to summarize those computational studies that are of greater significance to the development of CNT-MMCs. This chapter also focuses on proposing some of the computational methods for future research on CNT-MMCs.

10.1 OVERVIEW OF EARLY COMPUTATIONAL STUDIES ON CNT-MMCs

A computational study of SWNT-Ni interface for electronic transport was performed using Monte Carlo simulations [10], which showed that junctions between SWNT and nickel clusters exist on the cluster surface and not at a subsurface location. Such behavior was observed independent of nanotube chirality, temperature, and nature of docking. Figure 10.1 shows Ni55-SWNT structures from density functional theory (DFT) optimization and Monte Carlo simulations. It is evident from Figure 10.1 that Ni clusters adapt to the nanotube geometry and SWNT-metal interface forms on the cluster surface. This study can be very useful in predicting the electronic transport behavior of 1-D CNT-Ni composite synthesized by electrodeposition techniques.

In another study on modeling, the mechanical behavior of a graphene/polymer interface was studied using MD simulations [11]. The consistent valence force field (CVFF) method was adopted to simulate the atomistic interactions and nanoscale load transfer between polyethylene and a graphene sheet, which was considered analogous to the polymer matrix and CNT. Although this study is on a CNT-polymer system, it provides an excellent methodology to perform similar simulations on CNT-MMCs because atomistic simulations for metals are less complex due to their well-defined lattice and metallic bonding. Figure 10.2 shows the schematic that was used to model the mechanical behavior of CNT-polymer composites. Figure 10.2 shows the nanocomposite at the macro-scale, consisting of CNTs dispersed in the polymer, in aligned or random orientations, and the molecular detail of the CNT-polymer interface. A similar model for a CNT-metal interface could provide continuum length-scale micromechanical models that could predict the overall material behavior in tension, compression, and sliding modes.

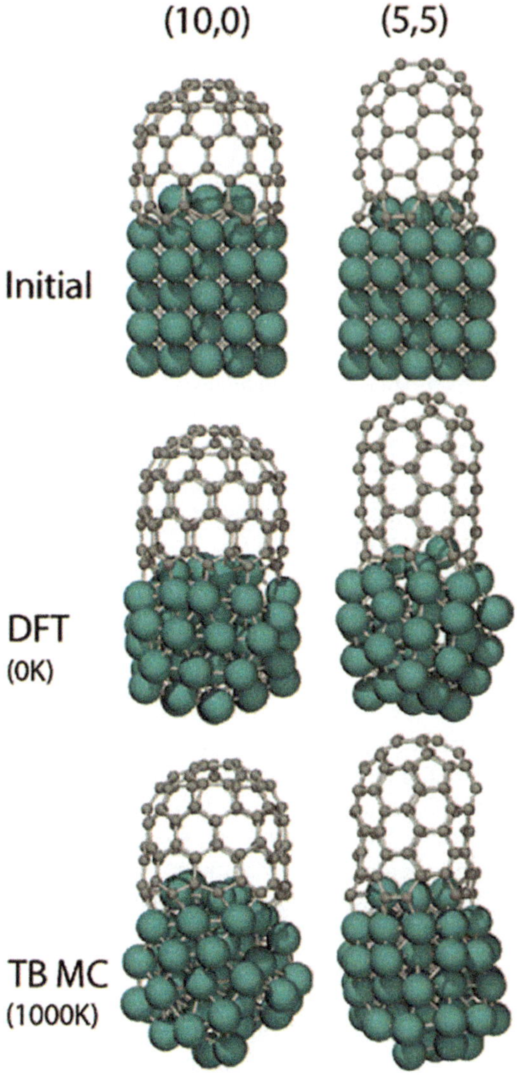

FIGURE 10.1 Initial (top), optimized (middle), and typical finite temperature (bottom) structures with Ni55 clusters and (10,0) (left) and (5,5) (right) SWNTs. The initial Ni55 geometry is FCC and the optimizations (middle) are done with DFT forces at 0 K. The typical structures at finite temperature are obtained from TB MC simulations at 1000 K [10]. (Reproduced with permission from American Chemical Society).

The mechanical behavior of nickel-coated SWNT embedded in gold (Au) matrix under axial tension was investigated using molecular dynamics (MD) simulation method [12]. A comparison between the elastic modulus of an Au-SWNT composite with parallel and vertically embedded nanotubes was made. The uncoated nanotubes improved the elastic modulus of the Au-SWNT composite under the condition of parallel loading, but such improvement disappears under vertical loading because the

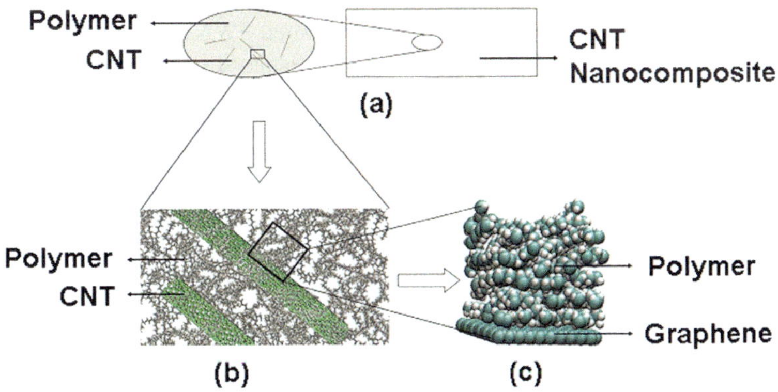

FIGURE 10.2 Modeling of mechanical behavior of CNT nanocomposites with interfacial effects: (a) nanocomposite at macroscale, (b) molecular detail of CNT and polymer, and (c) nanoscale interfacial RVE consisting of graphene and polymer chosen to perform simulations of separation [11]. (Reproduced with permission from IOP Publishing).

interaction between SWNT and the gold matrix was too weak for effective load transfer. The mechanical behavior of the nickel-coated and uncoated armchair (5, 5)

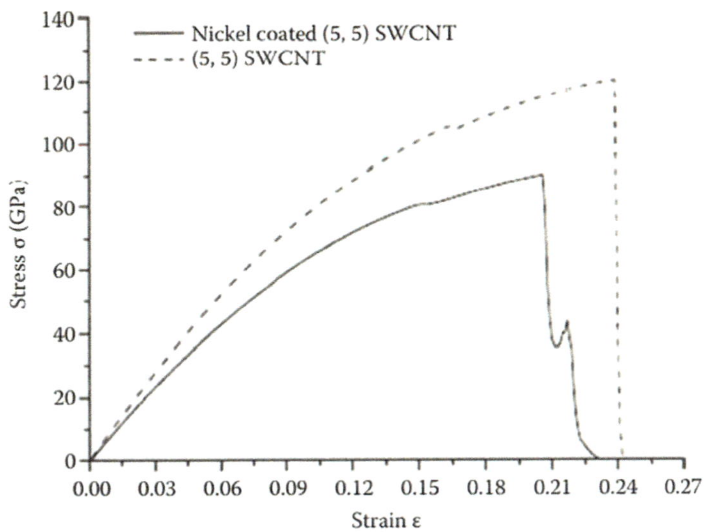

FIGURE 10.3 The stress–strain curve of the nickel-coated and uncoated (5, 5) SWNTs under tension [12]. (Reproduced with permission from Elsevier).

SWNTs under axial tension using the MD simulation method was also studied. Figure 10.3 shows the stress–strain curves of the nickel-coated and uncoated (5, 5) SWNTs. The strength of the nickel-coated SWNT is lower than that of uncoated SWNTs due to an increase in the cross-sectional area of the SWNT after being coated by nickel. Nickel-coated SWNT can significantly improve the composite behavior. Song and Zha also investigated the effect of Ni-coated SWNTs on mechanical behavior of Al matrix composite [13]. The simulation results showed Young's modulus of Ni-coated SWNT–Al composite was higher than uncoated SWNT-Al composite. In addition, the effects of the Ni coating density on the interfacial bonding strength of the composites were investigated by simulating the pullout behavior of nanotube from Al matrix. The results revealed that the initial pullout load for the Ni-coated SWNT from Al matrix about are 13.5, 16.9, and 21.7 times large for Ni coating density, 4.84, 9.68, and 19.36%, respectively. Hence, the Ni coating on CNT helped improve the interfacial bonding between the nanotube and the Al matrix to a great extent [13]. This is in accordance with the observation made in Chapter 6, where it was concluded that nickel-coated CNTs promote interfacial bonding.

Based on the previously mentioned studies, it can be concluded that computational techniques can be effectively utilized to study CNT-MMCs. Thermodynamic and kinetic computations can be made to analyze the chemical changes in the CNT-MMC systems during processing, which will affect the phase composition of the final product. Computational techniques can also be applied to study the microstructural evolution during processing of CNT-MMCs. The microstructure and phase composition have an effect on the final properties. Computational methods can also be adopted to estimate the properties (thermal, electrical, and mechanical) from the microstructure, which otherwise cannot be measured. These methodologies are presented in the following section, which will assist the readers and future researchers in developing CNT-MMCs.

10.2 THERMODYNAMIC PREDICTION OF CARBON NANOTUBE-METAL INTERFACE

The matrix composition can have a profound impact on the reactions occurring at the CNT-MMC interface [14,15]. As was discussed in Chapter 5, the type of reaction product forming at the interface will influence the load transfer and determine the fracture mechanism of the composite. For example, formation of high strength carbide at the metal/CNT interface will enable a large load transfer to the CNTs, while a low strength and brittle carbide may lead to interface failure and inefficient load transfer leading to CNT pullout. A model for the strength of the composite based on the shear strength of the interfacial layer has also been developed [16], which was utilized for the Al-Si-CNT composite [17]. The feasibility of formation of the reaction product (metal carbide) at the interface is governed by the thermodynamics of the CNT-MMC reaction, while the mechanism, rate, and amount of the carbide formation are governed by the kinetics. For example, in the case of an Al-Si alloy reacting with CNTs, whether Al_4C_3 or SiC that will form will depend on which carbide has a lower free energy of formation. The free energy depends on the activity of the Si and Al in the Al-Si alloy, which is dependent on the alloy composition and processing temperature. Since the carbide formation occurs at the processing temperature, it can be controlled by manipulating processing conditions and alloy composition Additionally, coating CNTs with preferential carbide-forming metals like chromium or titanium may help in avoiding formation of either Al_4C_3 or SiC in the case of Al-CNT composites.

Thermodynamics calculations can be performed with the help of software like FactSage™ [18] and Pandat™ (Computerm, LLC, Madison, WI) [19]. FactSage, one of the largest fully integrated database computing systems in chemical thermodynamics, was introduced in 2001 and is the result of over 20 years of collaborative efforts between Thermfact/CRCT (Montreal, Canada) [20] and GTT-Technologies (Aachen, Germany) [21]. FactSage is basically a fusion of the FACT-Win/F*A*C*T and ChemSage/SOLGASMIX thermochemical packages. Pandat is a similar thermodynamic engine and database, which can calculate the phase diagram and thermodynamic properties of multi-component systems. These softwares work on the principle of minimization of Gibbs free energy by optimization of the relative amount of different phases. The Gibbs free energy of different solid and liquid phases such as solid solutions and intermetallic compounds can be described using different thermodynamic models. Commonly used models include the random substitutional solution model, the stoichiometric compound model, the crystal electric field (CEF) model, the ionic liquid model, the associate solution model, the CEF with ionic species, and the more advanced cluster/site approximation (CSA). Using these computational tools, one can estimate the outcome in terms of phase composition of a reaction between metal matrix and CNTs. In lieu of the lack of the thermodynamic properties of CNTs, and the fact that CNTs contain sp^2 hybridized carbon, the thermodynamic properties of graphite can be utilized [15].

In the case of Al-Si alloys, the free energy of carbide formation per mole of carbon can be obtained for the formation of Al_4C_3 and SiC for a given activity of Si or Al in the alloy using FactSage. The activity of Al and Si in an Al-Si alloy can be obtained from the partial molar free energy values provided by the Pandat database. The variation of the free energy change for the formation of Al_4C_3 and SiC as a function of alloy composition was shown in Figure 7.9. From the data, pseudo phase diagrams as shown in Figure 7.9 can be obtained that would predict the carbide that would form for a given Al-Si alloy at a given processing temperature. By comparison of the energy values, it can be said which carbide will preferentially form. Figure 7.4 was also obtained using FactSage. The amount of SiC and Al_4C_3 formed from the complete reaction between CNTs and Al-Si matrix can be calculated through modules in FactSage. Hence, interfaces between CNT and a metal matrix can be engineered by addition of a suitable carbide former to the matrix. Thus, these thermodynamic software packages can be helpful in deciding process parameters for obtaining the desired interface in CNT-MMCs.

10.3 MICROSTRUCTURAL SIMULATION

The understanding of the formation of microstructure is essential in controlling the process. Modeling of microstructure is an active research area in the field of computational materials science. Phase field modeling has been developed for studying several phenomena such as microstructural evolution during solidification, grain growth and coarsening, precipitation reactions, spinodal decomposition, and other second-order transformations [22,23]. Some open-source software like the mesoscale microstructure simulation project (MMSP) is also available [24]. It may be possible to use some of these techniques to predict the effect of the presence of CNTs on the microstructure evolution.

It was shown that thermal spray techniques could be successfully adopted for large-scale synthesis of CNT-MMC coatings. A splat is a single unit of thermal-sprayed microstructure that dictates the overall properties of the coating. The molten droplets (formed from the powder feedstock) impinge the substrate and solidify to form disc-shaped splats that may have a regular geometry or may be fingered and even broken into smaller particles. The splat morphology and role of CNT on splat formation can be effectively modeled using Simulent Drop (Simulent Inc., Toronto, Canada) softwares. The following governing equations (Eqs. 10.1–10.3) can be solved to predict the shape of the splats under various conditions of pressure, velocity, and temperature:

$$\nabla.V = 0 \tag{10.1}$$

$$\frac{\partial V}{\partial t} + \nabla.(VV) = -\frac{1}{\rho}\nabla_p + \nabla^2 \upsilon V + \frac{1}{\rho} F_b \tag{10.2}$$

$$\frac{\partial h}{\partial t} + (V.\nabla)h = \frac{1}{\rho}\nabla.(k\nabla T) \tag{10.3}$$

where V represents the velocity vector, p is the pressure, ρ is the density, ν is the kinematic viscosity, F_b is the body forces acting on the fluid, h is the enthalpy of the fluid, k is the thermal conductivity, and T is the temperature. Simulent Drop is based on the volume of fluid (VOF) technique, which is essentially a robust and free surface modeling technique. The tracking of the droplet-free surface in this three-dimensional model is done by combining a fixed mesh discretization of the Navier-Stokes equations with a piece-wise linear volume tracking algorithm. Since the code allows the use of the dynamic contact angle as and when needed, the simulation of the liquid–solid interfaces is done accurately. Surface tension is modeled as a volume force, acting on fluid in the vicinity of the free surface. A contact angle is specified as a boundary condition along the contact line. The full Navier-Stokes equations in a three-dimensional Cartesian coordinate system are solved using a finite difference scheme. The algorithm has been written for laminar, incompressible flows. The effect of CNT addition is to increase the viscosity of the fluid because it acts as a second phase that is in the solid state. The viscosity of the molten droplet can be modeled by Eq. 10.4 [25]:

$$\mu_e = \frac{\mu}{1 - \varphi\left(1 + {R_c}/{R_p}\right)} \tag{10.4}$$

where μ_e is the effective dynamic viscosity, μ is the dynamic viscosity of the liquid phase, R_C is the radius of CNT clusters, R_P is the droplet size, and ϕ is the CNT fraction. Considering $R_C/R_P \ll 1$, Eq. 10.4 predicts a 6.5% and 13.8% increase in viscosity of molten Al-Si alloy due to addition of 5 and 10 wt.% of CNTs. The larger the viscosity of the material, the smaller will be the splat size and the smaller will be the

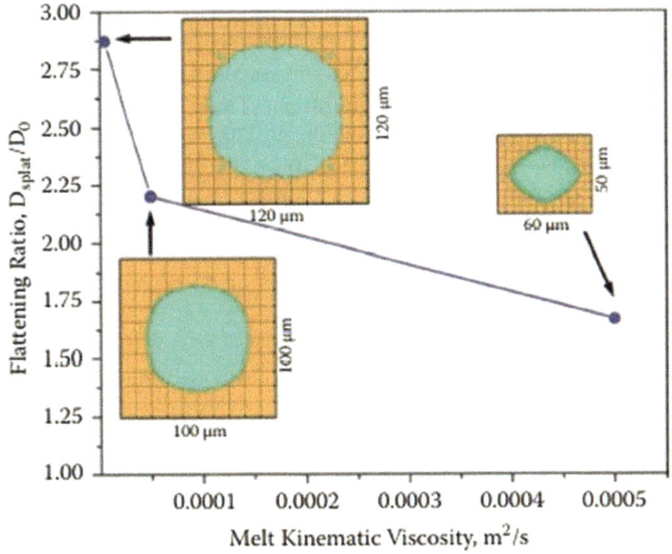

FIGURE 10.4 Variation of the ratio of splat diameter to the droplet diameter vs. the kinematic viscosity of the molten droplet as calculated using Simulent Drop software. The kinematic viscosity of the melt increases with the addition of CNTs. The images are the simulated images of the splats formed after solidification.

flattening ratio defined by the ratio of the diameter of the splat to that of the molten droplet. Figure 10.4 shows the variation of the flattening ratio with the melt viscosity. The simulated images of the splats are also shown. It can be seen that there is splashing and splat breakdown at low viscosity, while the splat size is smaller and irregular for higher viscosity. Figure 10.5a and 10.5b show the SEM images of single splats obtained by plasma spraying of the spray-dried agglomerated powder of Al-12 wt.% Si containing 5 wt.% CNT (Al-5 CNT) and 10 wt.% CNTs (Al-10 CNT),

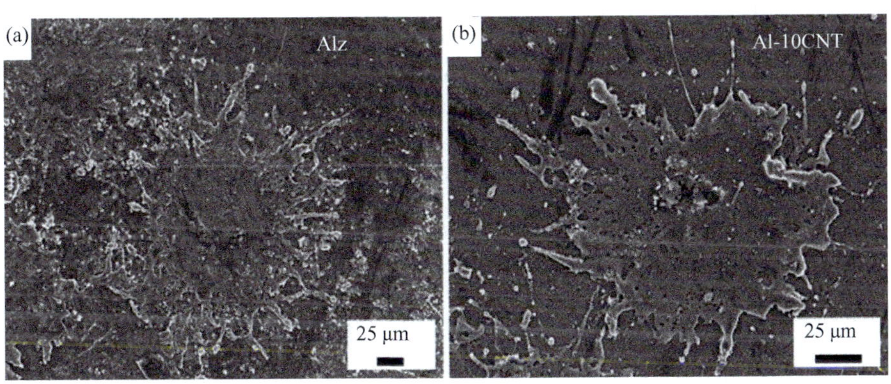

FIGURE 10.5 SEM images of single splats obtained on a glass substrate by plasma spraying of spray-dried Al-12 wt.% Si agglomerates containing (a) 5 wt.% CNT and (b) 10 wt.% CNT.

respectively. A uniform spreading and splashing in the case of Al-5 CNT was observed, which is expected to have a lower viscosity. The splat shape is irregular and smaller in the case of Al-10 CNT as predicted from the flattening ratio in Figure 10.4. The effect of surface roughness on spreading can also be incorporated in simulation to match experimental conditions. Thus, these simulations provide insights into the splat formation, which is the smallest unit of the microstructure of thermal spray coatings and structures. The effect of CNT addition can be modeled similarly in other processes, assuming CNT affects the flow stress and viscosity of the material.

10.4 MECHANICAL AND THERMAL PROPERTY PREDICTION BY THE OBJECT-ORIENTED FINITE ELEMENT METHOD

One of the main advantages of computational techniques is for predicting the properties when experimental data are unavailable or when measurement is difficult. For example, measuring the thermal conductivity of a metal matrix containing a single CNT is difficult experimentally. Similarly, measuring the mechanical properties for this structure is very challenging. Although advances in measurement techniques have made it possible to measure thermal [26] and mechanical properties [27] of small systems, computational prediction provides a theoretical understanding that can be verified by comparison with experimental values.

Finite element techniques are very popular to solve inhomogeneous systems. Finite element simulations can be carried out using commercially available software packages like ABAQUS. ABAQUS has been used to determine the heat transfer between CNTs in a matrix as a function of the separation distance [28]. It was observed that for very high values of the ratio of the thermal conductivity of the CNT to the matrix (>104), there was very small heat flow between the tubes leading to the conclusion that there could be a lack of thermal percolation in such composites. Another method that demonstrated adeptness for predicting thermal conductivity and mechanical properties of composites is the object-oriented finite element method (OOF) [29,30]. This method directly utilizes the microstructure, which can be divided into several pixel groups based on the color intensity and then different properties can be assigned to each pixel group. Software for analyzing microstructures, OOF2, has been developed by the Center for Theoretical and Computational Materials Science at the National Institute of Standards and Technology, Gaithersburg, MD, and is available for free [31]. The OOF2 software uses an adaptive mesh (skeleton) option that facilitates discretization of images such that it confines the pixel boundary. Once the FE mesh has been generated, it can be solved for a given set of boundary conditions and thermal and mechanical properties of the microstructure can be found as a function of the properties of the constituents.

OOF2 has been utilized for determining the effective thermal conductivity and elastic modulus of thermally sprayed yttria-stabilized zirconia coating [32] and overall thermal conductivity of solution precursor plasma-sprayed (SPPS) ZrO_2-7 wt.% Y_2O_3 coating [33] considering the effect of pores and interfaces. The result obtained was in good agreement with experimental values. The elastic property of lamellar Al-12 wt.% Si/Al_2O_3 composite produced by metal infiltration was determined by OOF2 and showed good agreement with experimental measured values [34]. OOF2 is a powerful tool for predicting the properties of CNT-MMCs. OOF2 can capture the effect of CNT curvature and CNT dispersion. Binary images of CNT-MMC microstructures are needed for the analysis. Figure 10.6 shows the case of Al-12 wt.% Si coating containing 10 wt.% CNTs whose thermal conductivity has been measured using OOF2. Figure 10.6a shows the optical micrograph, which indicates the presence of CNT clusters in the matrix. The microstructure can be made binary by considering the Al-Si-CNT matrix (white) and CNT clusters (black) as shown in Figure 10.6b. Figure 10.6c shows the finite element mesh developed using OOF2. The finite element problem has been solved with the boundary conditions as shown in Figure 10.6c with a value for the conductivity of the Al-Si-CNT matrix as 136 W m^{-1} K^{-1} and the CNT cluster as 2.5 W m^{-1} K^{-1}. Figure 10.6d shows the variation of the heat flux in the microstructure. The heat flux distribution shows that the majority of the heat transfer occurs through the matrix if CNT clusters have a very low conductivity. The overall thermal conductivity as calculated from the finite element method (FEM) is 37.9 W m^{-1} K^{-1}, which is close to the experimentally measured value of 25 W m^{-1}

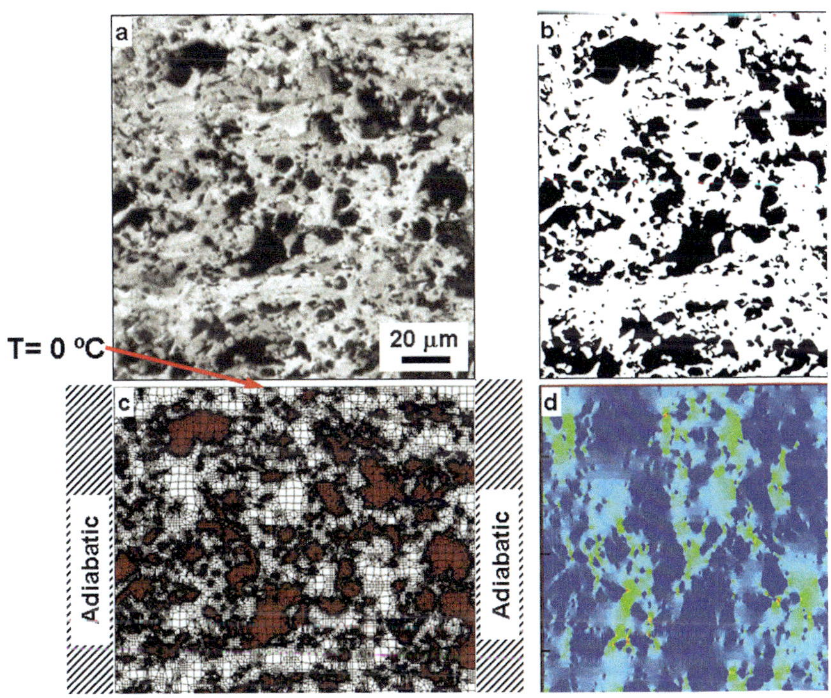

FIGURE 10.6 (a) Optical micrograph of Al-12 wt.% Si alloy coating containing 10 wt.% CNT, (b) the corresponding binary image used for OOF2 analysis, (c) the finite element mesh created using OOF2 and the boundary conditions applied, and (d) the heat flux distribution along the different regions of the microstructure.

K^{-1}. Thus, OOF2 can be a very useful technique for evaluation of the effective properties of the CNT-MMC.

By using the same technique on images such as those in Figure 2.19b, one can find conductivity of well-dispersed systems. OOF2 can also be used to find effective mechanical properties of CNT-MMCs at different length scales by considering microstructure at multiple magnifications.

10.5 CHAPTER HIGHLIGHTS

There are very few computational studies on CNT-MMCs. Most of the computational studies focus on the isolated nanotube and do not translate to predicting the properties of the bulk CNT-MMCs. Some of the studies on multi-scale modeling of polymer-CNT can also be adapted for CNT-MMCs. Computational techniques can be applied to interface design, processing, and property evaluation. Thermodynamic computations can be applied to study the chemistry of CNT-MMCs and the changes occurring at the interface due to the CNT-matrix reactions. Computational techniques may also be employed to study the microstructure evolution during processing. The effect of CNTs on microstructure evolution, as in the case of thermal spraying, can also be studied. Using the information obtained, the process parameters can be tailored to acquire the desired microstructure. Thermo-physical and mechanical properties of CNT-MMCs at multiple length scales can be obtained using microstructured-based computational methods such as OOF.

REFERENCES

1. Heo, S. J., and S. B. Sinnott. 2009. Computational investigation of the mechanical properties of nanomaterials. *Diamond Relat. Mater.* 18: 438–442.
2. Jayasekara, T., Landis, B. A., and J. W. Mintmire. 2008. First-principles simulations of chiral double-wall carbon nanotubes. *Int. J. Quantum Chem.* 108: 2943–2949.
3. Pozrikidis, C. 2009. Effect of the Stone–Wales defect on the structure and mechanical properties of single-wall carbon nanotubes in axial stretch and twist. *Arch. Appl. Mech.* 79: 113–123.
4. Feng, C., and K. M. Liew. 2009. A molecular mechanics analysis of the buckling behavior of carbon nanorings under tension. *Carbon* 47: 3508–3514.
5. Georgantzinos, S. K., Giannopoulos, G. I., and N. K. Anifantis. 2009. An efficient numerical model for vibration analysis of single-walled carbon nanotubes. *Comput. Mech.* 43: 731–741.
6. Ragab, T., and C. Basaran. 2009. A framework for stress computation in single-walled carbon nanotubes under uniaxial tension. *Comput. Mater. Sci.* 46: 1135–1143.
7. Li, Y-F., Li, B-R., and H-L. Zhang. 2009. The computational design of junctions between carbon nanotubes and graphene nanoribbons. *Nanotechnology* 20: 225202 (1–11).
8. Stoltz, G., Lazzeri, M., and F. Mauri. 2009. Thermal transport in isotopically disordered carbon nanotubes: a comparison between Green's functions and Boltzmann approaches. *J. Phys.: condens. Matter* 21: 245302 (1–11).
9. Tsai, P., and Y. Jeng. 2013. Experimental and numerical investigation into the effect of carbon nanotube buckling on the reinforcement of CNT/Cu composites. *Compos Sci Technol.* 79: 28–34.

10. Borjesson, A., Zhu, W., Amara, H., Bichara, C., and K. Bolton. 2009. Computational studies of metal-carbon nanotube interfaces for regrowth and electronic transport. *Nano Lett.* 9: 1117–1120.
11. Awasthi, A. P., Lagoudas, D. C., and D. C. Hammer. 2009. Modeling of graphene–polymer interfacial mechanical behavior using molecular dynamics. *Modelling Simul. Mater. Sci. Eng.* 17: 015002 (1–37).
12. Song, H-Y., and X-W. Zha. 2010. Mechanical properties of nickel-coated single-walled carbon nanotubes and their embedded gold matrix composites. *Phys. Lett. A* 374: 1068–1072.
13. Song, H. Y., and X-W. Zha. 2010. Influence of nickel coating on the interfacial bonding characteristics of carbon nanotube-aluminum composites. *Comput Mater Sci* [Internet] 49: 899–903.
14. Laha, T., Kuchibhatla, S., Seal, S., Li, W., and A. Agarwal. 2007. Interfacial phenomena in thermally sprayed multiwalled carbon nanotube reinforced aluminum nanocomposite. *Acta Mater.* 55: 1059–1066.
15. Bakshi, S. R., Keshri, A. K., Singh, V., Seal, S., and A. Agarwal. 2009. Interface in carbon nanotube reinforced aluminum silicon composites: thermodynamic analysis and experimental verification. *J. Alloys Compd.* 481: 207–213.
16. Coleman, J. N., Cadek, M., Blake, R., Nicolosi, V., Ryan, K. P., Belton, C., Fonseca, A., Nagy, J. B., Gunko, Y. K., and W. J. Blau. 2004. High-performance nanotube-reinforced plastics: understanding the mechanism of strength increase. *Adv. Funct. Mater.* 14: 791–798.
17. Laha, T., Chen, Y., Lahiri, D., and A. Agarwal. 2009. Tensile properties of carbon nanotube reinforced aluminum nanocomposite fabricated by plasma spray forming. *Composites A* 40: 589–594.
18. Bale, C. W., Chartrand, P., Degterov, S. A., Eriksson, G., Hack, K., Mahfoud, R. B., Melanqon, J., Pelton, A. D., and S. Petersen. 2002. FactSage thermochemical software and databases. *Calphad* 26: 189–228.
19. Cao, W., Chen, S-L., Zhang, F., Wu, K., Yang, Y., Chang, Y., Schmid-Fetzer, R., and W. A. Oates. 2009. PANDAT software with PanEngine, PanOptimizer and PanPrecipitation for multi-component phase diagram calculation and materials property simulation. *Calphad* 33: 328–342.
20. www.crct.polymtl.ca.
21. www.gtt-technologies.de.
22. Boettinger, W. J., Warren, J. A., Beckermann, C., and A. Karma. 2002. Phase-field simulation of solidification. *Annu. Rev. Mater. Res.* 32: 163–194.
23. Moelans, N., Blanpain, B., and P. Wollants. 2008. An introduction to phase-field modeling of microstructure evolution. *Comp. Coupling Phase Diag. Thermochem.* 32: 268–294.
24. http://matforge.org/cmu/wiki/mmsp.
25. Sobolev, V. V., and J. M. Guillemany. 2000. Formation of splats during thermal spraying of composite powder particles. *Mater. Lett.* 42: 46–51.
26. Itkis, M. E., Borondics, F., Yu, A., and R. C. Haddon. 2007. Thermal conductivity measurements of semitransparent single-walled carbon nanotube films by a bolometric technique. *Nano Lett.* 7: 900–904.
27. Peng, B., Locascio, M., Zapol, P., Shuyou, L., Mielke, S. L., Schatz, G. C., and Espinosa, H. D. 2008. Measurements of near-ultimate strength for multiwalled carbon nanotubes and irradiation-induced crosslinking improvements. *Nature Nanotech.* 3: 626–631.
28. Shenogina, N., Shenogin, S., Xue, L., and P. Keblinski. 2005. On the lack of thermal percolation in carbon nanotube composites. *Appl. Phys. Lett.* 87: 133106.
29. Langer, S. A., Fuller, Jr., E. R., and W. C. Carter. 2001. Image-based finite element mesh construction for material microstructures. *Comp. Sci. Eng.* 3: 15.

30. Reid, A. C. E., Langer, S. A., Lua, R. C., Coffman, V. R., Haan, S-I., and R. E. Garcıa. 2008. Image-based finite element mesh construction for material microstructures. *Comp. Mater. Sci.* 43: 989–999.
31. http://www.ctcms.nist.gov/oof/oof2/.
32. Wang, Z., Kulkarni, A., Deshpande, S., Nakamura, T., and H. Herman. 2003. Effects of pores and interfaces on effective properties of plasma sprayed zirconia coatings. *Acta Mater.* 51: 5319–5334.
33. Jadhav, A. D., Padture, N. P., Jordan, E. H., Gell, M., Miranzo, P., and E. R. Fuller, Jr. 2006. Low-thermal-conductivity plasma-sprayed thermal barrier coatings with engineered microstructures. *Acta Mater.* 54: 3343–3349.
34. Ziegler, T., Neubrand, A., Roy, S., Wanner, A., and R. Piat. 2009. Elastic constants of metal/ceramic composites with lamellar microstructures: finite element modelling and ultrasonic experiments. *Comp. Sci. Tech.* 69: 620–626.

11 Patents and Commercialization

Use of nanotechnology and nanomaterials-based products has been increasing. Figure 11.1 shows trends in the United States and worldwide in patents filed on carbon nanotube-reinforced composites. It definitively indicates an increasing trend of research in nanocomposites utilizing CNTs. The increasing amount of patents indicates continued innovation in potential applications of CNT composites, novel applications enabled by CNTs, and new ways of effectively creating CNT composites. While most of the early patent literature revolves around the incorporation of CNTs into polymeric materials, significant strides have been made in CNT-MMCs as well. This chapter presents an overview of patents and patent applications on CNT-MMCs in order to shed light on what the first wave of commercialization might look like. It is evidence of what is widely known, after decades of exciting research and substantial promising; significant CNT-MMC research is now being done by private industry and other organizations that do not readily report findings and advances in the open literature. By surveying the patent literature, it is possible to take a glimpse into the current and future state of commercial development of CNT-MMC technology.

11.1 PATENTS

Many patented inventions on CNT composites have revolved around the relatively "low hanging fruit" in terms of composite science – polymer composites. These composites can be processed using low temperatures and pressure, often not significantly greater than ambient conditions, and can be readily formed into complex shapes required for commercial products using a variety of methods. CNT-polymer composites (CNT-PMCs) patents have ranged from broad claims regarding CNT processing using any solvent and any matrix [1], to specific patents with claims specific to a specific product, such as the patent filed by Comeau [2] on CNT-polymer composites for the intermediate layer in golf balls. The CNT-MMC patent literature is thus far mostly in the former stage, with a few exceptions. This section covers salient patents and inventions filed in the US and abroad.

11.1.1 US Patents

CNT-reinforced metal patents span the fabrication life of composites. Starting with methods for CNT-MMC powder synthesis [3], generalized composite microstructures envisioned [4], CNT fiber-reinforced composites [5,6], surface engineered CNT-MMCs [7,8], hybrid dual reinforced composites [9], interphase design [10], and generalized processing methods applicable to metal and ceramic nanocomposites [11].

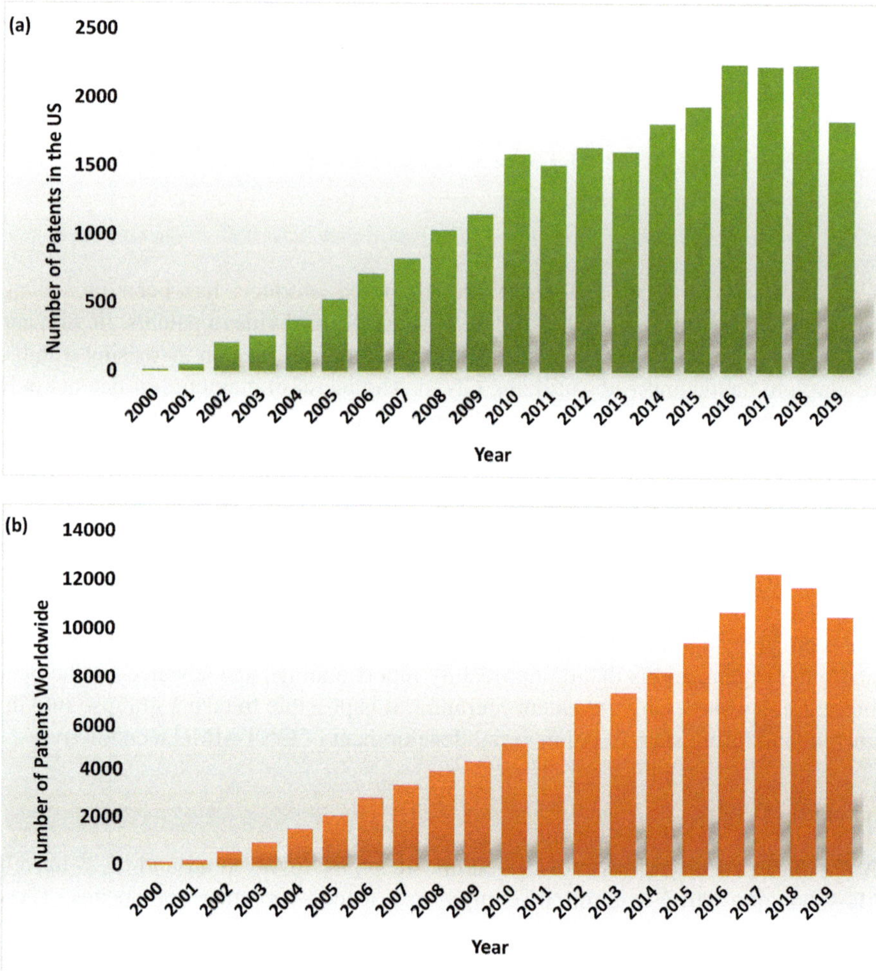

FIGURE 11.1 Trends in CNT-Composite patents. Data gathered from the European Patent Office (Espacenet search engine) using search term ["carbon nanotube" AND "Composite"] from 2000 to 2019 for (a) US Patents, and (b) Worldwide Patents.

Hong et al. [3] patented generalized techniques for processing nanocomposite CNT-MMC powders, as outlined in Figure 11.2. The patent claims their invention utilizes any solvent that is capable of functionalizing the CNTs to generate a repelling electrostatic charge on the surface of the CNTs to aid in uniform dispersion. CNTs are mixed in the solvent through ultrasonication, with the example parameters listed consisting of ultrasonic waves in the range of 40–60 kHz and sonication times of 2–10 h. The metal matrix introduced into the dispersed CNT solution in this invention consists of any metallic materials that form water-soluble metal salts or metal hydrates. This is followed by further ultrasonication using similar parameters as the first. The mixed solution is then dried (or calcined for some metal matrices) at

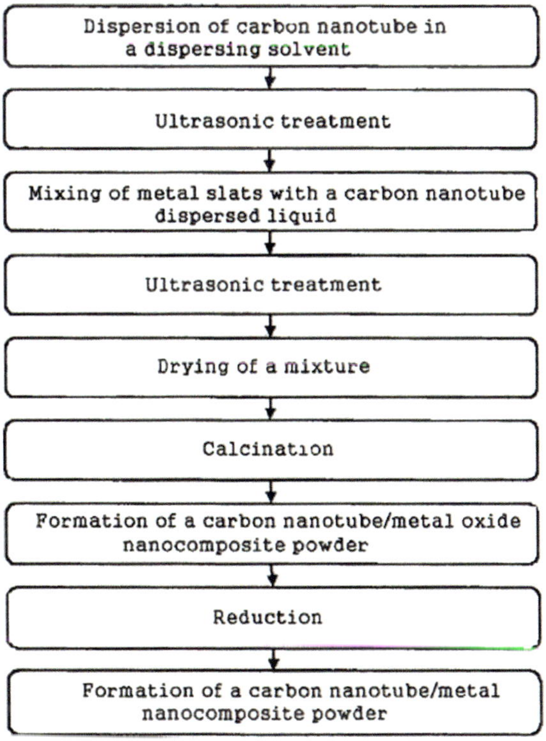

FIGURE 11.2 Flowchart for patented process for producing CNT-MMC composites [3].

temperatures typically below 100 °C for 6–12 h in either a vacuum or inert gas atmosphere. In the case of starting with a metal oxide precursor (e.g., CuO for Cu matrix), then a reduction step is needed to obtain CNT-MMC powder [3]. Chen et al. [4] patent generalized methods for producing CNT-MMCs with enhanced thermal and electrical conductivity, greater mechanical strength, and tailorable coefficient of thermal expansion (CTE). CTE is stated to be tailorable by varying the amount of CNTs in the composite. The patent states the use of SWCNT is preferred in the reported invention and is generally more suitable when enhanced electrical conductivity is the objective. While the patent broadly covers CNT-MMC materials, a preference is stated for electrochemical co-deposition processing in order to produce aligned CNT composites. As with the patent by Hong et al. [3] and widely practiced in scientific literature, CNTs are functionalized. Chen et al. [4] list cetyl trimethyl ammonium bromide (CTAB) and octyl trimethylammonium bromide (OTAB) as preferred surfactants. The electrolytc solution typically includes CNTs (in amounts varying from 0.5 to 80%), CTAB or OTAB, a metal ion source (e.g., $CuSO_4$ for a Cu-CNT composite) and is kept at a pH of 2–5. In addition, the intended matrix of the composite can be used as the anode material for a co-deposition process. The cathode will be the intended substrate to which the CNT-MMC will be co-deposited [4].

Shah et al. [5] filed a patent for composite utilizing CNT-infused fiber materials. Their invention revolves around the indirect use of CNTs as reinforcement, that is, they are a secondary phase within a more conventional fiber such as carbon fiber, ceramic fibers, or glass fibers, in either chopped or continuous form. The CNTs used in this invention are stated to be 40 μm +/− 20%. The CNT-reinforced fiber is then used to reinforce a metal matrix. The primary targeted properties for enhancement by this invention are electrical conductivity, thermal conductivity, and CTE. The CNT-infused fiber approach is not an approach otherwise reported in the literature. Most of the composite literature on CNT composites has attempted to scale up the CNTs and bulk CNT yarn sheets have been successfully utilized in polymer composites [6]. As discussed previously, one of the primary challenges associated with CNT-MMCs, as compared to CNT-PMCs, is the much higher processing temperatures needed. These higher temperatures are prone to inducing defects on CNTs. High-quality defect-free CNTs are especially critical for attaining enhancing transport properties such as electrical conductivity. As such, the approach by Shah et al. [5] aims to circumvent processing-induced damage on CNTs by embedding them within the fiber – effectively shielding them from high temperature – while maintaining a continuous CNT network capable of enhancing conductivity. Furthermore, the inventors state that there invention could utilize protective passivation layers on the CNTs to prevent reactions with the fibers. The inventors propose that the transport properties of interest are tailorable by varying the amount, diameter, length, and chirality of the infused CNTs [5]. In regards to scale-up, the individually CNT-infused fibers can be bundled into fiber tows or twisted into yarns or braids. Furthermore, CNTs can be infused onto both sides of a fiber weave or tape. The CNTs can be directly grown on the surface (via CVD) by doping the fiber with metal catalyst particles. Applying CNTs only to the surface of the fiber structure could reduce costs while attaining the desired conductivity enhancement at the surface.

Surface engineering is a logical approach for applications where surface properties are key (e.g., corrosion, wear), and can decrease costs due to small volume of coatings, and avoid challenges associated with processing of complex systems such as CNT-MMCs. Patents have been filed for producing CNT-MMCs on metal surfaces using friction stir welding [7] and thermal spraying [8]. The patent on friction stir welding describes a process of laying metal layers, followed by a deposit of CNTs on those layers, either directly or thru a CNT-based solution. Stacks of metal layers with CNTs on top of them are then subjected to the friction stir process that welds the layers together and mixes the CNTs within the metal matrix, while maintaining frictional heat below the melting point of the metal matrix. The inventors describe varying layer thickness of the layers and lay claim to composites utilizing up to 30 vol.% of CNTs. It should be noted that the patent by Brice et al. [7] lays claim to any nanomaterial-reinforced metal composite produced by friction stir welding; despite the well-known fact in the literature that different challenges are encountered during processing of different nanomaterials, depending on the composition (e.g., boron nitride vs. carbon nanotubes), or in morphology (e.g., carbon nanotubes vs. graphene sheets), for example. Buresch et al. [8] lay claim to all carbon-based nanomaterials used to reinforce thermally sprayed metal matrix composite coatings, to include plasma spraying high velocity oxyfuel spraying, and cold spraying.

A patent by Zhang et al. [9] is the first to lay claim to a type of hybrid or dual reinforced composite. CNTs are used to reinforce WC-Co, which can be considered (especially for high Co content compositions) as a metal matrix composite with WC particulate reinforcements. The patented invention focuses on the use of CNT (0.5–50 vol.%) reinforced WC-Co formed into cutting tools, including drill bits, mining picks, and core bits. In some embodiments of the invention, the authors cite the need for functionalization of CNTs or metallic coatings to enhance compatibility and bonding with the base phases of the cermet. The primary motivation for incorporating CNTs into the cermets is to enhance fracture toughness without an associated decrease in hardness [9]. This is typically not possible in conventional cermets as increasing toughness dictates an increase in the metallic binder, which in turn decreases hardness.

Karandikar et al. [10] patent the use of an interphase tailored to yield silicon carbide in silicon containing metallic alloys and resulting composites. The interphase is envisioned to be a carbonaceous precursor or other source of free carbon that could prevent direct reaction of silicon (molten or otherwise) from reacting with, and hence degrading the CNTs. The reaction is envisioned to form a sufficiently thick and uniform silicon carbide interphase so as to seal off the CNTs. The CNT-MMCs with these reactively formed SiC interphases exhibit enhanced electrical conductivity and low CTE (particularly in Si-CNT composites). A variation in the invention is to directly coat CNTs with a silicon-containing polymer such as polysilazane. With sufficient free carbon, the current invention can be used to fabricate CNT-SiC composites by fully reacting molten Si with free carbon (independent of CNTs). Another patent to encapsulate both metal and ceramic composites reinforced with CNTs (or other nanomaterials), and perhaps the broadest of the patents surveyed [11], includes any nanometric reinforcement and any metal, ceramic, or composite phase produced by powder metallurgy and other common techniques available in the literature and outlined earlier in Chapters 2 and 3. The broad claims by some inventors will be discussed briefly in Section 11.2.

11.1.2 International Patents

Patents abroad have also kept pace with the promising advancements in CNT-MMC technologies. A few salient inventions have led to international patents being filed with the World International Property Organization, including overlap with US patents, for example, the inventions regarding CNT-infused fiber composites by Shah et al. [12].

An international patent application by Barrera et al. [13] deals with using SWCNTs as reinforcements in carbide forming metal matrices without any reaction with the CNT, a critical step to ensuring survival of a SWCNT. The invention utilizes protective barrier layers to protect CNTs during high temperature processing using induction heating. Induction heating is an ideal technique due to its rapid heating that allows for reduced processing times and less time for CNTs to potentially react with the carbide-forming matrix. The invention utilizes a thick nickel coating on CNTs. A uniform and thick nickel coating is achieved by first fluorinating the CNTs; however, it is recognized that fluorine atoms form bonds with carbon and degrade the excellent thermal and electrical conductivity of CNTs [13].

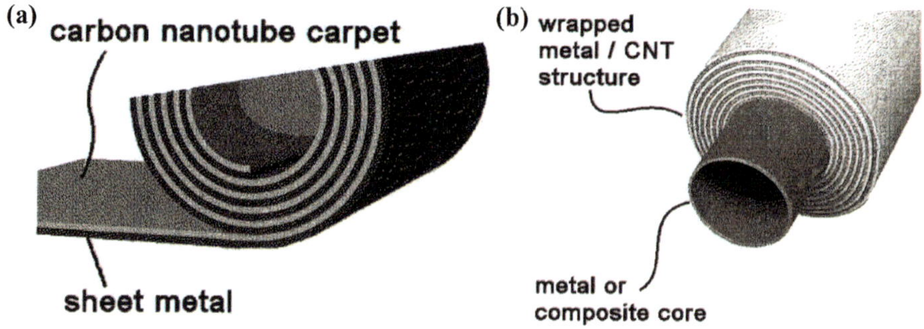

FIGURE 11.3 Schematics of CNT-infused electrical cables, (a) CNT carpet consisting of CNTs infused on a thin metal sheet that is then wrapped around itself to form a cable, and (b) configuration where CNT-infused thin metal sheet is wrapped around an electrically conductive metal core that provides structural support.

Another exciting invention cited in the international patent literature is that of a conductive CNT "carpet" for use in electrical cables [14], an idea not discussed in the scientific literature. Figure 11.3 shows schematics of two promising configurations envisioned for CNT-carpets formed into electrical cables. The idea is to roll up or fold a single or multiple metal sheets of high conductivity CNT-MMCs in order to form electrical cables with high electrical conductivity, high current density capacity, and low specific weight. The process begins by growing CNTs directly on a metal surface using catalyst particles and CVD, or a process utilizing vapor phase catalyst delivery. The metal sheet with in situ grown CNTs is then rolled to achieve the desired layer thickness. The thin CNT-infused metal sheet can be rolled onto itself to produce a wire, as shown in Figure 11.3a. Alternatively, the CNT-infused metal sheet can be rolled around a conductive metal rod that provides structural stability, as shown in Figure 11.3b. The inventors claim the resulting electrical cables are suitable for applications requiring high power, low losses, low weight, and high current density capacity.

Several patents from individual countries have also contributed noteworthy inventions related to CNT-MMCs. A Chinese patent [15] develops a method to produce highly dispersed CNTs in Al powder by first mixing CNTs with an adhesive that is then subjected to high shear mixing. Al powder is then introduced and the mixture is subjected to further high shear mixing. The mixture consisting of Al powders coated with CNTs, which are in turn coated with the adhesive, is then subjected to ball milling. The formed powders can be processed using conventional solid state and powder metallurgy processing routes.

Patents from several countries have focused on modified approaches to the consolidation of CNT-MMCs [16–19]. DeBladis [16] patented a process for producing CNT-MMCs thru a continuous extrusion process of CNT-MMC powder thru a movable extrusion wheel that imparts frictional heating to the powder. The invention is noteworthy as extrusion processing is typically done on formed solid parts, and not directly on powder mixtures, as is claimed in DeBladis's invention [16]. A Japanese patent application proposes an invention for developing layered CNT-MMCs [17]. The method proposes using composite powders to directly form layers of CNT-MMC upon a base layer, and the process can be repeated using subsequent layers of powder. The invention appears analogous to a selective laser melting additive manufacturing process, except that layers are not melted by a laser. Instead, powders are heated in a non-oxidizing state to yield either sintering, partial melting, or complete melting of the CNT-MMC powder. The invention is stated to work with 2–20 vol.% of CNTs [17]. A Chinese patent presents an invention for the ultrasonic welding of Al-CNT composites [18]. The process starts with an acid-based purification step (which presumably can be circumvented by starting with high purity CNTs) and proceeds with de-agglomeration steps consisting of both ultrasonication (referred to as ultrasonic shearing) and ball milling techniques. The CNTs are then dispersed on a prepared Al metal surface. Ultrasonic welding is then conducted on the CNT-coated Al surface using a frequency of 20 kHz, a welding pressure of 12.25–21 MPa, and an amplitude of 35 m. The primary advantages of this invention are the low temperatures used and quick welding times that prevent interfacial reactions of carbide products [18]. As discussed earlier, the science behind controlling the degree of interfacial reaction between Al and CNT has advanced to the point of harnessing carbide products for beneficial load transfer with a sufficiently thin carbide. A Japanese patent by Ishida et al. [19] proposes an invention for fabricating formable CNT-MMC rods or wires. The invention begins with a carbon rod that is subjected to DC discharges within a He atmosphere to yield carbon nanotubes and carbon "nanocapsule" structures. A metal matrix is then deposited, presumably on the cathode side where the carbon nanostructures formed. Then the rod can be worked by conventional means such as drawing or compression forming to manufacture CNT-MMC wires or transmission lines and cables [19].

As mentioned earlier in this book, the area of dual-reinforcement CNT-MMCs promises to be an emerging technological approach for harnessing multiple nanomaterials and tailoring properties and such approach is taken by a couple of inventions disclosed in Chinese patents [20,21]. One invention utilizes Al_2O_3 alongside CNTs [20], and the other utilizes graphene alongside CNTs [21]. The CNT-Al_2O_3 reinforcement is produced by uniformly growing CNTs on Fe/Al_2O_3 catalysts and then mixing this CNT-ceramic matrix composite phase into molten Mg. Using Al_2O_3 particles effectively as a "carrier" of CNTs allows for uniform dispersion of CNTs in Mg matrix, as the density of Al_2O_3 and Mg is similar so that stirring alone provides a uniform dispersion of CNT-Al_2O_3 particulates throughout the molten Mg phase. The resulting CNT-Al_2O_3/Mg mixture is then cast into the desired form, with the CNT-Al_2O_3 hybrid reinforcement constituting 1–15 wt.% of the hybrid MMC [20]. The CNT-graphene dual reinforcement composite yields a networked structure of interconnected nanostructures that enhanced mechanical, electrical, and thermal properties [21]. The invention uses graphene with up to 10 layers and a volume fraction of 0.1–1%, and a CNT volume fraction of 0.5–5%.

11.2 PROSPECTS & CHALLENGES FOR COMMERCIALIZATION

As was stated at onset of the chapter, the patent literature reveals those technologies that are approaching commercial relevance. A key to a successful material is one with the flexibility to have tailored properties. CNT-MMCs have a myriad of levers to pull for tailoring properties, from processing techniques and their associated parameters, to different CNT sizes and purities, to the dual-reinforcement approach much advocated for in this book and shown to have commercial relevance by several relevant US [9] and international patents [20,21]. Among the most promising uses of CNT-MMCs appear to be those requiring properties that cannot be enhanced by conventional fibers or at least not as easily, such as electrical and thermal conductivity, and tailoring of coefficient of thermal expansion. In addition, the use of nanometric reinforcements means that layering of property gradients can be done at much finer scales than with conventional fibers. In fact, with the advent of nanotechnologies such as MEMS/NEMS, there are geometric constraints that make conventional fiber-reinforced composite approaches untenable. There are length scales, where the nano-engineer will only have nano-reinforcements at their disposal to tailor the properties of metallic materials.

While many technical and scientific challenges lie ahead (and are discussed in Chapter 12), there are myriad issues related to patents that will affect the progress of CNT-MMC commercialization, which are broadly summarized in Table 11.1. Patent "floods" have previously occurred in the fields of molecular biology, superconductors, and petroleum refining and are thought to have led to low-quality and overlapping patents due to a strain in the governing patent office (e.g., US Patent and Trade Office) [23]. The issue of patent validity can come in several forms, including when claims made in patents may not have been fully validated or the underlying mechanisms not fully understood. For example, a schematic and description in a patent may claim that CNTs must become embedded in metal grains [3]. Without TEM evidence,

TABLE 11.1
Patent-Related Challenges to CNT-MMC Commercialization and Potential Solutions

Challenges	Potential Solutions
Patent Validity – Technological Rigor of Claims	• Increased fundamental understanding of CNT-MMC mechanisms
Patent Validity – CNT Size Ranges	• Development of better tools to quantify ranges and distributions • Further exploration of size- and morphology-dependent mechanisms in CNT-MMC processing and properties
Definitions of "*novelty*," "*Obviousness*," and "*one skilled in the art*"	• Increased participation of CNT-MMC experts in the patent granting process. • Better training for patent offices in determining what is set fundamental knowledge vs. current and evolving state-of-the-art understandings
Detection of Infringement and Patent Enforcement	• Assessment of costs associated with patent infringement and detection thereof • Conducting forums and conferences to come up with licensing schemes specific to nanotechnology [22]
Significant Holdings of Nanotechnology Patents by Universities	• Increase collaboration between government and industry with universities • Increased participation from universities in technology transfer and investment in start-up companies

it is difficult to verify this fact. In the early days of CNT-MMC development, some may have assumed such a microstructure to be preferred. However, if in fact the process described in the patent does not yield embedded CNTs, then one can imagine a scenario where the claims in the patent become invalid if a subsequent process proves it attains embedded CNTs and this resulting microstructure yields superior properties to the previous patent. A review of the CNT-MMC patent literature finds most claims are broad and examples given typically do not contain the rigor to mechanistically validate approaches or extrapolations of behavior. Broad CNT compositions are often stated with little evidence indicating that those ranges have indeed been explored and validated to yield useful properties with the given processing approach.

A related aspect that can invalidate patent claims is an issue broadly seen in nanotechnology related patents – the issue of sizes and size ranges. The field of CNT-MMCs is likely to come across the issue of size ranges listed in patents and exceptions to those ranges in new inventions, as has occurred in other nanotechnology-related patents and technologies. In short, what should happen if a patent has claimed its invention encompasses a certain range of sizes, when in fact not the entire range has been thoroughly investigated? The cases of *Smithkline Beecham Biologicals v Wyeth Holdings Corporation* [24] and *BASF v Orica Australia* [25] are highlighted by Watal and Faunce [26] as examples where new patents were granted for inventions that fell in the size range of a previous similar invention. The European Patent Office (EPO) held that a Smithkline patent on a lipid used in a Hepatitis B vaccine was valid even though its size (60–120 nm) overlapped a previous patent where a similar lipid with a wider size range (80–500 nm) was utilized [24]. The new patent using the narrower size ranges was deemed to be novel as it was in the lower end of the previous patent's range and the smaller lipids exhibited significantly enhanced performance (improved immune response induced by vaccine) [24]. One could imagine similar issues occurring with CNT-MMCs, for example with the size of CNTs used (short vs. long), or other characteristics such as chirality, purity, functionalization, or distribution of sizes. Nanotechnology, including CNT-MMCs, is still a relatively new field and the range of behaviors that occur at different length scales has not yet been mapped and prior claims could be invalidated or overlooked if the prior art did not fully characterize or understand the mechanisms (and resulting properties) that result in its claimed range of sizes or even compositions.

A similar issue was at stake in *BASF v Orica Australia* [25] in that the newer Orica patent was using polymer nanoparticles (<100 nm) only slightly smaller than a previous patent by BASF utilizing polymer nanoparticles >111 nm. BASF argued that the claims by Orica fell within their invention because further scaling down the particles would constitute an extension that is "obvious" to one "skilled in the art," two terms that hold significant legal weight in the patent world. The same language is used in patents on CNT-MMCs to claim rights over multiple nano reinforcements, or in some cases to lay claim to all nanomaterials reinforcements and/or all metal matrices possible. In the Orica case, it was deemed that their processing temperature of 40 °C was not covered or obvious extensions of the BASF patent that utilized temperatures below 50 °C. BASF argued that simple trial and error studies at ranges below 50 °C would have led to the synthesis of <100 nm particles. However, the EPO rejected that argument on the basis that BASF's patent implied temperatures below 50 °C were not needed and was specifically targeted for particles larger than 111 nm, hence their invention would not have predicted that synthesis of <100 nm particles. In truth, it can be expected that the terms "novel," "obvious," and "one skilled in the art" are difficult and perhaps impossible to define when a field is rapidly evolving, as CNT-MMC technology clearly is. What may not have been obvious a few years ago, may now be common knowledge or obvious to those working in the field. Does that change the validity of prior patents (that are typically active for 20 years) and how do patent reviewers get training to navigate and identify the current state of the art or the current understanding of next logical steps for maturing a technology?

All of these issues logically lead to questions of patent enforcement. Enforcement of CNT-MMC patents will again come upon similar issues broadly seen in enforcing nanotechnology patents. For example, could one argue that downsizing in of itself (e.g., Hall–Petch effect) would be obvious to one skilled in the art. The now known Inverse Hall–Petch effect shows us that trends in material properties and behavior do not often continue as initially thought and unexpected behavior is especially known to crop up in nanotechnology. Another example could be the previous avoidance of any carbide formation in CNT-MMCs. Many patents clearly claim their invention avoids carbide formation to avoid those deleterious effects. However, as discussed earlier in Chapter 7, precise control of the carbide interface (previously thought too fine to control) can serve to enhance load transfer and tailor properties. Will new understandings of CNT-MMCs invalidate previous claims in patents that depend on those prior held facts, either directly or indirectly? Many of those questions are likely to be resolved only in court cases, if and when they come up. That's brings up another factor in enforcing CNT-MMC patents, that is somewhat unique to nanotechnology – the difficult and resulting expense in determining whether ones patent rights or invention have been infringed upon. The nature of CNT-MMCs makes it so that it may be difficult to assess patent infringement unless detailed characterization is conducted. In come case, patents make reference to uniform distribution of CNTs, a characteristic notoriously difficult to quantify reliably. What if a patented invention relies on specific functionalization of CNTs. TEM and other expensive techniques would need to be used just to determine whether ones' patent was infringed upon. Inventors would have to weigh the cost of determining patent infringement and the resulting benefit thereof. Morris [23] points out that as related to nanotechnology broadly, the difficulty in inforcement has tacitly given researchers and engineers free reign to further develop new technologies without significant risk or concern for infringing upon others patents. This could change as technologies such as CNT-MMCs mature and enter a more commercial phase.

Morris [23] points out that the current landscape of patents has evolved, particularly since the Bayh-Dohl act. The motivation for obtaining and being granted a patent was previously to attain an "exclusive monopoly on an invention" [23]. The Bayh-Dohl Act encouraged federally funded researchers to patent the basic science findings, leading to a significant portion of nanotechnology patents to be held by universities. These basic findings are often referred to as "upstream patents," that hold little promise for direct use as a commercial product. One could imagine the creation and design of a specific interphase for CNT-MMCs as an example. The interphase in itself would rely on other processing techniques and possibly other patented nanoscale architectures or functionalization processes. Many upstream patents are possible due to lack of prior art (because of the novelty of the field as a whole) in order to determine the validity of a new patent. It should also be no surprise that many of these upstream patents are held by universities, who typically do not have "exclusive monopoly of an invention" as a primary motivation. There is some concern that a gamut of upstream patents, referred to as a "thicket" of patents could stunt the development of "downstream" patents that do directly lead to end-use products [22]. A proliferation of University-based patentees leads to multiple "owners" of a given technology. This historically leads to less utilization and underdevelopment of a technology due to difficulties in getting all owners to agree on its collective use, which in the patent literature is referred to as the tragedy of the anticommons [23]. However, Morris suggests that the "thicket" of patents held by universities is a minor impediment to developing nanotechnology as compared to lack of funds, lack of expertise, and public-safety concerns [23]. Many now mature technologies previously saw patent thickets of their own, including sewing machines in the 1850s and airplanes in the early 1900s [26].

It should be noted that many universities have sought to develop robust technology transfer offices that make use of the breadth of the patent portfolio held by many universities. If universities hold patents on critical technologies, then it seems logical that there should be increased collaborations between universities and government and industry partners. Universities may also feel compelled to directly participate in commercialization by backing start-up companies or being stakeholders in start-up companies headed by university inventors. What is clear is that transformative technologies such as CNT-MMCs have attracted broad enough attention that commercialization may be inevitable. Rao et al. [27] emphasize that the extraordinary properties of CNTs and other nanomaterials will allow them to overcome barriers such as the so-called "Trough of Disillusionment" and "Valley of Death," as well as warrant risk-taking in patent infringement and patent overlap in order to reap the benefits of CNT-derived technologies, such as CNT-MMC. Table 11.2 provides examples of specific

TABLE 11.2
Specific Potential Products and Uses of CNT-MMCs Mentioned in CNT-MMC Patent Literature

Sector or Industry	Prospective Products and Uses of CNT-MMCs
Aerospace	Nose cones, aircraft wing leading edges, flaps, airfoils, propellers, air brakes, fuselages, helicopter shells, rotor blades, floors, doors, seats, tanks, rocket nozzles, engine and motor components
Marine	Boat hulls, Propellers, rudders, decks, stringers
Sporting Goods	Water and snow skis, bicycles, kayaks and canoes, snowboards, golf clubs
Energy & Electrical	Wind turbine blades, solar structures, laptops, cell phones, electronic enclosures, light poles, power lines, circuit boards
Construction	Columns, pediments, domes, cornices, formwork, window profiles, reinforcing bars, pre-stressing tendons, roadside sign supports, guardrails, noise barriers, municipal pipes, storage tanks
Consumer Services	Pans, housings, gas pressure bottles, washers, dryers, air conditioners

commercial product and uses in several important fields including aerospace, marine, energy, sporting, construction, and consumer goods industries.

11.3 CHAPTER HIGHLIGHTS

This chapter has sought to summarize the current state of patent filings on CNT-MMC inventions and technologies and its impact on the commercialization of CNT-MMCs. US and international patents show that a significant portion have patented observations generally available in the open scientific literature; however, in some cases, novel approaches and uses have been proposed. The patent literature is the best way to catch a glimpse of research being done in government or private labs that may not engage the academic community substantially. The chapter then discusses aspects of patent law that may need to be addressed or overcome as CNT-MMC technology matures and issues of patent overlap and patent infringement enter the legal system. At this stage, the thicket of patents does not appear to be hindering innovation and development of CNT-MMCs and is still secondary to the need for increased funding and expertise in the area. The field is still evolving and ripe for further discoveries, which will constantly shift the boundaries of what is "obvious" to one "skilled in the art." It may be possible as knowledge of CNT-MMCs evolves, many claims in earlier patents will be deemed invalid as new mechanisms replace prior understanding. Regardless of patent rights challenges, it is clear from the patent literature that CNT-MMCs have the potential to proliferate nearly all industries affecting our daily lives, from the seas to the skies, to space, to the insides of our electronic devices.

REFERENCES

1. Shambaugh, R. L., "Nanotube/Matrix Composites and Methods of Production and Use", U.S. Patent 7,001,556 B1, Issued Feb. 21, 2006.
2. Comeau, B., "Carbon-Nanotube-Reinforced Composites for Golf Ball Layers", U.S. Patent Application Publication US2006/0270790 A1, Published Nov. 30, 2006.
3. Hong, S. H., Cha, S. I., Kim, K. T., Hong, S. H., "Method of Producing Metal Nanocomposite Powders Reinforced with Carbon Nanotubes and the Powder Prepared Thereby", U.S. Patent 7,217,311 B2, Issued May 15, 2007.
4. Chen, Q., "Carbon Nanotube Reinforced Metal Composites", U.S. Patent Application Publication US2007/0036978 A1, Published Feb. 15, 2007.
5. Shah, T. K., Malecki, H. C., Waicukauski, J. A., "Metal Matrix Composite Materials Containing Carbon Nanotube-Infused Fiber Materials and Methods for Production Thereof", U.S. Patent Application Publication US2012/0164429 A1, Published Jun. 28, 2012.
6. Zhang, M., Zhang, H., Wang, B., Liang, R., Zhang, C., "Composite Materials Reinforced with Carbon Nanotube Yarns", U.S. Patent Application Publication US2012/0085970 A1, Published Apr. 12, 2012.
7. Brice, C. A., "Metal Matrix Carbon Nanotube Composite Material and Method of Making Same", U.S. Patent Application Publication US2009/0258232 A1, Published Oct. 15, 2009.
8. Buresch, I., Kroemmer, W., "Process for Producing a Metal Matrix Composite Material", U.S. Patent Application Publication US2012/0077017 A1, Published Mar. 29, 2012.

9. Zhang, Y., Zhan, G., Sheng, X., Lockstedt, A. W., Shen, Y., Deng, H., Griffo, A., Keshavan, M. K., "Hybrid Carbon Nanotube Reinforced Composite Bodies", U.S. Patent Application Publication US2008/0210473 A1, Published Sep. 4, 2008.
10. Karandikar, P. G., "Nanotube-Containing Composites Bodies, and Methods for Making Same", U.S. Patent Application Publication US2005/0181209 A1, Published Aug. 18, 2005.
11. Seals, R. D., Ripley, E. B., Ludtka, G. M., "Nanostructured Composite Reinforced Material", U.S. Patent Application Publication US2012/0177905 A1, Published Jul. 12, 2012.
12. Shah, T. K., Malecki, H. C., Waicukauski, J. A., "Metal Matrix Composite Materials Containing Carbon Nanotube-Infused Fiber Materials and Methods for Production Thereof", International Application Publication Number WO2011/078934 A1, Published Jun. 30, 2011.
13. Barrera, E. V., Bayazitoglu, Y., Wilson, K., "Processing of Single-Walled Carbon Nanotube Metal Matrix Composites Manufactured by an Induction Heating Method", International Application Publication Number WO2007/118048 A2, Published Oct. 18, 2007.
14. Adams, H., "Carbon Nanotube Enhanced Electrical Cables", International Application Publication Number WO2013/127444 A1, Published Sep. 9, 2013.
15. Cai, W., Li, Q., Wang, L., Chen, M., Wu, H., Liao, J., Huang, Y., "Carbon-Nanotube-Reinforced Aluminum-Base Composite Material", Chinese Patent Application CN103602843A, Published Nov. 4, 2015.
16. Debladis, F., "Method for Manufacturing a Composite Material with Metal Matrix and Carbon Reinforcement", European Patent Application EP3083114A1, Published Oct. 26, 2016.
17. Kuno, M., Makuchi, Y., Tshushima, K., et al., "Carbon Nanotube-Containing Composite Material, and Method for Producing the Same", Japanese Patent Application JP2007016262A, Published Jan. 25, 2007.
18. Zhu, Z., Xiong, Z., "Method for Preparing Aluminum Composite Material Reinforced with Carbon Nano Tubes by Ultrasonic Welding", Chinese Patent Application CN101914735A, Published Jul. 4, 2012.
19. Ishida, Y., Ito, K., Kuzumaki, T., et al., "Carbon Nanotube-Reinforced Aluminum Composite Material", Japanese Patent JPH1088256A, Published Jan. 5, 2005.
20. Li, H., Zhang, K., Wang, H., Fan, J., Xia, X., "Preparation Method of Carbon Nanotube-Alumina Composite Reinforced Magnesium-Based Composite Material", Chinese Patent Application CN102206793A, Published Sep. 9, 2012.
21. Li, Z., Wang, J., Li, Z., Fan, G., Zhang, D., "Graphene and Carbon Nanotube Mixed Enhanced Metal-Matrix Composite Material and Preparation Method Thereof", Chinese Patent Application CN102719693A, Published Apr. 16, 2014.
22. Harris, D. L., "Carbon Nanotube Patent Thickets". In *Nanotechnology & Society*. F. Allhoff and P. Lin, Eds. Dordrecht: Springer, 2009.
23. Morris, E. M., "The Irrelevance of Nanotechnology Patents", *Connecticut Law Review* 49 (2016): 499–552.
24. *Smithkline Beecham Biologicals v Wyeth Holdings Corporation*, Boards of Appeal of the EPO, T-0552/00, Oct. 30, 2003.
25. *BASF v Orica Australia*, Boards of Appeal of the EPO, T-0547/99, Jan. 8, 2002.
26. Watal, A., Faunce, T. A., "Patenting Nanotechnology: Exploring the Challenges", *World Intellectual Property Organization Magazine*, April 2011.
27. Rao, R., Pint, C. L., et al., "Carbon Nanotubes and Related Nanomaterials: Critical Advances and Challenges for Synthesis toward Mainstream Commercial Applications", *ACS Nano* 12 (2012): 11756–11784.

12 Summary and Future Directions

The past two decades of research on CNT-MMCs have demonstrated that the novel properties that CNTs possess can be harnessed to design and make metallic composites with extraordinary combinations of properties. CNT-MMCs are potential next generation materials for high-performance applications. This chapter briefly summarizes the research on CNT-MMCs and provides the current state of the art to the readers. With continued research in this maturing technology, specific properties and techniques are changing day by day, new benchmarks are being set, and effectively dispersed CNTs and efficient strengthening levels are being achieved. The lessons learned from previous studies continue to be utilized in further developing novel techniques and improving on existing methods. The second important objective of this chapter is to provide the readers with new future research directions and a renewed roadmap for the technological mature of this revolutionary materials technology – CNT-MMCs.

12.1 SUMMARY OF CNT-MMC RESEARCH

Through the previous chapters, several key findings concerning CNT-MMCs were discussed. These are briefly summarized as follows.

1. **Composite Powder Processing:** The unique problems associated with processing of CNT-MMCs were outlined in detail in Chapters 2 and 3. The large surface area of the CNTs results in agglomeration and formation of clusters, which were found to be detrimental to the mechanical and thermal properties of the composite. Metal-infiltrated CNT clusters are less problematic as compared to CNT clusters that were not infiltrated, which effectively act as large pores [1]. Thus, improving the CNT distribution, especially at high CNT concentrations, has become the Holy Grail in this area. A majority of the processing has been through powder metallurgy techniques owing to the freedom it provides in terms of designing starting compositions and forming a diverse range of shapes and parts. Conventional blending processes were found to be ineffective for obtaining a good CNT dispersion in feedstock powders or materials. Novel methods like molecular level mixing, ball milling, flake powder metallurgy, spray drying, nanoscale dispersion, and even growing CNTs on individual particulates were developed and found effective for this purpose. Modeling and optimization studies have yielded an effective parameter range for ball milling techniques that results in both excellent CNT dispersion and strong CNT/metal adhesion. CNT dispersion quantification methods were described and the advantages and limitations of these studies were presented.

These methods help in quantifying the effect of different processes as well as changes in the processing parameters for a given process on the CNT distribution. CNT distribution affects properties and there are only a few studies that directly correlate CNT distribution with properties [2].

2. **Composite Consolidation:** Conventional consolidation processes like pressing and sintering or hot pressing have met with little to moderate success. Processes involving hot deformation like hot extrusion and hot rolling have been effective in attaining dense products with improved CNT/metal bonding. Deformation processes have the capability of breaking down the CNT clusters, densifying and aligning CNTs in the deformation direction thereby making use of the infiltrated clusters for strengthening [3,4]. Al-CNT composites prepared by deformation and severe plastic deformation based processing have shown significant strengthening efficiency [4–6]. Use of novel consolidation processes like spark plasma sintering (SPS) stood as one of the most popular and promising methods of the last decade of CNT-MMC research. The shorter sintering times required in SPS have the interesting prospect of retaining the CNT dispersion of powder feedstock and extraordinary strengthening has been reported [7,8]. Processes involving molten metals are less effective because infiltration of CNT clusters is difficult and unlikely to occur. Novel methods like disintegrated melt deposition (DMD) have been developed for Mg-CNT composites, which have shown some success [9]. Additive processing techniques such as plasma spraying, HVOF, and cold spraying have been utilized to synthesize Al-CNT composites as coatings and freestanding structures [10–12]. These techniques have the advantages of manufacturing large bulk CNT-MMC composites as coatings and near net shape products. Ni-CNT and Al-CNT coatings [13] have been shown to have good wear resistance properties over Ni coatings [14,15]. For applications in MEMS and nano-electromechanical systems (NEMS), CNT-MMCs can be extremely useful [16]. Such parts can be easily manufactured from bulk composites that contain areas of good CNT distribution. Ultrahigh frequency micromechanical resonators have been fabricated from Al-SWNT composites [17].
3. **Characterization:** Due to the unique structure and the small dimensions of CNTs, several interesting phenomena occur. Study of these phenomena requires advanced characterization techniques, some of which have obtained special attention; for example, Raman spectroscopy. Raman spectroscopy has gained popularity in studying the structural changes occurring within the CNTs during processing and deformation. Raman spectroscopy has been combined with other nano-characterization techniques such as AFM, to precisely map the type and damage of carbon-based materials with nanometric resolution. Use of HRTEM has made it possible to study the structural changes to CNTs due to processing, CNT fracture/deformation mechanisms, and interfacial product formation at the CNT/matrix interface. The last decade of CNT-MMC research saw great strides in understanding the make-up and role of interfacial products and was aided in large part by wide spread use of focused ion beam milling for sample prep and availability of HRTEM. Novel mechanical property characterization techniques like nano-indentation, nano-dynamic mechanical testing,

nano-scratch, and nano-dynamic modulus mapping techniques were discussed, which provide insight into the strengthening due to CNTs at the nano/micro-length scales. The frontier of characterization technology has migrated to in situ characterization using the aforementioned techniques.

4. **Stiffening and Strengthening Mechanisms:** The mechanisms of strengthening and stiffening were outlined in Chapter 6. It was observed that improvement in CNT dispersion and CNT-matrix bonding lead to increased strengthening [5,7,8]. Improvement of these factors can be brought about by manipulating processing parameters. A significant scatter in the data on strengthening for similar CNT concentrations is still observed. This is attributed to various powder processing and consolidation techniques used. Also, to be expected different specimen sizes will lead to non-conformity, even if standards are used. The micromechanical models available to predict the mechanical properties based on composites were analyzed to determine their efficacy for CNT-MMCs. The strengthening due to CNTs occurring at the nano/micro-length scales was found to result in increased properties of bulk samples. For Al-CNT composites, simple micro-mechanical models like rule-of-mixtures (ROM) were found to be applicable at lower CNT concentrations (<2 vol.% CNT) indicating possible alignment of CNTs, while the Halpin-Tsai equations and combined Voigt-Reuss models provided conservative values for the strengthening effect. The various mechanisms of strengthening were also discussed and compared. Micromechanical models developed specifically for micro- and nano-reinforcements are still in the early stages. Advanced in situ mechanical testing techniques are elucidating the unique mechanisms present in nanocomposites such as CNT-MMCs, providing the fundamental physics-based understanding needed to develop accurate and robust models.
5. **Interface in CNT-MMCs:** The importance of interface in transferring the stress to the CNTs was discussed in Chapter 7. Good bonding between the CNTs and the metal matrix can result in effective strengthening [5,8]. The reaction product forming at the interface is expected to influence the interfacial stress transfer and the mechanical properties of the composite. Thermodynamic and kinetic factors affect the type and amount of the metal carbide forming at the interface. The design of the interface through manipulation of the matrix composition, processing conditions, and application of coated CNTs was also discussed. Interfacial reactions between CNT and matrix have significant implications in determining the wettability, which has consequences in maintaining CNT dispersion and infiltration of CNT clusters.
6. **Thermal, Electrical, and Chemical Properties:** The improvement in thermal, electrical, and chemical properties due to addition of CNTs has been discussed in detail in Chapter 8. Apart from structural applications, CNT-MMCs have great potential as thermal management materials, wear resistant coatings, and in functional applications in Li-ion batteries and hydrogen storage. Addition of CNTs has been shown to improve the thermal conductivity and decrease the coefficient of thermal expansion, which is useful for the electronic packaging industry and space applications such as satellites that are thermally cycled from near-cryo temperatures to over 250 °C [18]. However, increase in

the thermal conductivity is subject to the presence of uniformly distributed CNTs in the matrix. CNT clusters have thermal conductivity 2–3 orders lower than individual CNTs and reduce the overall conductivity of CNT-MMCs. Addition of CNTs has also shown an improvement in the corrosion resistance of Ni-CNT and Zn-CNT composites, which has been attributed to the inert nature of CNTs and the fact that CNTs fill cracks and crevices inhibiting pitting corrosion [19,20]. The wear resistance of Cu-CNT and Ni-CNT composites also displayed improvement as compared to unreinforced material, which is attributed to matrix stiffening and strengthening, as well as a possible lubricating effect from sword-in-sheath sliding of nanotube walls within an MWCNT. The large surface area provided by CNTs is also beneficial for the use of CNT-MMCs as sensors and catalysts.

7. **CNT-MMCs with Dual Reinforcement:** CNT-MMCs are perhaps the most mature metallic nanocomposite; however, the past decade has also produced exciting results in metallic nanocomposites utilizing oxides and other carbon allotropes. Recent studies have sought to investigate the nature of dual reinforcements in a metal matrix, with particular interest on any possible synergism between different types of nanomaterials. The use of nano-oxides has shown to serve as a nanoscale mechanical agitator, capable in interspersing between CNT bundles and breaking them up. This leads to more rapid and effective cluster break-up and dispersion of CNTs. CNTs decorated with nanoparticles may exhibit other exciting functional properties and deformation mechanisms. The field is in the nascent stages and is likely to continue to branch out in many directions as the tools available for processing CNT-MMCs expand simultaneously.
8. **Computational Research in CNT-MMCs:** Use of computational techniques in CNT-MMCs is still lagging behind and only a few studies have been reported. Computational techniques used in compositional design of the CNT-MMC interface using thermodynamic databases like FactSage, microstructure evolution, and property correlation have been discussed. Microstructure-based software "OOF" has been discussed. This software can aid in predicting material properties from known microstructures, such as mechanical and thermal properties of CNT-MMC. The information can be used to predict new properties and the necessary levels of CNT dispersion and microstructural architectures needed to attain them.

Figure 12.1 shows the journey in the synthesis of CNT-MMC since the discovery of CNTs. The significant achievements or milestones are highlighted. CNT-MMCs can be potentially be applied in many applications, which are tabulated in Table 12.1 [21]. CNTs may be uniquely suited to reinforce fine structures, where even micro-particulates may be too large, such as in nano electromechanical systems (NEMS). Apart from use as structural materials, CNT-MMCs can be used for other functional applications for their thermal and electrochemical properties. CNTs have been found effective in making conductive fuel supply lines, which can dissipate all electrostatic charges that build up [22]. CNT-MMCs can also be used for batteries, hydrogen storage, and in thermal management applications. In these applications too, dispersion has a critical

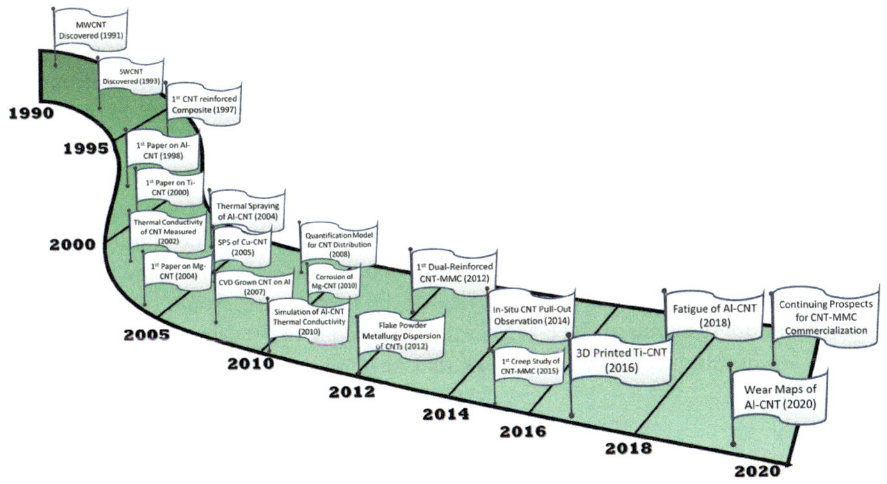

FIGURE 12.1 Roadmap showing significant milestones in the development of CNT-MMCs since the discovery of CNTs.

TABLE 12.1
Potential Applications of CNT-Reinforced Metal-Matrix Composites in Various Industries [19]

Industry	Application	Property Desired
Automobile	Break shoes, cylinder liners, piston rings, gears	High strength, wear resistance, good thermal conductivity, low density
Aerospace	Aircraft brakes, landing gears	Good wear resistance, good thermal conductivity, low density, high strength
Space	High gain antenna boom, structural radiators	Low density, high strength, low coefficient of thermal expansion, good electrical conductivity
Sports	Lightweight bicycles, tennis and badminton rackets	High strength, high elastic modulus
Electronic Packaging	Heat sinks for thermal management, solders	High thermal conductivity, low coefficient of thermal expansion, increased strength
MEMS and Sensors	Micro-beams, micro-gears	High elastic modulus, high surface area
Battery and Energy Storage	Anodes and anode coatings, hydrogen storage materials	Large surface area, high current density, reduced response times, increased H, adsorption-desorption area

role. The thermal conductivity of composites is significantly affected by CNT distribution and alignment. However, it was seen that there is a large scope and the extraordinary properties that the CNTs have to offer has been underutilized. The key to obtaining desired characteristics and properties from CNT-MMCs is intelligent design. Many studies are being undertaken to find solutions for the challenges. The scope and direction for further research in this area is summarized in the next section.

12.2 FUTURE DIRECTIONS

There is tremendous room for innovative ways of processing and characterizing the properties of CNT-MMCs. Great strides have been made since the first edition, but many of the same issues persist at varying degrees. Design strategies are being guided by new discoveries of CNT-MMC behavior and performance. Along with an improvement of properties and developing processing methods for bulk scale fabrication in existing applications, the aperture must continue to be widen for new applications, and for CNT-MMCs to serve key enablers of new technologies. The unique combinations of properties enshrined in CNT-MMCs makes it possible that they may set the foundation for revolutionary technologies of the near future. Some of the potential areas for future research in CNT-MMCs are included here, key findings since the first edition are highlighted, and new issues that have arisen during the last decade are discussed.

12.2.1 Purity & Quality of CNTs

Most of the studies carried out to date have used CNTs that have a wide range of length and diameter and are often curved in shape and entangled. Curled CNTs have been shown to be bad for thermal conductivity enhancement [23] because they have a lower "length efficiency." Use of straight CNTs of similar aspect ratio is expected to lead to improved properties. Only a few recent studies have utilized long and straight CNTs, due to their relative unavailability [24]. CNTs obtained by CVD techniques can yield a wide distribution in the CNT diameter and length. To be able to separate out or classify CNTs with a narrow range of diameters and lengths is a big challenge. Some success has been observed in separating metallic and semiconducting SWNTs having different diameters [25–27]. Such a separation technique for multi-walled CNTs will bring a new dimension to this field. Use of high-quality graphitized CNTs must be made and the effect on thermal conductivity and strengthening must be studied. The graphitization can be done by heat treating as-received CNTs at temperatures above 2000 °C. There are reports of very long CNTs being prepared by CVD technique [28,29]. Triple-walled CNTs up to 100 mm in length have been reported [29]. Researchers at Tianjin University in China have devised a method to spin continuous fibers of CNTs having a strength of 0.4–1.25 GPa [30]. Use of such long CNTs and fiber mats for synthesis of CNT-MMCs must be explored. In particular, the qualities and properties of CNTs within a CNT cloth or paper must be verified and assessed. The use of such papers or cloths would offer an excellent path forward toward mitigating issues with poor CNT dispersion.

Another issue related to CNT processing that directly affects resulting CNT-MMCs is the purity of the material, typically available in powder form. Impurities are typically other forms of carbon or catalysts used to synthesize the CNTs. Impurities of either form can have unexpected and detrimental effects on thermal and electrical conductivity. As discussed later in this chapter, metallic catalysts are also primarily responsible for the cytotoxic effects seen in CNTs, which has led to reduced interest in CNT-MMCs for biomedical applications over the last decade. Lastly, catalyst on CNTs could have possible adverse reactions as more complex, multi-reinforcement composites are developed. If secondary metallic particles are introduced to enhance functionality, they may experience micro galvanic corrosion with metallic catalysts such as Fe, leading to unexpected deterioration of the interface and subsequent degradation on composite performance. In highly corrosive environments, the metal matrix itself could form a galvanic couple with the catalyst and degrade the intimate interfacial characteristics needed in an effective CNT-MMC.

12.2.2 Challenges Related to Processing

The challenge for obtaining good CNT distribution in bulk samples as well as at higher loading of CNTs remains. From the discussion in Chapters 2 and 3, it is evident that processes involving shorter sintering times, higher pressures, and semi-molten state are desirable for achieving good dispersion. The dispersion of CNTs in the starting powders is a necessary condition. CNT clusters present in the powders are carried over to the consolidated product in most processing routes. Rotational ball milling modeling and early optimization studies have laid a road map for effective mixing that yields both adequate CNT dispersion and CNTs that are well adhered to the metal matrix powder [31]. Another possible approach to attaining good dispersion that has remained relatively unexplored is the use of nanometallic powders. The dispersion in this case would be in the fine nanoscale level. Use of nano powders smaller than the asperities between the CNT clusters will penetrate into the pores of any remaining CNT clusters and will result in well-infiltrated clusters upon processing. The lack of interest in using nanometallic powders involves the cost of the powders themselves and equipment to mitigate health and handling hazards, such as the use of glove boxes or other contained inert environments. However, other advanced MMC system go to great lengths to avoid oxygen contamination and utilize contained inert environments [32]. Fabrication of advanced CNT-MMCs for sensitive or high-performance applications may need such processes anyway.

12.2.3 Optimal CNT Dimensions & Aligned CNT-MMCs

It is well known that the best properties of CNTs are along their axial direction. Composites with aligned CNTs are expected to have better mechanical and thermal properties in the direction of alignment. This is another area that needs to be explored for CNT-MMCs. There are some studies on aligning CNTs during processing by using a magnetic field [33–35]. CNTs coated with magnetic nanoparticles like Fe_3O_4 or Fe are easier to align [36]. Use of magnetic fields in processing CNT-MMCs needs to be explored. Such alignment can be brought about by using aligned preforms or through processing. This will require development of meticulous processing techniques with provisions for in situ monitoring. For application in MEMS, sensors, and thermal management for strategic electronic components, the cost of fabricating these advanced composites can be aptly justified. Another area to be explored is the use of Bucky papers. Highly aligned Bucky papers that possess high thermal conductivities are available [37,38]. Use of these Bucky papers can result in high thermal conductivity composites for electrical applications.

Many applications may seek to avoid anisotropic properties. Structural applications may need isotropic material, but also highly dense material, which is often attained by high pressure processing techniques that inevitable align CNTs. Shorter CNTs could be utilized to mitigate that aspect, as such the effective length of CNTs to retain adequate reinforcement is needed. A material could be tailored by alternating the content, orientation, and size of CNTs across different layers of the material. Aligned CNT arrays to be laid in alternating patterns as is done with conventional polymeric composites [39]. Likewise, 3D or mesoscale ceramics with alternating ductile and hard layers could be tailored simply by alternating the length and size of CNTs used in each layer. Control of CNT sizes could be done by simple ball milling, with precise control over environment, milling materials, and milling time.

12.2.4 In Situ Characterization and Modeling

The high strength and elastic modulus of CNT reinforcement in the metal matrix is of no use unless the applied load is transferred to them. Therefore, understanding load transfer is crucial. In situ micromechanical testing studies have significantly broadened our understanding of CNT reinforcement mechanisms. In situ studies will continue to be needed to characterize different interfacial products, engineered interfaces, and variations in strengthening and deformation mechanisms as environment and temperature are varied. In situ testing facilities within an SEM or TEM environment have seen substantial growth in the last decade and are well suited for varying temperature, applying various strain loadings, and even simultaneous current and measurement of resistance [40,41]. The effect of using coated CNTs, surface modified CNTs, and the interfacial reaction between the matrix and the interface can be gauged thru in situ observation of sliding, pull-out, deformation, and fracture events. Such observations will prove critical as more nuanced approaches are developed, and as CNT-MMCs mature technologically and require only the materials design approaches to be implemented when CNT-MMCs are scaled up for specific applications. In situ observations will provide a strong knowledge base for establishing physics-based models that capture the unique properties and behavior of CNTs within the metal matrix environment. New models need to be formulated in order to take the geometry, composition, environment, and load transfer mechanisms of carbon nanotubes into account when predicting mechanical and functional properties.

12.2.5 Dynamic Mechanical Behavior

The vast majority of studies on CNT-MMCs have focused on static mechanic properties, namely elastic modulus and tensile strength. Only a few studies have focused on dynamic mechanical behavior, such as wear [14,15], fatigue [42,43], and creep [44,45]. Dynamic behavior more closely reflects engineering application conditions and hence demonstration of robust performance and reliability under these conditions is a prerequisite to CNT-MMCs being widely implemented in engineering applications. Deformation mechanisms over time may vary, especially for mechanisms unique to CNTs. CNT sliding and telescoping mechanisms may change under varying fatigue loadings and cycles. The quality of CNT may determine whether the efficacy is degraded over time. Other dynamic behavior of interest may be in impact and high strain rate conditions. Armor applications often rely on dissipating as much energy as possible and intrinsic energy dissipating mechanisms of CNTs and related 2D materials such as graphene nanoplatelets may be valuable for these applications. Early mechanical damping studies on CNT-polymer composites have shown promise [46], but damping behavior of CNT-MMCs is largely unexplored [47]. Composites with high mechanical damping properties could be used a shock absorbers or applications where exposure to high frequency vibrations could occur.

12.2.6 Interphase Development

As CNT-MMC technology matures, properties may need to be tailorable and perhaps only tweaked to meet application requirements. Interphases are a well-known approach in conventional composite science to enhanced mechanical properties by enhancing load transfer, or for protecting fiber reinforcements from environmental degradation. Interphases in CNT-MMCs have been limited largely to the utilization of sputtered metals coatings to enhance interfacial bonding with the metal matrix. Typically, a metal compatible with the matrix (either the same base material or common alloying element) is sputtered prior to a hot consolidation process. Thicker interphases have yet to be explored and could be well suited for protection or as reactive or sacrificial phases during processing. While other nanotubes such as boron nitride nanotubes (BNNT) have received attention due to higher intrinsic ambient stability under elevated temperatures, no studies have attempted to develop interphases to protect CNTs from oxidation under elevated temperature. Oxide-based interphases could perhaps significantly push the operating envelope of CNTs considerably, considering the intrinsic thermodynamic stability of the sp^2 bonded carbon. The presence of an interphase could potentially be used to mitigate surface defects on lesser quality industrial CNTs.

12.2.7 Extreme Environment Stability

Nanomaterials are often considered poorly suited for extreme environment applications due to their nanometric dimensions that may make them unstable or overly reactive, especially in elevated temperature environments. However, with development of aforementioned interphases, understanding of dynamic and intrinsic properties, and optimization of CNTs and microstructures of CNT-MMCs, these materials may provide unique properties suited for extreme environments. The sp^2 bonded structure is inherently inert and could be suitable for use in corrosive environments provided that a non-conductive interphase is present to eliminate possibility of electron transfer and hence any galvanic mismatch. Many high-strength high-performance materials suffer from corrosion due to localized microgalvanic or intergranular corrosion. A deliberate thick carbide layer could be used to electrochemically isolate CNTs from a metal matrix. Similarly, as mentioned earlier, an protective oxide layer could render CNTs ideal for high-temperature oxidation environments. Of particular interest could be bond coats for thermal and environmental barrier coatings (T/EBCs). Bond coats require high strength and toughness that could be imparted by CNTs. Bonds coats in EBCs must prevent moisture infiltration to protect sensitive CMCs such as SiC-SiC fiber-based composites. The hydrophobic nature of CNTs could be used to develop a moisture barrier either as a bulk layer or at grain boundaries of a silicon- or

NiCoCrAlY-based bond coat. Early studies [48,49] have demonstrated great potential for CNTs to provide electromagnetic shielding. Light metals such as Al and Mg that are increasing being utilized in aerospace applications could be loaded with CNTs to impart aircraft skin layers with electromagnetic shielding.

12.2.8 Environmental and Toxicity Aspects of CNT-MMCs

Discussion about CNTs and their composites is not complete without mention of their potential toxic effect on biological systems, such as the human body. The wear of CNT-MMC structures is expected to generate CNT containing debris, which may be disposed of in the atmosphere, or within the human body if CNT-MMC implants or prosthetics are utilized. The issue related to biocompatibility or cytotoxicity of CNTs is highly debated and is yet to merge into universal agreement. However, the last decade has seen scientists and practitioners err on the side of caution, and interest in biomedical applications of CNTs has waned. This is partially attributed to the increased prominence of lower risk nanomaterial reinforcements such as graphene and boron nitride, which have shown more promise in cytotoxicity studies [50–52]. Many early studies showed that CNTs are non-cytotoxic to living bodies [53–56], and actually are helpful in biomedical applications, even in certain in vivo applications [56–59]. Upon exposure to mouse bloodstream by injection, CNTs are not found to be retained by any of the reticuloendothelial system organs (liver or spleen) and are rapidly cleared from the bloodstream through the renal excretion route [53]. Multiwall CNTs also exhibited good bone-tissue compatibility without any measurable toxic effect in those studies. When implanted on bones alongside collagen, CNT can assist in bone repair by accelerating bone growth [56]. In addition, CNTs inhibit bone resorption by inhibiting osteoclast proliferation and hence are potential candidates for treatment of diseases like arthritis [57]. CNTs are also promising candidates for antiseptic bandages [59]. Another very interesting study by Khodakovskaya et al. shows that the presence of CNTs significantly helps seed germination and plant growth [60].

Likewise, several studies have hinted at negative effects of CNTs on human health [61–63]. Pulmonary [64], hepatic [65], and splenic [66] cytotoxicity in rodents has been reported as an effect of the presence of CNTs. Furthermore, multiwall CNTs may be toxic to macrophages, the security guard of blood, as they could cause incomplete phagocytosis or mechanically pierce through the plasma membrane and result in oxidative stress and cell death [67]. However, another study shows that CNT is successfully ingested by macrophage without cytotoxic effect [55]. The cytotoxic effect of CNTs also depends on their surface treatment [65].

Toxicity of carbon nanotubes has been actively debated and summarized in several review articles [68–70]. Insights from these publications reveal that the impurities present in CNTs and their clustering are some of the main causes of their cytotoxicity [68,71]. Moreover, it has also been noticed that CNTs embedded in the matrix of a composite structure do not show toxic effect. In addition, many reports showing CNT cytotoxicity are for CNTs that are suspended in a fluid medium [69]. Perhaps the greatest barrier to CNT acceptance is that nano materials in general are not fully understood and some changes in behavior of human cells or organs can be induced by the presence of any type of nano-particles. Thus, Ren et al. [68] have very correctly stated that, "assessment of effect of carbon nanotube on the cells, organ, or whole organism should also be standardized" before predicting the cytotoxicity of CNTs. Guidelines must be developed for safe use of CNTs and other nanomaterials, to include acceptable levels as nanocomposites become ubiquitous, just as safe amounts of other hazards such as radiation and air pollution can be established. A better understanding will foster better handling and manufacturing protocols for ensuring safety of researchers and workers involved in CNT-MMC production.

12.2.9 CNTs as Key Technology Enablers

A majority of research focuses on the development of CNT-MMCs for commonly known structural applications listed in Table 12.1. Studies must be carried out to find novel applications for CNT-MMCs. For example, the pores within a partially infiltrated CNT cluster might have very fine dimensions that can be used for filtration. CNT-MMCs could be a key enabler for metallic membranes and sensors. The advent of nanotechnology and micro-fabrication techniques has led to the invention of many novel devices, actuators, and sensors; most of which have dimensions in the microscale (MEMS) or nanoscale (NEMS). High-strength micro- and nano-scale materials such as CNT-MMCs could be key enablers of new high-performance MEMS and NEMS. Silicon has been the choice material for fabrication of these devices. Some studies on the use of CNT-MMCs for structural applications at micro- and nano-length scales have shown great promise and must be explored [16,17].

The possibilities with CNT-MMCs are many and exciting. However, in order to realize these goals, more research is required to continue to discover the exciting properties of CNT-MMCs and to enable solutions to CNT-MMC challenges that

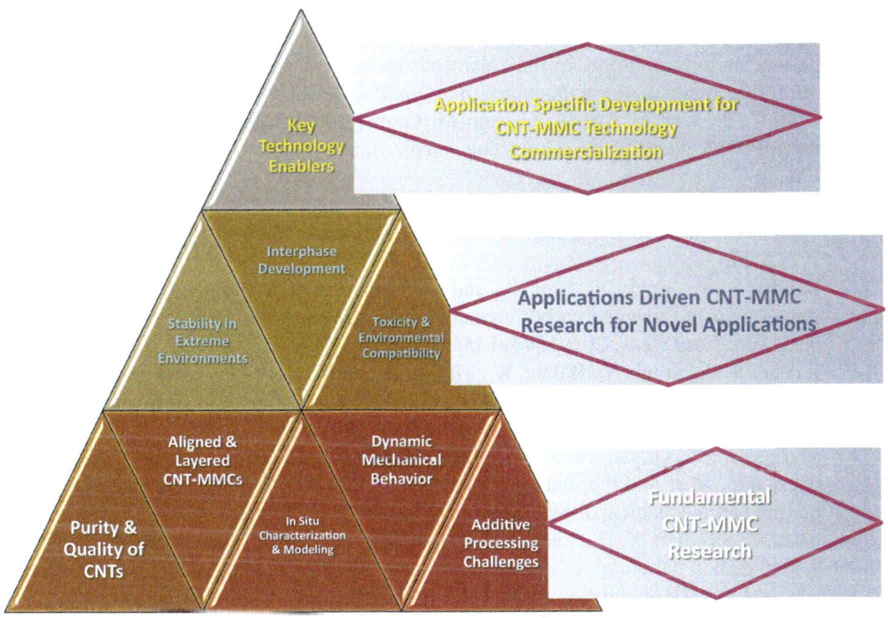

FIGURE 12.2 Directions for future research in CNT-MMCs, spanning fundamental research to applications driven research to enable commercialization of CNT-MMC technology.

will continue to promote the maturation and commercialization of these novel materials. Figure 12.2 shows the future directions for research in CNT-MMCs. Application-oriented research with sound understanding of the underlying scientific fundamentals is the north star to advancing CNT-MMCs technologies for myriad applications.

REFERENCES

1. Bakshi, S. R. 2009. Plasma and cold sprayed aluminum carbon nanotube composites: quantification of nanotube distribution and multi-scale mechanical properties. PhD Dissertation. Florida International University. FIU Electronic Theses and Dissertations. Paper 97. http://digitalcommons.fiu.edu/etd/97/.
2. Bakshi, S. R., Batista, R. G., and A. Agarwal. 2009. Quantification of carbon nanotube distribution and property correlation in nanocomposites. *Composites A* 40: 1311–1318.
3. Kim, K. T., Cha, S. I., Hong, S. H., and S. H. Hong. 2006. Microstructures and tensile behavior of carbon nanotube reinforced Cu matrix nanocomposites. *Mater. Sci. Eng. A* 430: 27–33.
4. Kwon, H., Estili, M., Takagi, K., Miyazaki, T., and A. Kawasaki. 2009. Combination of hot extrusion and spark plasma sintering for producing carbon nanotube reinforced aluminum matrix composites. *Carbon* 47: 570–577.
5. Choi, H., Shin, J., Min, B., Park, J., and D. Bae. 2009. Reinforcing effects of carbon nanotubes in structural aluminum matrix nanocomposites. *J. Mater. Res.* 24: 2610–2616.
6. Pérez-Bustamante, R., Gómez-Esparza, C. D., Estrada-Guel, I., Miki-Yoshida, M., Licea-Jiménez, L., Pérez-García, S. A., and R. Martínez-Sánchez. 2009. Microstructural and mechanical characterization of Al–MWCNT composites produced by mechanical milling. *Mater. Sci. Eng. A* 502: 159–163.
7. Cha, S. I., Kim, K. T., Arshad, S. N., Mo, C. B., and S. H. Hong. 2008. Extraordinary strengthening effect of carbon nanotubes in metal-matrix nanocomposites processed by molecular-level mixing. *Adv. Mater.* 17: 1377–1381.
8. Kim, K. T., Cha, S. I., Gemming, T., Eckert, J., and S. H. Hong. 2008. The role of interfacial oxygen atoms in the enhanced mechanical properties of carbon nanotube-reinforced metal matrix nanocomposites. *Small* 4: 1936–1940.
9. Goh, C. S., Wei, J., Lee, L. C., and M. Gupta. 2006. Simultaneous enhancement in strength and ductility by reinforcing magnesium with carbon nanotubes. *Mater. Sci. Eng. A* 423: 153–156.
10. Laha, T., Liu, Y., and A. Agarwal. 2007. Carbon nanotube reinforced aluminum nanocomposite via plasma and high velocity oxy-fuel spray forming. *J. Nanosci. Nanotech.* 7: 515–524.
11. Bakshi, S. R., Singh, V., Seal, S., and A. Agarwal. 2009. Aluminum composite reinforced with multiwalled carbon nanotubes from plasma spraying of spray dried powders. *Surf. Coat. Tech.* 203: 1544–1554.
12. Bakshi, S. R., Singh, V., Balani, K., McCartney, D. G., Seal, S., and A. Agarwal. 2008. Carbon nanotube reinforced aluminum composite coating via cold spraying. *Surf. Coat. Tech.* 202: 5162–5169.
13. Murugesan, R., Gopal, M., and G. Murali. 2019. Effect of Cu, Ni addition on the CNTs dispersion, wear and thermal expansion behavior of Al-CNT composites by molecular mixing and mechanical alloying. *Applied Surface Science* 495: 143542.
14. Chen, C. S., Chen, X. H., Yang, Z., Li, W. H., Xu, L. S., and B. Yi. 2006. Effect of multiwalled carbon nanotubes as reinforced fibres on tribological behaviour of Ni–P electroless coatings. *Diamond Relat. Mater.* 15: 151–156.
15. Chen, W. X., Tu, J. P., Wang, L. Y., Gan, H. Y., Xu, Z. D., and X. B. Zhang. 2003. Tribological application of carbon nanotubes in a metal-based composite coating and composites. *Carbon* 41: 215–222.

16. Shen, G. R., Cheng, Y. T., and L. T. Tsai. 2005. Synthesis and characterization of Ni–P-CNT's nanocomposite film for MEMS applications. *IEEE Trans. Nanotech.* 4(5): 539–547.
17. Bak, J. H., Kim, Y. D., Hong, S. S., Lee, B. Y., Lee, S. R., Jang, J. H., Kim, M., Char, K., Hong, S., and Y. D. Park. 2008. High-frequency micromechanical resonators from aluminium–carbon nanotube nanolaminates. *Nature Mater.* 7: 459–463.
18. Liu, J., Xiong, D-B., Su, Y., Guo, Q., Li, Z., and D. Zhang. 2019. Effect of thermal cycling on the mechanical properties of carbon nanotubes reinforced copper matrix nanolaminated composites. *Materials Science & Engineering A* 739: 132–139.
19. Chen, X. H., Chen, C. S., Xiao, H. N., Cheng, F. Q., Zhang, G., and G. J. Yi. 2005. Corrosion behavior of carbon nanotubes–Ni composite coating. *Surf. Coat. Technol.* 191: 351–356.
20. Praveen, B. M., Venkatesha, T. V., Naik, Y. A., and K. Prashantha. 2007. Corrosion studies of carbon nanotubes–Zn composite coating. *Surf. Coat. Technol.* 201: 5836–5842.
21. Bakshi, S. R., Lahiri, D., and A. Agarwal. 2010. Carbon nanotube reinforced metal matrix composites—a review. *Int. Mater. Rev.* 55: 41–64.
22. Baughman, R. H., Zakhidov, A. A., and W. A. de Heer. 2002. Carbon nanotubes—the route toward applications. *Science* 297: 787–792.
23. Song, P. C., Liu, C. H., and S. S. Fan. 2006. Improving the thermal conductivity of nanocomposites by increasing the length efficiency of loading carbon nanotube. *Appl. Phys. Lett.* 88: 153111.
24. Peng, T., and I. Chang. 2015. Uniformly dispersion of carbon nanotube in aluminum powders by wet shake-mixing approach. *Powder Technology* 284: 32–39.
25. Heller, D. A., Mayrhofer, R. M., Baik, S., Grinkova, Y. V., Usrey, M. L., and M. S. Strano. 2004. Concomitant length and diameter separation of single-walled carbon nanotubes. *J. Am. Chem. Soc.* 126: 14568–14573.
26. Krupke, R., and F. Hennrich. 2005. Separation techniques for carbon nanotubes. *Adv. Eng. Mater.* 7: 111–116.
27. An, K. H., Yang, C-M., Lee, J. Y., Lim, S. C., Kang, C., Son, J-H., Jeong, M. S., and Y. H. Lee. 2006. A diameter-selective chiral separation of single-wall carbon nanotubes using nitronium ions. *J. Electronic Mater.* 35: 235–242.
28. Yun, Y., Shanov, V., Tu, Y., Subramaniam, S., and M. J. Schulz. 2006. Growth mechanism of long aligned multiwall carbon nanotube arrays by water-assisted chemical vapor deposition. *J. Phys. Chem. B* 110: 23920–23925.
29. Wen, Q., Qian, W., Niw, J., Cao, A., Ning, G., Wang, Y., Hu, L., Zhang, Q., Huang, J., and F. Wei. 2010. 100 mm long, semiconducting triple-walled carbon nanotubes. *Adv. Mater.* 22: 1–5.
30. Zhong, X-H., Li, Y-L., Liu, Y-K., Qiao, X-H., Feng, Y., Liang, J., Jin, J., Zhu, L., Hou, F., and J-Y. Li. 2010. Continuous multilayered carbon nanotube yarns. *Adv. Mater.* 22: 692–696.
31. Aiken, B. J. M., Courtney, T. H., and D. R. Maurice. 1991. Reaction rates during mechanical alloying. *Materials Science and Engineer A* 147: 229–237.
32. Yang, H., Topping, T. D., Wehage, K., Jiang, L., Lavernia, E. J., and J. M. Schoenung. 2014. Tensile behavior and strengthening mechanisms in a submicron B_4C-reinforced Al trimodal composite. *Materials Science and Engineering A* 616: 35–43.
33. Garmestani, H., Al-Haik, M. S., Dahmen, K., Tannenbaum, R., Li, D., Sablin, S. S., and M. Y. Hussaini. 2003. Polymer-mediated alignment of carbon nanotubes under high magnetic fields. *Adv. Mater.* 15: 1918–1921.
34. Shi, D., He, P., Lian, J., Chaud, X., Bud'ko, S. L., Beaugnon, E., Wang, L. M., Ewing, R. C., and R. Tournier. 2005. Magnetic alignment of carbon nanofibers in polymer composites and anisotropy of mechanical properties. *J. Appl. Phys.* 97: 0643e12 (1–5).
35. Camponeschi, E., Vance, R., Al-Haik, M., Garmestani, H., and R. Tannenbaum. 2007. Properties of carbon nanotube–polymer composites aligned in a magnetic field. *Carbon* 45: 2037–2046.

36. Correa-Duarte, M. A., Grzelczak, M., Salgueiriño-Maceira, V., Giersig, M., Liz-Marzán, L. M., Farle, M., Sierazdki, K., and R. Diaz. 2005. Alignment of carbon nanotubes under low magnetic fields through attachment of magnetic nanoparticles. *J. Phys. Chem. B* 109: 19060–19063.
37. Hone, J., Llaguno, M. C., Nemes, N. M., Johnson, A. T., Fisher, J. E., Walters, D. A., Casavant, M. J., Schmidt, J., and R. E. Smalley. 2000. Electrical and thermal transport properties of magnetically aligned single wall carbon nanotube films. *Appl. Phys. Lett.* 77: 666–668.
38. Wang, D., Song, P., Liu, C., Wu, W., and S. Fan. 2008. Highly oriented carbon nanotube papers made by aligned carbon nanotubes. *Nanotechnology* 19: 075609(1–6).
39. Demasi, L., Biagini, G., Vannucci, F., Santarpia, E., and R. Cavallaro. 2017. Equivalent single layer, zig-zag, and layer wise theories for variable angle tow composites based on the generalized unified formulation. *Composite Structures* 177: 54–79.
40. Yu, Q., Legros, M., and A. M. Minor. 2015. In situ TEM nanomechanics. *Materials Research Society Bulletin* 40: 62–70.
41. Bernal, R. A., Filleter, T., Connell, J. G., Sohn, K., Huang, J., Lauhon, L. J., and H. D. Espinosa. 2014. In situ electron microscopy four-point electromechanical characterization of freestanding metallic and semiconducting nanowires. *Small* 10: 725–733.
42. Goh, C. S., Wei, J., Lee, L. C., and M. Gupta. 2008. Ductility improvement and fatigue studies in Mg-CNT nanocomposites. *Composites Science and Technology* 68: 1432–1439.
43. Shin, S. E., and D. H. Bae. 2018. Fatigue behavior of Al2024 alloy-matrix nanocomposites reinforced with multi-walled carbon nanotubes. *Composite Part B: Engineering* 134: 61–68.
44. Alizadeh, A., Abdollahi, A., and H. Biukani. 2015. Creep behavior and wear resistance of Al 5083 based hybrid composites reinforced with carbon nanotubes (CNTs) and boron carbide (B_4C). *Journal of Alloys and Compounds* 650: 783–793.
45. Park, S. H., Cho, D. H., Cho, K. M., and I. M. Park. 2018. Creep properties of squeeze-infiltrated carbon nanotube and aluminum borate whisker reinforced AS52 Mg metal matrix composites. *Metals and Materials International* 24: 1162–1171.
46. Agrawal, R., Nieto, A., Chen, H., Mora, M., and A. Agarwal. 2013. Nanoscale damping characteristics of boron nitride nanotubes and carbon nanotubes reinforced polymer composites. *ACS Applied Materials & Interfaces* 5: 12052–12057.
47. Carvalho, O., Miranda, G., Buciumeanu, M., Gasik, M., Silva, F. S., and S. Madeira. 2016. High temperature damping behavior and dynamic Young's modulus of AlSi-CNT-SiC_p hybrid composite. *Composite Structures* 141: 155–162.
48. Micheli, D., Vricella, A., Pastore, R., Delfini, A., Giusti, A., Albano, M., Marchetti, M., Moglie, F., and V. M. Primiani. 2016. Ballistic and electromagnetic shielding behavior of multifunctional Kevlar fiber reinforced epoxy composites modified by carbon nanotubes. *Carbon* 104: 141–156.
49. Yang, K., Mei, H., Han, D., and L. Cheng. 2018. Enhanced electromagnetic shielding property of C/SiC composites via electrophoretically-deposited CNTs onto SiC coating. *Ceramics International* 44: 20187–20191.
50. Merlo, M., Mokkapati, V. R. S. S., Pandit, S., and I. Mijakovic. 2018. Boron nitride nanomaterials: biocompatibility and bio-applications. *Biomaterials Science* 6: 2298–2311.
51. Girao, A. F., Sousa, J., Dominguez-Bajo, A., Gonzalez-Mayorga, A., Bdikin, I., Pujades-Otero, E., Casan-Pastor, N., Hortiguela, M. H., Otero-Irurueta, G., Completo, A., Serrano, M. C., and P. A. D. A. P. Marques. 2020. 3D reduced graphene oxide scaffolds with a combinatorial fibrous-porous architecture for neural tissue engineering. *ACS Applied Materials & Interfaces.* doi:10.1021/acsami.0c10599.
52. Ruiz, A., Lucherelli, A., Murera, D., Lamon, D., Menard-Moyon, C., and A. Bianco. 2020. Toxicological evaluation of highly water dispersible few-layer graphene in vivo. *Carbon.* doi:10.1016/j.carbon.2020.08.023.

53. Singh, R., Pantarotto, D., Lacerda, L., Pastorin, G., Klumpp, C. D., Prato, M., Bianco, A., and K. Kostarelos. 2006. Tissue biodistribution and blood clearance rates of intravenously administered carbon nanotube radiotracers. *Proc. Natl. Acad. Sci. USA* 103(9): 3357–3362.
54. Cherukuri, P., Bachilo, S. M., Litovsky, S. H., and R. B. Weisman. 2004. Near-infrared fluorescence microscopy of single-walled carbon nanotubes in phagocytic cells. *J. Am. Chem. Soc.* 126: 15638–15639.
55. Kalbacova, M., Kalbac, M., Dunsch, L., and U. Hempel. 2007. Influence of single-walled carbon nanotube films on metabolic activity and adherence of human osteoblasts. *Carbon*. 45: 2266–2272.
56. Usui, Y., Aoki, K., Narita, N., Murakami, N., Nakamura, I., Nakamura, K., Ishigaki, N., Yamazaki, H., Horiuchi, H., Taruta, S., Kato, H., Ahm Kim, Y., Endo, M., and N. Saito. 2008. Carbon nanotubes with high bone-tissue compatibility and bone-formation acceleration effects. *Small* 4(2): 240–246.
57. Narita, N., Kobayashi, Y., Nakamura, H., Maeda, K., Ishihara, A., Mizoguchi, T., Usui, Y., Aoki, K., Simizu, M., Kato, H., Ozawa, H., Udagawa, N., Endo, M., Takahashi, N., and N. Saito. 2009. Multiwalled carbon nanotubes specifically inhibit osteoclast differentiation and function. *Nano. Lett.* 9(4): 1406–1413.
58. Lobo, A. O., Corat, E. M. A., Antunes, F. F., Palma, M. B. S., Pacheco-Soares, C., Garcia, E. E., and E. J. Corat. 2010. An evaluation of cell proliferation and adhesion on vertically aligned multi-walled carbon nanotube films. *Carbon* 48: 245–254.
59. Simmons, T. J., Lee, S. H., Park, T. J., Hashim, D. P., Ajayan, P. M., and R. J. Linhardt. 2009. Antiseptic single wall carbon nanotube bandages. *Carbon* 47: 1561–1564.
60. Khodakovskaya, M., Dervishi, E., Mahmood, M., Xu, Y., Li, Z., Watanabe, F., and A. S. Biris. 2009. Carbon nanotubes are able to penetrate plant seed coat and dramatically affect seed germination and plant growth. *ACS Nano* 3(10): 3221–3227.
61. Shvedova, A. A., Pietroiusti, A., Fadeel, B., and V. E. Kagan. 2012. Mechanisms of carbon nanotube-induced toxicity: focus on oxidative stress. *Toxicology and Applied Pharmacology* 261: 121–133.
62. Lanone, S., Andujar, P., Kermanizadeh, A., and J. Boczkowski. 2013. Determinants of carbon nanotube toxicity. *Advanced Drug Delivery Reviews* 65: 2063–2069.
63. Dong, J., and Q. Ma. 2015. Advances in mechanisms and signaling pathways of carbon nanotube toxicity. *Nanotoxicology* 9: 658–676.
64. Muller, J., Huaux, F., and D. Lison, 2006. Respiratory toxicity of carbon nanotubes: how worried should we be? *Carbon* 44: 1048–1056.
65. Ji, Z., Zhang, D., Li, L., Shen, S., Deng, X., Dong, L., Wu, M., and Y. Liu. 2009. The hepatotoxicity of multi-walled carbon nanotubes in mice. *Nanotechnology* 20: 445101.
66. Deng, X., Wu, F., Liu, Z., Luo, M., Li, L., Ni, Q., Jiao, Z., Wu, M., and Y. Liu. 2009. The splenic toxicity of water soluble multi-walled carbon nanotubes in mice. *Carbon* 47: 1421–1428.
67. Cheng, C., Müller, K. H., Koziol, K. K. K., Skepper, J. N., Midgley, P. A., Welland, M. E., and A. E. Porter. 2009. Toxicity and imaging of multi-walled carbon nanotubes in human macrophage cells. *Biomaterials* 30: 4152–4160.
68. Ren, H. X., Chen, X., Liu, J. H., Gu, N., and X. H. Huang. 2010. Toxicity of single-walled carbon nanotube: how we were wrong? *Mater Today* 13(1–2): 6–8.
69. Hussain, M. A., Kabir, M. A., and A. K. Sood. 2009. On the cytotoxicity of carbon nanotubes. *Current Sci.* 96(5): 664–673.
70. Lu, F., Gu, L., Meziani, M. J., Wang, X., Luo, P. G., Veca, L. M., Cao, L., and Y. P. Sun. 2009. Advances in bioapplication of carbon nanotubes. *Adv. Mater.* 21: 139–152.
71. Fiorito, S. 2008. *Carbon Nanotube: Angels or Demons?*. Singapore: Pan Stanford Publishing.

Index

Page numbers in *italic* indicate figures.

ABAQUS, 310
Ag-CNT, *see* silver-carbon nanotube composites
Al-CNT, *see* aluminum-carbon nanotube composites
aluminum-carbon nanotube composites
  Al4C3 in, 125, 132, 133, 241, 243
  aluminum *vs.*, 28, 32, 122
  bulk, 32
  CNT cluster in, 196
  cold rolled, 60, 131
  cold sprayed, 69, 70, 96, 100, 192
  computational studies, 307, 310, 312
  conventional sintering, 45
  from CVD, 133
  deformation processing, 51
  determining grain size of Al in, 92
  dislocation density in, 92
  elastic modulus, 63, 94, 133, 191, 217, 293
    CNT content and, 77, 100, 101, 189
  electrical resistivity, 111
  flowability of blended powders, 32, 63, 64
  friction stir processing, 77
  hardness, 50, 60, 67, 122, 133, 155, 161
  hot pressed, 50, 77, 199
  hot rolled, 205
  HVOF, 63, 66, 67
  mandrels, 65, 66
  mechanical milling, 18, 30, 122
  mechanical properties, 45, 51, 67, 77
  nanoindentation, 64, *101*, 192
  NSD, 16, 27, 50, 53, 122, 133, 201
  plasma sprayed, 62, 94, 128, 133, 190, 207, 255, 260, 262
  powder metallurgy technique, 24, 30, 45, 155, 164
  sandwich processing, 76
  SEM images, 18, 27, 32, 62, 64, 65, 67, 163, 309, *309*
  shock wave consolidation, 77
  SPS, 16, 27, 49, 50, 53, 133, 217, 235, 243
  strength, 27, 30, 53, 106, 108
  TEM image, 29, 70, 78, 97, 206, 211, 242, 244, 246
  tensile test specimens, 107
  tensile tests, 21, 122, 267
  tribological studies, 255
  wear rate, 58, 294
  X-ray diffraction spectrum, 52
  yield strength, 16, 51, 77, 199, 204, 205, 212, 213
associate solution model, 307
Au-CNT, *see* gold-carbon nanotube composites

ball milling, 54, 262, 266, 287, 293, 320, 329
  of Al-CNT, 62, 122, 206
  blending and, 62
  cryomilling, 19
  CNT dispersion and, 54, 71, 122, 133
  of Cu-CNT, 133
  CVD *vs.*, 158, 162
  disadvantages, 19, 335
  effectiveness, 156, 164, 336
  function, 21, 23, 162, 212, 220
  high energy ball milling, 17
  rotational milling modeling, 21–23
  SEM images, 18, 19
  shift-speed ball milling, 23
  solution ball milling, 20
ball-to-powder ratio (BPR), 17, 19, 21, 133, 156
biomimetics, 1
blending, 162
  ball milling and, 19
  composite, 16, 17, 122, 329
  flowability, 64
  plasma spraying and, 62
  sintering and, 45
BMG, *see* bulk metallic glass
boron, 2, 95, 285, 318, 337, 338
BPR, *see* Ball-to-Powder Ratio
bulk metallic glass (BMG), 56
  Ti-based, 169, 172
  Zr-based, 172

carbide formation
  aluminum, 234, 290
  detection, 63, 244
  Gibb's free energy of, 230, 232, 233, 241, 244
  kinetics, 214, 227, 234, 241
  loss of CNTs through, 43
  values, 237
  variation of critical thickness for, 232
carbon fiber(s), 53
  development of, 3, 9
  physical properties, 3, 5, 206
carbon nanotube (CNT)
  as composite paste detectors, 277
  composites

with metal matrices (CNT-MMCs), 10, 24, 54, 91, 109, 151, 164, 196, 232, 277, 290, 307, 317, 330
with various matrices, 7, 15, 30, 319
cytotoxicity, 338
discovery, 3, 7, 96, 332
dispersion, 41, 50, 82, 121, 164, 261, 278, 285, 329, 331
electrode, 74, 170, 240, 277
electrodeposition, 74, 82
future, 9, 54, 303, 306, 315, 329
interactions
energetics of, 230
in various systems, 233
interfacial phenomena, 227, 337
lateral heat conduction, 258
magnetic susceptibility, 3
mechanical behavior, modeling, 304
mechanical properties, 4
prediction of, 310
metal interactions, 230–246
energetics of, 230–233
in various systems, 234–246
multi-walled, 3
physical properties, 5
plant growth and, 338
publications, 7
quality, improvement in, 335
seed germination and, 338
single-walled, *see* single-walled carbon nanotube
specific heat capacity, 5
stability, 230
structure, 230
SWNT *vs.*, 7
synthesis, 6
therapeutic potential, 338
thermal conductivity, 5
thermal resistance, 258
thermodynamic prediction, 306
thermoelectric power, 5
toxicity, 338
world-wide interest, 7–9
catalytic chemical vapor deposition (CCVD), 3
CEF, *see* crystal electric field
chemical vapor deposition (CVD)
Al-CNT from, 16, 30, 122, 133, 219, 243
bounding and dispersion in CNTs from, 292, 299, 318, 320
Cu-CNT from, 299
dispersion, 15
purity, 243, 244, 334
Si-CNT from, 29, 30, 270
SWNT from, 6, 334
cluster/site approximation (CSA), 307
clustering parameter (CP), 34, 36, 37, 39
Co-CNT, *see* Cobalt-carbon nanotube composites
Co-deposition, 73, 317
bath for, 71, 73
by electrochemical and electroless processes, 71, 72
Ni-CNT fabricated by, 73, 143, 272
Cobalt-carbon nanotube composites, 166, 240
coefficient of friction (CoF), 254, 255
CNT and lowering of, 253, 294
dispersion and, 257
nanowear and, 255
wear resistance and decrease in, 253, 254, 257
coefficient of thermal expansion (CTE), 53, 262, 317
CNT content and, 262, 319
Cu-CNT, 270
measurement, 263, 264
thermal stresses and, 270
CoF, *see* coefficient of friction
cold spraying, 32, 60, 67, 69, 96, 122, 206, 318, 330
composite material(s)
fiber reinforced, 1
publications, 1
ubiquity, 1, 2
uses, 1, 10
consistent valence force field (CVFF), 303
copper-carbon nanotube composites, 10, 133, 134
ball milling, 48
CNT cluster in, 53
cold rolling, 134
conventional sintering, 45, 53
deformation processing, 53, 134
hardness, 134
hot pressed, 48
mechanical properties, 28, 173
microstructures, 50
MLM, 16, 28
powder metallurgy technique for, 45
SEM images, 76
SPS, 49, 53, 134, 143
strength, 28
studies, 133
TEM images, 28
tribological properties, 254
ultrasonication, 49
variation of microhardness, 28
wear resistance, 28, 53
yield strength, 50
Cox model, 188, 191, 192, 197
CP, *see* clustering parameter
crystal electric field (CEF), 307
CSA, *see* cluster/site approximation
CTE, *see* coefficient of thermal expansion
Cu-CNT, *see* copper-carbon nanotube composites
CVD, *see* chemical vapor deposition
CVFF, *see* consistent valence force field

Delaunay triangulation, 34, 36, 38
density functional theory (DFT), 303, *304*
diamond, 5, 10, 98, 285, 294
differential scanning calorimeter (DSC), 109, 110, 194, 234, 235
disintegrated melt decomposition (DMD), 56, 151, 330
dispersion parameter (DP), 34, 36, 37
dual-reinforcement, 285–300
  ceramic, 285–294
  nanocarbon, 294–299
dynamic modulus analysis (DMA), 99, 102, 103, 114, 144

ECAP, *see* equal channel angular pressing
EDM, *see* electrical discharge machining
EDS, *see* energy dispersive spectroscopy
EELS, *see* electron energy loss spectroscopy
effective medium approach (EMA), 260
EFTEM, *see* energy filtered transmission electron microscopy
elastic modulus, 1, 3, 6, 16, 63, 133, 172
  models
    comparisons of, 33, 108, 189, 193
    Cox, 191, 192, 197
    dispersion-based, 194
    Eshelby, 193
    Halpin-Tsai, 192
    Hashin-Shtrikman, 192
    Mori-Tanaka, 33
    Rule of Mixtures, 190, 285
  sintering, 288
electrical conductivity, 1, 5, 10, 72, 111, 114, 163, 270, 272, 273, 276–278, 317, 319, 333
electrical discharge machining (EDM), 107
electrochemical and electroless deposition, 30, 56, 71, 72, 133, 143, 173, 241, 272
electrodeposition, 71, 73
  co-deposition and, 143
  electronic transport behavior and, 143, 303
  hardness and, 143
  pulsed reverse, 143
  SWNT preparation, 111
electron energy loss spectroscopy (EELS), 98, 114
EMA, *see* effective medium approach
energy dispersive spectroscopy (EDS), 48, 95, 98, 234, 240, 241
  with SEM, 95
energy-filtered transmission electron microscopy (EFTEM), 98
equal channel angular pressing (ECAP), 51, 53, 139
Eshelby model, 193

Factsage, 230, 234, 307, 332
FEM, *see* finite element method
field assisted sintering technology (FAST), 49
finite element method (FEM), 222, 311
foams, 155–160
friction stir processing (FSP), 77, 79, 80, 160
friction stir welding (FSW), 77
full width at half maxima (FWHM), 92

GFRP, *see* glass fiber reinforced plastics
Gibb's free energy, 230, 231, 291, 307
glass fiber reinforced plastics (GFRP), 2
gold-carbon nanotube composites, 170, 304, *305*
grain refinement, 51, 53, 80, 212, 214
graphite, 3, 5, 7, 48, 65, 81, 91, 161, 172, 230, 231, 233, 234, 238–240, 257, 258, 275, 307

HAADF, *see* high angle annular dark field
Hall-Petch relation, 134
Halpin-Tsai model, 192
Hardness
  Ag-CNT, 16
  Al-CNT, 30, 50, 60, 67, *101*, 102, 122, 133, 212
  CNT from MLM powders, 50
  Cu-CNT, 28, 50, 134, 213, 248
  electrodeposition and, 143
  Mg-CNT, 151
  nano-indentation and, 99, 100
  Ni-CNT, 143
  SPS and, 50
  Ti-CNT, 48, 172
  Zi-based composites, 172
Hashin-Shtrikman model, 192
high angle annular dark field (HAADF), 97
high velocity oxy-fuel (HVOF), 63, 66, 67, 70, 122, 124, 330
HIP, *see* hot isostatic pressing
hot isostatic pressing (HIP), 44, 48, 51
hot pressing, 44, 45, 47–49, 151, 162, 172, 234, 245, 265, 287, 294, 298, 330
hot rolling, 17, 51, 53, 94, 133, 162, 163, 190, 192, 205, 212, 265, 330
HVOF, *see* high velocity oxy-fuel

in-situ mechanical testing, 108–109
inter-fiber spacing, 34
interfacial phenomena, 246–250, 264, 306
  effect of CNT defects, 241
  effect of temperature on carbide formation, 243
  formation and effect of carbides, 233–240
  significance of, 10, 15, 227, 246, 262, 331
  strengthening and, 29, 52, 109, 158, 204, 206, 214, 245, 265

studies, 91, 155, 164, 245, 256, 262, 306
TEM study, 96, 219, 248
ionic liquid model, 307

Kevlar fiber, 2

laser deposition, 81, 241
Laser engineered net shaping (LENS), 81, 143, 150

macroscale/bulk mechanical testing, 105
tensile/compression test, 106–107
tribological property evaluation, 107–108
magnesium-carbon nanotube composites
casting in synthesis of, 56, 151
CTE decrease, 110
deformation processing, 53, 56
DMD for, 151
friction stir processing, 77, 151
hydrogen storage properties, 151, 275
preparation, 48, 58, 151
studies, 48, 58, 151
tensile strength and ductility, 56
magnetic susceptibility, 5
mandrels, 65, 66
MD, *see* molecular dynamics
melt processing, 54, 58, 173
casting, 55
induction melting, 59
melt infiltration, 57
MEMS, *see* micro-electro-mechanical systems
mesoscale microstructure simulation project (MMSP), 307
Mg-CNT, *see* magnesium-carbon nanotube composites
micro-electro-mechanical systems (MEMS), 99, 105, 270, 322, 330, 333, 335, 339
enhancing performance of, 143, 270
mechanical properties, 99
microstructure simulation, 307–310
MLM, *see* molecular level mixing
MMSP, *see* mesoscale microstructure simulation project
modulus mapping, 99, 103, 114, 331
molecular dynamics (MD), 187, 188, 303, 304
molecular level mixing (MLM), 16, 28, 54, 83, 133–139, 209, 256, 276, 329
Monte Carlo simulations, 303
Mori-Tanaka model, 33

nano dynamic modulus analysis (Nano DMA), 99, 102, 330
nano-electro-mechanical system (NEMS), 322, 330, 332, 339
nano-indentation, 64, *101*, 192, 197
nanolaminates, 161–163
nanoscale dispersion (NSD)
Al-CNT, 16, 27, 50, 53, 122, 133, 201
SEM images, 27
nanoscale mechanical testing
modulus mapping, 99, 103, 114, 331
nano DMA, 99, 102, 330
nano-indentation, 64, *101*, 192, 197
nanoscratch, 99, 105
natural rubber, 27, 123, 201
NEMS, *see* nano-electro-mechanical system
nickel-carbon nanotube composites
carbide formation and, 143
co-deposition process, 143, 272
corrosion resistance, 273, 332
Cu-CNT *vs.*, 121, 254
electrodeposition, 72, 73
electronic transport behavior of 1-D, 272
field-emission performance, 272
free-standing, 143
laser deposition, 81
studies, 72, 121
tribological properties, 143, 253, 254
wear resistance, 254
Ni-CNT, *see* nickel-carbon nanotube composites
NSD, *see* nanoscale dispersion

Object-oriented finite element method (OOF), 310, 312, 322
Oliver and Pharr method, 100
Orowan strengthening, 211, 215

PAS, *see* plasma assisted sintering
Patents, 315–326
challenges, 321
international patents, 319–320
prospects, 321
US patents, 315
PECS, *see* pulse electric current sintering
physical vapor deposition (PVD), 25, 26, 31, 166
plasma assisted sintering (PAS), 49
plasma spray forming (PSF), 66, 124, 278
plasma spraying, 32, 41, 62, 65–67, 85, 100, 115, 122, 133, 174, 201, 207, 214, 235–237, 278, 309, 310, 318, 330
polyacrylonitrile, 3
polymer matrix composite (PMC), 7, 9, 108, 214, 318
polyvinyl alcohol (PVA), 24–26, 31, 64, 162
powder metallurgy techniques, 43, 57
advantages, 43, 122, 133, 164
conventional sintering, 45
deformation processing, 51
flake powder metallurgy, 24, 245, 263, 265, 297

hot pressing, 44, 45, 47–49, 151, 162, 172, 234, 245, 265, 287, 294, 298, 330
SPS, 49
uses, 43, 45, 70, 82
PSF, *see* plasma spray forming
Pulse electric current sintering (PECS), 49
PVA, *see* polyvinyl alcohol
PVD, *see* physical vapor deposition

radial breathing mode (RBM), 93
Raman spectroscopy, 6, 21, 49, 92–94, 163, 290, 307
random substitutional solution model, 307
RBM, *see* radial breathing mode
reactive wetting, 232, 233
Rule of Mixtures (ROM), 190, 215, 216, 264, 290, 331

SAED, *see* Selective area electron diffraction
sandwich processing, 76, 134
strengthening and, 131, 210
TEM images, 211
scanning electron microscopy (SEM), 6, 15
with EDS, 94, 157
images
Al-CNT, 18, 26, 36, 65, 67, 80, 157, 163, 207
ball milling, 18, 19
blending, 62
Co-CNT, 75
CNTs grown on SiC, 292
coating scratch, 257, 267
Cu-CNT, 28, 196
fracture surface, 36, 52, 65, 80, 207, 221
Ni-CNT, 74
NSD, 27
sandwich processing, 76
showing gold layer deposited due to sputter coating, 95
single splats, 309, *309*
splat finger, 229
spray-dried agglomerate, 32, 64, 229
wear-track, 267
in situ nano-indentation, 6, 218, 221
scanning probe microscope (SPM), 104, *106*, 218
in conjunction with SEM, 257
of indents in Al-CNTs, *101*
Scanning transmission electron microscopy (STEM), 97
Scherrer formula, 92
Selective area electron diffraction (SAED), 96
silicon-carbon nanotube composites, 10, 30, 164, 198, 270, 276, 306, *311*, 319
silver-carbon nanotube composites, 111, 272
Simulent Drop, 308, *309*

single-walled carbon nanotube (SWNT)
alloy, 169
aluminum, 48, 77, 78
CNT *vs.*, 6
cobalt, 6
copper, 46, 76, 111, 134
costs, 9
diameter, 4
discovery, 4
electrical resistivity, 5, 270
epoxy, 195
formation, 4, 6, 9
gold, 75, 304
hardness, 134
heat capacity, 5, 110
historical perspectives, 4
interaction between metal and, 75, 228, 230
magnetic susceptibility, 5
mechanical properties, 6
nickel, 46, 303–306
physical properties, 4, 6
Raman spectra, 6, 93
silicon, 165
specific heat capacity, 5
thermal conductivity, 5, 258
titanium, 75
solution precursor of plasma-sprayed (SPPS), 310
spark plasma sintering (SPS)
Al-CNT from, 16, 27, 49, 133, 233, 243
of ball-milled powders, 16
Cu-CNT from, 28, 49, 50, 53, 208, 254
hardness, 50, 143
MLM followed by, 28, 209, 240, 256
Ni-CNT from, 143
strengthening and, 16, 50, 53, 208, 214, 217
tensile strength and, 16, 133, 218
specific heat capacity, 5, 109–110
SPM, *see* solution precursor of plasma-sprayed
sputtering, 75, 82
gold layer deposited during, 75
STEM, *see* Scanning transmission electron microscopy
stoichiometric compound model, 307
Strengthening
clustering, 208
degrees of, 133
in-situ observation, 214–220
interfacial phenomena and, 29
by interphase, 206
mechanisms, 53, 94, 95, 109, 158, 187, 199
orientation of fibers and, 191, 196
Orowan, 211
predicting, 187–191, 205, 299
sandwich processing and, 131, 210

SPS and, 28
TEM images, 221
SWNT, *see* single-walled carbon nanotube

tensile/compression test, 106–107
tensile strength, 121, 253
Al-CNT, 16, 122, 291
in ball milling *vs.* CVD, 46, 133
clustering and, 99
interfacial strength and, 52
Mg-CNT, 56, 151
Micro-voids and, 52
Ni-CNT, 143
pressure-sintering and, 16
SPS and, 16, 50, 53, 208, 214, 217
SWNT, 54
vapor deposition and, 30
tensile tests, 106–107, 199
thermal conductivity, 5
thermal spraying, 60–70
freestanding cylinders prepared by, 63–66
gas temperatures and particle velocities, 60
HVOF, 66
thermodynamic prediction, 306
models, 307
associate solution, 307
cluster/site approximation, 307
crystal electric field, 307
ionic liquid, 307
random substitutional solution, 307
stoichiometric compound, 307
thermoelectric power, 5
Ti-CNT, *see* titanium-CNT composites
titanium-CNT composites
hot pressed, 47, 172, 240
microstructure, 48
preparation, 47
production, 47
TiC formation and, 172
TNT, *see* trinitro-toluene
torsion/friction processing, 77, 78, 106
transmission electron microscopy (TEM)
images
Al-CNT, 29, 70, 78, 97, 206, 211, 242, 244, 246
Cu-CNT, 28
interfacial phenomena, 96, 219, 248
sandwich processing, 211
scanning, 97
tribological properties, 104, 107–108, 253–264, 294
trinitro-toluene (TNT), 277

vapor deposition techniques, 3, 6, 15, 31, 162
vapor grown carbon fibers (VGCF), 3
volume of fluid (VOF), 308

X-ray diffraction (XRD), 91, 92, 114, 158, 172, 219, 239, 317
X-ray photoelectron spectroscopy (XPS), 98, 99, 114

Yield strength
Al-CNT, 16, 51, 199
compressive, 16, 27, 51, 213
computation of, 204, 209
Cu-CNT, 28, 50, 53
grain refinement and, 212
Mg-CNT, 53, 177
nano-indentation and, 99
predicting, 204, 210

Zinc-carbon nanotube composites, 169, 274, 332
Zirconium, 56, 75, 164, 172

9780367274924